Advances in Applied Mechanics

Volume 23

Editorial Board

T. Brooke Benjamin
Y. C. Fung
Paul Germain
Rodney Hill
L. Howarth
C.-S. Yih (Editor, 1971–1982)

Contributors to Volume 23

Robert J. Asaro
M. F. Ashby
Cornelius O. Horgan
James K. Knowles
A. Libai
Gérard A. Maugin
J. G. Simmonds

ADVANCES IN
APPLIED MECHANICS

Edited by

John W. Hutchinson

DIVISION OF APPLIED SCIENCES
HARVARD UNIVERSITY
CAMBRIDGE, MASSACHUSETTS

Theodore Y. Wu

ENGINEERING SCIENCE DEPARTMENT
CALIFORNIA INSTITUTE OF TECHNOLOGY
PASADENA, CALIFORNIA

VOLUME 23

1983

ACADEMIC PRESS
A Subsidiary of Harcourt Brace Jovanovich, Publishers
New York   London
Paris   San Diego   San Francisco   São Paulo   Sydney   Tokyo   Toronto

COPYRIGHT © 1983, BY ACADEMIC PRESS, INC.
ALL RIGHTS RESERVED.
NO PART OF THIS PUBLICATION MAY BE REPRODUCED OR
TRANSMITTED IN ANY FORM OR BY ANY MEANS, ELECTRONIC
OR MECHANICAL, INCLUDING PHOTOCOPY, RECORDING, OR ANY
INFORMATION STORAGE AND RETRIEVAL SYSTEM, WITHOUT
PERMISSION IN WRITING FROM THE PUBLISHER.

ACADEMIC PRESS, INC.
111 Fifth Avenue, New York, New York 10003

United Kingdom Edition published by
ACADEMIC PRESS, INC. (LONDON) LTD.
24/28 Oval Road, London NW1 7DX

LIBRARY OF CONGRESS CATALOG CARD NUMBER: 48-8503

ISBN 0-12-002023-8

PRINTED IN THE UNITED STATES OF AMERICA

83 84 85 86   9 8 7 6 5 4 3 2 1

Contents

LIST OF CONTRIBUTORS   vii

PREFACE   ix

Micromechanics of Crystals and Polycrystals

Robert J. Asaro

I. Introduction	2
II. Micromechanics of Crystallographic Slip	4
III. Constitutive Laws for Elastic–Plastic Crystals	36
IV. Applications to Elastic–Plastic Deformation of Single Crystals	67
V. Some Outstanding Problems	81
References	111

Mechanisms of Deformation and Fracture

M. F. Ashby

I. Introduction	118
II. Mechanisms of General Plasticity and Creep	121
III. Deformation Mechanism Maps	130
IV. Mechanisms of Fracture by General Damage	147
V. Fracture Mechanism Maps	155
VI. Conclusions	171
References	172

Recent Developments Concerning Saint-Venant's Principle

Cornelius O. Horgan and James K. Knowles

I. Introduction	180
II. A Model Problem	182
III. Linear Elastostatic Problems	213
IV. Principles of Saint-Venant Type in Other Contexts	250
V. Concluding Comments	261
References	262

Nonlinear Elastic Shell Theory

A. Libai and J. G. Simmonds

I. Introduction	272
II. Cylindrical Motion of Infinite Cylindrical Shells (Beamshells)	274
III. The Equations of General Nonlinear Shell Theory	302
IV. Nonlinear Plate Theory	332
V. Static One-Dimensional Strain Fields	337
VI. Approximate Shell Theories	344
Appendix. The Equations of Three-Dimensional Continuum Mechanics	365
References	365

Elastic Surface Waves with Transverse Horizontal Polarization

Gérard A. Maugin

I. Introduction	373
II. Some General Features of Surface Waves	380
III. The Notion of Resonance Coupling between Modes	385
IV. A Typical SH Surface Mode: Bleustein–Gulyaev Wave	391
V. SH Surface Waves in Nonhomogeneous Elastic Media	395
VI. Electroacoustic SH Surface Waves in Ferroelectrics	407
VII. Magnetoacoustic SH Surface Waves in Magnetoelasticity	421
References	431

AUTHOR INDEX	435
SUBJECT INDEX	443

List of Contributors

Numbers in parentheses indicate the pages on which the authors' contributions begin.

ROBERT J. ASARO (1), Division of Engineering, Brown University, Providence, Rhode Island 02912

M. F. ASHBY (117), Department of Engineering, Cambridge University, Trumpington Street, Cambridge CB2 1PZ, England

CORNELIUS O. HORGAN (179), College of Engineering, Michigan State University, East Lansing, Michigan 48824

JAMES K. KNOWLES (179), Division of Engineering and Applied Science, California Institute of Technology, Pasadena, California 91125

A. LIBAI (271), Department of Aeronautical Engineering, Technion, Israel Institute of Technology, Haifa, Israel

GÉRARD A. MAUGIN (373), Laboratoire de Mécanique Théorique, Associé au C.N.R.S., Université Pierre-et-Marie-Curie, Paris, France

J. G. SIMMONDS (271), Department of Applied Mathematics and Computer Science, University of Virginia, Charlottesville, Virginia 22901

Preface

Volume 23 of the series reflects the broad span of solid mechanics today with articles on elasticity, plasticity, and fracture, with applications to structures and materials. I am indebted to the authors for their outstanding contributions. I join with my fellow Editor, Theodore Y. Wu, in thanking our predecessor, Chia-Shun Yih, for his long and distinguished service to the Advances. We will not let his experience and skill go unused, since Professor Yih has joined the Editorial Board.

JOHN W. HUTCHINSON

Micromechanics of Crystals and Polycrystals

ROBERT J. ASARO

Division of Engineering
Brown University
Providence, Rhode Island

I. Introduction	2
Notation	3
II. Micromechanics of Crystallographic Slip	4
A. Early Observations	4
B. Dislocations	13
C. Other Strengthening Mechanisms	24
D. Measurements of Latent Hardening	26
E. Observations of Slip in Single Crystals and Polycrystals at Modest Strains	32
III. Constitutive Laws for Elastic–Plastic Crystals	36
A. Kinematics of Crystalline Deformation	36
B. Constitutive Laws	39
C. Strain-Hardening Laws for Rate-Insensitive Crystals	42
D. Strain Rate-Dependent Flow Laws	53
E. The Hardening Modulus $h_{\alpha\beta}$	63
IV. Applications to Elastic–Plastic Deformation of Single Crystals	67
A. Kink Band Formation	67
B. Coarse Slip Band Formation	77
C. Macroscopic Shear Bands	81
V. Some Outstanding Problems	97
A. Polycrystalline Models	97
B. Deformation Shear Bands	107
References	111

I. Introduction

This article is primarily concerned with the incorporation of microstructure, crystallinity, and micromechanics into the continuum description of finite strain plasticity. The subject has received a great deal of attention during the past 100 years, with the result that a sound physical and mathematical basis for it now exists. Experimental and theoretical landmarks in the development can be identified and these are discussed in some detail mainly in Section II; in the context of the present article, however, one of the more significant of these can be attributed to the work of Taylor and co-workers (1923, 1925, 1934, 1938). In this the discovery of the crystal dislocation provided a clear atomistic interpretation of the slip process and strain hardening. Furthermore, and of vital significance, Taylor was able to show how micromechanics could be incorporated into macroscopic analyses of plastic flow. His 1938 model for polycrystals, for example, discussed in Section V, is still widely used to predict the development of crystallographic texture.

An intent of the present article is likewise to illustrate that it is indeed possible to incorporate important micromechanical features of plastic flow into macroscopic analyses. Furthermore, it is shown that when this is done, accurate descriptions of rather complex phenomena can be obtained. Illustrations of such analyses are taken from recent studies of nonuniform plastic deformation of crystalline metals and alloys.

In Section II a brief outline of only some of the important features of the micromechanics of crystalline plasticity is given. There is no intention here of providing a complete discussion of the phenomena involved. For example, the discussion is confined to plastic flow caused by dislocation slip, and face-centered-cubic crystals are used in the examples of dislocation mechanisms. Particular attention is paid to kinematics and to the phenomenology of strain hardening, since these are shown to play dominant roles in macroscopic response.

In Section III constitutive laws for elastic–plastic crystals are developed. The framework draws heavily on Hill's analysis of the mechanics of elastic–plastic crystals, but the theory is extended by incorporating the possibility of deviations from the Schmid rule of a critical resolved shear stress for slip. Deviations from the Schmid rule are motivated by micromechanical models for dislocation motion and are shown to lead to deviations from the "normality flow rule" of continuum plasticity. The implications of these "non-Schmid effects" regarding the stability of plastic flow are brought out via some examples of models for kinks bands and shear bands in Section IV. Emphasis is also placed on some apparent limitations of rate-independent constitutive laws for the analysis of finite strain plasticity and a very preliminary alternative rate-dependent constitutive law is suggested for isothermal deformation.

In Section IV some examples of analyses of elastic–plastic deformation in crystals are discussed. The examples involve predictions and descriptions of various common modes of nonuniform deformation, including necking and shear bands. These examples illustrate the important role that crystal kinematics play in establishing deformation patterns and therefore the loads and stresses. For example, the phenomenon of "geometrical softening," which involves lattice rotations into orientations more favorable to plastic slip, is described. It is shown by way of an analysis of localized shearing that geometrical softening can in effect compete with material strain hardening and produce a less stiff response of the crystal to certain deformation modes.

The article is concluded with some suggestions for fruitful research. These involve extensions of the theory to finite-strain rate-dependent polycrystalline models.

NOTATION

Standard notations are used throughout. Bold-faced symbols denote tensors, matrices, or dyads, the order of which is indicated by the context. The magnitude of a vector such as $\mathbf{b}$ is denoted as b, and similarly for the elements of higher-order tensors. The summation convention is used for Latin indices unless otherwise indicated. Superscripts and subscripts pertaining to crystal slip systems are written in Greek letters and summations over these are indicated explicitly. Dots and double dots are used to indicate the following products:

$$\mathbf{P} \cdot \mathbf{n} = P_{ij}n_j, \qquad \mathbf{Q} : \boldsymbol{\sigma} = Q_{ji}\sigma_{ij},$$

also

$$\mathbf{L} : \mathbf{P} = L_{ijkl}P_{lk}, \qquad \boldsymbol{\sigma} \cdot \mathbf{P} = \sigma_{ij}P_{jk}.$$

The cross product between $\mathbf{t}$ and $\mathbf{b}$ is expressed as $\mathbf{t} \otimes \mathbf{b}$. Time derivatives are often expressed by superposed dots, e.g.,

$$\partial \sigma_{ij}/\partial t = \dot{\sigma}_{ij}$$

or

$$\partial \boldsymbol{\sigma}/\partial t = \dot{\boldsymbol{\sigma}}.$$

Finally, the inverses to a second-order tensor $\mathbf{A}$ or a fourth-order tensor $\mathbf{B}$ are denoted $\mathbf{A}^{-1}$ and $\mathbf{B}^{-1}$ so that

$$\mathbf{A} \cdot \mathbf{A}^{-1} = \mathbf{A}^{-1} \cdot \mathbf{A} = \mathbf{I}$$

or

$$B:B^{-1} = B^{-1}:B = I$$

where **I** is the unit tensor.

II. Micromechanics of Crystallographic Slip

A. EARLY OBSERVATIONS

In a series of papers published between 1898 and 1900 Ewing and Rosenhain summarized their metallographic studies of deformed polycrystalline metals. The conclusions they reached concerning the mechanisms of plastic deformation provided a remarkably accurate picture of crystalline plasticity. Figure 1 is a schematic diagram, including some surrounding text, taken from their 1900 overview article. Figure 2 is one of their many excellent optical micrographs of deformed polycrystalline metals; the particular micrograph in Fig. 2 is of polycrystalline lead. They identified the steps a–e in Fig. 1 as "slip steps" caused by the emergence of "slip bands," which formed along crystallographic planes, at the specimen surfaces (thereby coining these two well-known phrases). Traces of the crystallographic slip

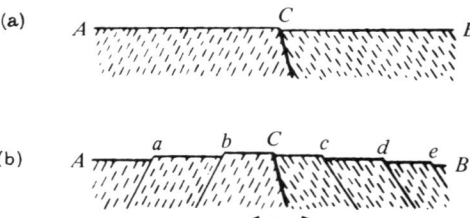

FIG. 1. Schematic diagram taken from Ewing and Rosenhain (1900) indicating slip steps on the surface of plastically deformed metal. (a) Before straining. (b) After straining. Slip was concluded to occur along crystallographic planes. Ewing and Rosenhain say, "The diagram, fig. 15, is intended to represent a section through the upper part of two contiguous surface grains, having cleavage or gliding places as indicated by the dotted lines, AB being a portion of the polished surface, C being the junction between the two grains.

"When the metal is strained beyond its elastic limit, as say by a pull in the direction of the arrows, yielding takes place by finite amounts of slips at a limited number of places, in the manner shown at a, b, c, d, e. This exposes short portions of inclined cleavage or gliding surfaces, and when viewed in the microscope under normally incident light these surfaces appear black because they return no light to the microscope. They consequently show as dark lines or narrow bands extending over the polished surface in directions which depend on the intersection of the polished surface with the surfaces of slip."

Fig. 2. Optical micrograph of deformed polycrystalline lead taken from Ewing and Rosenhain (1900).

planes were indicated by the dashed lines. The line labeled C was indicated by them to be a grain boundary separating two grains; the grains, they concluded, were *crystals* with a more or less homogeneous crystallographic orientation. Slip steps corresponding to the diagram of Fig. 1 are clearly visible in the micrograph of Fig. 2. Ewing and Rosenhain had not only concluded from their slip-line studies that metals and alloys were crystalline and composed of aggregates of crystallites (i.e., grains), but also that plastic deformation took place by simple shearing caused by the sliding of only certain families of crystal planes over each other in certain crystallographic directions lying in the planes. They also noted that certain metals deformed by "twinning" in addition to slip, which is also a crystallographic phenomenon.

Ewing and Rosenhain's (1900) early slip-line studies further indicated that the particular crystalline structure of the metals was preserved during plastic straining. This view was consistent with their conclusion that the simple shearing process of plastic flow involved only *crystallographic planes* sliding in *crystallographic directions*, but included the further assumption that slip progressed in whole multiples of lattice spacings so that during slip atoms were transported to equivalent lattice sites within the crystal structure. They recognized, however, that grains deformed inhomogeneously, i.e., only certain planes in the family of possible slip planes underwent slip, and this they attributed to a random distribution of microscopic imperfections of a nonspecified type that triggered the slipping process. They further argued that slip would disrupt these imperfections, which in turn would make it more difficult to activate further slip and thus cause strain hardening.

A remarkable aspect of these early studies of metal plasticity is that it was not until 1912 that Von Laue first diffracted X-rays from copper sulfate crystals, and not until 1913 that W. H. Bragg and W. L. Bragg made the first crystal structure determinations for ionic crystals (Bragg, 1933). In 1919 Hull published his results on the crystal structures of various common metals that provided proof and documentation of the crystalline structure of metals. As examples, Fig. 3a–c shows the "unit cells" of the face-centered-cubic (fcc), body-centered-cubic (bcc), and hexagonal-closed-packed (hcp) crystal structures. Aluminum, copper, nickel, and gold, along with γ-iron (austenite), are

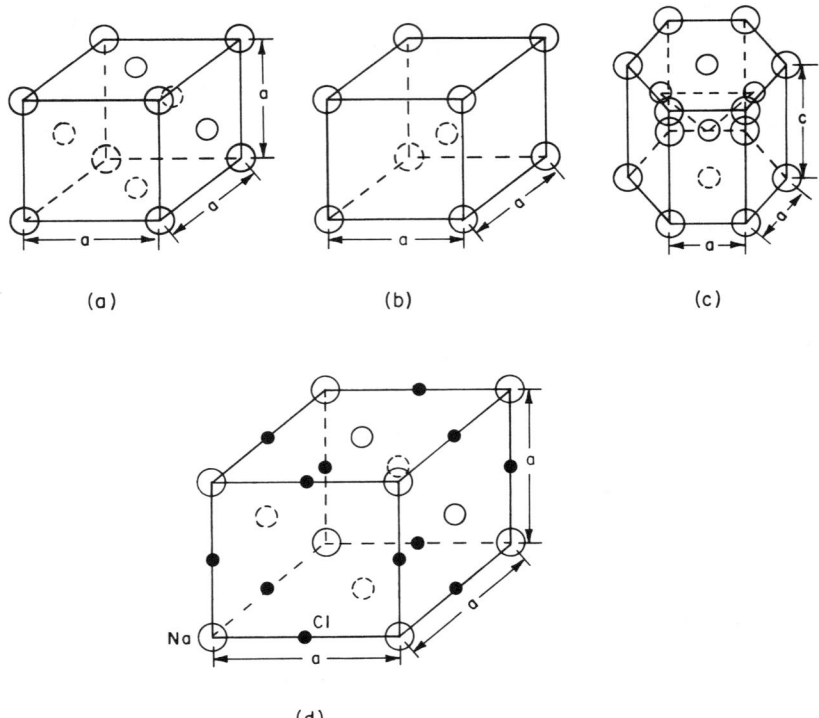

FIG. 3. Unit cells for four common crystal structures: (a) fcc, (b) bcc, (c) hcp, and (d) NaCl type.

fcc crystals; niobium, molybdenum, chromium, and α-iron (ferrite) are bcc; whereas zinc, magnesium, and cadmium are hcp. As a fourth example, Fig. 3d shows the crystal structure of a typical ionic crystal such as NaCl or LiF. It is easy to confirm that this "NaCl crystal structure" is based on the fcc point lattice but that each lattice site such as shown in Fig. 3a has four atoms associated with it, one (e.g., Na) placed on the site and three (Cl) placed at positions $(0, \frac{1}{2}a, 0)$, $(\frac{1}{2}a, 0, 0)$ and $(0, 0, \frac{1}{2}a)$ with respect to it.

A good deal of the present quantitative understanding of plastic deformation in crystalline materials is due to Taylor and coworkers, in particular to Taylor and Elam (1923, 1925). Their pioneering experiments carried out in the 1920s again firmly established the crystallographic nature of slip, but then with the aid of X-ray diffraction. They made a detailed study of aluminum single crystals and interpreted the kinematics of deformation in terms of the crystallography. For fcc aluminum single crystals they identified the slip planes as the family of octahedral $\{111\}$ planes and the slip directions as the particular $\langle 110 \rangle$ type directions lying in the $\{111\}$ planes. Figure 4 illustrates the (111) plane and the $[10\bar{1}]$ direction, one of the three crystallographically identical $\langle 110 \rangle$ type directions lying in that plane. Note that the Ewing and Rosenhain (1900) conclusion that crystal structure is preserved by plastic shearing requires that slips occur in units of "perfect lattice vectors" along the $\langle 110 \rangle$ directions. In Fig. 4 the shortest of these along $[10\bar{1}]$ is $a/2$ $[10\bar{1}]$, as indicated. Furthermore, as there are 4 independent (but crystallographically identical) $\{111\}$ planes and 3 $\langle 110 \rangle$ type directions in each plane, there are 12 distinct "slip systems" of the type $\{111\}\langle 110 \rangle$ in fcc crystals.

Taylor and Elam (1923, 1925) had also established for their aluminum crystals what is now commonly referred to as the Schmid law (1924) of a critical resolved shear stress for plastic yielding. They phrased it as follows: "Of the twelve crystallographically similar possible modes of shearing, the one for which the component of shear-stress in the direction of shear was

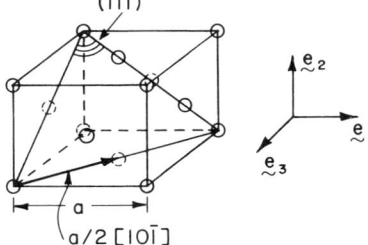

FIG. 4. Unit cell and one of the 12 crystallographically identical slip systems for fcc crystals.

greatest was the one which actually occurred" (1925, p. 28). Taylor and Elam had therefore identified the component of shear stress resolved in the slip plane and in the direction of slip, the "resolved shear stress," as the important component of stress in promoting slip. The term *mode of shearing* referred to the choice of active "slip systems" defined, as above, as the combination of slip plane and slip direction. We shall adopt this nomenclature and refer to a particular set of "active" slip systems as the "slip mode."

During this same period (1920–1930) research in Germany on crystalline plasticity was conducted by Polanyi, Masing, Schmid, and Orowan, among others. They had reached many of the conclusions during their studies of zinc (with a hcp crystal structure) that Taylor and Elam had with fcc aluminum. In particular, in 1922 Polanyi had not only observed crystallographic slip in zinc but had also noted in polycrystalline zinc that grains with initially random crystallographic orientations tended to assume preferred orientations after finite straining, i.e., they developed texture. In 1924 Schmid, using data on zinc single crystals subject to tension, suggested that plastic yield would begin on a slip system when the resolved shear stress reached a critical value, independent of the orientation of the tensile axis and thus of other components of stress resolved on the lattice. This was a clear statement of the Schmid law. It should be pointed out, however, that although experiments conducted in uniaxial tension or compression often yield an approximate confirmation of Schmid's Law, deviations from it are likely and have been found, as later discussion will show.

The early work of the 1920s included some important observations of strain-hardening behavior in crystalline plasticity. Taylor and Elam (1923, 1925) in particular noted that the individual slip systems hardened with strain and that slip on one slip system hardens other slip systems, even if the latter are not active. This is known as "latent hardening," and since it plays an important role in crystal mechanics the Taylor and Elam (1925) results will now be briefly discussed.

Figure 5a illustrates a single crystal in tension, oriented for single slip, i.e. the resolved shear stress is highest on the slip system $(\bar{1}11)$ $[110]$. The vectors **s** and **m** are unit vectors lying in the $[110]$ and $[111]$ directions, respectively, so that $(\mathbf{s}, \mathbf{m})$ now defines the slip system that is denoted the "primary" slip system. For the present the crystal is modeled as rigid–plastic, since the points being illustrated are not substantially affected by elasticity. Figure 5d is Taylor and Elam's stereographic plot of the orientation of the tensile axis, with respect to the crystal axes, as a function of the extensional strain; the point marked 0 represents the initial orientation. Now as the crystal slips, the orientation of the tensile axis *AB* changes with respect to the crystal axes and therefore to the slip direction **s** and the slip plane normal

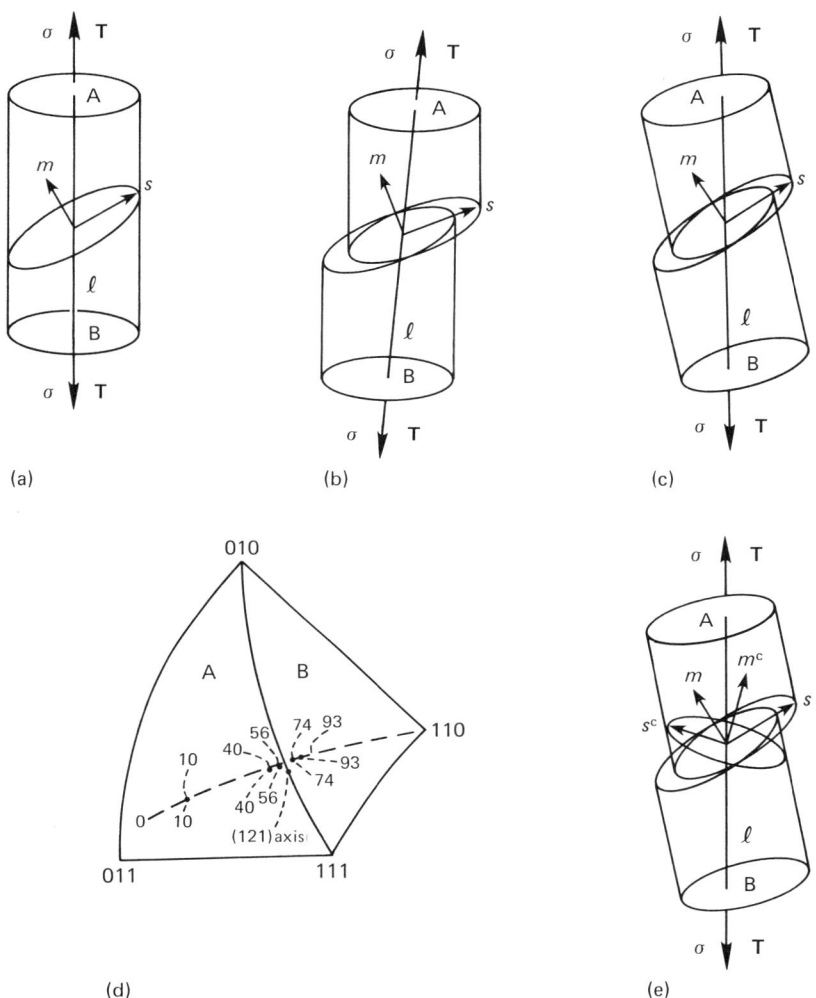

FIG. 5. Schematic diagrams for a crystal undergoing single slip in tension along the material fiber ℓ. (a) ℓ and AB are constrained to remain vertical, which causes a rigid lattice rotation (b to c). (d) Taylor and Elam's (1925) result for the change in crystallographic orientation of ℓ. (e) The relative orientation of the so-called conjugate slip system.

m, as shown in Fig. 5b. In fact, since most tensile machines constrain the loading axis to remain fixed in orientation relative to the laboratory (e.g., vertical in Fig. 5), the crystal undergoes a rigid rotation as shown in Fig. 5c. The result is that the slip direction rotates toward the tensile axis about an axis **r** orthogonal to both, where

$$\mathbf{r} = \mathbf{s} \otimes \mathbf{T}/[1 - (\mathbf{s} \cdot \mathbf{T})^2]^{1/2}. \tag{2.1}$$

On the stereographic projection the pole of the tensile axis appears to rotate along the great circle common to **s** and **T**. Now consider increments of extension, shear, and rotation starting at some current stage taken as reference, as indicated in Fig. 5e. The slip system that is said to be "conjugate" to the primary system, $(11\bar{1})[011]$, is also indicated. Conjugacy refers to the fact that single slip on the system $(11\bar{1})[011]$ will induce rotations during tensile straining that lead to large resolved shear stresses and inevitable slip on the system $(11\bar{1})[011]$. The rate of rotation $\dot{\beta}$ about **r** is easily shown to be

$$\dot{\beta} = \dot{\gamma}(\mathbf{T} \cdot \mathbf{m})[1 - (\mathbf{s} \cdot \mathbf{T})^2]^{1/2}, \tag{2.2}$$

where $\dot{\gamma}$ is the plastic shearing rate on the primary system. The resolved shear stress τ is given by

$$\tau = \sigma(\mathbf{m} \cdot \mathbf{T})(\mathbf{s} \cdot \mathbf{T}) = S\sigma, \tag{2.3}$$

with a similar relation holding for the conjugate system. In Eq. (2.3) S is defined as the "Schmid factor." Taylor and Elam (1925) noted that if the strain-hardening rate on the active slip system (call this the "self-hardening rate") were equal to the latent-hardening rate of the conjugate slip system, conjugate slip would become active when the two resolved shear stresses were equal. This occurs when the tensile axis rotates to any point on the $[111]$–$[010]$ symmetry boundary in Fig. 5d. The dotted arc in Fig. 5d is the great circle on the stereographic projection whose pole is **r** and is the path of rotation for the tensile axis. The "tick" marks indicate the crystallographic orientation of the tensile axis expected from pure single slip, and the dots the orientations measured by Taylor and Elam (1925) at the specified extensional strains (in percent). Until the tensile axis approaches the symmetry boundary, the predictions of single slip are closely matched by experiment. The X-ray measurements indicate, however, that the tensile axis "overshoots" the symmetry boundary, which leads to a larger resolved shear stress on the conjugate, i.e., latent, system. The measurements also indicate that the rotations are nonetheless much less than expected from single slip, which indicates that conjugate slip does occur, but at a lower rate than primary slip. If the double mode of slip were symmetric once the symmetry boundary were reached, the tensile axis would rotate to the $[121]$ pole and remain there. In this case the rotations caused by primary and conjugate slip exactly cancel. Such behavior was observed in some of their other tests. Taylor and Elam (1925) had thus demonstrated two very important aspects of crystal strain hardening: (1) Slip systems are hardened by slip on other systems (whether they themselves are active or not) and (2) this latent hardening is at least comparable in magnitude to self-hardening. The observations of "overshooting" indicated that latent-hardening rates are often somewhat larger than self-hardening rates, so that a slightly larger shear stress is required on the previously latent system to activate it. Taylor and Elam summarized the hardening behavior as follows: "It is found that though the double slipping does, in fact begin when the two planes get to the position in which they make equal angles with the axis, the rate of slipping on the original slip-plane is sometimes greater than it is on the new one" (1925, p. 29). They went on to note that "the process cannot be followed very far, however, because the specimen usually breaks when only a comparatively small amount of double slipping has occurred" (1925, p. 29). As will be discussed later in Section IV,C, crystals typically undergo necking and intense localized shearing after double slip begins.

In passing, we note an interesting consequence of the kinematics of crystalline slip, again using the rigid plastic crystal model of Fig. 5 for a crystal in tension. According to the Schmid rule, the resolved shear stress τ must remain at the critical yield value τ_c for slip to continue; τ_c increases with shear strain at a rate given by $\dot{\tau}_c = h\dot{\gamma}$, where h is the current strain-hardening (self-hardening) rate of the active slip system. Differentiating Eq. (2.3) with respect to time, evaluating $\dot{\mathbf{s}}$ and $\dot{\mathbf{m}}$ by noting that

$$\dot{\mathbf{s}} = \mathbf{\Omega}^* \cdot \mathbf{s} \quad \text{and} \quad \dot{\mathbf{m}} = \mathbf{\Omega}^* \cdot \mathbf{m}, \tag{2.4}$$

where

$$\mathbf{\Omega}^* = -(\mathbf{sT} - \mathbf{Ts})\dot{\gamma}(\mathbf{T} \cdot \mathbf{m}) \tag{2.5}$$

is the rigid lattice spin rate, and setting the result equal to $h\dot{\gamma}$ yields

$$\dot{\sigma} = \left(\frac{h}{\cos^2\phi \cos^2\theta} + \frac{\sigma \cos 2\phi}{\cos^2\phi}\right)\dot{\varepsilon}. \tag{2.6}$$

Here $\dot{\varepsilon}$ is the current rate of extension along the tensile direction, and we recall that these rates are taken from the current state, which is also taken as the reference state of strain. The first term, due to strain hardening, leads to an increase in true tensile stress with extension (i.e., if $h > 0$), whereas the second term is due to the change in τ, at fixed σ, caused by rotation of the lattice with respect to the fixed tensile axis. If $\cos 2\phi < 0$ and $h/\sigma < |\cos 2\phi|/\cos^2\theta$, the instantaneous modulus governing extension is negative, and the crystal softens. This softening is associated with purely geometrical effects and not material strain softening and for this reason is called "geometrical softening." Evidently, if lattice rotations that cause geometrical softening occurred locally within the crystal's gauge section, they would promote nonuniform and perhaps localized deformation. This is true for single crystals as well as for individual grains of polycrystals. We will show later that geometrical softening does indeed play a vital role in the phenomenology of crystalline slip.

Thus by 1930 much of the macroscopic phenomenology of crystalline plasticity, aside from strain rate effects, had been documented. The fundamental connection between the resolved shear stress and resolved shear strain on the slip systems had been clearly noted, along with the important notions of self-hardening and latent hardening of slip systems. Temperature had long been known to have an important influence on strength, as shown, for example, in Fig. 6, which is taken from Mitchell's 1964 review. A tenable micromechanistic theory, however, was missing. For example, in 1926 Frenkel made a simple but plausible calculation of the theoretical shear strength of a perfect crystal. The result indicated that the shear stress required to slide an entire unit area of crystal plane over the underlying plane by one lattice spacing is not less than about $\frac{1}{10}$ the elastic shear modulus, a value many orders of magnitude higher than observed yield strengths (see Fig. 6). As Orowan (1963) pointed out much later, it was also realized that Becker's thermal fluctuation theory could not resolve this large discrepancy. The satisfactory resolution of this problem led to the theoretical discovery of crystal dislocation in 1934.

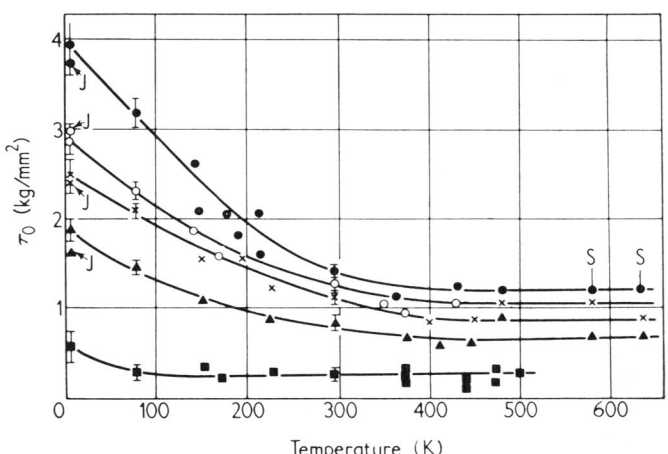

FIG. 6. Data for the critical resolved shear stress for initial yield versus temperature for various materials. ●, 70:30; ×, 90:10; ○, 80:20; ▲, 95:5; ■, copper. J. Jameson and Sherrill; S. Suzuki. Taken from Mitchell (1964).

B. Dislocations

Figures 7–9 were all published in 1934 and they describe essentially the same crystal defect known as an "edge" dislocation. In all three cases the dislocation is drawn in a simple cubic crystal and was used by Taylor (1934), Orowan (1934), and Polanyi (1934) to explain the micromechanics of slip. As is easily appreciated from the Taylor model in Fig. 7, slip is caused by the glide of the dislocation across the slip plane one lattice spacing at a time. The result is to displace the material on either side of the plane by the unit lattice spacing. Taylor (1934) argued that the shear stress required to cause incremental dislocation motion would be very low and thus propagation of such defects would result in shear strengths consistent with those observed. Crystal symmetry is preserved after glide and, in fact, even if the dislocation is trapped within the crystal, as in Fig. 7b or 7e, since the disregistry in the crystal structure is highly localized in the dislocation "core" region.

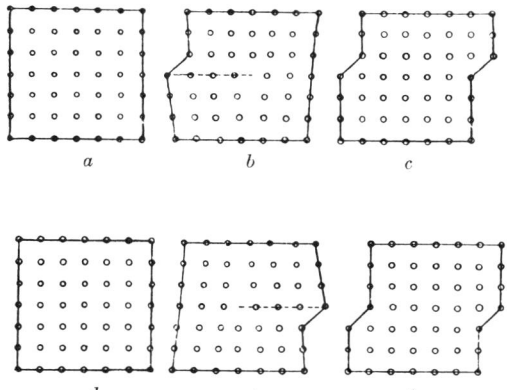

FIG. 7. Taylor's 1934 model picture for a crystal dislocation. Glide of the dislocation causes material on either side of the slip plane to slip by one lattice spacing. a, b, c, positive dislocation; d, e, f, negative dislocation.

FIG. 8. Orowan's 1934 model for a crystal dislocation.

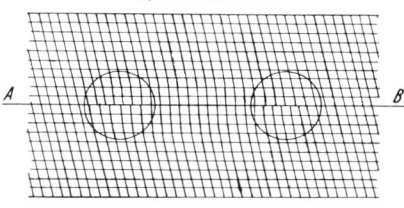

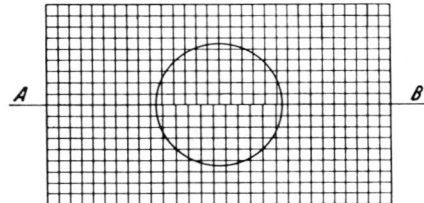

Fig. 9. Polanyi's 1934 model for a crystal dislocation.

The precise specification of a dislocation involves both the displacement vector for the material and the dislocation line, including its sense; in Fig. 7 the dislocation line is directed into the plane and orthogonal to the displacement vector. In 1939 Burgers proposed a dislocation for which the displacement vector would lie parallel to the line. The two dislocations are known as "edge" and "screw" dislocations, respectively, and actually coexist as shown in Fig. 10. Figure 10 illustrates that a glide dislocation can be formed within a continuum by making a cut over the surface S, bounded by the line C, and displacing material on either side of the cut by the vector **b**. The cut is then rejoined so that the material across the cut surface has been permanently slipped by **b**; **t** is the unit tangent vector to C and thus the segments of C for which $\mathbf{t} \cdot \mathbf{b} = 0$ are "edge" segments and those for which $\mathbf{t} \cdot \mathbf{b}/b = \pm 1$ are "screw" segments. The vector **b** is known as the Burgers vector and for crystal dislocations it must be one of the perfect lattice vectors if slip is to preserve crystal symmetry. The Schmid rule can now be reinterpreted in terms of a critical force required either to move or to generate dislocations. This force is defined to be work conjugate to the gliding motion and hence is equivalent to the resolved shear stress multiplied by the magnitude of **b**.

Taylor's 1934 analysis also provided a basis for a quantitative understanding of strain hardening. He recognized that dislocations were sources of internal elastic strain and stress and that dislocations would interact with each other through their mutual stress fields. Figure 11 shows two of Taylor's (1934) interaction problems involving dislocations of opposite and similar signs. The stress fields for these dislocations had already been calculated in linear isotropic elastic media by Timpe (1905), Volterra (1907), and Love

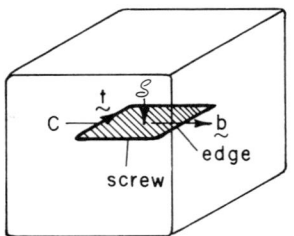

Fig. 10. A dislocation loop in a continuum is produced by first making a cut on surface S and displacing material across it by **b**.

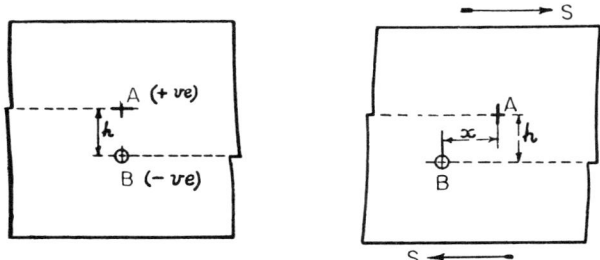

FIG. 11. Taylor's 1934 model for the elastic interaction of two edge dislocations gliding by each other.

(1927). The interactions depicted in Fig. 11 can be either attractive or repulsive, depending on the relative sign of the dislocations, but in either case the maximum resolved shear stress acting on the dislocations required to slip them by each other occurs at $|x| = h$ and is equal to $Gb/2\pi h$, where G is the elastic shear modulus. Thus the shear stress required to move dislocations within a "substructure" of other dislocations is considerably larger than that required to move an isolated dislocation. Furthermore, once dislocations and slip have been generated on a plane, it becomes easier to generate slip on planes further removed from those currently active. However, this leads to an increase in dislocation density, a decrease in the average spacing between dislocations, and hence to smaller values of h and an increase in the value of the "passing stress."

One model picture envisioned by Taylor for the dislocation arrangement is shown in Fig. 12. To see how this leads to a specific strain-hardening relation we consider the following dimensional arguments. Taylor assumed that dislocations would be generated at one surface of the crystal and would move a distance x along the slip plane; the maximum value x may have is L, which is either the dimension of the crystal or the distance to a boundary. The average distance dislocations move is then $L/2$. The average spacing between active slip planes in Fig. 12 is $d/2$, the spacing between dislocations

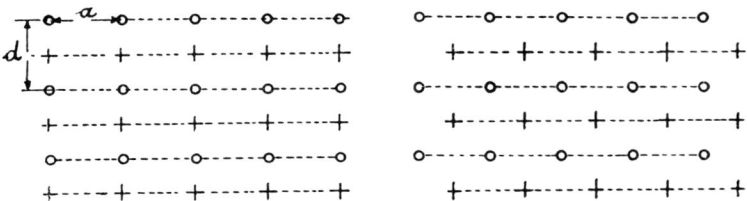

FIG. 12. Taylor's 1934 idealized arrangement for the dislocation substructure in a finitely deformed crystal.

is a, and so the average "dislocation density" ρ is $2/ad$[†]; the shear strain accumulated in this substructure, γ, is $L^2 \frac{1}{2} b/L = \rho L b/2 = Lb/ad$. Now since the passing stress is proportional to h^{-1} and h scales with $\rho^{-1/2}$ if the dislocation distribution is regular, it follows from dimensional considerations alone that the passing stress is proportional to $\rho^{1/2}$ or actually to $Gb\rho^{1/2}$, and hence to $G(\gamma b/L)^{1/2}$. Finally, if τ_i represents the shear stress required to move an isolated dislocation, then the stress–strain law derived above has the form

$$\tau_c = \tau_i + \kappa G(b\gamma/L)^{1/2} = \tau_i + \eta Gb\rho^{1/2}, \qquad (2.7)$$

where κ and η are constants determined by the geometry of the distribution.

Although Taylor's model arrangement for the distribution of dislocations is oversimplified, it serves to focus on an important cause of strain hardening—the elastic interaction between dislocations. The dimensional form of his strain-hardening law reflects the experimentally observed fact that, in metals at least, strain-hardening rates generally decrease with strain at finite strains. Furthermore, the proportionality between the flow stress and the square root of dislocation density is also well documented, especially in fcc metals, although, as we will see, this dimensional rule follows from other micromechanical models for dislocation interaction as well. Finally, we note that the indicated dependence of τ_c on the microstructural dimension L reflects the well-established fact that yield and flow strength increase with decreases in grain (or subgrain) size, or with refinement of the microstructure in general.

1. *Some Basic Properties of Dislocations in Crystals*

A linear elastic theory of dislocations, notably due to Volterra (1905), was available at the time Taylor published his 1934 analysis of dislocation interactions. This theory has since been extended to include elastic anisotropy and the development of a variety of techniques for solving for the elastic fields of complex arrays of dislocations (see, for example, Hirth and Lothe, 1968; Asaro *et al.*, 1973; and Asaro and Barnett, 1976). For an infinitely long and straight dislocation, for example, it follows from this linear theory that the elastic energy per unit length of dislocation line can be expressed as

$$E = K_{ij} b_i b_j \ln(R/r_0), \qquad (2.8)$$

where $\mathbf{K}$ is an energy factor matrix that depends on the direction of the dislocation line with respect to the crystal axes and R and r_0 are outer and inner cutoff radii not specified in the linear theory. The quadratic form in the

---

[†] Dislocation density is often measured as the number of dislocation lines penetrating a unit area of plane, as here, or as the total length of dislocation line per unit volume.

TABLE I

Crystal structure	Burgers vector	Slip plane (at 20°C)
fcc	$a/2\langle 110\rangle$	$\{111\}$
bcc	$a/2\langle 111\rangle$	$\{110\}$ and $\{112\}$
NaCl	$a/2\langle 110\rangle$	$\{110\}$

components of the Burgers vector indicates that stable dislocations take on the shortest Burgers vector possible—the energetics of crystal bonding, along with the evidence that slip does not affect crystal structure, indicate that **b** is a perfect lattice vector. Thus, with reference to Fig. 4, $\mathbf{b} = a/2\langle 110\rangle$ for fcc crystals. The slip systems for fcc metals crystals are then of the type $\{111\}$ $a/2\langle 110\rangle$, where a is the lattice parameter shown in Fig. 3a. Table I indicates some common slip systems for the cubic crystal types illustrated in Fig. 3.

Although the Burgers vector of a crystal dislocation is a perfect lattice vector, it is possible for dislocations to "dissociate" to form "partial dislocations," as illustrated in Fig. 13 for a glide dislocation in a fcc crystal. The extent of dissociation depends not only on the crystal goemetry, but also on the chemistry. A simple analysis, using linear isotropic elasticity, follows.

The glide force acting on a segment of dislocation line is by definition the force that is work conjugate to the gliding motion. Then if τ is the shear stress resolved in the plane of glide at the dislocation line and in the direction of the Burgers vector, the force is τb per unit length of line.[†] In general, the force per unit length, due to stress σ, acting on a segment of dislocation with unit tangent **t** is (Peach and Koehler, 1950)

$$\mathbf{f} = \mathbf{t} \otimes (\mathbf{b} \cdot \boldsymbol{\sigma}). \tag{2.9}$$

For a general discussion of forces on elastic singularities the reader is referred to Eshelby (1951, 1958). Now it can be shown that the in-plane tractions

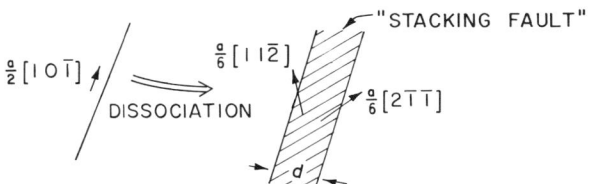

FIG. 13. Screw dislocation dissociating in an fcc crystal. Partial dislocations are produced by the dissociation of perfect dislocations.

[†] A more precise definition of τ accounting for finite lattice elasticity is given in Section III.

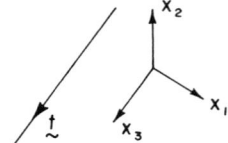

Fig. 14. Coordinate frame used to reference the elements of the **K** matrix.

acting on any plane containing an infinitely long and straight dislocation line are given by $\mathbf{T} = 2\mathbf{K} \cdot \mathbf{b}/d$, where d is the distance from the line that is taken to be positive if it falls to the right of the dislocation line (see, for example, Barnett and Asaro, 1972). With reference to Fig. 14, x_1, x_2, x_3 is a mutually orthogonal triad with x_3 along the dislocation line. The point in question lies along x_1, which is to the right. In this coordinate frame, the energy factor matrix **K** is diagonal, with $K_{11} = K_{22} = G/[4\pi(1-v)]$ and $K_{33} = G/4\pi$; v is Poisson's ratio. It is now easy to show that any two parallel dislocations *repel* each other on their common plane with a force given by $2\mathbf{b}^{(1)} \cdot \mathbf{K} \cdot \mathbf{b}^{(2)}/d$; if this quantity is negative, they attract each other. However, as the partial dislocations dissociate according to the reaction

$$a/2[10\bar{1}] = a/6[11\bar{2}] + a/6[2\bar{1}\bar{1}], \tag{2.10}$$

a "faulted" region is created across which the equilibrium atomic stacking sequence is disrupted. It is convenient to express the energy associated with this "stacking fault" as a surface tension, i.e., the reversible isothermal work required to create a unit area of fault, γ_s. Hirth and Lothe (1968) quoted some values for these as follows: aluminum, 200 ergs/cm^2; silver, 17 ergs/cm^2; and copper, 73 ergs/cm^2. The total force including the fault tension acting between these two partial dislocations per unit length is then

$$2\mathbf{b}^{(1)} \cdot \mathbf{K} \cdot \mathbf{b}^{(2)}/d - \gamma_s,$$

and at equilibrium this yields

$$d = [2\mathbf{b}^{(1)} \cdot \mathbf{K} \cdot \mathbf{b}^{(2)}]/\gamma_s. \tag{2.11}$$

If the undissociated dislocation in Fig. 13 were a perfect screw dislocation, for example, the estimated equilibrium spacings d would be on the order of 2 Å for aluminum and 20 Å for copper. In practical terms this suggests that partials are not "extended" in aluminum, but may well be in copper and almost certainly would be in silver.

An important feature of dissociated, or extended, dislocations is they may not readily undergo such micromechanical processes as "cross-slip." Figure 15 is an illustration of such a process in a fcc crystal adapted from Asaro and Rice (1977). Dislocations in fcc crystals may glide in any {111} plane containing the Burgers vector and the dislocation line. However, the individual

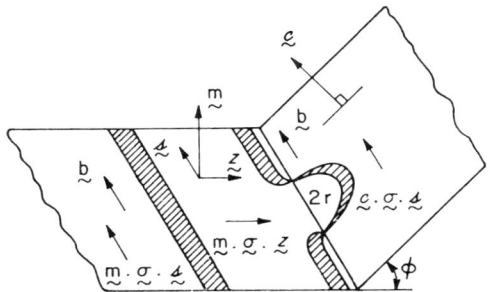

FIG. 15. Idealized model for the cross-slip of an extended screw dislocation. Note that whereas the perfect dislocation Burgers vector lies in the cross-slip plane, neither of the partial Burgers vectors do. The partial dislocations must first constrict before they can slip on the cross-slip plane.

partial dislocations do not lie in the cross-slip plane and so the extended dislocation on the primary slip system $(\mathbf{s}, \mathbf{m})$ must first develop a constricted segment that can bow out on the cross-slip plane. Once the dislocation has bowed by a critical amount on the cross-slip plane, it may then continue to glide on an adjacent primary plane. Micromechanical processes such as cross-slip lead to deviations from the Schmid law, since stress components other than the resolved shear stress τ_{ms} affect the constriction and bowing process.

2. Strain Hardening, Dislocation Interactions, and Dislocation Multiplication

Taylor's (1934) original analysis of strain hardening focused on the elastic interaction of essentially straight nonintersecting dislocations belonging to one slip system. An entirely different and revealing picture of dislocation substructure emerges when interactions between dislocations belonging to two or more systems are accounted for. Figure 16 illustrates an interaction for fcc crystals proposed by Lomer (1951) between dislocations having a primary–conjugate relationship. Suppose that the conjugate slip system, $(1\bar{1}1)a/2[\bar{1}\bar{1}0]$, is not active but that a certain initial "grown-in" density of its dislocations exists in the crystal. When the primary system, taken to be $(11\bar{1})a/2[101]$, yields, its dislocations necessarily intersect conjugate dislocations along the line common to both slip planes, [011]. At the juncture the two dislocations merge according to the reaction

$$a/2[101] + a/2[\bar{1}\bar{1}0] = a/2[0\bar{1}1], \quad (2.12)$$

where $\mathbf{b}^L = a/2[0\bar{1}1]$ is the Burgers vector of the product Lomer dislocation. The basis for the reaction lies in the fact that the elastic energies of all three

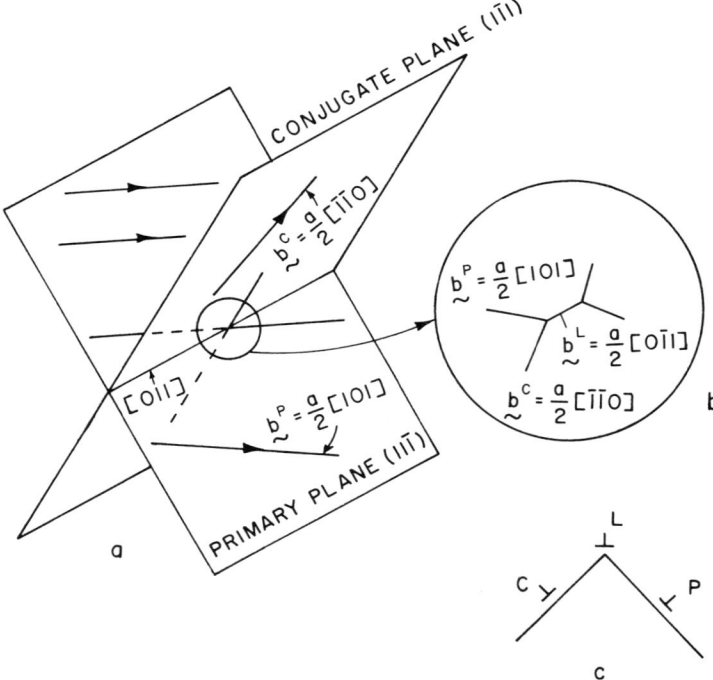

FIG. 16. Interaction between dislocations in a primary–conjugate relationship, leading to Lomer dislocations that act as barriers to the motion of both primary and conjugate slip.

dislocations involved in reaction (2.12) are equal and so the formation of the Lomer dislocation leads to a net decrease in the total elastic energy. A more rigorous method of analyzing dislocation reactions of this type has been given by Asaro and Hirth (1974). The Lomer dislocation is an edge dislocation, since $\mathbf{t}^L = 1/\sqrt{2}\,[011]$ and $\mathbf{t}^L \cdot \mathbf{b}^L = 0$, and if it were to glide, it would have to do so on the (100) plane containing $\mathbf{t}^L$ and $\mathbf{b}^L$. Since glide on {100} type planes is very difficult in fcc crystals, the Lomer dislocation is immobile and thus impedes further glide of the original primary dislocation. What is also evident is that glide of the conjugate dislocations would also be more difficult should the conjugate system become highly stressed. Figure 16c shows a view of the two slip planes along the [011] direction. Subsequent primary *and* conjugate dislocations are impeded in their motion, since they are both *repelled* from the Lomer dislocation with a force equal to $Ga^2/[8\pi(1-v)d]$. Both the primary and conjugate slip systems are then hardened in a roughly symmetrical way, which helps explain Taylor and Elam's (1923, 1925) observations of latent hardening.

Another type of dislocation reaction leading to strain hardening and latent hardening is illustrated in Fig. 17. The primary slip system and primary dislocations are again taken to be $(11\bar{1})a/2[101]$ and the other system, whose dislocations are referred to as "forest" dislocations, is $(111)a/2[\bar{1}10]$. When the two dislocations meet, they undergo the reaction

$$a/2[101] + a/2[\bar{1}10] = a/2[011], \qquad (2.13)$$

i.e., they react to form a product segment lying in both planes along the $[110]$ direction with the Burgers vector along $[011]$. $a/2[011]$ lies in the $(11\bar{1})$ plane and so it may glide in the primary plane. However, since the "nodes" shown in Fig. 17 are energetically stable, they essentially pin, and thus impede, the motion of the original primary *and* forest dislocations. If adjacent primary

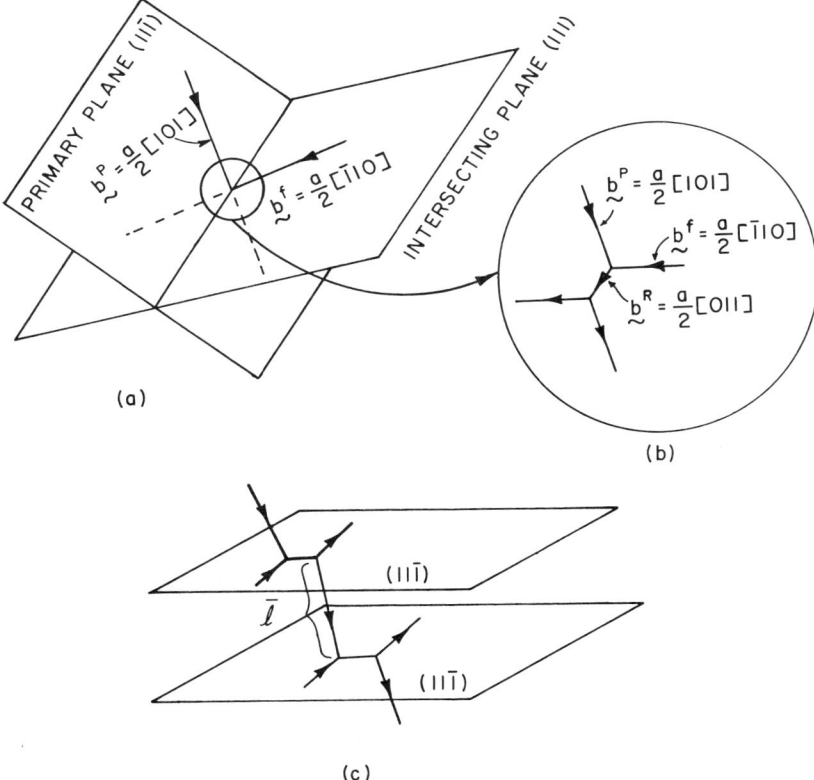

FIG. 17. (a, b) Interaction between primary dislocations and "forest" dislocations, leading to the formation of nodes and a segmented dislocation substructure. This also leads to barriers to the motion of both the primary and forest dislocations. (c) Process by which a dislocation network is produced from such interactions.

planes are considered [and then adjacent (111) planes], the effect is to produce a three-dimensional network, as suggested in Fig. 17c. As a consequence, dislocations are not arranged in the ideal fashion envisioned in the Taylor model of Fig. 12, but in a segmented distribution such as suggested by the idealized model of Fig. 17c.

The determination of the crystal's strength depends on the resistance of the network to the motion of dislocation *segments*. If we assume for a moment that the nodes are rigid and act as pinning sites, the segments move by bowing out in the manner illustrated in Fig. 18a. The simplest analyses of this follows if the dislocation is taken to be a line with average tension $\mathscr{L}$ and if the bowing is assumed to proceed in circular arcs. Then at any stage that is stable, $\tau b = \mathscr{L}/r$, where τ is the resolved shear stress for the dislocation and r is the radius of curvature of the bowing segment. If the shear stress exceeds a critical value $\tau_c > 2\mathscr{L}/b\bar{l}$ corresponding to $r = \bar{l}/2$, the bowing becomes unstable and the loop closes on itself, as shown in Fig. 18b. The result is that a new dislocation loop is produced and the original segment restored, whereupon it may bow again; the critical stress required for this is given by $\tau_c = 2\mathscr{L}/b\bar{l}$.

The mechanism just described constitutes one possible source of dislocations and represents a type of source introduced by Frank and Reed (1950). In well-annealed crystals dislocation densities can be on the order 10^6 cm^{-2} and even less in certain cases. However, after modest plastic strains of, say, 10%, this may increase to 10^{10}, as indicated in Fig. 19 taken from Ashby (1971). Dislocation sources then play a vital role in maintaining plastic flow. Another dislocation source involving double cross-slip was suggested by Koehler (1952). As shown in Fig. 20 (see Fig. 15), once a dislocation segment has cross-slipped onto the cross-slip plane, it may then cross-slip again onto another primary plane. This segment may then bow out and operate as a dislocation source.

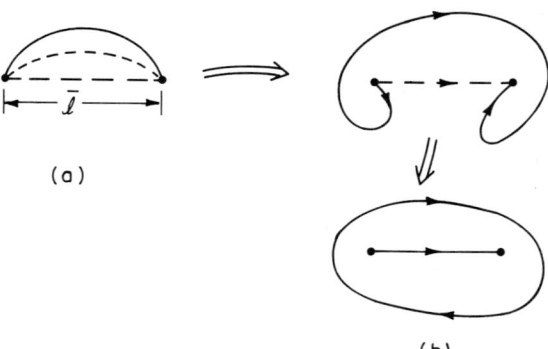

FIG. 18. (a) Idealized model for the movement of dislocation segments pinned at nodes. (b) Unstable bowing of the segments resulting in the production of new mobile dislocations.

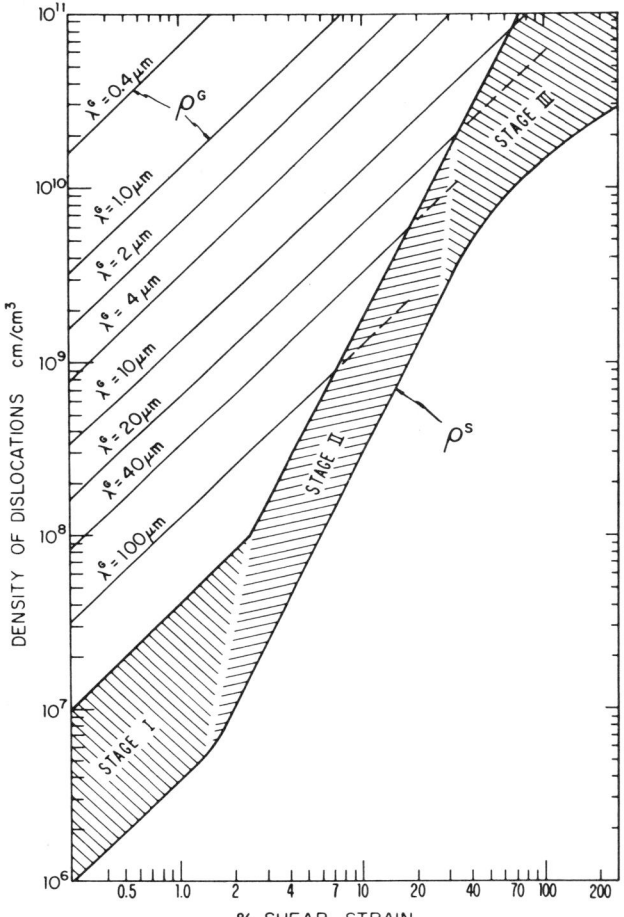

FIG. 19. Plot showing how the dislocation density increases by several orders of magnitude during plastic straining (Ashby, 1971).

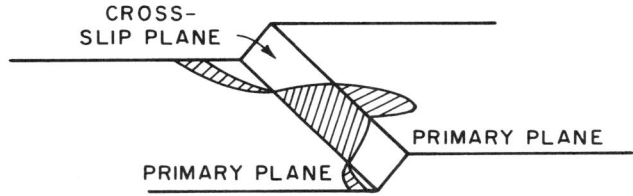

FIG. 20. Koehler's model of the so-called double cross-slip mechanism for dislocation multiplication. Note that cross-slip itself results in straining of the cross-slip system, but that the ratio of strain produced on this system to that on the primary system is very small.

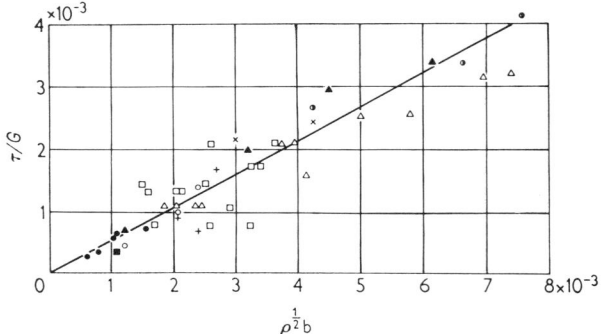

FIG. 21. Collection of data for single crystals for the flow stress on a slip system versus the average dislocation density. ●, copper; ×, copper–0.8% aluminium; △, copper–2.2% aluminum; □, copper–4.5% aluminum; ■, nickel; ○, nickel–40% cobalt; +, nickel–60% cobalt; ◐, silver; ▲, copper. From Mitchell (1964).

Figure 19 shown earlier indicates how dislocation density ρ depends on shear strain. Figure 21 shows some corresponding relationships for shear stress and $\rho^{1/2}$. As mentioned earlier, most models for strain hardening result in such relationships. For example, if the flow stress is determined by the stress τ_c, which is proportional to $\bar{l}^{-1}$, then by dimensional arguments $\bar{l}^{-1}$ should be proportional to $\rho^{1/2}$.

C. Other Strengthening Mechanisms

Most crystalline materials derive their strength from a combination of the intrinsic mechanisms described previously and the interaction of dislocations with other microstructural features. The yield strength and strain-hardening characteristics of polycrystals depend in an important way upon grain size, as shown by the two examples in Fig. 22. In both these cases a close correspondence with the well-known Hall–Petch (1951, 1953) relation between yield strength σ_y and average grain diameter d is found, $\sigma_y = \sigma_i + k_y d^{-1/2}$, at least for the range of grain sizes included by the data. Although the Hall–Petch relation bears a superficial resemblance to Taylor's flow stress equation (2.7), the physical underpinning of it is different. The yield strength σ_y is determined in large part by the process by which slip is transferred from grains that yield at the smallest plastic strain to the surrounding grains, and not solely by the interaction of dislocations and strain-hardening processes occurring in the grain interiors. As pointed out by Embury (1971), the local stress necessary to propagate slip may be determined by the critical conditions for (1) the unpinning of existing dislocations or the operation of

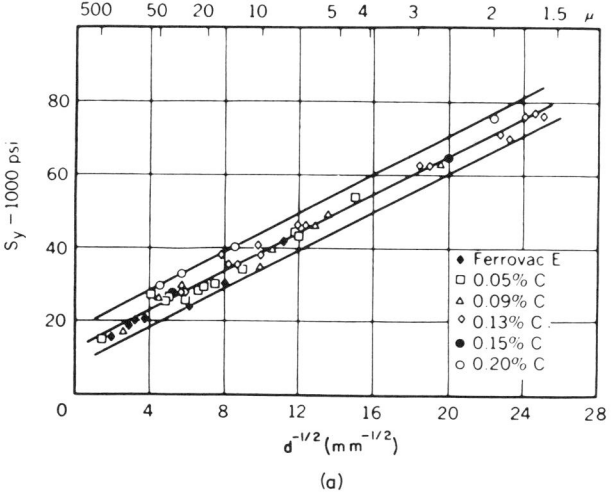

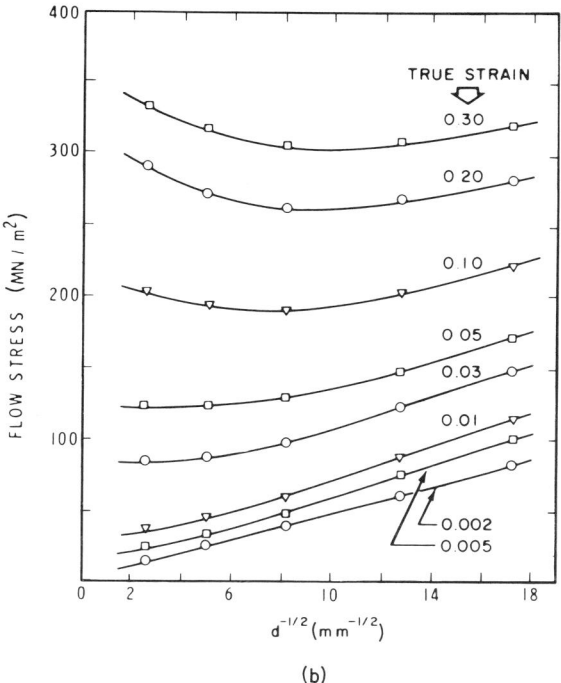

FIG. 22. (a) Morrison's (1966) data for the dependence of yield strength on grain diameter in polycrystalline iron. (b) Thompson and Baskes' (1973) data for the grain size dependence of flow stress in polycrystalline copper. Note that at small strains the data fits the Hall–Petch relationship, whereas at larger strains it does not.

dislocation sources in unyielded grains or (2) the creation of dislocations in the grain boundaries that glide into the unyielded grain. Furthermore, as is evident from the Thompson and Baskes (1973) data, the Hall–Petch relation holds, even in the range of grain sizes shown, only for proof stresses defined at plastic strains less than 0.01. The relation breaks down at higher strains, which indicates that grain size has a significant influence on strain-hardening behavior, especially at the finer grain sizes. Thompson (1975) correctly pointed out the implications of such grain size dependence regarding the development of continuum polycrystalline models, all of which have to date ignored grain size per se. In short, the data indicate that grains in polycrystals, especially grains less than 10^{-2} mm in diameter, do not strain harden as if they were single crystals subjected to comparable strains.

Finally, we note that the development of a segmented network of dislocations as described in the previous section typically evolves into the formation of dislocation cells after finite strain. An example of cell structure in single-crystal aluminum, taken from the work of Chiem and Duffy (1981), is shown in Fig. 23a. The cell walls are characterized by a high dislocation density, as described, for example, by Embury (1971) and Kuhlman-Wilsdorf (1976), and by the existence of dislocations of two or three different Burgers vectors in them. Cell size in turn plays an important role in determining strength, as shown in Fig. 23b, which is also taken from Chiem and Duffy (1981).

The present discussion does not begin to cover the wide range of strengthening and strain-hardening phenomena in crystalline materials and the importance of chemistry and microstructure. For example, topics such as solute strengthening and precipitation hardening have not been mentioned and the reader is encouraged to look elsewhere (e.g., Kelly and Nicholson, 1971). The intent rather is to present a very basic picture that serves both to identify important phenomena and construct continuum constitutive laws which allow for a more precise analysis of them. As it happens, one of the more important of these phenomena is latent hardening, and so some of the available data is discussed next.

D. Measurements of Latent Hardening

Latent hardening has been measured in essentially two ways. One method involves deforming large crystals in single slip and then machining specimens aligned for slip on another slip system from them. Yield stresses on previously latent systems are then measured and compared to the flow stress reached on the original, primary, system. The results are typically described by a variation of the "latent-hardening ratio," the ratio of these two flow stresses. A second method involves measurements of lattice rotations in single crystals

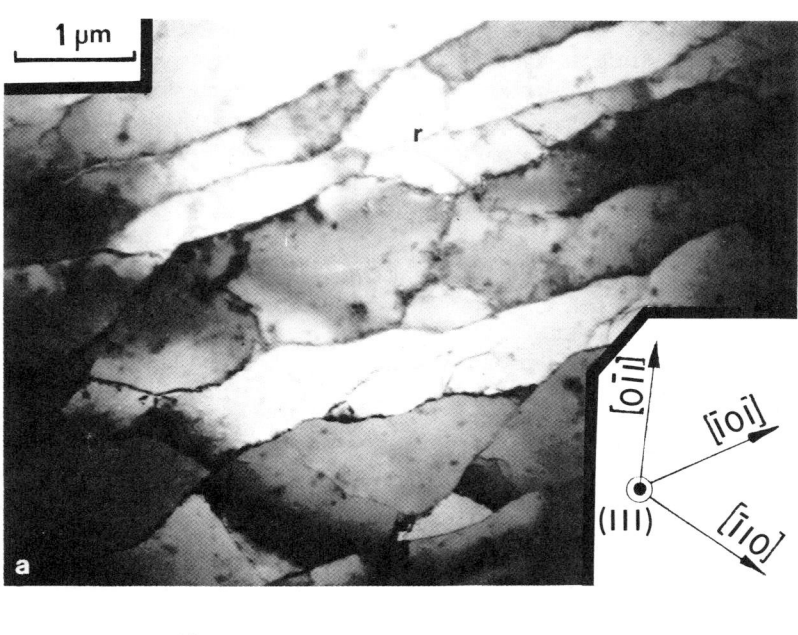
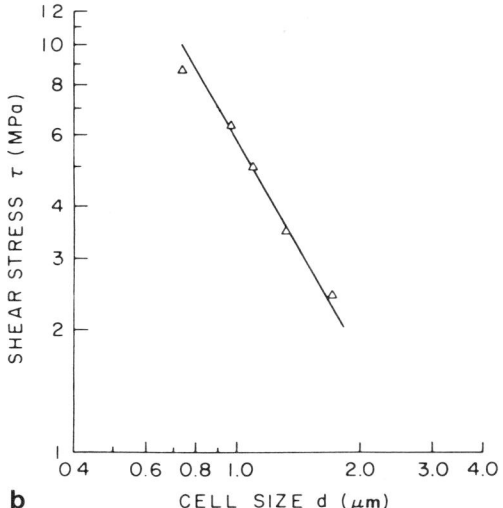

FIG. 23. (a) Dislocation cell structure in aluminum single crystals and (b) the relationship between flow strength and cell diameter (Chiem and Duffy, 1981).

that occur during tensile straining in single slip, as described in connection with Fig. 5. The rate of rotation with strain can be measured, and discrepancies with the single-slip predictions used to detect yielding on latent systems. The extent of overshoot can also be measured and used as an indication of the magnitude of latent hardening. These two techniques are not equivalent, since they subject the specimen to different strain histories; in the first only one system is active at any stage, whereas in the second at least two systems are simultaneously active at the point where yield stresses are measured for the initially latent system. To describe the experimental results quantitatively it is helpful to assume for the moment a rate-independent hardening rule of the form

$$d\tau_c^{(\alpha)} = \sum_\beta h_{\alpha\beta} d\gamma^{(\beta)}, \tag{2.14}$$

where $\tau_c^{(\alpha)}$ is the current yield strength on the α slip system, and $h_{\alpha\beta}$ the hardening rates. The off-diagonal terms in the matrix **h** represent latent hardening.

In his original analysis of polycrystals, Taylor (1938a,b) assumed that the latent-hardening rates were equal to self-hardening rates, i.e., $h_{\alpha\beta} = h$ for all α and β. He was undoubtedly motivated to assume such isotropic hardening by his experiments with Elam (1923, 1925) presented in Section II,A. It is important to note, however, that overshoots of the symmetry positions, where equal shear stresses existed on the conjugate system, of $2-3°$ were common in those experiments, indicating that slip on the initially latent conjugate system required a slightly larger resolved shear stress than on the active primary system. The Taylor isotropic rule was evidently meant to be approximate, since its background experiments actually indicated that the latent-hardening rate was slightly larger than the self-hardening rate.

A brief review of the measurements made by 1970 was given by Kocks (1970), who concluded that the average ratio of latent-hardening to self-hardening rates is nearly unity for coplanar slip systems (i.e., systems sharing the same plane) and between 1 and 1.4 for noncoplanar systems. This appears to be a very reasonable approximate range, especially for pure metals with intermediate or high stacking fault energies (e.g., Al and Cu) at finite strain. Work by Franciosi *et al.* (1980) on copper and aluminum is consistent with the earlier results in terms of the average hardening rates after finite strains, although the authors reported the variation of latent-hardening ratios with strain to be more complicated than described by Jackson and Basinski (1967) for copper and by Kocks and Brown (1966) for aluminum. Franciosi *et al.* (1980) found that the latent-hardening ratios first increase from unity to peak values of 1.6–2.2 for aluminum and to peak values nearly twice as high for copper. It is important to note, though, that these peak values occur at prestrains of only 0.2% or so. The ratios rapidly decrease at larger strains and level off at values near 1.3 and 1.5 for aluminum and copper, respectively.

The experimental study of Piercy et al. (1955) on α-brass crystals is an interesting example of earlier work on latent hardening. Their crystals were initially oriented for single slip in tension, and lattice rotations as described in connection with Fig. 5 were measured by X-ray diffraction. The point where conjugate slip began was determined "by continual microscopical examination (at 15×) of the specimen during the period of linear hardening, after easy glide" (1955, p. 332). Figure 24 shows a micrograph of the gauge section of one of their crystals. The crystallographic orientation of the tensile axis was then determined at the point where activity on the conjugate system was detected and the resolved shear stress on the conjugate and primary systems calculated. Piercy et al. (1955) found the latent-hardening ratio to be essentially constant and equal to $\tau_c^{(c)}/\tau_c^{(p)} = 1.28$, thus indicating that the average rate of latent hardening on the conjugate system was approximately 28% greater than the self-hardening rate on the primary system. Note that their experiments are of the second type described previously and involve simultaneous activity on the primary and conjugate systems. The amount of overshoot they observed, 4–7°, was rather large but consistent with the measurements made earlier on copper–aluminum by Elam (1927). She also found overshoot of several degrees. α-Brass and copper–5 at percent aluminum alloys have low stacking fault energies, and this appears to contribute to strong latent hardening. Mitchell and Thornton (1964) have also reported large amounts of overshoot in α-brass of up to 7°.

The large amounts of overshoot reported in α-brass and Cu–Al alloys and the high latent hardening they imply do not appear typical of other fcc

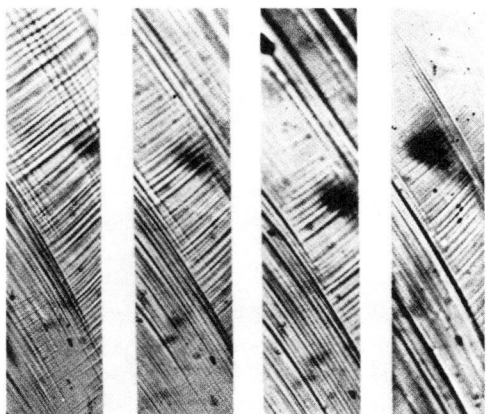

FIG. 24. Optical micrographs of the gauge section of α-brass crystals deformed in tension. The crystals were deformed in single slip, and the onset of conjugate slip determined by the appearance of slip lines belonging to the conjugate system.

crystals. Taylor was able to interpret his experiments conducted with Elam on aluminum in terms of isotropic hardening, although it is important to note that up to 2–3° of overshoot were reported (Taylor and Elam 1925). Conjugate slip was definitely observed, though, even as the symmetry line was approached. Mitchell and Thornton (1964) also reported that conjugate slip began in copper crystals before the symmetry boundary was reached. They nonetheless found that the tensile axis rotated past the symmetry position and followed a path that fell between the predictions of single and symmetric double slip. Ramaswami et al. (1965) reported very little or no overshoot in pure silver crystals and between 2 and 3° in alloy crystals of silver–10% gold. They estimated the ratio of latent-hardening to self-hardening rates to be 0.95 for silver and 1.05 for the silver–gold alloy. Their results for silver are interesting, considering its low stacking fault energy, but the trend toward larger overshoots with alloying, and therefore lower stacking fault energy, is consistent with the results on α-brass and Cu–Al alloys. Chang and Asaro (1980) found that in age-hardened aluminum–copper alloys overshoot ranged from 0 to 4° at most. As in many other reported cases, conjugate slip was often observed to begin before the tensile axis reached the symmetry line.

Linear relationships between flow (or yield) stresses on latent systems, $\tau_c^{(L)}$, and flow stresses on active primary systems, $\tau_c^{(p)}$, of the form $\tau_c^{(L)} = A\tau_c^{(p)} + B$ have been reported in several of these studies *after a strain of several percent:* Piercy et al. (1955) found $A = 1.28$ and $B = 0$ for α-brass; Jackson and Basinski (1967) found $A = 1.36$ and $B = 1.49$ MPa for copper; for silver and silver–10% gold Ramaswami et al. (1965) found (where the latent system was what they called the "half-critical" system) $A = 1.4$ and 1.2, respectively, with $B = 0$ in both cases. Thus Kocks' suggestion that the average ratio of latent-hardening to self-hardening rates lies between 1 and 1.4 is consistent with available measurements.

It must be realized that the direct measurements of latent hardening are limited to only a few studies in which the hardening of latent systems is measured following single slip. Furthermore, the measurements described above have all been interpreted as if plastic flow were rate insensitive, and so the influence of strain rate is uncertain. In a study on aluminum crystals Joshi and Green (1980) found, for a few crystals, that larger strain rates were accompanied by larger amounts of overshoot—for other crystals no systematic correlation between strain rate and overshoot was detected. In other studies, such as those of Mitchell and Thornton (1964) on copper, as well as those of Joshi and Green (1980), the observed lattice rotations during the transition from single to double slip are not readily described in detail by the simple relation suggested above with A constant. Figure 25 shows data from these two studies of the observed lattice rotation compared in the one

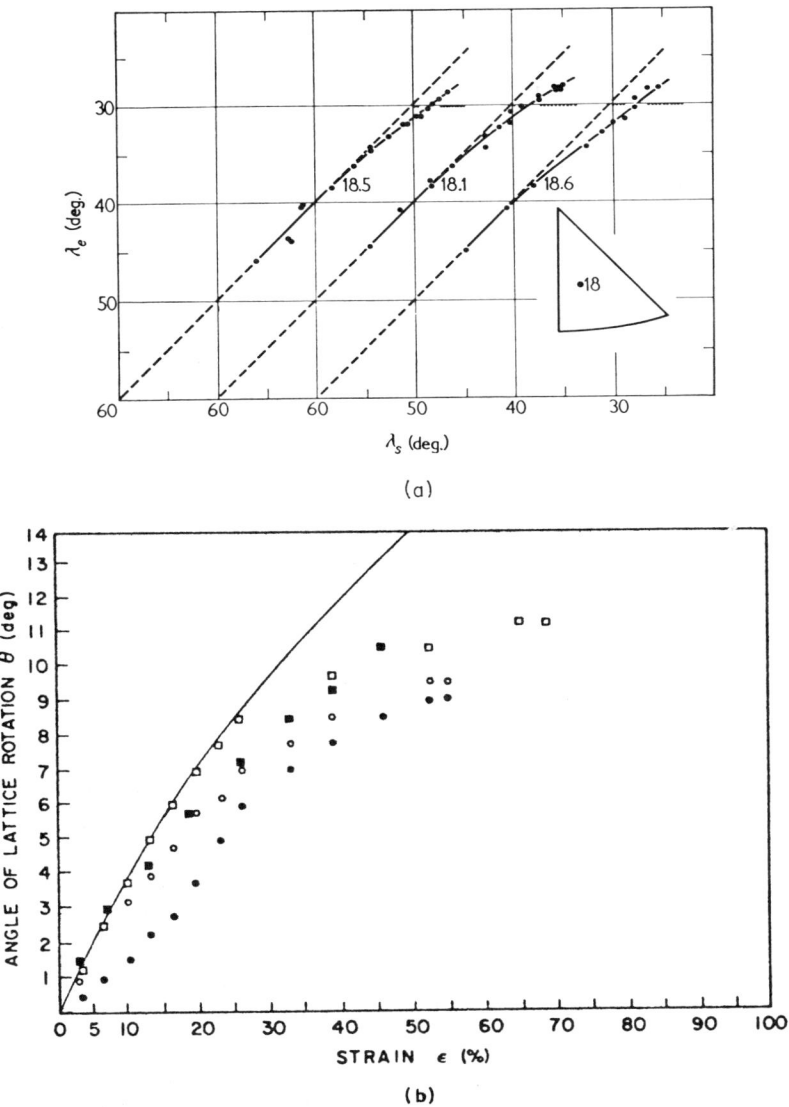

FIG. 25. (a) Mitchell and Thornton's (1964) data for the angle λ_e between the tensile axis and primary slip direction and λ_s, the angle expected from single slip. Note that deviations from the single-slip predictions occur before the symmetry position is reached, but that the tensile axis overshoots the symmetry position. The dotted lines indicate the values for λ_e, assuming equal slipping on the primary and conjugate systems once symmetry position is reached. (b) Joshi and Green's (1980) data for the angle of lattice rotation versus tensile strain. The solid line is based on the prediction of single slip.

case to the predicted rotation assuming single slip, and in the other to the extensional strain; Fig. 25 also shows, by dotted lines, the rotation expected if the hardening were isotropic and symmetric double slip began when the tensile axis reached the symmetry line. Figure 25 indicates that although a constant A (larger than 1) accounts for the observed average ratio of hardening rates, it cannot account for both the apparent premature conjugate slip *and* subsequent overshoot. In a later section an alternative description of overshoot is suggested based upon a rate-dependent constitutive law, which may help to explain these observations.

Finally we note that the limited experimental data available suggests that the rate of latent hardening may also be influenced by the ratio of strain rates on the active systems. For example, the high latent-hardening ratios of 1.36 and 1.5 in copper reported by Jackson and Basinski (1967) and Franciosi *et al.* (1980), respectively, are not consistent with the lattice rotations and conjugate slip reported by Mitchell and Thornton (1964). The latter found that conjugate slip began even before the symmetry line corresponding to equal shear stresses on the primary and conjugate systems was reached. These high latent-hardening ratios would have required large amounts of overshoot to activate the conjugate system. The two measurements differ in that the former maintains the crystal in a single-slip mode, first on the primary system, then on one of the latent systems, whereas the latter measurement allows the crystal to deform in multiple slip. The inclusion of rate sensitivity as discussed in Section III,D may account for some of these differences, but there may still be influences having to do more with the ratio of shearing rates that will only be sorted out by further experiment.

E. Observations of Slip in Single Crystals and Polycrystals at Modest Strains

When slip is confined to one system, i.e., single slip, macroscopic deformation often occurs uniformly over the gauge section of uniformly stressed crystals. When more than one system is active, however, observations suggest that this is generally not the case. For example, Fig. 24 shows the "patchy" slip-line pattern that develops in α-brass crystals when they are stretched in tension so that double primary–conjugate slip eventually occurs after primary single slip. The patches consist of regions of single slip on one or the other system contiguous with regions of double slip. Piercy *et al.* (1955) attributed this nonuniform slip mode to latent hardening. They argued that "these results prove the reality of latent-hardening, in the sense that the

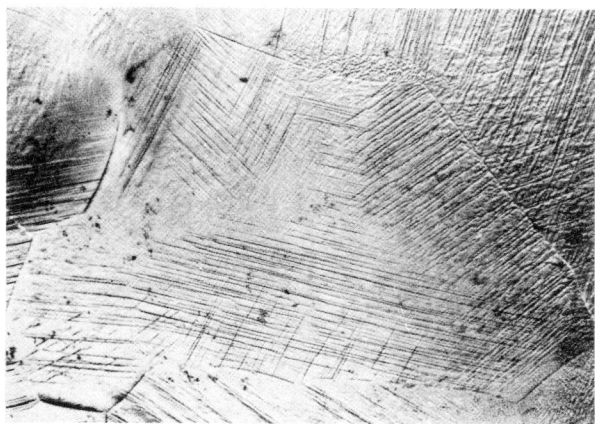

FIG. 26. Optical micrograph of the slip-line pattern in polycrystalline aluminum. Note the discontinuous slip mode leading to a pattern of "patchy slip" (Boas and Ogilvie, 1954).

slip lines of the one system experience difficulty in breaking through the active slip lines of the other one" (1955, p. 337). A very similar kind of patchy slip is observed within the grains of polycrystals, as shown, for example, in Fig. 26 taken from the work of Boas and Ogilvie (1954) on aluminum. In both these examples the loading was uniform, and in the polycrystalline case the macroscopic deformation field that included the grain in question was also uniform. Nonetheless, the slip mode was highly nonuniform.

One significance of patchy slip is that the slip mode, and therefore the lattice rotation with respect to the material, becomes difficult to specify for an entire grain. Thus grain, or crystal, reorientation becomes ambiguous. At small strains this may not be important, but at finite strains, where crystallographic texture develops owing to finite amounts of grain reorientation, specification of the slip mode and thus the relative rotation of the lattice with respect to the material is vital to the evolution of constitutive behavior.

Initiation of a nonuniform slip mode may occur owing to the initiation of slip at grain boundaries, as shown for an iron–silicon alloy in Fig. 27a. On the other hand, a nonuniform slip mode may be triggered by the impingement of slip bands at grain boundaries, resulting in stress concentrations, as shown in Fig. 27b for a copper–aluminum alloy. A third example of a nonuniform slip mode is shown in Fig. 27c, which shows an aluminum bicrystal. This latter example is taken from the work of Miura and Saeki (1978), who argued that the stress concentrations at the bicrystal (grain) boundary, caused by the impinging slip bands belonging to the slip systems

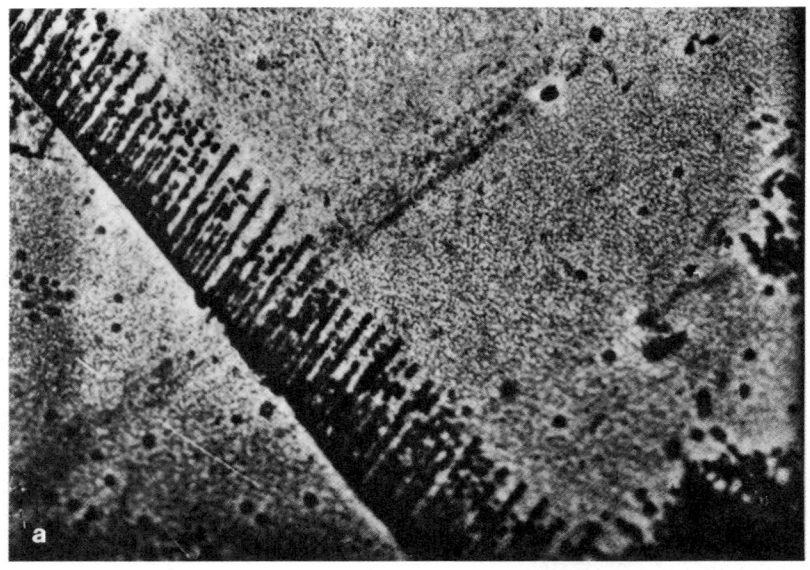
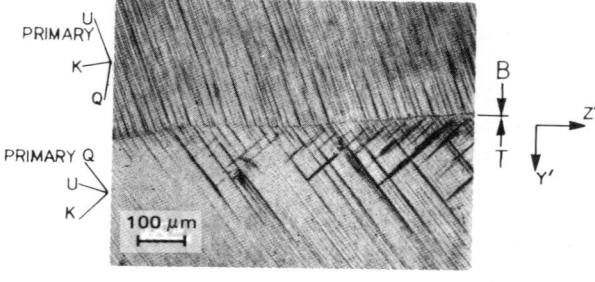
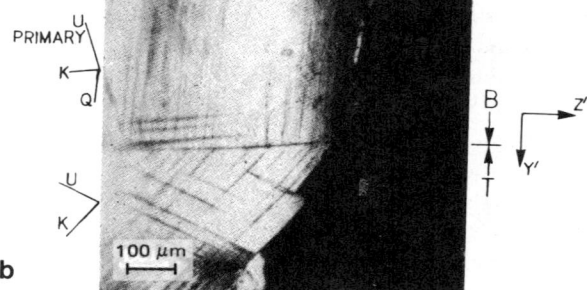

FIG. 27. Slip bands initiated at (a) a grain boundary of iron–silicon polycrystals (Worthington and Smith, 1964), (b) a grain boundary in a copper–4 wt. % aluminum alloy (Swearengen and Taggart, 1971), and (c) multiple slip induced at the bicrystal boundary due to the impingement of slip bands. The asterisk indicates an induced slip system. Strain 4.2% (Miura and Saeki, 1978).

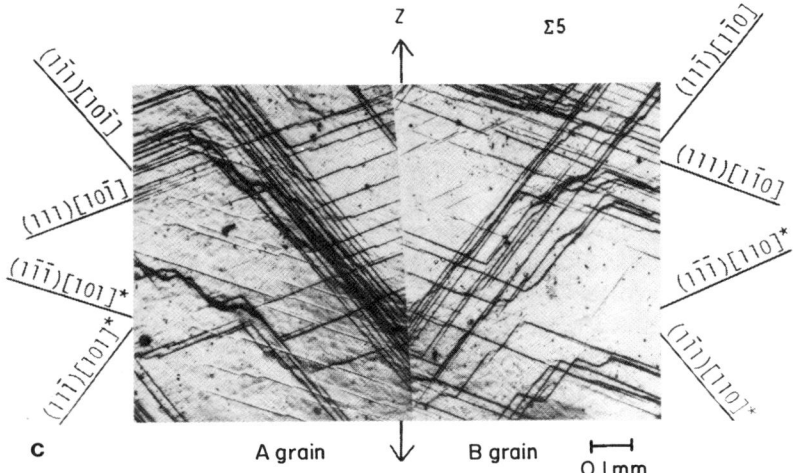

FIG. 27. (continued)

highly stressed by the applied loads, "induced" slip activity on other systems. In this way full compatibility between the crystals was maintained. On the other hand, Peirce *et al.* (1982) have shown that small spatial variations in stress coupled with relatively strong latent hardening ($A \approx 1.4$) would bring about patchy slip in single crystals. Whatever the mechanism, the result is to produce a highly nonuniform mode of slip throughout the grain, as well as a nonuniform dislocation density. It is common that the dislocation density becomes much larger in regions adjacent to the grain boundaries, where complex multiple slip modes are generated, than in the grain interiors. Remarkably, Miura and Saeki (1978) found that provided that the slip mode within the grains was multiple slip, there was very little difference in the strain-hardening properties of their bicrystals and the component single crystals comprising them. When the grain slip mode was single slip, however, they reported that the grain boundaries caused multiple slip, even at very small strains, and this affected the rate of strain hardening. Since complex multiple slipping tends to be concentrated near the grain boundaries, it is likely its effect on the overall strain hardening of the grain will be negligible at extremely large grain sizes. When the grains are fine and the more highly dislocated boundary zones occupy a proportionately larger volume, the influence of the boundary-induced slip modes should be more significant. This may well be an important factor contributing to the grain size dependence of polycrystalline strain hardening evidenced in Fig. 22b. As yet these effects have not been included in continuum polycrystalline models.

III. Constitutive Laws for Elastic–Plastic Crystals

A. Kinematics of Crystalline Deformation

The development of constitutive laws for the elastic–plastic deformation of crystals at finite strain began with Taylor's (1938a,b) analysis of polycrystalline deformation. More comprehensive formulations have since been presented by Hill (1966), Hill and Rice (1972), Asaro and Rice (1977), and Peirce *et al.* (1982); an interesting reformulation of the Hill and Rice (1972) work has been given by Hill and Havner (1982). A basic tenet of these analyses is that material flows through the crystal lattice via dislocation motion as described previously, whereas the lattice itself undergoes elastic deformations. Thus there are two physically different mechanisms for deforming and reorienting material fibers—plastic slip and lattice deformation. Crystallites, i.e., single crystals or polycrystalline grains, are generally subjected to rigid body rotations owing to boundary constraints or compatibility requirements, and it may on occasion be convenient (although arbitrary) to consider this a third mechanism.

As an example of some important principles and as a preface to a general kinematical formulation it is helpful to reconsider the simple problem introduced in Fig. 5. Figure 28 illustrates a simpler model of Fig. 5 for a single crystal undergoing single slip on the slip system $(\mathbf{s}^{(p)}, \mathbf{m}^{(p)})$; for later reference the conjugate (latent) system is also indicated. This model is idealized as planar in that $(\mathbf{s}^{(p)}, \mathbf{m}^{(p)})$ and $(\mathbf{s}^{(c)}, \mathbf{m}^{(c)})$, along with the tensile axis $\mathbf{T}$, are in the plane of the drawing. The $(\mathbf{s}^{(\alpha)}, \mathbf{m}^{(\alpha)})$, that define the slip

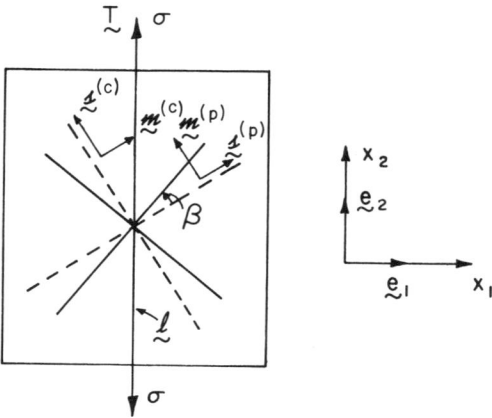

Fig. 28. Plane model for a crystal undergoing single slip or double primary-conjugate slip in tension.

system α are taken to be orthogonal unit vectors in the reference state. The total deformation and rotation of the crystal when it is under load is given by the deformation gradient **F**, which can be decomposed into three operations as follows. First the material undergoes plastic slip through the *undeformed* crystal lattice according to $\mathbf{F}^P$; in single slip on system $(\mathbf{s}, \mathbf{m})$, for example, $\mathbf{F}^P$ can always be written as

$$\mathbf{F}^P = \mathbf{I} + \gamma \mathbf{sm}, \tag{3.1}$$

where γ is the shear strain measured with respect to the undeformed lattice. Next, the crystal is given a rigid rotation prescribed by $\mathbf{F}^R$, which, from the discussion concerning Fig. 5, is imposed by the special constraint of the tension test that the fiber **l** remain vertical. Finally, *under load* the lattice and material deform together according to $\mathbf{F}^e$ and again **l** must remain vertical. The total deformation gradient is then

$$\mathbf{F} = \mathbf{F}^e \cdot \mathbf{F}^R \cdot \mathbf{F}^P, \tag{3.2}$$

where the residual deformation gradient remaining on unloading is $\mathbf{F}^{e-1} \cdot \mathbf{F}$ or $\mathbf{F}^R \cdot \mathbf{F}^P$. The residual deformation $\mathbf{F}^P$ is due to plastic slip alone, and $\mathbf{F}^R$ is the residual rotation of the lattice that accounts for its reorientation with respect to the fixed (vertical) tensile axis. When the axial orientation with respect to the lattice is determined by X-ray Laue back-diffraction (as in Fig. 5d), the incident X-ray beam is directed along **l** and the crystal direction parallel to **l** recorded. The lattice is then found to have rotated with respect to the laboratory frame, and **l**, and so develops a sort of "texture." We will continue the analysis of this model later.

In general, the deformation gradient can be written with reference to Fig. 29 as

$$\mathbf{F} = \mathbf{F}^* \cdot \mathbf{F}^P, \tag{3.3}$$

where $\mathbf{F}^P$ is again the deformation due solely to plastic shearing on crystallographic slip systems and $\mathbf{F}^*$ is caused by stretching and rotation of the crystal lattice. The deformation gradient remaining after elastic unloading *and upon returning the lattice to its orientation in the reference state is* $\mathbf{F}^{*-1} \cdot \mathbf{F} = \mathbf{F}^P$.

As the crystal deforms, vectors connecting lattice sites are stretched and rotated according to $\mathbf{F}^*$, i.e., they convect with the lattice. The slip direction vectors are regarded as such vectors, as are other lattice vectors lying in the slip planes. In the elastically deformed configuration the slip direction of the slip system α is then given by

$$\mathbf{s}^{*(\alpha)} = \mathbf{F}^* \cdot \mathbf{s}^{(\alpha)}, \tag{3.4}$$

where each $\mathbf{s}^{(\alpha)}$ is a unit vector. The normal to the slip plane $\mathbf{m}^{*(\alpha)}$, however, is taken as the cross-product of two orthogonal vectors in the slip plane

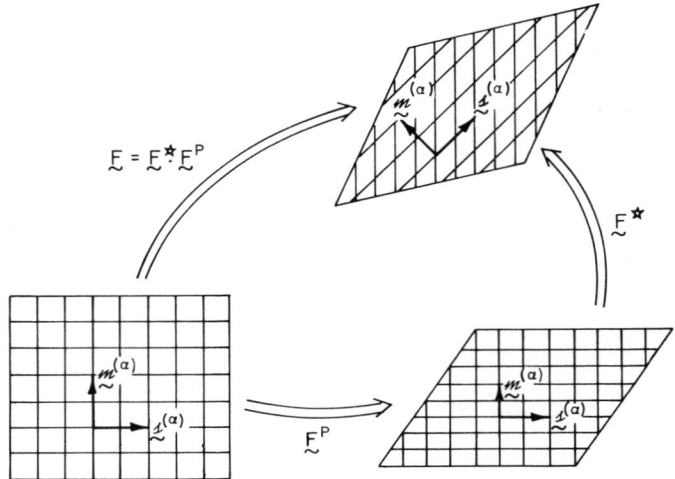

FIG. 29. Kinematics of elastic–plastic deformation in crystalline media. Plastic deformation via dislocation slip deforms and rotates the material, but does not affect the lattice. The material and lattice deform and rotate together under $\mathbf{F}^*$.

and is given by

$$\mathbf{m}^{*(\alpha)} = \mathbf{m}^{(\alpha)} \cdot \mathbf{F}^{*-1}, \tag{3.5}$$

where each $\mathbf{m}^{(\alpha)}$ is also a unit vector. $\mathbf{s}^{*(\alpha)}$ and $\mathbf{m}^{*(\alpha)}$ are in general not unit vectors, but remain orthogonal, since $\mathbf{s}^{*(\alpha)} \cdot \mathbf{m}^{*(\alpha)} = \mathbf{s}^{(\alpha)} \cdot \mathbf{m}^{(\alpha)} = 0$.

The velocity gradient in the current state is given by a standard formula:

$$\mathbf{L} = \mathbf{v}\nabla = \dot{\mathbf{F}} \cdot \mathbf{F}^{-1} = \dot{\mathbf{F}}^* \cdot \mathbf{F}^{*-1} + \mathbf{F}^* \cdot \dot{\mathbf{F}}^\mathrm{P} \cdot \mathbf{F}^{\mathrm{P}-1} \cdot \mathbf{F}^{*-1} \tag{3.6}$$

where $\mathbf{L}$ can also be written as

$$\mathbf{L} = \mathbf{D} + \mathbf{\Omega}. \tag{3.7}$$

The terms $\mathbf{D}$ and $\mathbf{\Omega}$ are the symmetric rate of stretching tensor and the rate of spin tensor, respectively. They are then decomposed from Eq. (3.6) and Fig. 29 into parts due to plastic slip $(\mathbf{D}^\mathrm{P}, \mathbf{\Omega}^\mathrm{P})$ and lattice deformation $(\mathbf{D}^*, \mathbf{\Omega}^*)$ as follows:

$$\mathbf{D} = \mathbf{D}^* + \mathbf{D}^\mathrm{P}, \qquad \mathbf{\Omega} = \mathbf{\Omega}^* + \mathbf{\Omega}^\mathrm{P} \tag{3.8a}$$

where

$$\mathbf{D}^\mathrm{P} + \mathbf{\Omega}^\mathrm{P} = \mathbf{F}^* \cdot \dot{\mathbf{F}}^\mathrm{P} \cdot \mathbf{F}^{\mathrm{P}-1} \cdot \mathbf{F}^{*-1}. \tag{3.8b}$$

Furthermore, since plastic deformation occurs by dislocation slip, we find in the current state

$$\mathbf{D}^P + \mathbf{\Omega}^P = \sum_{\alpha=1}^{n} \dot{\gamma}^{(\alpha)} \mathbf{s}^{*(\alpha)} \mathbf{m}^{*(\alpha)}, \quad (3.9)$$

where the sum runs over the active slip systems and $\dot{\gamma}^{(\alpha)}$ is the shearing rate on the system α measured relative to the lattice. In the reference configuration we note, however, that

$$\dot{\mathbf{F}}^P \cdot \mathbf{F}^{P-1} = \sum_{\alpha=1}^{n} \dot{\gamma}^{(\alpha)} \mathbf{s}^{(\alpha)} \mathbf{m}^{(\alpha)}, \quad (3.10)$$

which means that the magnitude of the material displacement and shear strain due to slip is different in the reference and current configurations. The product $\dot{\mathbf{F}}^P \cdot \mathbf{F}^{P-1}$ is also the residual increment of deformation gradient after elastic unloading and returning the lattice to its orientation in the reference state. It is also worth noting that because of definitions (3.4) and (3.5), the decompositions in Eqs. (3.3) and (3.8) are exact.

Finally, we note that the plastic parts of the rate of stretching and the rate of spin are given by the symmetric and skew parts of Eq. (3.9). If we define

$$\mathbf{P}^{(\alpha)} = \tfrac{1}{2}(\mathbf{s}^{*(\alpha)} \mathbf{m}^{*(\alpha)} + \mathbf{m}^{*(\alpha)} \mathbf{s}^{*(\alpha)}) \quad (3.11a)$$

and

$$\mathbf{W}^{(\alpha)} = \tfrac{1}{2}(\mathbf{s}^{*(\alpha)} \mathbf{m}^{*(\alpha)} - \mathbf{m}^{*(\alpha)} \mathbf{s}^{*(\alpha)}), \quad (3.11b)$$

then

$$\mathbf{D}^P = \sum_{\alpha=1}^{n} \mathbf{P}^{(\alpha)} \dot{\gamma}^{(\alpha)} \quad (3.12a)$$

and

$$\mathbf{\Omega}^P = \sum_{\alpha=1}^{n} \mathbf{W}^{(\alpha)} \dot{\gamma}^{(\alpha)}. \quad (3.12b)$$

B. Constitutive Laws

If we assume that the crystal's elasticity is unaffected by slip, then, following Hill and Rice (1972), we can write the elastic law as

$$\tau^{\nabla *} = \mathbf{L} : \mathbf{D}^*, \quad (3.13)$$

where $\mathbf{L}$ is the tensor of elastic moduli (with components formed on coordinates fixed on the crystal lattice) and $\tau^{\nabla *}$ is the Jaumann rate of Kirchhoff stress formed on axes that spin with the *lattice*,

$$\tau^{\nabla *} = \dot{\tau} - \mathbf{\Omega}^* \cdot \tau + \tau \cdot \mathbf{\Omega}^*, \tag{3.14}$$

with $\dot{\tau}$ the material rate of Kirchhoff stress. The Kirchhoff stress τ is defined as $(\rho_0/\rho)\sigma$, where σ is the Cauchy stress and ρ_0 and ρ are material densities in the reference and current states. On the other hand, τ^∇ is the Jaumann rate of Kirchhoff stress formed on axes that rotate with the *material*,

$$\tau^\nabla = \dot{\tau} - \mathbf{\Omega} \cdot \tau + \tau \cdot \mathbf{\Omega}, \tag{3.15}$$

and the difference between these two rates is

$$\tau^{\nabla *} - \tau^\nabla = \sum_{\alpha=1}^{n} \boldsymbol{\beta}^{(\alpha)} \dot{\gamma}^{(\alpha)}, \tag{3.16a}$$

where

$$\boldsymbol{\beta}^{(\alpha)} = \mathbf{W}^{(\alpha)} \cdot \tau - \tau \cdot \mathbf{W}^{(\alpha)}. \tag{3.16b}$$

When there is no relative rotation of the material and lattice or only elastic deformation, the right-hand side of Eq. (3.16a) vanishes and $\tau^{\nabla *} = \tau^\nabla$.

When Eqs. (3.12), (3.13), and (3.16) are combined, the resulting constitutive law is

$$\tau^\nabla = \mathbf{L}:\mathbf{D} - \sum_{\alpha=1}^{n} [\mathbf{L}:\mathbf{P}^{(\alpha)} + \boldsymbol{\beta}^{(\alpha)}] \dot{\gamma}^{(\alpha)}, \tag{3.17}$$

where it remains to specify the shearing rates $\dot{\gamma}^{(\alpha)}$; that is, the $\dot{\gamma}$'s must be related to either the stress rates, the deformation rates, or the current stress state.

Before developing relations for the $\dot{\gamma}$'s, we introduce a particular definition of the resolved shear stress or Schmid stress. It has already been noted that the resolved shear stress plays a vital role in promoting slip, since it produces a force on dislocations. Asaro and Rice (1977) have discussed various ways in which this may be accomplished, but for now we will define $\tau^{(\alpha)}$ as the "Schmid stress" on the slip system α so that $\tau^{(\alpha)} \dot{\gamma}^{(\alpha)}$ is precisely the rate of working due to slip on system α per unit reference volume. Then from the general expression for this rate of working,

$$\tau:\mathbf{D}^P = \sum_{\alpha=1}^{n} \tau:\mathbf{P}^{(\alpha)} \dot{\gamma}^{(\alpha)},$$

we identify

$$\tau^{(\alpha)} = \mathbf{P}^{(\alpha)}:\tau. \tag{3.18}$$

The rate of change of $\tau^{(\alpha)}$ can be calculated from material time derivatives of $\mathbf{P}^{(\alpha)}$ and τ,

$$\dot{\tau}^{(\alpha)} = \dot{\mathbf{P}}^{(\alpha)}:\tau + \mathbf{P}^{(\alpha)}:\dot{\tau} \tag{3.19}$$

or from the two corotational derivatives introduced previously, namely,

$$\dot{\tau}^{(\alpha)} = \mathbf{P}^{\nabla*(\alpha)}:\tau + \mathbf{P}^{(\alpha)}:\tau^{\nabla*} \tag{3.20a}$$

and

$$\dot{\tau}^{(\alpha)} = \mathbf{P}^{\nabla(\alpha)}:\tau + \mathbf{P}^{(\alpha)}:\tau^{\nabla}. \tag{3.20b}$$

It is interesting to note that in particular

$$\mathbf{P}^{\nabla*(\alpha)}:\tau = \boldsymbol{\beta}^{(\alpha)}:\mathbf{D}^* \tag{3.21}$$

and that therefore $\dot{\tau}^{(\alpha)}$ may also be calculated from

$$\dot{\tau}^{(\alpha)} = \mathbf{P}^{(\alpha)}:\tau^{\nabla*} + \boldsymbol{\beta}^{(\alpha)}:\mathbf{D}^*. \tag{3.22}$$

Equation (3.21) can also be derived from the more general "pivotal" relation given by Hill and Havner [1982, Eq. (3.3)]. From Eqs. (3.22) and (3.13) we also have that

$$\dot{\tau}^{(\alpha)} = (\mathbf{P}^{(\alpha)}:\mathbf{L} + \boldsymbol{\beta}^{(\alpha)}):\mathbf{D}^* = \boldsymbol{\lambda}^{(\alpha)}:\mathbf{D}^*, \tag{3.23}$$

which allows us to again construct Eq. (3.17) as

$$\tau^{\nabla} = \mathbf{L}:\mathbf{D} - \sum_{\alpha=1}^{n} \boldsymbol{\lambda}^{(\alpha)}\dot{\gamma}^{(\alpha)}, \tag{3.24}$$

provided that L_{ijkl} is symmetric in the indicial pairs ij and kl. Furthermore, with the expectation that $\mathbf{L}^{-1}$ exists, operating with it on Eq. (3.24) yields

$$\mathbf{D} = \mathbf{L}^{-1}:\tau^{\nabla} + \sum_{\alpha=1}^{n} \boldsymbol{\mu}^{(\alpha)}\dot{\gamma}^{(\alpha)}, \tag{3.25}$$

with

$$\boldsymbol{\mu}^{(\alpha)} = \mathbf{L}^{-1}:\boldsymbol{\lambda}^{(\alpha)} = \mathbf{P}^{(\alpha)} + \mathbf{L}^{-1}:\boldsymbol{\beta}^{(\alpha)}. \tag{3.26}$$

Finally, from Eqs. (3.23) and (3.22) and the symmetries in $\mathbf{P}^{(\alpha)}$, $\boldsymbol{\beta}^{(\alpha)}$, and $\mathbf{L}^{-1}$, $\dot{\tau}^{(\alpha)}$ is alternatively expressed as

$$\dot{\tau}^{(\alpha)} = \boldsymbol{\mu}^{(\alpha)}:\tau^{\nabla*} \tag{3.27}$$

or, as in Asaro and Rice (1977) and Peirce et al. (1982),

$$\dot{\tau}^{(\alpha)} = \mathbf{m}^{*(\alpha)} \cdot (\tau^{\nabla*} - \mathbf{D}^* \cdot \tau + \tau \cdot \mathbf{D}^*) \cdot \mathbf{s}^{*(\alpha)}. \tag{3.28}$$

C. Strain-Hardening Laws for Rate-Insensitive Crystals

1. The Schmid Law and Plastic Normality

As mentioned earlier, the Schmid law of a critical resolved shear stress has approximate experimental confirmation in fcc crystals. It is important to note, however, that such verification has been obtained for the most part from crystals strained at quasi-static rates in tension or compression. The micromechanics of slip, on the other hand, suggest that deviations from the Schmid law should exist, even if they may be small. Nevertheless, Schmid's law provides a valuable starting point in the formulation of strain-hardening laws or kinetic relations for dislocation motion that can then be amended to account for dislocation cross-slip, climb, or other micromechanical processes that are influenced by components of stress other than the resolved shear stress.

To adapt the Schmid rule to plastic flow we may simply state that for yield to begin on the slip system α, $\tau^{(\alpha)}$ must reach a critical value $\tau_c^{(\alpha)}$ that is determined by the current dislocation density and substructure. The set of systems for which $\tau^{(\alpha)} = \tau_c^{(\alpha)}$ is said to be the set of potentially active or critical systems; we assume there are n of these. For the system α to remain active, $\tau^{(\alpha)}$ must increase to (and remain at) the critical value $\tau_c^{(\alpha)}$, which changes according to

$$\dot{\tau}^{(\alpha)} = \dot{\tau}_c^{(\alpha)} = \sum_{\beta=1}^{n} h_{\alpha\beta}\dot{\gamma}^{(\beta)}, \qquad \dot{\gamma}^{(\alpha)} > 0. \tag{3.29a}$$

If a critical system is inactive,

$$\dot{\tau}^{(\alpha)} \leq \dot{\tau}_c^{(\alpha)} = \sum_{\beta=1}^{n} h_{\alpha\beta}\dot{\gamma}^{(\beta)}, \qquad \dot{\gamma}^{(\alpha)} = 0. \tag{3.29b}$$

For noncritical systems there is only the inequality

$$\tau^{(\alpha)} < \tau_c^{(\alpha)}, \qquad \dot{\gamma}^{(\alpha)} = 0. \tag{3.29c}$$

The $h_{\alpha\beta}$ are the instantaneous slip plane hardening rates; $h_{\alpha\alpha}$ is the self-hardening rate on system α and $h_{\alpha\beta}$ ($\alpha \neq \beta$) is the latent-hardening rate of system α (whether it is active or not) caused by slip on system β. To include reverse plastic flow, it is common to define separate positive and negative systems, i.e., $(\mathbf{s}^{(\alpha)}, \mathbf{m}^{(\alpha)})$ and $(-\mathbf{s}^{(\alpha)}, \mathbf{m}^{(\alpha)})$, so that positive (or zero) $\dot{\gamma}^{(\alpha)}$'s are allowed. For an fcc crystal there are then 24 systems. n is the number of potentially active systems for which $\tau^{(\alpha)} = \tau_c^{(\alpha)}$, and for an fcc crystal $n \leq 12$ at *any point* in the crystal. It has also been common to assume that the $h_{\alpha\beta}$

Micromechanics of Crystals and Polycrystals 43

do not depend upon the current set of active systems or on the ratio of the positive $\dot{\gamma}$'s. This must be viewed as a model assumption, whose accuracy is not yet clear.

The static picture, then, in stress space under load is the set of planes $\tau^{(\alpha)} = \tau_c^{(\alpha)}$ defining the yield locus. When the loads change or the crystal deforms and rotates, $\tau^{(\alpha)}$ changes as prescribed previously. A normality structure accompanies this version of the Schmid law as follows. Suppose the current stress were on the boundary on a yield plane $\tau^{(\alpha)} = \tau_c$ or at a corner common to several planes. Stress changes that would invoke a purely *elastic* response from system α are those that tend to decrease $\tau^{(\alpha)}$, i.e.,

$$\dot{\tau}^{(\alpha)} = \lambda^{(\alpha)} : \mathbf{D}^* = \boldsymbol{\mu}^{(\alpha)} : \boldsymbol{\tau}^{\nabla *} \leq 0. \quad (3.30)$$

The vector $\boldsymbol{\mu}^{(\alpha)}$ is then the outward-pointing normal to the yield plane of system α. Since, according to Eq. (3.25), the plastic part of the strain rate contributed by slip system α is given by $\boldsymbol{\mu}^{(\alpha)} \dot{\gamma}^{(\alpha)}$, it is necessarily directed along the normal $\boldsymbol{\mu}^{(\alpha)}$. Alternatively, by rewriting Eq. (3.30) as

$$\boldsymbol{\mu}^{(\alpha)} : \mathbf{L} : \mathbf{D}^* = \mathbf{L} : \boldsymbol{\mu}^{(\alpha)} : \mathbf{D}^* \leq 0, \quad (3.31)$$

it is seen, as noted by Hill and Rice (1972), that any pure plastic strain increment has a nonpositive scalar product with any elastic strain increment. Thus, as originally shown by Rice (1971) and Hill and Rice (1972), the Schmid law using $\tau^{(\alpha)}$ as defined in Eq. (3.18) leads precisely to normality in work-conjugate variables.

Combining Eq. (3.29) with Eqs. (3.27) and (3.23) then yields for the critical systems

$$\lambda^{(\alpha)} : \mathbf{D}^* = \boldsymbol{\mu}^{(\alpha)} : \boldsymbol{\tau}^{\nabla *} = \sum_{\beta=1}^{n} h_{\alpha\beta} \dot{\gamma}^{(\beta)}, \qquad \dot{\gamma}^{(\alpha)} > 0, \quad (3.32a)$$

$$\lambda^{(\alpha)} : \mathbf{D}^* = \boldsymbol{\mu}^{(\alpha)} : \boldsymbol{\tau}^{\nabla *} \leq \sum_{\beta=1}^{n} h_{\alpha\beta} \dot{\gamma}^{(\beta)}, \qquad \dot{\gamma}^{(\alpha)} = 0. \quad (3.32b)$$

Alternatively, following Hill and Rice (1972), Eq. (3.32) may be expressed in terms of the total material deformation rate $\mathbf{D}$, using Eqs. (3.8a) and (3.12a), as

$$\lambda^{(\alpha)} : \mathbf{D} = \sum_{\beta=1}^{n} g_{\alpha\beta} \dot{\gamma}^{(\beta)}, \qquad \dot{\gamma}^{(\alpha)} > 0, \quad (3.33a)$$

$$\lambda^{(\alpha)} : \mathbf{D} \leq \sum_{\beta=1}^{n} g_{\alpha\beta} \dot{\gamma}^{(\beta)}, \qquad \dot{\gamma}^{(\alpha)} = 0, \quad (3.33b)$$

where

$$g_{\alpha\beta} = h_{\alpha\beta} + \lambda^{(\alpha)} : \mathbf{P}^{(\beta)} = h_{\alpha\beta} + \mathbf{P}^{(\alpha)} : \mathbf{L} : \mathbf{P}^{(\beta)} + \boldsymbol{\beta}^{(\alpha)} : \mathbf{P}^{(\beta)}. \quad (3.34)$$

In terms of the material stress rates τ^V, Eq. (3.32) can also be written as

$$\boldsymbol{\mu}^{(\alpha)} : \boldsymbol{\tau}^V = \sum_\beta k_{\alpha\beta} \dot{\gamma}^{(\beta)}, \qquad \dot{\gamma}^{(\alpha)} > 0, \tag{3.35a}$$

$$\boldsymbol{\mu}^{(\alpha)} : \boldsymbol{\tau}^V \leq \sum_\beta k_{\alpha\beta} \dot{\gamma}^{(\beta)}, \qquad \dot{\gamma}^{(\alpha)} = 0, \tag{3.35b}$$

where

$$k_{\alpha\beta} = h_{\alpha\beta} - \boldsymbol{\mu}^{(\alpha)} : \boldsymbol{\beta}^{(\beta)} = h_{\alpha\beta} - \mathbf{P}^{(\alpha)} : \boldsymbol{\beta}^{(\beta)} - \boldsymbol{\beta}^{(\alpha)} : \mathbf{L}^{-1} : \boldsymbol{\beta}^{(\beta)}. \tag{3.36}$$

The hardening moduli **g** and **k** were introduced in this context by Hill and Rice (1972). If Eq. (3.33) were uniquely invertible with prescribed **D**, or Eq. (3.35) invertible with prescibed τ^V, the $\dot{\gamma}^{(\alpha)}$ could be computed and substituted into Eq. (3.24) or (3.25), respectively.

Sufficiency conditions for unique specification of the $\dot{\gamma}$'s have been derived by Hill and Rice (1972) and here we list only their result. The $\dot{\gamma}$'s are unique with prescribed deformations if **g** is positive definite; the $\dot{\gamma}$'s are unique with prescribed τ^V if **k** is positive definite. If **g** is positive definite, $\mathbf{g}^{-1}$ exists and, for example, with prescribed **D**,

$$\dot{\gamma}^{(\alpha)} = \sum_{\beta=1}^n g_{\alpha\beta}^{-1} \boldsymbol{\lambda}^{(\beta)} : \mathbf{D} \tag{3.37}$$

and

$$\boldsymbol{\tau}^V = \mathbf{C} : \mathbf{D}, \tag{3.38}$$

with

$$\mathbf{C} = \mathbf{L} - \sum_{\alpha=1}^n \sum_{\beta=1}^n \boldsymbol{\lambda}^{(\alpha)} g_{\alpha\beta}^{-1} \boldsymbol{\lambda}^{(\beta)}. \tag{3.39}$$

Uniqueness, however, is not generally guaranteed, and therefore neither is the relative rotation of the lattice with respect to the material. Conditions that guarantee uniqueness depend sensitively on the hardening rates, the stress state (especially with respect to the lattice), and the number and relative orientation of the active systems. General conditions are not derived here, since they would be cumbersome and uninstructive, but since the question is important, a brief discussion follows.

For illustration we take the lattice elasticity as linear isotropic; G is the elastic shear modulus. In forming the elements of **g** and **k** we note that

$$\mathbf{P}^{(\alpha)} : \mathbf{L} : \mathbf{P}^{(\beta)} = 2G \mathbf{P}^{(\alpha)} : \mathbf{P}^{(\beta)}, \tag{3.40}$$

whereby the elements of **g** are given by

$$g_{\alpha\beta} = \underbrace{h_{\alpha\beta} + \boldsymbol{\beta}^{(\alpha)} : \mathbf{P}^{(\beta)}}_{\mathbf{g}^h} + \underbrace{2G \mathbf{P}^{(\alpha)} : \mathbf{P}^{(\beta)}}_{\mathbf{g}^e}. \tag{3.41}$$

Since $\mathbf{P}^{(\alpha)}:\mathbf{P}^{(\beta)} = \frac{1}{2}$ when $\alpha = \beta$ and is no more than $\frac{1}{3}$ when $\alpha \neq \beta$, at least for fcc crystals, the matrix $\mathbf{g}^e$ can be positive definite as long as the number of slip systems included (i.e., potentially active) does not exceed 5. If $n > 5$, the $\mathbf{P}^{(\alpha)}$ are not linearly independent and $\mathbf{g}^e$ becomes singular. The matrix $\mathbf{g}^h$ depends on the stress state resolved on the lattice through the $\boldsymbol{\beta}$'s and on the slip plane hardening rates $h_{\alpha\beta}$. The latter tend to be negative definite, since, as discussed in Section II,D, $h_{\alpha\beta} > h_{\alpha\alpha}$. The remaining terms in $\mathbf{g}^h$ depend on the resolved stress state, but in general we expect that $|g^h_{\alpha\beta}| \ll |g^e_{\alpha\beta}|$. Thus when $n \leq 5$ and the included systems are linearly independent (three coplanar systems in a fcc crystal are not independent), then $\mathbf{g}$ is almost assured to be positive definite by virtue of the $\mathbf{g}^e$ component. It is far more difficult to make general statements regarding $\mathbf{k}$, except that since it contains no analog to $\mathbf{g}^e$, positive definiteness is more difficult to anticipate. However, to provide some helpful perspective on the trends we conclude this discussion with a specific example for a physical problem analyzed in much more detail in Section IV,C.

We consider Fig. 30, which shows a fcc single crystal aligned for double primary–conjugate slip in tension. Figure 30a shows the actual geometry and the tensile axis somewhere along the $\langle 111 \rangle$–$\langle 100 \rangle$ symmetry boundary. Figure 30b shows an idealized plane version of this crystal that will serve as the basis for a discussion of necking, localized shearing, and fracture in Section IV,C. The stress state with respect to the laboratory frame is one of simple tension along the x_2 axis. For the model crystal

$$g_{\alpha\beta} = h_{\alpha\beta} - \sigma/2 \cos 2\phi + G, \qquad \alpha = \beta, \tag{3.42a}$$

$$g_{\alpha\beta} = h_{\alpha\beta} + \sigma/2 \cos 2\phi - G \cos 4\phi, \qquad \alpha \neq \beta \tag{3.42b}$$

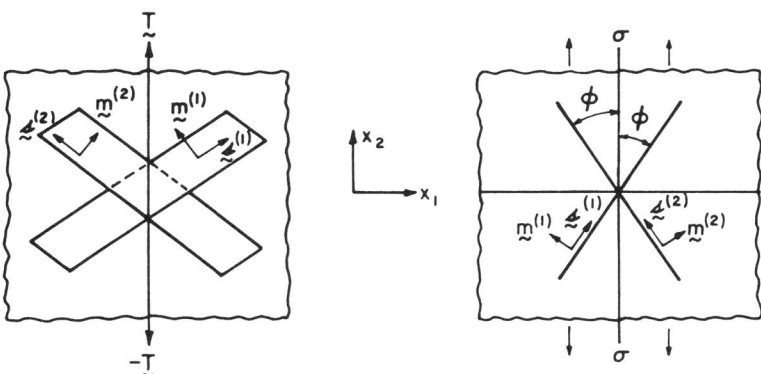

FIG. 30. Model for a crystal undergoing double primary–conjugate slip in tension. (a) The actual geometry of an fcc crystal. (b) An idealized plane model where both slip plane normals and slip directions lie in the plane of the drawing (Asaro, 1979).

and

$$k_{\alpha\beta} = h_{\alpha\beta} + \sigma/2\cos 2\phi + O(\sigma^2/G), \qquad \alpha = \beta, \tag{3.43a}$$

$$k_{\alpha\beta} = h_{\alpha\beta} - \sigma/2\cos 2\phi + O(\sigma^2/G), \qquad \alpha \neq \beta. \tag{3.43b}$$

For this case we tacitly assume that $\sigma/G \ll 1$. The values of these components for the actual geometry are very similar if ϕ is set equal to 30°; in fact, if $\mathbf{T}\|\langle 211\rangle$, the only changes required are to replace $\sigma/2\cos 2\phi$ in Eqs. (3.42a) and (3.42b) with 0.264, along with a similar change in the last term in Eq. (3.42b). Now if the model is applied to a fcc crystal where $\phi = 30°$ and $\cos 4\phi = -0.5$, then $\mathbf{g}$ is positive definite so long as σ and the $h_{\alpha\beta}$ are much smaller in magnitude than G. This is the case for most values of ϕ. However, a notable but very special situation is where $\phi = 45°$ and $\cos 4\phi = -1$, in which case $g_{11} = h_{11} + G$, $g_{22} = h_{22} + G$, $g_{12} = h_{12} + G$, and $g_{21} = h_{21} + G$. If we take as an example $h_{11} = h_{22} = h$ and $h_{12} = h_{21} = h_1$, as the model's orthotropic symmetry might imply, then positive definiteness depends on $h > h_1$, which is effectively ruled out by the experimental estimates of latent hardening. The situation for $\mathbf{k}$ is much more difficult to state in general. For example, if we again take $h_{11} = h_{22} = h$ and $h_{12} = h_{21} = h_1$ and neglect the terms of order σ^2/G, then $\mathbf{k}$ is positive definite if and only if

$$h - h_1 + \sigma\cos 2\phi > 0 \tag{3.44a}$$

or

$$\cos 2\phi > (h/\sigma)(q - 1), \qquad q \equiv h_1/h. \tag{3.44b}$$

As we noted earlier, q is expected to lie in the range $1 \leq q \leq 1.4$, and if the model is to represent a fcc crystal in tension, $\phi \approx 30°$. These two specifications require that for the inequality in Eq. (3.44) to hold $h/\sigma < 1.25$, or since $\sigma \approx 2.3\tau$, where τ is the resolved shear stress on either slip system, $h/\tau < 2.875$. The number of cases where this inequality would fail for real metal crystals are legion—failure would be common, for example, at initial yield and during the immediately subsequent 1 or 2% strain in aluminum alloy crystals, as discussed by Peirce et al. (1982). After strains of 1 or 2%, h/τ would drop below 2 or 3, insuring that Eq. (3.44) is satisfied.

The importance of latent hardening regarding uniqueness of slip mode is made clear in the previous example, since if $q < 1$, uniqueness in the slip mode is always assured as long as $\cos 2\phi > 0$. However, the main point is that rate-independent hardening laws will not lead to a unique slip mode in general, and especially when physically realistic descriptions of latent hardening are included. Whether a unique slip mode *should* be demanded on physical grounds is another matter that must be evaluated in view of the actual rate dependence of slip and such phenomena as "patchy slip."

2. Non-Schmid Effects and Deviations from Plastic Normality in Single Slip

Deviations from normality arise whenever $\tau_c^{(\alpha)}$ in Eq. (3.29) is interpreted to be other than the Schmid stress $\tau^{(\alpha)}$ defined in Eq. (3.18). This has been discussed by Asaro and Rice (1977) in particular, who illustrated that other definitions of τ_c are generated, depending on how **s** and **m** are chosen to convect and rotate with the lattice; the relations (3.4) and (3.5) are only one possible choice. Since all the physically appropriate definitions constrain **s** to remain in the slip plane (and thus to rotate with the lattice), the differences in the various choices for the $\tilde{\tau}$'s are of the order of the elastic *strain* rates and are thus usually negligible, as shown below. On the other hand, deviations from the Schmid law are inherent in the micromechanics of dislocation slip itself, regardless of how τ is related to the lattice elasticity. For example, consider Fig. 31 (see Fig. 15), which shows idealized models for cross-slip of a screw dislocation and "climb" of an edge dislocation. In both cases these localized micromechanical processes represent a means for the dislocations to overcome obstacles to their continued motion. In both cases

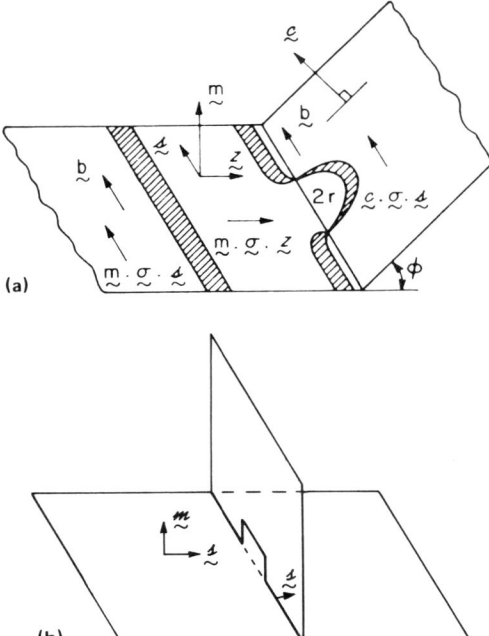

FIG. 31. Idealized models for (a) cross-slip of screw dislocations and (b) climb of edge dislocations.

the critical events are influenced by components of stress other than the Schmid stress τ_{ms}.

In the cross-slip process the extended dislocation must first constrict over a segment followed by a critical amount of bowing of the constricted segment on the cross-slip plane. Constriction is influenced by the shear stress τ_{ms}, whereas the bowing process is partially driven by τ_{cs}, which has a component arising from τ_{zs}. In single slip this leads to a flow law of the type

$$\dot{\gamma}_{ms} = \frac{1}{h}(\dot{\tau}_{ms} + \alpha \dot{\tau}_{zs} + \beta \dot{\tau}_{mz}), \tag{3.45}$$

where h is the current strain-hardening rate of the system $(\mathbf{s}, \mathbf{m})$. Here α and β, for example, give the decrement in the Schmid stress τ_{ms}, consistent with maintaining plastic flow, per unit increase in the non-Schmid stresses τ_{zs} and τ_{mz}. The climb process for the edge segment shown in Fig. 31b is, on the other hand, influenced by the normal stress τ_{ss}, which by the Peach–Koehler relation (2.9) causes a climb force in a plane orthogonal to the Burgers vector (i.e., $\mathbf{s}$) and containing the dislocation line. A flow law including this effect would have the form

$$\dot{\gamma}_{ms} = \frac{1}{h}(\tau_{ms} - \varepsilon \tau_{ss}). \tag{3.46}$$

In a section of stress space resolved on the lattice the related yield surfaces might then look as depicted in Fig. 32, which also illustrates the nonnormality of the plastic strain increment. Normality would be restored in the example of Eq. (3.45) and Fig. 31a if relations of the type $d\gamma_{zs} = \alpha \, d\gamma_{ms}$ and $d\gamma_{mz} = \beta \, d\gamma_{ms}$ held. This is quite unlikely in general, and in analyzing single-slip phenomena on the system $(\mathbf{s}, \mathbf{m})$ it is reasonable to take $d\gamma_{zs} = d\gamma_{mz} = 0$. To the extent that the Schmid law is verified, we expect the parameters α, β, or ε to be significantly less than unity, but direct measurements of these non-Schmid effects are at present limited.

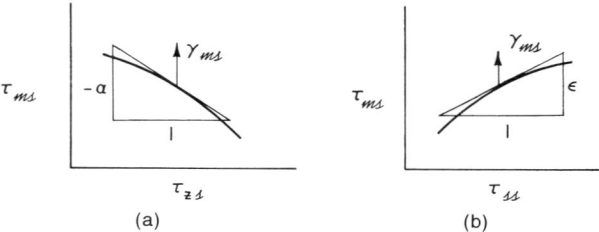

FIG. 32. Hypothetical yield surfaces constructed as if (a) cross-slip were the critical event in maintaining primary slip and (b) climb were the critical event. In both cases plastic strain rates would not be directed normal to the yield surfaces.

Micromechanics of Crystals and Polycrystals

To incorporate non-Schmid effects into a crystal flow law of the type given earlier in Eq. (3.32), it is necessary to relate all the $\dot{\tau}$'s in relations such as Eqs. (3.45) and (3.46) to $\tau^{\nabla*}$. However, given the current uncertainties in the magnitudes of these effects, it is difficult to do this with precision. Thus it seems prudent to readopt the simple approach originally suggested by Asaro and Rice (1977).

To begin, we recall that when τ is defined as in Eq. (3.18) with **P** given by Eq. (3.11), $\dot{\tau}_{ms}$ is given by Eq. (3.28),

$$\dot{\tau}_{ms} = \mathbf{m}^{*(\alpha)} \cdot (\boldsymbol{\tau}^{\nabla*} - \mathbf{D}^* \cdot \boldsymbol{\tau} + \boldsymbol{\tau} \cdot \mathbf{D}^*) \cdot \mathbf{s}^{*(\alpha)}. \tag{3.47}$$

If for the sake of clarity we now take the current state of deformation coincident with the reference state,[†] $\boldsymbol{\tau} = \boldsymbol{\sigma}$, $\mathbf{s} = \mathbf{s}^*$, $\mathbf{m} = \mathbf{m}^*$, and

$$\tau_{ms} = \mathbf{m} \cdot \boldsymbol{\sigma} \cdot \mathbf{s}. \tag{3.48}$$

If τ is again taken to be the Schmid stress defined by Eq. (3.18), with **P** defined by Eq. (3.11), then we note, in computing $\dot{\tau}$, that

$$\dot{\mathbf{s}} = (\mathbf{D}^* + \boldsymbol{\Omega}^*) \cdot \mathbf{s} \tag{3.49a}$$

and

$$\dot{\mathbf{m}} = -\mathbf{m} \cdot (\mathbf{D}^* + \boldsymbol{\Omega}^*) + \mathbf{m}\,\mathrm{tr}(\mathbf{D}^*) \tag{3.49b}$$

and that

$$\dot{\tau}_{ms} = \mathbf{m} \cdot [\boldsymbol{\sigma}^{\nabla*} + \boldsymbol{\sigma}\,\mathrm{tr}(\mathbf{D}^*) - \mathbf{D}^* \cdot \boldsymbol{\sigma} + \boldsymbol{\sigma} \cdot \mathbf{D}^*] \cdot \mathbf{s}. \tag{3.49c}$$

This is equivalent to Eq. (3.47), since $\tau^{\nabla*} = \sigma^{\nabla*} + \sigma\,\mathrm{tr}(\mathbf{D})$ with the reference so chosen. However, another definition of τ and $\dot{\tau}$ results if in Eq. (3.48) we take **s** and **m** to remain *unit* orthogonal vectors that rotate with the lattice so that **s** remains in the slip plane. Then

$$\dot{\mathbf{s}} = (\mathbf{D}^* + \boldsymbol{\Omega}^*) \cdot \mathbf{s} - \mathbf{s}(\mathbf{s} \cdot \mathbf{D}^* \cdot \mathbf{s}), \tag{3.50a}$$

$$\dot{\mathbf{m}} = -\mathbf{m} \cdot (\mathbf{D}^* + \boldsymbol{\Omega}^*) + \mathbf{m}(\mathbf{m} \cdot \mathbf{D}^* \cdot \mathbf{m}), \tag{3.50b}$$

and

$$\dot{\tau}_{ms} = \mathbf{m} \cdot [\boldsymbol{\sigma}^{\nabla*} - \mathbf{D}^* \cdot \boldsymbol{\sigma} + \boldsymbol{\sigma} \cdot \mathbf{D}^* + \boldsymbol{\sigma}(\mathbf{m} \cdot \mathbf{D}^* \cdot \mathbf{m} - \mathbf{s} \cdot \mathbf{D}^* \cdot \mathbf{s})] \cdot \mathbf{s}. \tag{3.50c}$$

A third definition results by taking **s** and **m** in Eq. (3.48) to be orthogonal unit vectors that simply rotate at the lattice spin rate $\boldsymbol{\Omega}^*$. Then

$$\dot{\mathbf{s}} = \boldsymbol{\Omega}^* \cdot \mathbf{s}, \tag{3.51a}$$

$$\dot{\mathbf{m}} = \boldsymbol{\Omega}^* \cdot \mathbf{m}, \tag{3.51b}$$

[†] Another reason for doing this here is to establish some relations required in a later analysis.

and

$$\dot{\tau}_{ms} = \mathbf{m} \cdot \boldsymbol{\sigma}^{\nabla *} \cdot \mathbf{s}. \qquad (3.51c)$$

As shown by Asaro and Rice (1977), all such interpretations of the resolved shear stress lead to $\dot{\tau}$'s of the form

$$\dot{\tau}_{ms} = \mathbf{m} \cdot [\boldsymbol{\sigma}^{\nabla *} + \boldsymbol{\sigma}\,\mathrm{tr}(\mathbf{D}^*)] \cdot \mathbf{s} + \boldsymbol{\sigma}:\mathbf{H}:\mathbf{D}^* = \mathbf{m} \cdot \boldsymbol{\tau}^{\nabla *} \cdot \mathbf{s} + \boldsymbol{\sigma}:\mathbf{H}:\mathbf{D}^*, \qquad (3.52)$$

where $\mathbf{H}$ is a fourth-rank tensor with elements of $O(1)$ that depend on the precise way in which the vectors $\mathbf{s}$ and $\mathbf{m}$ deform with the lattice. By inverting Eq. (3.13), Eq. (3.52) can also be written in the form

$$\dot{\tau}_{ms} = (\mathbf{sm} + \boldsymbol{\sigma}:\mathbf{H}:\mathbf{L}^{-1}):\boldsymbol{\tau}^{\nabla *} = (\mathbf{P} + \boldsymbol{\sigma}:\mathbf{H}:\mathbf{L}^{-1}):\boldsymbol{\tau}^{\nabla *}. \qquad (3.53)$$

This illustrates the point made earlier, that as long as the vector $\mathbf{s}$ is chosen to rotate with the lattice, or preferably to remain in the slip plane by convecting with the lattice, the various $\dot{\tau}$'s differ *only* by terms of $O(\sigma/L)$, which are typically negligible. Certainly experimental verification of the Schmid rule, not to mention the current less well documented understanding for the non-Schmid effects, is not known to an accuracy involving terms of this order.

In view of this, it seems entirely adequate to write expressions for $\dot{\tau}_{mz}$, $\dot{\tau}_{zs}$, etc., in laws like Eq. (3.45), of the type $\mathbf{m} \cdot \boldsymbol{\tau}^{\nabla *} \cdot \mathbf{z}$, $\mathbf{z} \cdot \boldsymbol{\tau}^{\nabla *} \cdot \mathbf{s}$, etc., and to then write a more general flow law incorporating the non-Schmid effects of the form

$$\dot{\gamma}_{ms} = \frac{1}{h}\mathbf{Q}:\boldsymbol{\tau}^{\nabla *}, \qquad (3.54)$$

where

$$\mathbf{Q} = \mathbf{P} + \boldsymbol{\sigma}:\mathbf{H}:\mathbf{L}^{-1} + \boldsymbol{\alpha} \qquad (3.55)$$

and

$$\boldsymbol{\alpha} = \begin{bmatrix} \alpha_{ss} & 0 & \alpha_{sz} \\ 0 & \alpha_{mm} & \alpha_{mz} \\ \alpha_{sz} & \alpha_{mz} & \alpha_{zz} \end{bmatrix}. \qquad (3.56)$$

On the other hand, when the Schmid stress of Eq. (3.18) is again used, Eq. (3.54) could alternatively be written as

$$\dot{\gamma}_{ms} = \frac{1}{h}(\boldsymbol{\mu} + \boldsymbol{\alpha}):\boldsymbol{\tau}^{\nabla *}. \qquad (3.57)$$

In any case, aside from the probably negligible effects of lattice elasticity, for example, $\alpha_{zs} = \alpha_{sz} = 1/2\alpha$ and $\alpha_{zm} = \alpha_{mz} = 1/2\beta$ in Eq. (3.45) and $\alpha_{ss} = -\varepsilon$ in Eq. (3.46).

Some Estimates of Non-Schmid Terms

As mentioned above, direct measurements of these effects are very limited, but some data have been reported. For example, Barendreght and Sharpe (1973) performed an interesting experiment on hcp single crystals of zinc, which they reported slipped only on the basal plane (see. Fig. 3c). Their crystals were loaded in biaxial tension so that the critical resolved shear stresses were reached on the basal slip system, with widely varying normal stresses acting on the slip plane. Their results are shown in Fig. 33, which indicates that the critical resolved shear stress for initial yield decreases by as much as 30% of the value measured in uniaxial tension. This can be expressed using the procedure introduced in this context by Asaro and Rice (1977) as follows: If $(\tau_{ms})_{onset}$ is the critical value of the resolved shear stress to initiate slip on the $(\mathbf{s}, \mathbf{m})$ system,

$$(\tau_{ms})_{onset} = \tau_c - \eta \tau_{mm}. \tag{3.58}$$

Then the Schmid stress rate $\dot{\tau}_{ms}$ is replaced by

$$\dot{\tau}_{ms} + \eta \dot{\tau}_{mm},$$

and comparing with Eq. (3.49),

$$\alpha_{mm} = \eta. \tag{3.59}$$

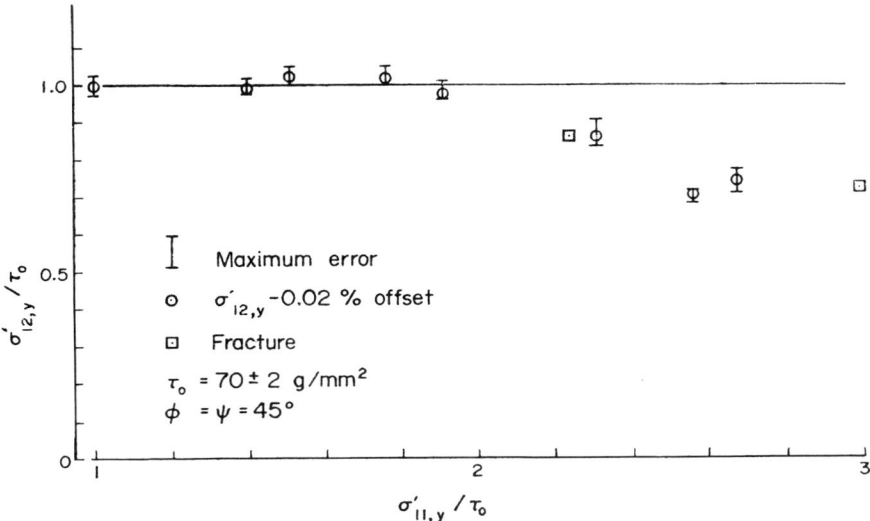

FIG. 33. Barendreght and Sharpe's (1975) data for the dependence of yield strength on the state of biaxial tension in zinc crystals.

The data in Fig. 33 suggest that η may actually be as large as $1/6$, at least when the normal tension on the slip plane is greater than twice the value for τ in uniaxial tension.

In analogous fashion we can consider the possibility of pressure sensitivity of the critical Schmid stress. Barendreght and Sharpe's data could be interpreted this way as well. For example, suppose that

$$(\tau_{ms})_{onset} = \tau_c + \kappa p, \qquad (3.60)$$

where $p = -\frac{1}{3}\sigma_{kk}$. Now the Schmid stress rate $\dot{\tau}_{ms}$ is replaced by

$$\dot{\tau}_{ms} + \tfrac{1}{3}\kappa(\dot{\tau}_{mm} + \dot{\tau}_{ss} + \dot{\tau}_{zz}),$$

and comparing with Eq. (3.48),

$$\alpha_{ij} = \tfrac{1}{3}\kappa\delta_{ij}. \qquad (3.61)$$

As noted by Asaro and Rice (1977), κ is readily interpreted in terms of the difference between yield strengths in uniaxial tension σ_t and in uniaxial compression σ_c. If the slip system is oriented for a maximum resolved shear stress, i.e., both the slip plane normal and slip direction are at $45°$ to the tensile axis so that $\tau_{ms} = \tfrac{1}{2}\sigma$, then Eq. (3.60) leads to

$$\frac{1}{2}(\sigma_c - \sigma_t) = \frac{\kappa}{3}(\sigma_c + \sigma_t) \qquad (3.62a)$$

or

$$\text{SD} \equiv \frac{\sigma_c - \sigma_t}{(1/2)(\sigma_c + \sigma_t)} = \frac{4\kappa}{3}, \qquad (3.62b)$$

where SD is the "strength differential." Values of SD are reported to be quite low (or zero) for pure single-phase fcc crystals, but have been reported to be as high as 0.07–0.1 in high-strength martensitic steels (Spitzig et al., 1975).

Finally we add that theoretical estimates of the terms α_{zs} and α_{mz} for cross-slip have been given by Asaro and Rice (1977) following the procedure outlined below.

One approach to the development of kinetic laws for dislocation motion and constitutive laws rests on the premise that micromechanisms such as the two shown previously are thermally activated events. What is then required is a calculation of the reversible work ΔG required to drive the local system to the transition state. When the flow is idealized as *explicitly* rate insensitive, ΔG, which depends on τ_{ms}, τ_{ss}, etc., is taken to be a yield function in the sense that plastic flow only occurs at a measurable rate when ΔG falls to a critically low (essentially zero) value. This leads to a relation among the arguments of

ΔG of the form

$$\Phi(\tau_{ms}, \tau_{mz}, \tau_{zs}, \ldots \rho) = 0, \qquad (3.63)$$

where ρ is a structure parameter related no doubt to dislocation density that increases with strain due to strain hardening, e.g.,

$$d\rho = \mu \, d\gamma. \qquad (3.64)$$

Since Φ must maintain its critical value during plastic flow, there is also the consistency condition

$$(\partial\Phi/\partial\tau_{ms}) d\tau_{ms} + (\partial\Phi/\partial\tau_{mz}) d\tau_{mz} + (\partial\Phi/\partial\tau_{zs}) d\tau_{zs} + \cdots = -\mu(\partial\Phi/\partial\rho) d\gamma. \qquad (3.65)$$

This has the same form as Eq. (3.45), provided that we make the identities

$$h \equiv -\mu(\partial\Phi/\partial\rho)(\partial\Phi/\partial\tau_{ms})^{-1} \qquad (3.66a)$$

$$\alpha = (\partial\Phi/\partial\tau_{zs})(\partial\Phi/\partial\tau_{ms})^{-1}, \qquad (3.66b)$$

$$\beta = (\partial\Phi/\partial\tau_{mz})(\partial\Phi/\partial\tau_{ms})^{-1}, \qquad (3.66c)$$

$$\vdots$$

Using this framework, Asaro and Rice (1977) proposed several versions of the cross-slip model depicted in Fig. 31a. The models differed in the detail concerning the piling up of primary dislocations behind the lead dislocation undergoing cross-slip. They estimated values for α of between 0.05 and 0.08, and values for β of between 0.01 and 0.06. The details of their calculation are quite lengthy and are not reproduced here. However, the procedures are applicable to other kinds of micromechanical processes and so the reader is referred to their work for a full discussion.

D. Strain Rate-Dependent Flow Laws

In the preceding section it was noted that the resistance to dislocation motion, and therefore the shear strength of a slip system, depends upon the existence of obstacles to dislocation motion. Impurity or solute atoms, "forest" intersections, second-phase particles or precipitates, Lomer locks, etc., are examples of such barriers. It was also noted that obstacles are overcome by micromechanisms such as cross-slip and climb, among others, that require gliding dislocations to achieve some higher free-energy critical state. The reversible work required to achieve this state was designated ΔG. When plastic flow is idealized as rate insensitive, the applied local stresses are presumed to be sufficient to drive ΔG to an essentially zero value so that the obstacles are surmounted in a purely mechanical fashion. At strain rates less than 10^3 or so, however, thermal fluctuations supply part of ΔG so that dislocation motion and straining at finite rates occur at stresses below the mechanical threshold. Within the framework of transition state theory (see, for example, Krausz and Erying, 1975) the probability of a region experiencing a critically large thermal fluctuation is taken to be proportional to

$$e^{-\Delta G/kT}, \tag{3.67}$$

where k is Boltzmann's constant and T is the absolute temperature. The rate at which dislocations are presumed to overcome the obstacles is then given by

$$v e^{-\Delta G/kT}, \tag{3.68}$$

where v is a characteristic "frequency" at which attempts are made to reach the critical state. This last statement is, of course, highly idealized, in that v evidently stands for a kind of average frequency that is strongly dependent upon the details of the transition state (i.e., the dislocation arrangement), which in turn changes with strain. When the average dislocation velocity v is determined by this sort of "thermal release" from obstacles, as opposed to the rate at which dislocations glide between obstacles for example, v can be expressed as

$$\bar{l} v e^{-\Delta G/kT}, \tag{3.69}$$

where $\bar{l}$ is an average spacing between obstacles. Then v depends very sensitively on stress state via ΔG and also on the temperature T. As discussed earlier, ΔG depends in general on the resolved shear stress τ acting on the slip system, as well as other components of stress. In the discussion to follow, however, non-Schmid effects are neglected.

Experimental measurements, we note, measure the macroscopic response of a slip system to applied stresses that involve the motion of many dislocations. This is usually analyzed using the Orowan relation

$$\dot{\gamma} = \rho_m b v, \tag{3.70}$$

where v is again the average dislocation velocity, b the magnitude of the Burgers vector, and ρ_m the density of *mobile dislocations* that are able to slip. For a given dislocated state of a crystal it must be assumed that ρ_m also depends on stress. Now to characterize the strain rate sensitivity of the material (and its current internal structure) the parameter m is defined as

$$m \equiv \frac{\partial \ln \dot{\gamma}}{\partial \ln \tau} \approx \frac{\partial \ln \rho_m}{\partial \ln \tau} + \frac{\partial \ln v}{\partial \ln \tau}, \tag{3.71}$$

where the second term can be expressed as

$$\frac{\partial \ln v}{\partial \ln \tau} = \frac{\partial \ln(v\bar{l})}{\partial \ln \tau} - \frac{\tau}{kT}\frac{\partial \Delta G}{\partial \tau} = \frac{\partial \ln(v\bar{l})}{\partial \ln \tau} + \frac{\tau v^*}{kT}. \qquad (3.72)$$

The magnitude of m is strongly influenced (among other things) by the stress dependence of ΔG; when this is large, so is m and the material shows relatively low rate sensitivity. In essence, this comes about as follows: v^*, which has the dimensions of volume and is called the "activation volume," is effectively the volume through which the generalized force τ acts in affecting (decreasing) ΔG. When v^* is large or, in other words, when ΔG is strongly stress dependent, the transition state must be very nearly overcome mechanically before thermal fluctuations become important. Krausz and Eyring (1975) listed some approximate values for v^* corresponding to common dislocation mechanisms:

- Climb $\qquad v^* = 1b^3$,
- Cross-slip $\qquad v^* = 10-10^2 b^3$,
- Dislocation intersection $\qquad v^* = 10^2-10^4 b^3$.

Byrne et al. (1961), as another example, measured v^* to be in the range $10^2-10^3 b^3$ in precipitation-hardened aluminum–copper alloys, where dislocations must "cut through" precipitates. When v^* is this large, the rate sensitivity and corresponding temperature dependence of plastic flow is found to be relatively small. In general, we note that m will depend upon internal structure and should change with plastic straining.

As a specific example some recent experimental results on strain rate effects in aluminum single crystals are discussed. The results, taken from the recent work of Chiem and Duffy (1981), utilized novel experimental techniques that allow for accurate determinations of τ versus γ relations at strain rates as large as 2×10^3 sec^{-1}.

Figure 34 illustrates τ versus γ over the range of strain rates 4×10^{-5}–1.6×10^3 sec^{-1}. The data indicate both an influence of strain rate on the level of the flow stress, as well as on the rate of strain hardening. Figure 35 shows the material response to sudden changes in strain rate; for the cases shown the strain rate was *increased* from 5×10^{-4} to 850 sec^{-1} after various strains achieved at the lower rate.

Chiem and Duffy (1981) interpreted their results using thermal rate analysis and correlated them with the dislocation substructure they observed using electron microscopy. Figure 36 shows their results for the total dislocation density versus shear strain at the strain rate 5×10^{-4} sec^{-1}. At quasi-static rates well-defined dislocation cells formed, and the relation between the average cell size and flow strength is shown in Fig. 37. The analytical form for this was reported to be

$$\rho_{\text{total}} = 2.7 \times 10^8 + 1.15 \times 10^{10} \gamma_s \text{ cm}^{-2} \qquad (3.73)$$

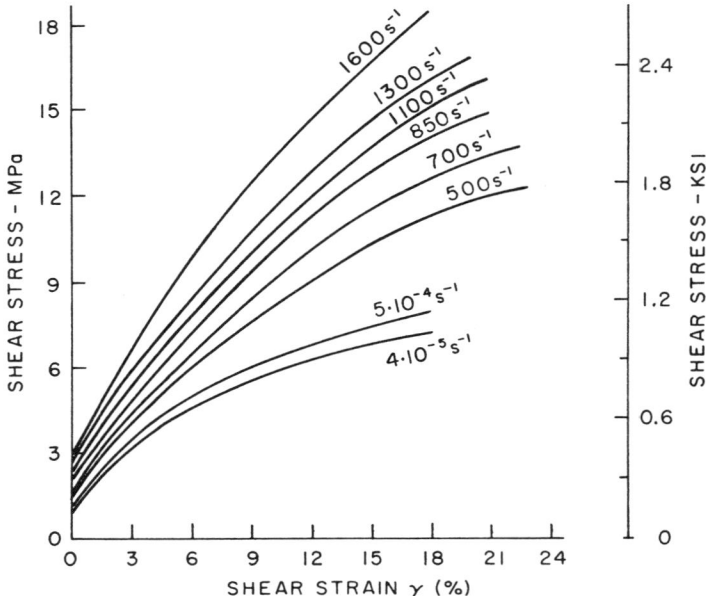

FIG. 34 Resolved shear stress versus shear strain curves obtained at constant strain rates rates ranging from 4×10^{-5} to 1.6×10^{-3} s^{-1} at 20°C (Chiem and Duffy, 1981).

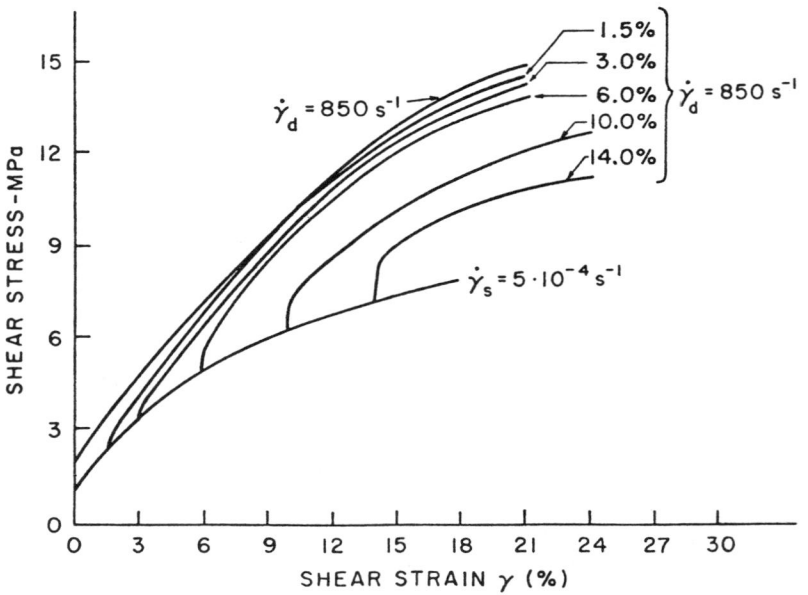

FIG. 35. Strain rate "jump" tests obtained by rapidly increasing the strain rate after various prestrains at 20°C (Chiem and Duffy, 1981).

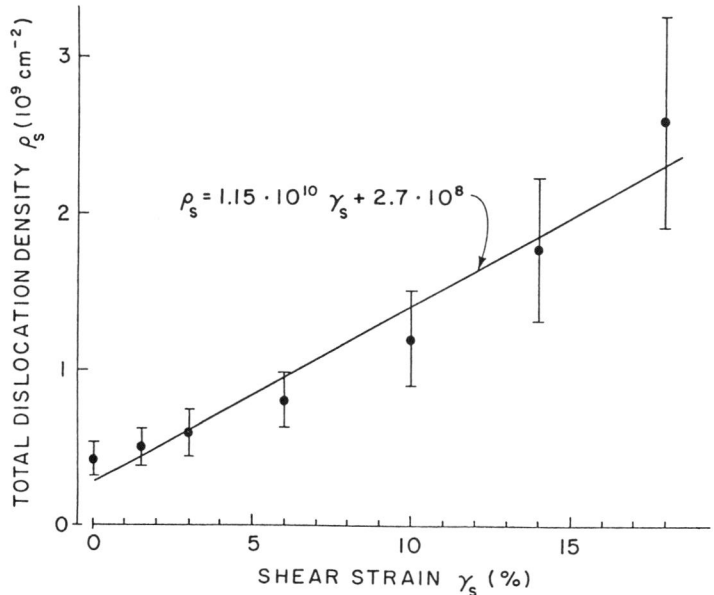

FIG. 36. Dislocation density versus shear strain following quasi-static deformation at a strain rate of 5×10^{-4} s^{-1} at 20°C (Chiem and Duffy, 1981).

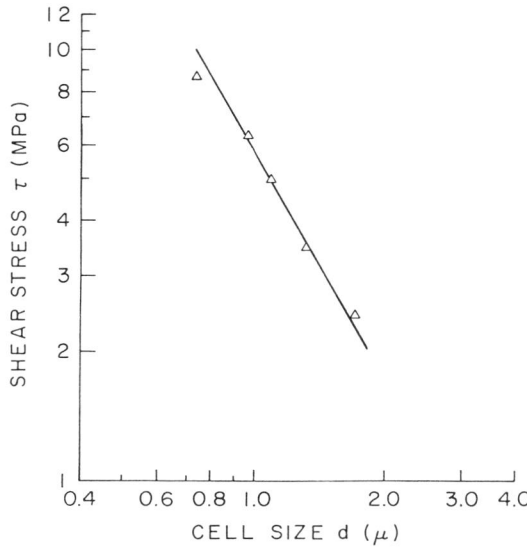

FIG. 37. Relation between flow strength and dislocation cell diameter at 20°C (Chiem and Duffy, 1981).

and

$$\tau = 7.45d^{-1} - 1.84. \tag{3.74a}$$

Chiem and Duffy (1981) noted, however, that a Hall–Petch dimensional form

$$\tau = 14.66d^{-1/2} - 8.92 \tag{3.74b}$$

also fit the τ versus d data with nearly equal precision. They reported that cells also formed at the higher strain rates and that they were smaller in diameter, in accordance with the higher strengths. Several important parameters that enter the thermal analysis were computed, namely, m and v^* defined previously, and the quantities β and v^* defined as

$$\beta \equiv \left(\frac{\partial \ln \dot{\gamma}}{\partial \tau}\right)_{T,\rho} \quad \text{and} \quad v^* \equiv -\left(\frac{\partial \Delta G}{\partial \tau}\right)_{T,\rho}. \tag{3.75}$$

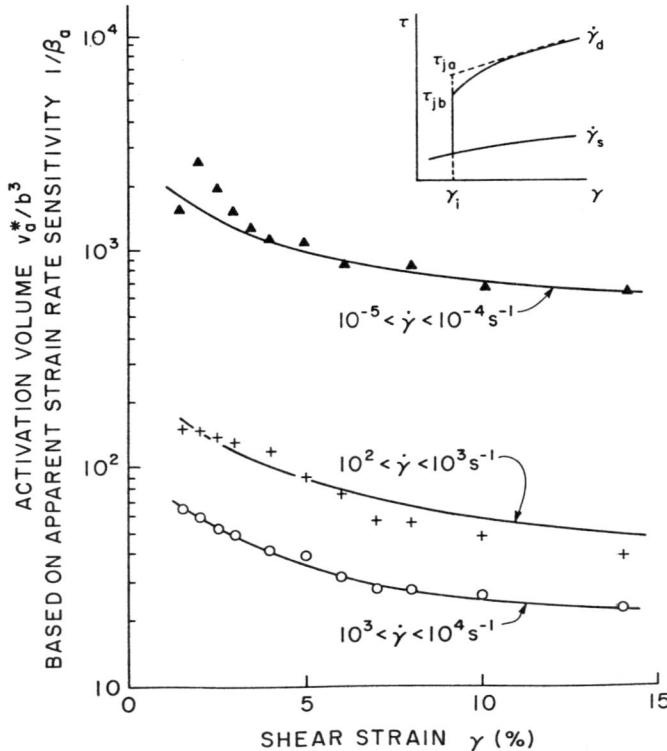

FIG. 38. Apparent activation volume v_a^* versus shear strain at 20°C (Chiem and Duffy, 1981).

The two derivatives that define β and v^* are taken in principle at fixed *internal structure*, i.e., at a fixed density and local configuration of dislocations. In practice it is often assumed that the derivatives in Eqs. (3.71) and (3.72), $\partial \ln \rho_m/\partial \ln \tau$ and $\partial \ln(v\bar{l})/\partial \ln \tau$, are small in comparison to those involving ΔG and v. In that case there is a simple relation among β, v^*, and m:

$$m \approx \beta\tau \approx v^*\tau/(kT). \qquad (3.76)$$

When m, β, and v^* are evaluated from constant strain rate data (Fig. 34), they are denoted "apparent" values; when they are evaluated from the strain rate jump tests (Fig. 35), they are called "true" values. These are designated with subscripts "a" and "t," respectively.

Values for v_a^* as a function of shear strain are shown in Fig. 38. The results for v_t^* computed from the data of Fig. 35 are shown in Fig. 39. It is evident that v^* depends upon both strain and strain rate. Similarly, the strain rate sensitivity parameter β computed at the strain rate 5×10^{-4} sec^{-1} is

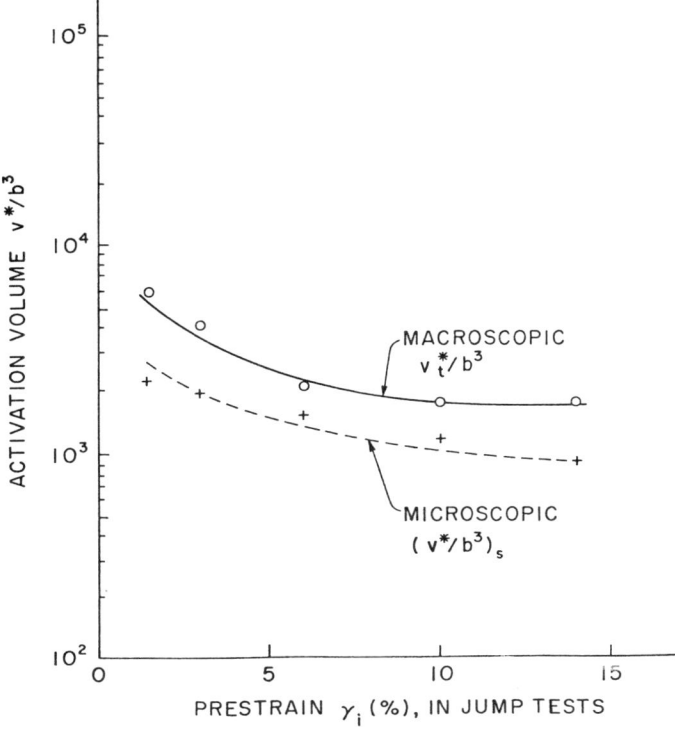

FIG. 39. True activation volume v_t^* versus shear strain at 20°C (Chiem and Duffy, 1981).

shown in Fig. 40; β is defined in Eqs. (3.75) and (3.76) and, as shown by Chiem and Duffy's plot, it depends upon the dislocation substructure and therefore on the strain. When m is computed from Eq. (3.76) using Fig. 37 and 40, it is found to vary in the range 70–100. It is interesting to note that when m is computed using the "apparent" values of β derived from constant strain rate data, it is found to be some two to three times as large. Chiem and Duffy (1981) have further shown that the strain rate sensitivity parameter m tends to decrease at larger strain rates, which indicates a stronger rate sensitivity at high rates. Nevertheless, for pure fcc materials like aluminum m is relatively large, on the order of 10^2, and the material response often displays an apparent rate insensitivity especially when the nominal strain rates are in the quasi-static regime 10^{-6}–10^{-4} sec^{-1}.

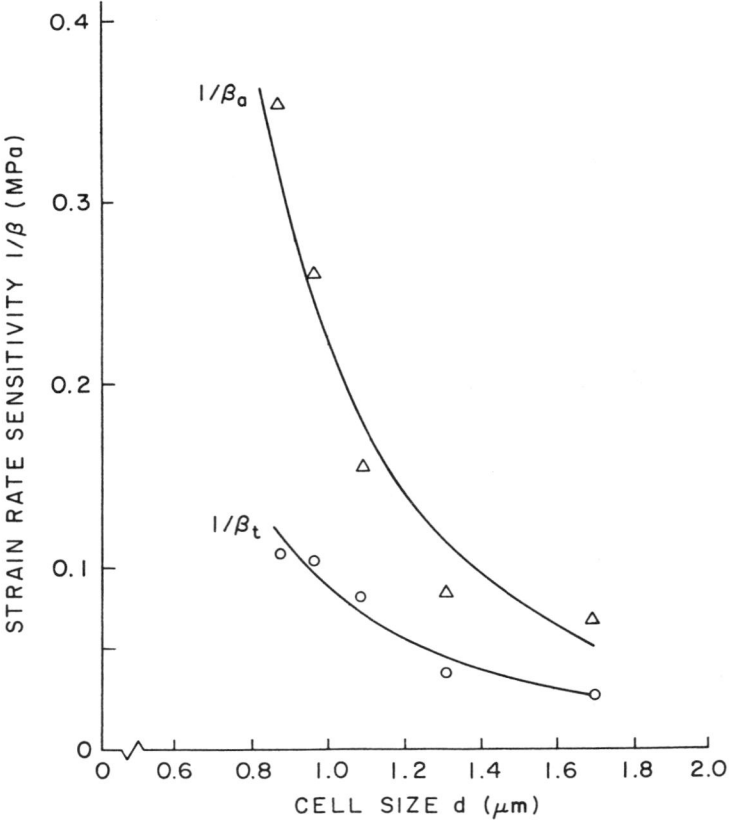

FIG. 40. Strain rate sensitivity parameter β versus dislocation cell size (Chiem and Duffy, 1981).

Body-centered-cubic metals tend to be more rate sensitive than fcc metals. For example, Fig. 41 shows some typical results of strain rate jump tests and for the calculated true rate sensitivity parameter m for a mild steel (0.125 wt. % C, 1.15 wt. % Mn) tested at 20°C. The phenomenology following sudden increases in strain rate is quite different than that for fcc metals, in that the flow stresses at the higher rate jump to values *above* those that would have prevailed had the strain rate been constant at the higher rate. With continued straining after the jump, however, the flow stress tends to drop to values corresponding to the strain-hardening curve at the constant higher strain rate. Here m is correspondingly larger than for fcc metals (some two to three times as large), especially in the temperature range 100–300 K.

A simple isothermal relation that has computational attractiveness and that accounts for a part of the phenomenology just described is the nonlinear form

$$\dot{\gamma} = \dot{a}(\tau/g)^m, \tag{3.77}$$

where the value of g is determined by the current internal structure and $\dot{a}$ is a parameter defined so that $\dot{\gamma} = \dot{a}$ when the resolved shear stress $\tau = g$. Strain hardening is accounted for by allowing g to change with strain according to a kinetic law, possibly of the form

$$\dot{g} = h\dot{\gamma}, \tag{3.78}$$

where h is chosen to depend on γ (or $\dot{\gamma}$), as may seem appropriate. The above constitutive law in fact encompasses the often used law

$$\sigma = (\text{const})\varepsilon^n \dot{\varepsilon}^{1/m}, \tag{3.79}$$

which describes power-law hardening. The nonlinear power law suggested has a straightforward generalization to multiple slip that was used by Pan and Rice (1982). For example, we can extend Eq. (3.77) to imply that for each slip system

$$\begin{aligned}\dot{\gamma}^{(\alpha)} &= \dot{a}^{(\alpha)}(\tau^{(\alpha)}/g^{(\alpha)})^m, & \tau^{(\alpha)} &> 0, \\ \dot{\gamma}^{(\alpha)} &= 0, & \tau^{(\alpha)} &\leq 0,\end{aligned} \tag{3.80a}$$

so that each $\dot{\gamma}^\alpha \geq 0$ and we continue to define positive and negative slip systems. Alternatively, we may allow both positive and negative shearing on the systems, in which case Eq. (3.80a) can be rewritten as

$$\dot{\gamma}^{(\alpha)} = \dot{a}^{(\alpha)}|\tau^{(\alpha)}/g^{(\alpha)}|^{m-1}\tau^{(\alpha)}/g^{(\alpha)}. \tag{3.80b}$$

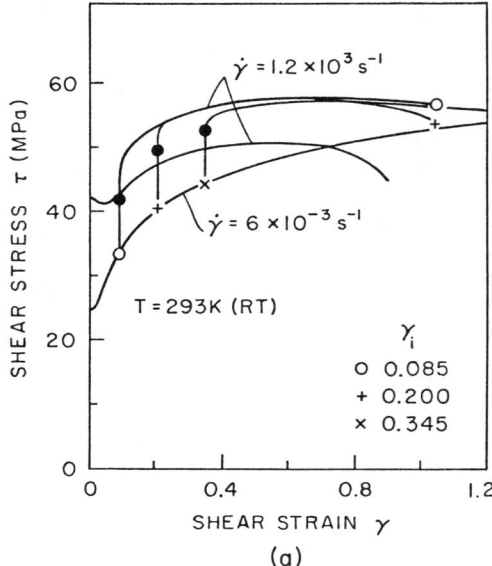

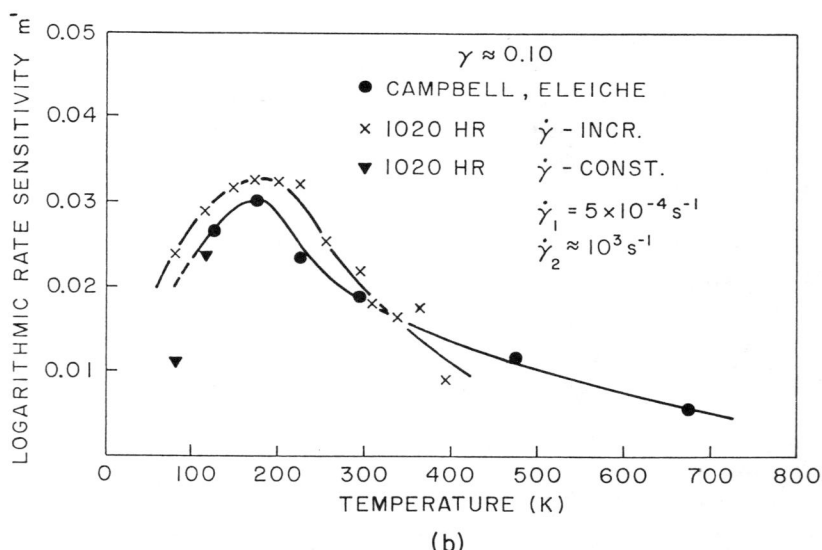

Fig. 41. (a) Strain rate jump tests for a mild steel and (b) the calculated rate sensitivity m' (Campbell *et al.*, 1977, and Klepaczko, 1981; they define $m' = 1/m$).

Which form (or some other) is more appropriate will depend on what feature of the plastic flow is to be modeled. It may, for instance, be easier to incorporate Bauschinger effects with Eq. (3.80a). On the other hand, kinematic hardening can also be included by reinterpreting $\tau^{(\alpha)}$ as $\tau^{(\alpha)} - y^{(\alpha)}$, where $y^{(\alpha)}$ might represent an "internal stress" that also changes with strain. In any event, the hardness parameters $g^{(\alpha)}$ are positive and change according to

$$\dot{g}^{(\alpha)} = \sum_{\beta=1}^{n} h_{\alpha\beta}|\dot{\gamma}^{(\beta)}| \qquad (3.81)$$

so that latent hardening is accounted for. A rate-dependent constitutive law results if Eqs. (3.80) and (3.81) are used to calculate the $\dot{\gamma}$'s in Eq. (3.17). In this case there is no indeterminacy in the choice of the $\dot{\gamma}^{(\alpha)}$ so long as the current values of $g^{(\alpha)}$ and $\tau^{(\alpha)}$ are known. In a sense all slip systems are active, with their shearing rates calculated from Eq. (3.80). For relatively rate-insensitive materials, i.e., those for which $m \geq 100$, the $g^{(\alpha)}$ will be nearly equal to the τ versus γ relations measured during quasi-static deformation. However, as an analysis of latent hardening presented in Section III,E shows, the description of plastic flow is still quite different.

E. The Hardening Modulus $h_{\alpha\beta}$

Latent hardening as discussed in Sections II,D and III,C,1 has a vital influence on the elastic—plastic response of crystalline materials. When plasticity is modeled as explicitly rate insensitive, for example, strong latent hardening can easily lead to a loss of uniqueness in the slip mode. Furthermore, as the plane double-slip model worked out in Section III,C,1 illustrates, the range of relative magnitude of the $h_{\alpha\beta}$ suggested by experiment often brackets the criteria for loss of uniqueness. When rate dependence is accounted for, uniqueness may be restored, provided, of course, that a complete enough description of the current material and stress states is available to construct a $\dot{\gamma}^{(\alpha)}$ versus $\tau^{(\alpha)}$ relation. Nevertheless, if straining on a slip system tends to increase the hardness on other systems more so than on itself, homogeneous multiple slip is unlikely and a patchy slip pattern is expected. Various physical mechanisms that may initiate nonuniform slip modes were discussed in Section II,E. Because of this important role of the $h_{\alpha\beta}$, we conclude this discussion of constitutive laws by providing more perspective on them.

Specifically, the plane model for a crystal in undergoing tensile straining in either single primary slip or double primary–conjugate slip is analyzed.

The main objective of the analysis is to calculate the amount of lattice rotation that occurs when the crystal is initially oriented for single slip and later undergoes double slip, and to compare this with the experimental findings discussed earlier. The model used was shown earlier in Fig. 28, where β now denotes the angle by which the lattice has rotated with respect to the fixed tensile axis along **l**.

To include both rate-dependent and essentially rate-independent material behavior, the flow law is taken as in Eq. (3.80):

$$\dot{\gamma}^{(\alpha)} = \dot{a}^{(\alpha)}(\tau^{(\alpha)}/g^{(\alpha)})^m. \tag{3.80a}$$

By letting $m \to \infty$, rate-independent behavior is approached. For a specific calculation we also take $\dot{g}^{(p)}$ and $\dot{g}^{(c)}$ to have the form of Eq. (3.81):

$$\begin{aligned} \dot{g}^{(p)} &= h_{pp}\dot{\gamma}^{(p)} + h_{pc}\dot{\gamma}^{(c)}, \\ \dot{g}^{(c)} &= h_{cp}\dot{\gamma}^{(p)} + h_{cc}\dot{\gamma}^{(c)}. \end{aligned} \tag{3.82}$$

Further, we make the model assumption that

$$\begin{aligned} h_{pc} &= qh_{pp}, & h_{cp} &= qh_{cc}, \\ h_{pp} &= h(\gamma^{(p)}), & h_{cc} &= h(\gamma^{(c)}). \end{aligned} \tag{3.83}$$

Here q is taken to be a *constant* ratio of latent-hardening to self-hardening rates, which our earlier discussion indicated was in the range $1 \leq q \leq 1.4$. Finally, we choose, for illustration,

$$h(\gamma) = h_0 \operatorname{sech}^2\left(\frac{h_0\gamma}{\tau_s - \tau_0}\right), \tag{3.84}$$

a form that Peirce *et al.* (1982) have recently found useful in modeling aluminum alloy crystals. In Eq. (3.84) $h_0 = 8.9\tau_0$ and $\tau_s = 1.8\tau_0$ was found to fit the strain-hardening data of Chang and Asaro (1980) for aluminum–2.8 wt. % copper alloys.

Now, returning to the model of Fig. 28, we note that when the crystal is taken to be elastically rigid, Eq. (3.2) yields

$$\mathbf{F} = \mathbf{F}^* \cdot \mathbf{F}^P, \tag{3.85}$$

where $\mathbf{F}^*$ corresponds to a rigid rotation of the crystal by β and $\mathbf{F}^P$ is calculated by integrating the expression [derived from Eq. (3.10)]

$$\dot{\mathbf{F}}^P = [\dot{\gamma}^{(P)}\mathbf{s}^{*(P)}\mathbf{m}^{*(P)} + \dot{\gamma}^{(c)}\mathbf{s}^{*(c)}\mathbf{m}^{*(c)}] \cdot \mathbf{F}^P.$$

If e_2 is a *unit* vector aligned along the fiber **l**, the requirement that **l** remain vertical (i.e., along the x_2 axis) implies that

$$\mathbf{F} \cdot \mathbf{e}_2 = \lambda \mathbf{e}_2, \tag{3.86}$$

from which we conclude that

$$F_{12} = 0 \tag{3.87}$$

and

$$\tan \beta = F^P_{12}/F^P_{22}. \tag{3.88}$$

An expression for $\dot\beta$ follows by differentiating Eq. (3.88):

$$\dot\beta = \cos^2\beta\left(\frac{\dot F^P_{12}}{F^P_{22}} - \frac{F^P_{12}\dot F^P_{22}}{(F^P_{22})^2}\right). \tag{3.89}$$

We also note that the two resolved shear stresses required in Eq. (3.80) are calculated at any stage from

$$\tau^{(P)} = \sigma/2 \sin(2\phi - 2\beta) \tag{3.90a}$$

and

$$\tau^{(c)} = \sigma/2 \sin(2\psi + 2\beta). \tag{3.90b}$$

The system of equations (3.80)–(3.90) can be employed in a straightforward time integration scheme to calculate β, $\gamma^{(P)}$, $\gamma^{(c)}$, and their time derivatives as a function of the stretch ratio λ.

Figure 42 shows the results of one such calculation in which $\dot\lambda = \dot l/l_0 = 10^{-2}$ and $\dot a^{(P)} = \dot a^{(c)} = 10^{-2}$. Two levels of strain rate sensitivity were considered corresponding to $m = 200$ and $m = 20$, and three levels of latent hardening corresponding to $q = 1$, 1.2, and 1.4. In addition, an initial orientation was taken for which $\phi = 40°$ and $\psi = 20°$; $\phi + \psi = 60°$ corresponds to the angle between the primary and conjugate slip directions in fcc crystals.

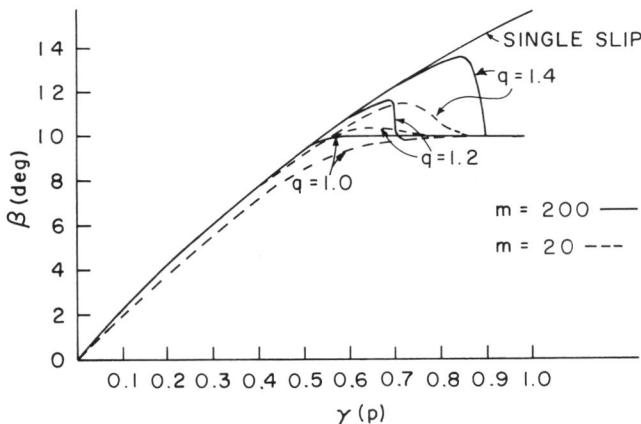

FIG. 42. Lattice rotation β versus primary shear strain for various rate-dependent constitutive laws. Note that with $q > 1$ overshoot is observed, even though noticeable conjugate slip begins before the symmetry position is reached.

When $m = 200$, the effect of $q > 1$ is to promote essentially single slip until the conjugate system becomes active and causes reverse rotation. Conjugate slip appears to develop very abruptly, as is borne out by the calculated strain rates (not shown). In this case of rather weak rate sensitivity, $1 \le q \le 1.4$ leads to overshoot of between 0 and 3.6°. When $m = 20$ the phenomenology is rather different, in that even when q is as large as 1.4, perceptible conjugate slip occurs before the symmetry position of $\beta = 10°$ is reached. Nonetheless, overshoot of up to 1.5° develops with $q = 1.4$. It appears, then, that rate sensitivity, in conjunction with strong latent hardening, offers a possible means of explaining the lattice rotations reported by Mitchell and Thornton (1964) and Joshi and Green (1980) discussed earlier in reference to Fig. 25. Until more data of this type is available, especially data that includes rate effects, it will be difficult to describe latent hardening in more detail. Nonetheless, it does appear that Kocks's (1970) suggestion that q lies in the range 1–1.4 provides a reasonable estimate of latent hardening in fcc crystals.

Finally, we note that other specific forms for rate-independent hardening laws have been proposed. For example, as mentioned by Havner and coworkers (1977, 1978) and Peirce *et al.* (1982), there are distinct computational advantages in imposing symmetry of the matrix $\mathbf{C}$ in Eq. (3.38). Symmetry of $\mathbf{C}$, however, requires symmetry in the matrix $g_{\alpha\beta}$, which in turn implies an asymmetry in $h_{\alpha\beta}$. Havner and Shalaby (1977) have suggested a hardening law of the form

$$h_{\alpha\beta} = H_{\alpha\beta} + \mathbf{P}^{(\alpha)} : \boldsymbol{\beta}^{(\alpha)}. \tag{3.91}$$

In their "simple theory" they take $H_{\alpha\beta}$ to be symmetric and set all the elements $H_{\alpha\beta} = H$. This has the effect of introducing a very strong latent hardening, which they have used to describe overshoot. Peirce *et al.* (1982) have suggested a modified form of such a law,

$$h_{\alpha\beta} = H_{\alpha\beta} + \tfrac{1}{2}(\boldsymbol{\beta}^{(\beta)} : \mathbf{P}^{(\alpha)} - \boldsymbol{\beta}^{(\alpha)} : \mathbf{P}^{(\beta)}), \tag{3.92}$$

which they showed leads to less severe latent hardening more in line with experiments but also preserves the symmetric form of $g_{\alpha\beta}$. These forms may indeed be convenient for specific purposes, but it must be remembered that the off-diagonal Eqs. (3.91) and (3.92) have the effect of coupling elements of $h_{\alpha\beta}$ to the current stress level via the $\boldsymbol{\beta}$'s. This can introduce latent hardening too strong to match experimental observations, which can lead to a premature loss in uniqueness for the slip mode. Furthermore, since the ratio of stress to slip plane hardening moduli becomes significantly larger than unity (of order 10–100 in fcc crystals after finite strains; see, e.g., Mitchell, 1964, or Chang and Asaro, 1981), these stress terms can dominate to the extent that the experimental or physical description of material hardening, which enters via the H's, is lost.

IV. Applications to Elastic–Plastic Deformation of Single Crystals

As the discussion in Section II indicated, plastic deformation in crystals is inherently nonuniform in both a microscopic and a macroscopic sense. For example, it is common for plastic flow in crystals deformed to finite plastic strains to become highly localized, even though the materials are strain hardening and are not undergoing fracture. Examples of nonuniform deformation include "deformation bands" (Barrett, 1952, Brown, 1972; Dillamore *et al.*, 1979), "kink bands" (Orowan, 1942; Honeycombe, 1950; Cahn, 1951), and "shear bands" (Price and Kelly, 1964; Saimoto *et al.*, 1965; Chang and Asaro, 1981). All of these modes have important influences on the plastic flow behavior of crystals that have yet to be fully incorporated in macroscopic continuum modeling of plastic deformation.

Although these phenomena have been documented empirically, progress toward a theoretical understanding of them has only been recently attained, with much more work still required. In the next three sections some recent analyses of nonuniform plastic flow in crystals are reviewed. The analyses of kink bands and coarse slip bands are at present speculative, whereas the analysis of shear bands has been shown to be consistent with the results of rather detailed experimental studies. However, all three cases illustrate that much of the micromechanical phenomenology of crystalline slip can be addressed using the constitutive framework discussed in Sections II and III.

A. Kink Band Formation

Figure 43 shows two examples of deformation kink bands in aluminum single crystals (Cahn, 1951). In both cases the bands formed during the first 1 or 2% extensional strain in crystals deformed in uniaxial tension. The slip mode was predominantly single slip and in fact it was noted by Honeycombe (1950), who also studied kink band formation in aluminum crystals, that the bands did not occur when there was double primary–conjugate slip from the beginning of plastic flow. Honeycombe (1950) further noted that in crystals that displayed conjugate slip *after* a small amount of deformation in single slip, kink bands formed but they "did not develop to the same degree as those found in crystals deforming on one primary set of slip planes" (Honeycombe, 1950, p. 52).

The material plane of these deformation kink bands is reported to lie nearly orthogonal to the primary slip direction when they initiate, but their orientation varies with plastic strain. Figure 44, taken from Cahn's (1951) article, illustrates the orientation of these bands with respect to the underlying

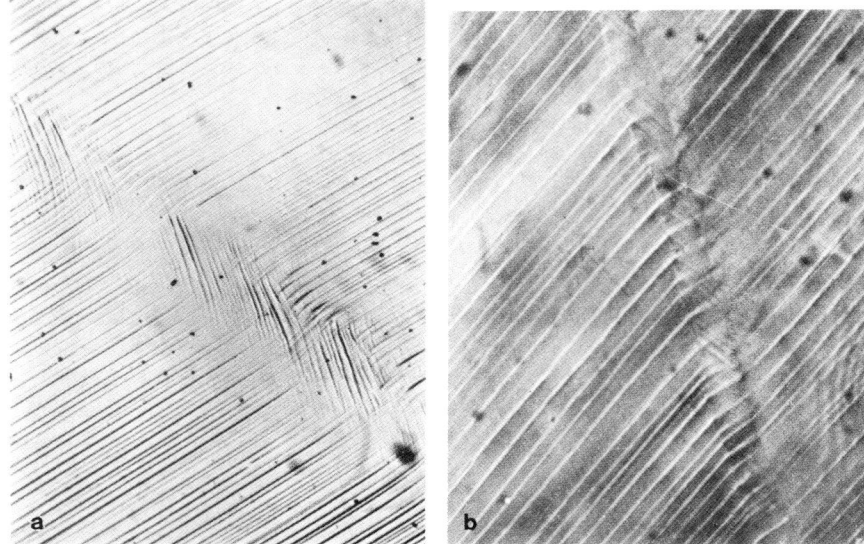

Fig. 43. Kinks bands formed in aluminum single crystals during the first 1–2% plastic strain. The slip mode was single slip and the bands are nearly orthogonal to the primary slip direction.

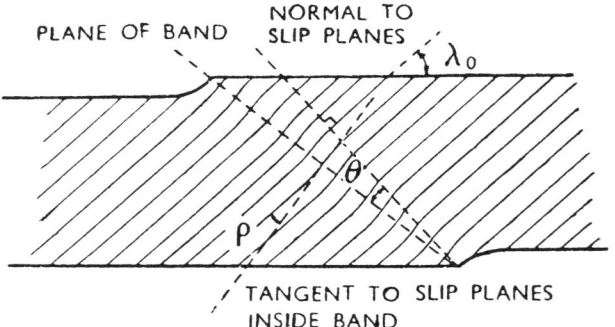

Fig. 44. Cahn's 1951 schematic illustration of the orientation of a kink band with respect to the crystal lattice. Here ρ is the angle between the slip direction inside and outside the band and θ is the angle between the band normal and the slip direction.

crystal lattice. As Fig. 44 shows, the bands involve lattice bending, with particle motion occurring in the plane of the band essentially collinear with the active slip plane normal. Cahn (1951) reported that the angle θ was nearly zero (within a few degrees) when the bands first appeared, but in more highly extended crystals θ increased to values in the range 12–20°. If **n** is the unit normal to the plane of the band, then its relation to the (single) slip system vectors (**s**, **m**) is given by

$$\mathbf{n} = \cos\theta \mathbf{s} + \sin\theta \mathbf{m}, \tag{4.1a}$$

which for later reference we write as

$$\mathbf{n} = \mathbf{s} + \theta \mathbf{m} + O(\theta^2). \tag{4.1b}$$

Figure 44 also indicates that the lattice within the bands is oriented differently than it is outside them. In particular, the trace of the slip plane in the band makes an angle ρ with respect to the corresponding trace outside, where ρ was reported to be less than θ by 1–4°.

Cahn (1951) suggested that kink bands play an important role in crystal strain hardening, especially at small strains. The bands are composed of dipolar arrays of edge dislocations aligned perpendicular to the primary slip direction. Cahn argued that these arrays can act as efficient dislocation traps and thus contribute to strength and strain hardening.

Whereas the geometry of kink bands is now documented, a completely satisfactory theory for their formation is not yet available. There have been suggestions that they form in response to asymmetries in loading, e.g., flexure or bending due to grip constraints (Cahn, 1951), and this may indeed be an important aspect of the overall initiation process. However, there are reasons, to be made clear in what follows, to suppose that such modes are actually inherent in the mechanics of crystalline slip.

Asaro and Rice (1977) analyzed a wide class of localization phenomena occurring in *single slip*. They assumed the material behavior to be explicitly rate insensitive and viewed the *onset* of localized deformation as a bifurcation of an otherwise homogeneous elastic–plastic flow field. Specifically, the general bifurcation mode they considered was a localized band of deformation bounded by characteristic planes across which the material velocity gradients are discontinuous. Non-Schmid effects as described in Section III,C,2 were included in their constitutive description, and it was found that this especially led to the prediction of localized plastic flow in strain-hardening crystals, with homogeneous material properties, as shown by numerous experiments. Some of these modes resemble kink bands and some resemble the coarse slip bands considered in the next section.

a. Constitutive Laws

For the analysis of these bifurcation modes it is convenient to take the current state of deformation coincident with the reference state when formulating the constitutive laws. Then, with reference to Section III,B, Eq. (3.17) becomes

$$\tau^V = \mathbf{L}:\mathbf{D} - (\mathbf{L}:\mathbf{P} + \boldsymbol{\beta})\dot{\gamma}, \tag{4.2}$$

with

$$\mathbf{P} = \tfrac{1}{2}(\mathbf{sm} + \mathbf{ms}), \tag{4.3a}$$

$$\boldsymbol{\beta} = \boldsymbol{\omega}\cdot\boldsymbol{\sigma} - \boldsymbol{\sigma}\cdot\boldsymbol{\omega}, \tag{4.3b}$$

and

$$\mathbf{W} = \tfrac{1}{2}(\mathbf{sm} - \mathbf{ms}). \tag{4.3c}$$

No superscripts are required, since we are now considering only single slip. The hardening law used by Asaro and Rice (1977) was already developed in Section III,C,2, namely, that given by Eq. (3.54) repeated here as

$$\dot{\gamma} = \frac{1}{h}\mathbf{Q}:\tau^{V*}, \tag{4.4}$$

where

$$\mathbf{Q} = \mathbf{P} + \boldsymbol{\alpha} + \boldsymbol{\sigma}:\mathbf{H}:\mathbf{L}^{-1}. \tag{4.5}$$

We recall that $\mathbf{H}$ is a fourth-rank tensor with elements of $O(1)$ that depend on the manner in which the vectors $\mathbf{s}$ and $\mathbf{m}$ *deform* with the lattice, while the elements of $\boldsymbol{\alpha}$ are the coefficients of the non-Schmid terms in the flow law. Combining Eq. (4.4) with Eq. (4.2) yields the Asaro and Rice (1977) constitutive law,

$$\tau^V = \left(\mathbf{L} - \frac{(\mathbf{L}:\mathbf{P} + \boldsymbol{\beta})(\mathbf{Q}:\mathbf{L})}{h + \mathbf{Q}:\mathbf{L}:\mathbf{P}}\right):\mathbf{D}. \tag{4.6}$$

b. *Conditions for Localization*

Hill (1962) has presented a general theory of bifurcation of a homogeneous elastic–plastic flow field into a band of localized deformation. For this to occur, there are two general conditions that must be met. The first is the kinematic restriction that for localization in a thin planar band bounded by planes with unit normal $\mathbf{n}$ in the current state (Fig. 45), the velocity gradient field $\partial v_i/\partial x_j$ inside the band can differ from that outside, $\partial v_i^0/\partial x_j$, only by an expression of the form

$$\partial v_i/\partial x_j - \partial v_i^0/\partial x_j = g_i n_j. \tag{4.7}$$

In addition, there is the equilibrium requirement that the traction rate be continuous across the band or, equivalently, given Eq. (4.7),

$$\mathbf{n}\cdot(\dot{\boldsymbol{\sigma}} - \dot{\boldsymbol{\sigma}}^0) = 0. \tag{4.8}$$

FIG. 45. Orientation of a band of localized deformation in the current state with respect to the slip system vectors.

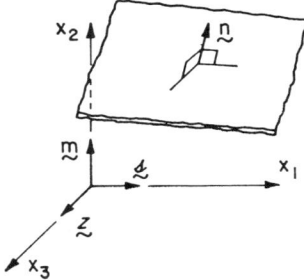

If Eq. (4.6) is rewritten in the form

$$\dot{\sigma}_{ij} = \hat{C}_{ijkl} \partial v_k/\partial x_l \tag{4.9}$$

and if the same constitutive moduli are presumed to apply both inside and outside the incipient band,[†] then Eqs. (4.7) and (4.8) are both satisfied if

$$(n_i \hat{C}_{ijkl} n_l) g_k = 0. \tag{4.10}$$

Thus the critical condition for localization on a plane with normal $\mathbf{n}$ is

$$\det(\mathbf{n} \cdot \hat{\mathbf{C}} \cdot \mathbf{n}) = 0. \tag{4.11}$$

Here $\hat{\mathbf{C}}$ is readily identified by rewriting Eq. (4.6) as

$$\dot{\boldsymbol{\sigma}} = \left(\mathbf{L} - \frac{(\mathbf{L}:\mathbf{P} + \boldsymbol{\beta})(\mathbf{Q}:\mathbf{L})}{h + \mathbf{Q}:\mathbf{L}:\mathbf{P}}\right):\mathbf{D} + \boldsymbol{\Omega} \cdot \boldsymbol{\sigma} - \boldsymbol{\sigma} \cdot \boldsymbol{\Omega} - \boldsymbol{\sigma} \operatorname{tr}(\mathbf{D}). \tag{4.12}$$

Then to construct Eqs. (4.10) and (4.11) we substitute

$$\tfrac{1}{2}(\mathbf{gn} + \mathbf{ng}) \quad \text{and} \quad \tfrac{1}{2}(\mathbf{gn} - \mathbf{ng})$$

for $\mathbf{D}$ and $\boldsymbol{\Omega}$ in Eq. (4.12) and then multiply from the left by $\mathbf{n}$. The result is

$$0 = \left((\mathbf{n} \cdot \mathbf{L} \cdot \mathbf{n}) - \frac{(\mathbf{n} \cdot \mathbf{L}:\mathbf{P} + \mathbf{n} \cdot \boldsymbol{\beta})(\mathbf{Q}:\mathbf{L} \cdot \mathbf{n})}{h + \mathbf{Q}:\mathbf{L}:\mathbf{P}}\right) \cdot \mathbf{g} + \mathbf{A} \cdot \mathbf{g}, \tag{4.13}$$

where

$$\mathbf{A} = \tfrac{1}{2}[(\mathbf{n} \cdot \boldsymbol{\sigma} \cdot \mathbf{n})\mathbf{I} - \boldsymbol{\sigma} - (\mathbf{n} \cdot \boldsymbol{\sigma})\mathbf{n} - \mathbf{n}(\boldsymbol{\sigma} \cdot \mathbf{n})]. \tag{4.14}$$

Now, rather than attempting to set the determinant of the matrix multiplying $\mathbf{g}$ in Eq. (4.13) to zero, the procedure used by Rice (1977, pp. 214–215) and Asaro and Rice (1977, pp. 318–319) may be again invoked. We assume that the inverse matrix $(\mathbf{n} \cdot \mathbf{L} \cdot \mathbf{n})^{-1}$ exists. If Eq. (4.13) is multiplied by this

---

[†] This would not be the case if the location where the band formed contained imperfections that might lower the strength or strain-hardening rates.

inverse, we obtain

$$0 = \left([\mathbf{I} + (\mathbf{n} \cdot \mathbf{L} \cdot \mathbf{n})^{-1} \cdot \mathbf{A}] - \frac{\{(\mathbf{n} \cdot \mathbf{L} \cdot \mathbf{n})^{-1} \cdot [\mathbf{n} \cdot (\mathbf{L}:\mathbf{P} + \boldsymbol{\beta})]\}(\mathbf{Q}:\mathbf{L} \cdot \mathbf{n})}{h + \mathbf{Q}:\mathbf{L}:\mathbf{P}} \right) \cdot \mathbf{g}. \quad (4.15)$$

Furthermore, since $\mathbf{A}$ of Eq. (4.14) has elements of $O(\sigma)$, the bracketed tensor

$$[\mathbf{I} + (\mathbf{n} \cdot \mathbf{L} \cdot \mathbf{n})^{-1} \cdot \mathbf{A}]$$

differs from the unit tensor by terms of $O(\sigma/L)$, which we will assume are very much smaller than unity; in representative cases they would be on the order of 10^{-3} or less. Thus we may assume that this tensor also has an inverse that can be computed from the expansion

$$[\mathbf{I} + (\mathbf{n} \cdot \mathbf{L} \cdot \mathbf{n})^{-1} \cdot \mathbf{A}]^{-1} \approx \mathbf{I} - (\mathbf{n} \cdot \mathbf{L} \cdot \mathbf{n})^{-1} \cdot \mathbf{A} + [(\mathbf{n} \cdot \mathbf{L} \cdot \mathbf{n})^{-1} \cdot \mathbf{A}]$$
$$\cdot [(\mathbf{n} \cdot \mathbf{L} \cdot \mathbf{n})^{-1} \cdot \mathbf{A}] + \cdots . \quad (4.16)$$

When Eq. (4.15) is multiplied by this inverse, we obtain

$$\left(\mathbf{I} - \frac{\mathbf{af}}{h + \mathbf{Q}:\mathbf{L}:\mathbf{P}} \right) \cdot \mathbf{g} = \mathbf{0}, \quad (4.17)$$

where

$$\mathbf{a} = [\mathbf{I} + (\mathbf{n} \cdot \mathbf{L} \cdot \mathbf{n})^{-1} \cdot \mathbf{A}]^{-1} \cdot (\mathbf{n} \cdot \mathbf{L} \cdot \mathbf{n})^{-1} \cdot [\mathbf{n} \cdot (\mathbf{L}:\mathbf{P} + \boldsymbol{\beta})] \quad (4.18a)$$

and

$$\mathbf{f} = \mathbf{Q}:\mathbf{L} \cdot \mathbf{n}. \quad (4.18b)$$

Now if Eq. (4.17) is multiplied from the left with $\mathbf{f} \cdot$, we obtain

$$\left(1 - \frac{\mathbf{fa}}{h + \mathbf{Q}:\mathbf{L}:\mathbf{P}} \right) (\mathbf{f} \cdot \mathbf{g}) = 0. \quad (4.19)$$

As noted by Rice (1977) and Asaro and Rice (1977), $\mathbf{f} \cdot \mathbf{g}$ cannot vanish for nonzero $\mathbf{g}$ unless the bifurcation mode does not involve plastic strain. Therefore the relevant condition is that the bracketed quantity in Eq. (4.19) vanishes, which leads to a critically *low* value for h,

$$h_{cr} = \mathbf{f} \cdot \mathbf{a} - \mathbf{Q}:\mathbf{L}:\mathbf{P}$$

or

$$h_{cr} = -\mathbf{Q}:\mathbf{L}:\mathbf{P} + (\mathbf{Q}:\mathbf{L} \cdot \mathbf{n}) \cdot [\mathbf{I} + (\mathbf{n} \cdot \mathbf{L} \cdot \mathbf{n})^{-1} \cdot \mathbf{A}]^{-1} \cdot [\mathbf{n} \cdot (\mathbf{L}:\mathbf{P} + \boldsymbol{\beta})]. \quad (4.20)$$

It is also easy to verify from Eq. (4.19) that $\mathbf{g}$ has the form

$$\mathbf{g} \propto \mathbf{a}. \quad (4.21)$$

Before evaluating Eq. (4.20), it is important to note the order of magnitude for the various terms appearing in it. The leading terms are of $O(L)$, since the elements of **P** are of $O(1)$. There are then terms of $O(\alpha L)$ and of $O(\sigma)$. In the cases of interest to us here we suppose that the order of ranking of the magnitudes of these terms is

$$O(L) > O(\alpha L) > O(\sigma). \tag{4.22}$$

This is so since the various elements of **α** in Eq. (4.5) are thought to be less than unity, but when they are at all significant, they are in the range 10^{-1}–10^{-2}. In any event, the discussion in Section III,C,2 showed that the elements of **α** are certainly not known to an accuracy of terms involving *elastic strains* that are of $O(\sigma/L)$.

On the other hand, the order of stress/(elastic moduli) is, except in extraordinarily hard materials like martensitic steels, not more than 10^{-3} (see Figs. 6 and 21). In view of this, Eq. (4.22) seems justified and as a consequence all terms of order

$$\alpha\sigma, \alpha\sigma^2/L, \alpha\sigma^3/L^2, \ldots \quad \text{and} \quad \sigma^2/L, \sigma^3/L^2, \ldots$$

may be deleted from expansions of Eq. (4.20). When this is done, Eq. (4.20) can be arranged as

$$\begin{aligned}h_{cr} = &[-\mathbf{P}:\mathbf{L}:\mathbf{P} + (\mathbf{P}:\mathbf{L}\cdot\mathbf{n})\cdot(\mathbf{n}\cdot\mathbf{L}\cdot\mathbf{n})^{-1}\cdot(\mathbf{n}\cdot\mathbf{L}:\mathbf{P})] \\ &+ [-\boldsymbol{\alpha}:\mathbf{L}:\mathbf{P} + (\boldsymbol{\alpha}:\mathbf{L}\cdot\mathbf{n})\cdot(\mathbf{n}\cdot\mathbf{L}\cdot\mathbf{n})^{-1}\cdot(\mathbf{n}\cdot\mathbf{L}:\mathbf{P})] \\ &+ [-\boldsymbol{\sigma}:\mathbf{H}:\mathbf{P} + (\boldsymbol{\sigma}:\mathbf{H}\cdot\mathbf{n})\cdot(\mathbf{n}\cdot\mathbf{L}\cdot\mathbf{n})^{-1}\cdot(\mathbf{n}\cdot\mathbf{L}:\mathbf{P}) \\ &+ (\mathbf{P}:\mathbf{L}\cdot\mathbf{n})\cdot(\mathbf{n}\cdot\mathbf{L}\cdot\mathbf{n})^{-1}\cdot\mathbf{n}\cdot\boldsymbol{\beta} \\ &- (\mathbf{P}:\mathbf{L}\cdot\mathbf{n})\cdot(\mathbf{n}\cdot\mathbf{L}\cdot\mathbf{n})^{-1}\cdot\mathbf{A}\cdot(\mathbf{n}\cdot\mathbf{L}\cdot\mathbf{n})^{-1}\cdot(\mathbf{n}\cdot\mathbf{L}:\mathbf{P})]. \end{aligned} \tag{4.23}$$

The bracketed quantities in Eq. (4.23) place h_{cr} in the form

$$h_{cr} = LF_0(\mathbf{n}) + \alpha L F_1(\mathbf{n}) + \sigma F_2(\mathbf{n}), \tag{4.24}$$

where all the F functions are of $O(1)$ and can be assessed bearing in mind Eq. (4.22).

c. Some Specific Results

If the non-Schmid effects represented by the **α** terms are neglected for the moment, and likewise the terms in σ, it may be shown that h_{cr} is either *negative* or *zero*, the latter occurring when **n** is equal to either **m** or **s**. Thus to localize *in the slip plane* **m** or what we will now dub *the kink plane* **s**, an ideally plastic state must be achieved. This is at variance with the observations reviewed previously. On the other hand, when the non-Schmid effects are taken into account, localization becomes possible on planes slightly removed from **m** or **s**, with positive values for h. Specially, Asaro and Rice (1977) showed that when $\mathbf{n}$ is perturbed from $\mathbf{m}$ as

$$\mathbf{n} \approx \mathbf{m} + \tfrac{1}{2}(\mathbf{s} \cdot \mathbf{M} \cdot \mathbf{s})^{-1} \cdot (\mathbf{s} \cdot \mathbf{M} : \boldsymbol{\alpha}) + O(\alpha^2, \sigma/L), \qquad (4.25)$$

where

$$\mathbf{M} \equiv \mathbf{L} - (\mathbf{L} \cdot \mathbf{m}) \cdot (\mathbf{m} \cdot \mathbf{L} \cdot \mathbf{m})^{-1} \cdot (\mathbf{m} \cdot \mathbf{L}), \qquad (4.26)$$

then

$$h_{cr} \approx \tfrac{1}{4}(\boldsymbol{\alpha} : \mathbf{M} \cdot \mathbf{s}) \cdot (\mathbf{s} \cdot \mathbf{M} \cdot \mathbf{s})^{-1} \cdot (\mathbf{s} \cdot \mathbf{M} : \boldsymbol{\alpha}) + O(\alpha\sigma, \sigma^2/L, \alpha^3 L). \qquad (4.27)$$

A similar perturbation about $\mathbf{n} \approx \mathbf{s}$,

$$\mathbf{n} \approx \mathbf{s} + \tfrac{1}{2}(\mathbf{m} \cdot \boldsymbol{\zeta} \cdot \mathbf{m})^{-1} \cdot (\mathbf{m} \cdot \boldsymbol{\zeta} : \boldsymbol{\alpha}) + O(\alpha^2, \sigma/L), \qquad (4.28)$$

with

$$\boldsymbol{\zeta} \equiv \mathbf{L} - (\mathbf{L} \cdot \mathbf{s}) \cdot (\mathbf{s} \cdot \mathbf{L} \cdot \mathbf{s})^{-1} \cdot (\mathbf{s} \cdot \mathbf{L}), \qquad (4.29)$$

yields

$$h_{cr} \approx \tfrac{1}{4}(\boldsymbol{\alpha} : \boldsymbol{\zeta} \cdot \mathbf{m}) \cdot (\mathbf{m} \cdot \boldsymbol{\zeta} \cdot \mathbf{m})^{-1} \cdot (\mathbf{m} \cdot \boldsymbol{\zeta} : \boldsymbol{\alpha}) + O(\alpha\sigma, \sigma^2/L, \alpha^3 L). \qquad (4.30)$$

To evaluate the predictions in Eqs. (4.27) and (4.30) we now suppose that $\mathbf{L}$ has the isotropic form

$$L_{ijkl} = G(\delta_{ik}\delta_{jl} + \delta_{il}\delta_{jk}) + \Lambda\delta_{ij}\delta_{kl}, \qquad (4.31)$$

where G and Λ are the usual Lamé moduli.
Then

$$\boldsymbol{\alpha} : \mathbf{L} \cdot \mathbf{s} = 2G\boldsymbol{\alpha} \cdot \mathbf{s} + \Lambda \mathbf{s}\,\mathrm{tr}(\boldsymbol{\alpha}), \qquad (4.32\mathrm{a})$$

$$\mathbf{m} \cdot \boldsymbol{\zeta} \cdot \mathbf{m} = G(\mathbf{I} + \mathbf{mm}) + \Lambda \mathbf{mm}, \qquad (4.32\mathrm{b})$$

$$\mathbf{m} \cdot \mathbf{L} \cdot \mathbf{s} = G\mathbf{sm} + \Lambda\mathbf{ms}, \qquad (4.32\mathrm{c})$$

and

$$(\mathbf{m} \cdot \mathbf{L} \cdot \mathbf{m})^{-1} = G^{-1}[\mathbf{zz} + \mathbf{ss}/(4\xi)], \qquad \text{where } \xi = (\Lambda + G)/(\Lambda + 2G). \qquad (4.32\mathrm{d})$$

With these expressions we also find that

$$(\mathbf{s} \cdot \mathbf{M} \cdot \mathbf{s}) = G(\mathbf{zz} + 4\xi\mathbf{ss}) \qquad (4.32\mathrm{e})$$

$$(\mathbf{s} \cdot \mathbf{M} \cdot \mathbf{s})^{-1} = G^{-1}(\mathbf{zz} + \mathbf{ss}/4\xi), \qquad (4.32\mathrm{f})$$

$$\boldsymbol{\alpha} : \mathbf{M} \cdot \mathbf{s} = 2G\{\alpha_{zs}\mathbf{z} + [(2\xi - 1)\alpha_{zz} + 2\xi\alpha_{ss}]\mathbf{s}\}. \qquad (4.32\mathrm{g})$$

The expressions in Eq. (4.32) can now be used to evaluate Eqs. (4.25)–(4.30), and when this is done [neglecting the terms of $O(\sigma\alpha)$], the following two results are found.

Case (i) Perturbation about the slip plane **m**:

$$\mathbf{n} \approx \mathbf{m} + \alpha_{sz}\mathbf{z} + \frac{1}{4\xi}[(2\xi - 1)\alpha_{zz} + 2\xi\alpha_{ss}]\mathbf{s} + O(\alpha^2, \sigma/G), \quad (4.33a)$$

with a corresponding h_{cr},

$$h_{cr} \approx G\left\{\alpha_{sz}^2 + \frac{1}{4\xi}[(2\xi-1)\alpha_{zz} + 2\xi\alpha_{ss}]^2\right\} + O(\alpha\sigma, \sigma^2/G, \alpha^3 G). \quad (4.33b)$$

Case (ii) Perturbation about the slip direction **s**:

$$\mathbf{n} \approx \mathbf{s} + \alpha_{mz}\mathbf{z} + \frac{1}{4\xi}[(2\xi - 1)\alpha_{zz} + 2\xi\alpha_{mm}]\mathbf{m} + O(\alpha^2, \sigma/G), \quad (4.34a)$$

with

$$h_{cr} \approx G\left\{\alpha_{mz}^2 + \frac{1}{4\xi}[(2\xi-1)\alpha_{zz} + 2\xi\alpha_{mm}]^2\right\} + O(\alpha\sigma, \sigma^2/G, \alpha^3 G). \quad (4.34b)$$

We may also now express Eq. (4.21) as

$$\mathbf{g} \propto \mathbf{n} \cdot \mathbf{P} - \xi\mathbf{n}(\mathbf{n} \cdot \mathbf{P} \cdot \mathbf{n}), \quad (4.35)$$

from which it may be shown that $\mathbf{g} \cdot \mathbf{n}$ is itself of $O(\alpha)$ and that the bifurcation mode is very nearly a simple shearing mode.

d. Some Particular Cases

A localized mode resembling a kink or deformation band results if the possibility of normal stress dependence of yielding is considered. For example, as discussed in Section III,C,2, Barendreght and Sharpe's (1973) observations on zinc crystals suggested that α_{mm} may be on the order of 0.1 or so. Taking this into account, Eq. (4.34b) yields

$$h_{cr} \approx G\xi\alpha_{mm}^2, \quad (4.36)$$

or if $\xi \approx 2/3$ (if $G = \Lambda$, for example) and $\alpha_{mm} \approx 0.1$,

$$h_{cr}/G \approx 0.66\alpha_{mm}^2 \approx 0.0066 \approx \tfrac{1}{150}. \quad (4.37)$$

The orientation of the band is, from Eq. (4.34a),

$$\mathbf{n} \approx \mathbf{s} + 1/2\alpha_{mm}\mathbf{m}, \quad (4.38)$$

which places its normal some $1/2\alpha_{mm}$ radians from **s** toward **m** or, given the supposed value for α_{mm}, about 3° from **s** toward **m**. The expected value for $(h/G)_{cr}$ of $\tfrac{1}{150}$ (again given the order of magnitude estimate for α_{mm}) is well within the typical range for strain-hardening rates quoted for fcc or hcp crystals (Mitchell, 1964). Furthermore, the kinematics of this band are consistent with that described earlier in connection with Cahn's (1951) observations. In fcc crystals values for α_{mm} as large as 0.1 are not readily justified from available data. On the other hand, the rate of strain hardening in so-called "easy glide," which occurs at small strains in fcc metals, is often on the order of only $10^{-4}G$. If this is instead substituted into Eq. (4.37), the resulting value of α_{mm} that would account for critical conditions is of the order 0.01. Such a small normal stress dependence of yielding could easily go unnoticed in simple tension or compression tests and yet have this significant consequence regarding the uniformity of plastic flow.

A similar result follows if the possibility of pressure sensitivity of yielding is considered. Using the definitions found in Eq. (3.62b) for the strength differential effect SD, we find that *both* Eqs. (4.33b) *and* (4.34b) yield

$$h_{cr} \approx 0.07 \text{SD}^2 G, \tag{4.39a}$$

with

$$\mathbf{n} \approx \mathbf{m} + 0.16 \text{SDs} \tag{4.39b}$$

or

$$\mathbf{n} \approx \mathbf{s} + 0.16 \text{SDm}, \tag{4.39c}$$

depending on whether the band normal is chosen to lie nearly coincident with $\mathbf{m}$ or the slip direction $\mathbf{s}$. Now since strength differentials SD have been reported as large as 0.08–0.1 only for certain martensitic steels (Spitzig *et al.*, 1975), with much lower values suggested for fcc crystals, both the values of h_{cr} and the orientations of the kink bands predicted by Eq. (4.39) are further removed from observational values than those given earlier using Barendreght and Sharpe's zinc data. Nonetheless, both kink- and deformation-type bands are predicted with positive h, along with shear bands aligned close to the active slip system.

Finally, we note that a third way in which deformation- and kink-type modes can be explained is by considering the possibility that the parameter α_{mz} is nonzero. A possible physical mechanism that accounts for such an effect is the constriction of extended screw dislocations as they undergo cross-slip. Direct experimental documentation of this sort of non-Schmid effect has not yet been attempted. However, Asaro and Rice (1977) have made theoretical estimates of the parameters involved, as mentioned in Section III,C,2, and have suggested that α_{mz} may lie in the range 0.005–0.03.

From Eq. (4.34a) we obtain a mode for which

$$\mathbf{n} \approx \mathbf{s} + \alpha_{mz} \mathbf{z} \tag{4.40a}$$

corresponding to

$$h_{cr} \approx \alpha_{mz}^2 G. \tag{4.40b}$$

This is kinematically different than those described previously, but does correspond approximately to kink bands that lie nearly orthogonal to **s**. The critical values of $h_{cr}/G \approx 10^{-5}$–10^{-3} are again in the range commonly observed during the so-called easy glide regime at small strains and again after finite strains have accumulated.

At the present time it is impossible to say which, if any, of the above modes should be used to describe kink band initiation in fcc or hcp crystals. Evidently much more careful experimentation aimed at documenting the precise conditions of stress state, strain-hardening rates, etc., required for their inception is needed. However, the preceding analyses provide two important bits of theoretical background: In the first place, kink band modes can be explained in a manner consistent with experiment if non-Schmid effects are included, and the bands appear as an entirely natural part of the mechanics of crystalline slip. In this regard it should be re-emphasized that the predictions outlined do not involve either material or geometrical imperfections or asymmetries in the applied loads. The latter, especially, may be important in the actual initiation process, but the analysis indicates a more general underlying cause.

B. COARSE SLIP BAND FORMATION

Figure 46 shows another example of localized flow taken from Chang and Asaro's (1981) experimental studies of coarse slip band and macroscopic shear band formation in precipitation-hardened aluminum–copper alloy crystals. The coarse slip bands are labelled CSB in Fig. 46. Unlike kink bands, coarse slip bands are closely aligned with one of the active slip systems. They appear to form in either single or multiple slip, in tension or compression, and while the materials are strain hardening, as shown by Chang and Asaro (1981) and Price and Kelly (1964). The significance of the latter observation is that it illustrates that the bands are not merely the result of a loss in strain-hardening capacity of the slip systems or of any fractures that may have been in progress; rather, the bands form in damage-free material, often in clusters, and later serve as preferred sites for *macroscopic* shearing, which then leads to fracture. Clusters of coarse slip bands and macroscopic shear bands (MSB) are also evident in Fig. 46. A possibly related phenomenon is the formation of coarse slip lines, as shown in Fig. 47 taken from Cahn's (1951) work on pure aluminum crystals. The offsets in

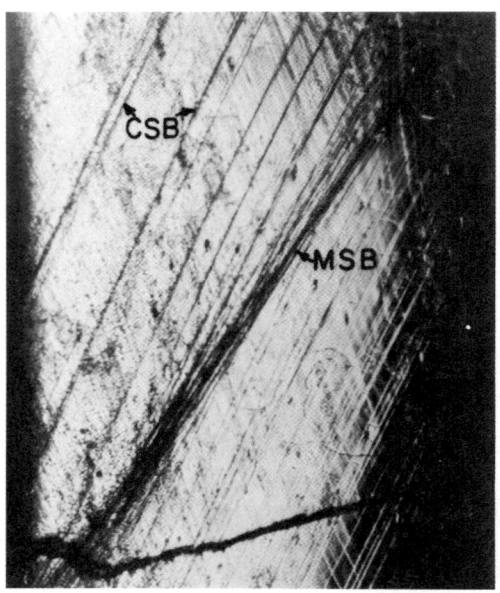

FIG. 46. Coarse slip band (CSB) formation in aluminum–2.8 wt. % copper alloys. Note also how macroscopic shear bands (MSB) form within clusters of CSBs.

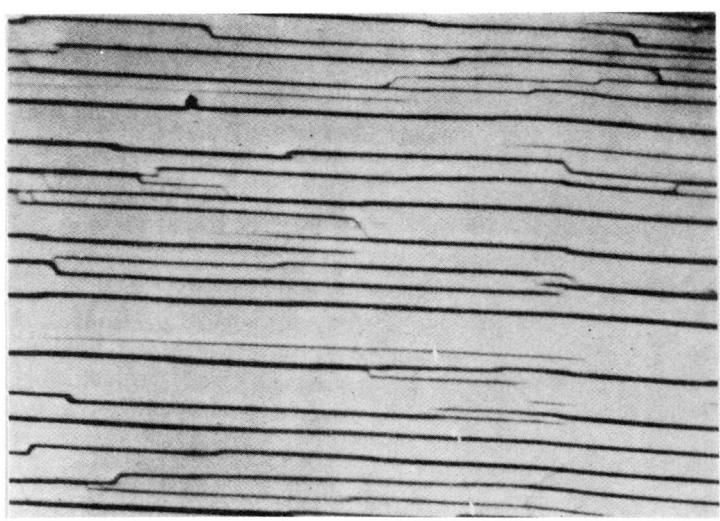

FIG. 47. Coarse slip lines in aluminum single crystals (Cahn, 1951).

the slip lines are associated with cross-slip. Cross-slip had been associated with CSB formation in aluminum alloy crystals by Price and Kelly (1964), and this in particular led Asaro and Rice (1977) to examine the possibility that the phenomenon could be explained as a constitutive instability. The relevant analysis has been developed in Section IV,A.

We recall that localized shearing in imperfection-free and positively strain-hardening material is predictable from pressure-sensitive yielding behavior. The (essentially) shearing mode consists of the plane **n** given by Eq. (4.33a) and **g**, which, as shown earlier, will lie almost entirely in **n** parallel to **s**. However, another way to explain this mode is to consider the possibility that cross-slip is important and to evaluate the role of the parameter α_{zs} in Eq. (3.56). Defining $\alpha = 1/2\alpha_{zs} = 1/2\alpha_{sz}$, we find from Eq. (4.33) that

$$\mathbf{n} \approx \mathbf{m} + 1/2\alpha\mathbf{z} \tag{4.41a}$$

and

$$h_{cr} \approx 1/2\alpha^2 G, \tag{4.41b}$$

where we recall that Asaro and Rice (1977) had estimated α to lie in the range 0.05–0.1. If we take the middle of this range for discussion's sake, we find that

$$(h/G)_{cr} \approx 10^{-3},$$

which, although small, is well within the experimentally observed range of strain-hardening rates after finite strains.

These predictions led Chang and Asaro (1981) to document *experimentally* the conditions prevailing at the inception of coarse slipping; their results are shown in Fig. 48. The nomenclature S.S., GP(II), and GP(I) stands for well-known age-hardening treatments and microstructures formed in this alloy, explained more fully in their paper, but the outcome of the experiments was a good correlation of the onset of coarse slip bands with a critical ratio of the slip plane strain-hardening rate to elastic shear modulus. Also shown in Fig. 48 is an attempted correlation with the ratio of the resolved shear stress to the modulus. The values of the measured critical strain-hardening rates are consistent with the values for the non-Schmid parameters α_{zs} estimated by Asaro and Rice (1977), and so it appears that the constitutive-type instability predictable by including non-Schmid effects and the associated deviations from plastic normality offers a way to explain these localized deformation modes.

The preceding discussion of kink bands and coarse slip bands is at the present time speculative and a good deal more careful critical experimentation is required before a precise explanation of their causes will be available. For example, the geometry of kink bands indicates that they are composed

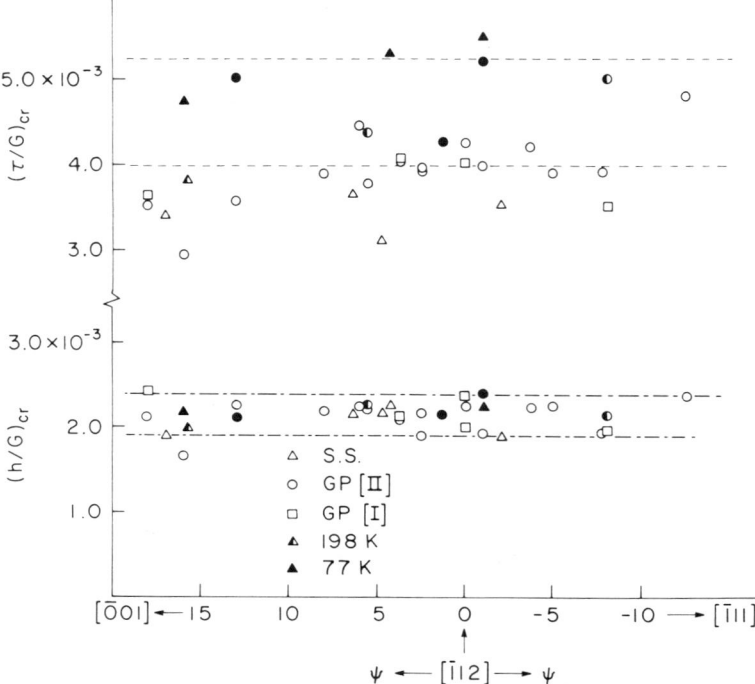

FIG. 48. Correlation of the onset of coarse slip band formation with strain-hardening rate and resolved shear stress.

of edge dislocation dipoles aligned so that the dislocations of each sign are arrayed as tilt boundaries (Cahn, 1951; Honeycombe, 1951). How such an arrangement happens to form is not clear on the dislocation mechanistic level, but if the sort of constitutive description suggested previously is pertinent, consistency with the dislocation substructure must be demonstrated. On the other hand, there are other kinds of explanations possible for coarse slip bands, for example, those that involve temporary breakdown of barriers to slip (Price and Kelly, 1964). Since such occurrences might be considered to be "stress driven," a correlation in terms of the resolved shear stress prevailing at CSB inception would be more natural than one involving the current rate of strain hardening. As Fig. 48 indicates, such a correlation is possible, especially if the data are evaluated at fixed temperature. However, coarse slip bands are not observed to be softer than the surrounding crystal parts (Price and Kelly, 1964; Chang and Asaro, 1981), and in fact they appear to continue to strain harden after forming. Thus even if slip barriers do "break down," this does not appear to lead to a state of strain softening.

The analyses outlined clearly indicate the important influence that rather small deviations from the Schmid rule have in promoting nonuniform deformation. The non-Schmid effects are themselves clearly suggested by dislocation mechanics, and because of their important influence should be evaluated more thoroughly by experiment.

C. Macroscopic Shear Bands

As a third and final example of nonuniform deformation in crystals we consider the phenomenon of macroscopic localized shearing. Figure 49 shows four cases of this, where uniaxial tensile straining of single crystals was terminated by intensely localized shearing and fracture. These four examples span a wide range of crystal structures and chemistries and thus demonstrate the generality of this type of localized deformation mode. Macroscopic shearing occurred in these examples, as well as in other examples reported in the literature, only after the slip mode involved multiple slip. In particular, the slip mode was primary–conjugate slip in the examples shown in Fig. 49.

Figure 49a,c is of the *same* fcc aluminum–2.8 wt.% copper alloy heat treated to contain two different microstructures (Chang and Asaro, 1981). The microstructure of Fig. 49a contained GP(II) zones and displayed coarse slip bands prior to macroscopic shear bands. Macroscopic shear bands formed within clusters of the CSBs, and only then did the accumulated strains become so large as to cause ductile fracture. The microstructure of Fig. 49c contained θ' precipitates and did not display coarse slip bands at any stage in the deformation. In this case macroscopic shearing directly followed necking. Crystals age hardened to contain θ' precipitates underwent considerably more necking than did those containing coherent zones such as GP(II) zones, and this, along with the absence of coarse slip bands in the former, indicates some important microstructural influences on the phenomenology of localized deformation.

Figure 49d taken from the work of Lisiecki *et al.* (1981) shows a very similar phenomenology of macroscopic shearing in pure single-phase fcc copper single crystals. As in the aluminum alloy crystals containing θ' precipitates (Fig. 49c), these crystals underwent substantial necking prior to shear band initiation. The bands in this case are not as sharply delineated as they are in the stronger aluminum alloy crystals, and this apparent connection between strength and the strain-hardening level and shear band pattern will be discussed later. Finally, Fig. 49b shows an example of macroscopic shearing in a single crystal of pure bcc niobium taken from the work

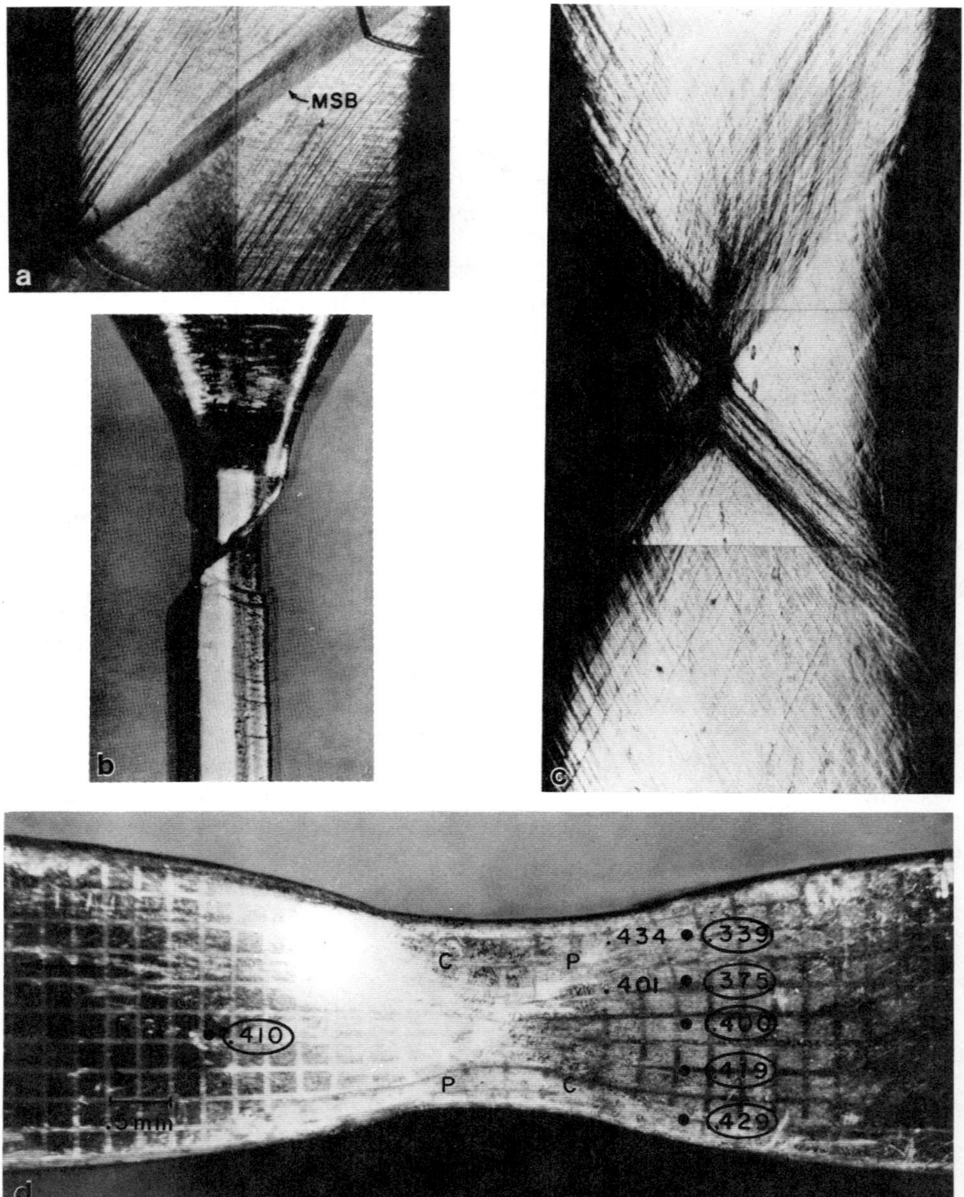

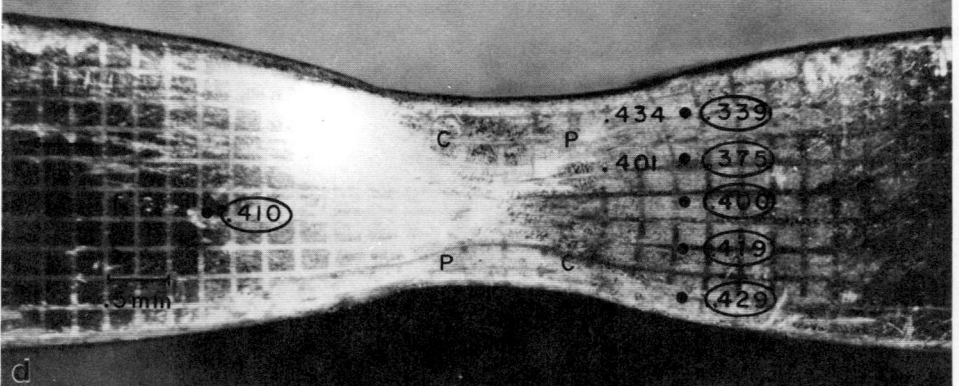

FIG. 49. Four examples of localized shearing in single crystals: (a) aluminum–copper age hardened to contain coherent GP(II) zones, (b) pure niobium (Reid et al., 1966), (c) aluminum–copper with θ' precipitates (Chang and Asaro, 1981), (d) pure single-phase copper (Lisiecki et al., 1981). In all cases shear bands appeared only after a double mode of slip prevailed.

of Reid et al., (1966). In this case the bands formed very abruptly after very little straining and led to fracture.

Chang and Asaro (1981) demonstrated two major differences in the kinematics of macroscopic shear bands and coarse slip bands in their aluminum–copper alloy crystals. In the first place the *material* planes of the macroscopic bands are not aligned with either of the two active slip systems in adjacent parts of the crystal, but are misoriented with respect to them by some 4–12°. This is clearly indicated in Fig. 46, which shows the misorientation between macroscopic and coarse slip bands, where the latter, we recall, are closely aligned with one of the slip systems. In addition, the *lattice* within the bands is also misoriented with respect to the lattice outside. Figure 50, an X-ray micrograph, illustrates the "orientation contrast" produced by this abrupt change in lattice orientation across the band. For both the material planes and the lattice, the misorientation is the result of rotations *away* from the tensile axis. Chang and Asaro (1981) further showed that the lattice rotations in the band increased the Schmid factor [see Eq. (2.3)] and hence the resolved shear stress on that slip system to which the band was nearly parallel; in other words, the bands became "geometrically soft." They noted that further straining in the bands became more single slip in character on the geometrically softened system.

In all the cases that have been studied to date, it has been reported that the bands form while the materials are strain hardening. Thus, like coarse slip bands, they cannot be attributed to an ideally plastic or strain-softening state or to "fractures in progress." In fact, in their aluminum alloy crystals Chang and Asaro (1981) showed that the *material* in the bands continued to strain harden after strains of the order of unity had accumulated in them.

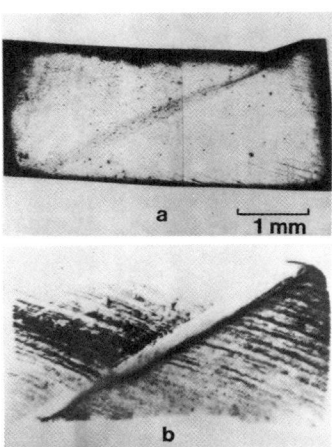

FIG. 50. X-ray micrograph of a shear band in an aluminum–2.8 wt. % copper alloy single crystal. The orientation contrast illustrates the lattice misorientation across the band.

The analysis of macroscopic shearing in crystals deformed in tension in multiple slip has received a good deal of attention recently, and comparisons with the experimental phenomenology just described are very encouraging. The model shown in Fig. 30b for a fcc or bcc crystal subjected to uniaxial tension has served as a basis for the bifurcation analysis of Asaro (1979), which focused on shear band initiation, and the finite element calculations of nonuniform and localized deformation of Peirce et al. (1982). In the preceding the material response was taken to be explicitly rate insensitive. Non-Schmid effects were not included, so that the Schmid law in its simplest form is adopted as the flow rule. Some of the analyses and results are described below.

a. Bifurcation and Uniqueness

We begin by recalling the work of Hill and Hutchinson (1975) on bifurcation in the plane strain tensile test. Their material model was a plane orthotropic incompressible solid that can be applied to the plane model of the single crystal, provided that the elastic behavior of the crystal is taken to be incompressible. We also note that this plane model possesses orthotropic symmetry, since both slip systems are inclined by the angle ϕ to the tensile axis with the load applied. The homogeneous prebifurcation stress state is $\sigma_{22} = \sigma$ and $\sigma_{11} = \sigma_{12} = 0$ and, as shown by Hill and Hutchinson (1975) (see also Biot, 1965), the incremental deformation of an incompressible solid with this symmetry is characterized by the following constitutive laws:

$$\sigma_{22}^V - \sigma_{11}^V = 2\mu^*(D_{22} - D_{11}), \tag{4.42a}$$

$$\sigma_{12}^V = 2\mu D_{12}, \tag{4.42b}$$

$$D_{11} + D_{22} = 0. \tag{4.42c}$$

In Eq. (4.42) the two instantaneous moduli μ^* and μ are defined such that μ governs shearing parallel to the coordinate axes and μ^* governs shearing at 45° to them.

To evaluate μ and μ^* in the prebifurcation state we appeal to the model symmetry and take the hardening moduli $h_{\alpha\beta}$ of Eq. (3.29) in the form $h_{11} = h_{22} = H$ and $h_{12} = h_{21} = qH$. Then, using Eqs. (3.38) and (3.39), we find that (Asaro, 1979; Peirce et al., 1982)

$$\mu = \frac{G[h^*(1-q) + \sigma \cos 2\phi] - 1/2\sigma^2}{h^*(1-q) + 2G \cos^2 2\phi - \sigma \cos 2\phi} \tag{4.43a}$$

and

$$\mu^* = \frac{Gh^*(1+q)}{h^*(1+q) + 2G \sin^2 2\phi}. \tag{4.43b}$$

In Eq. (4.43)

$$h^* = H/\lambda_{ss}^{*4}, \qquad (4.44a)$$

and

$$\tan\phi/\tan\phi_0 = \lambda^*. \qquad (4.44b)$$

where λ_{ss}^* and λ^* are the *elastic* stretch ratios in either slip direction and in the tensile direction, respectively. Now, in Eq. (4.43) we recognize that terms of $O(H)$ and $O(\sigma)$ are typically comparable in magnitude, for example, H/σ ranges from 10 to 5×10^{-2} for the crystals shown in Fig. 49a,c,d. On the other hand, σ/G rarely attains values as large as 10^{-2} in single crystals and is typically no larger than 10^{-3} in the aforementioned cases. Thus it seems reasonable to delete the last term in Eq. (4.43a) and set $\lambda_{ss}^* = \lambda^* = 1$. This yields the moduli given by Asaro (1979),

$$\mu = \frac{G[H(1-q) + \sigma\cos 2\phi_0]}{H(1-q) + 2G\cos^2 2\phi_0} \qquad (4.45a)$$

and

$$\mu^* = \frac{GH(1+q)}{H(1+q) + 2G\sin^2 2\phi_0} \qquad (4.45b)$$

or their rigid–plastic limits

$$\mu = \frac{H(1-q) + \sigma\cos 2\phi_0}{2\cos^2 2\phi_0} \qquad (4.46a)$$

and

$$\mu^* = \frac{H(1+q)}{2\sin^2 2\phi_0}. \qquad (4.46b)$$

To model a fcc crystal in essentially symmetric primary–conjugate slip, we take $\phi_0 = 30°$, since this is the angle that either slip direction makes with the $\langle 112 \rangle$ axis lying roughly midway along the $\langle 100 \rangle$–$\langle 111 \rangle$ symmetry line (see Fig. 5d).

Asaro (1979) used the relations (4.45) and (4.46) to show that bifurcation into a shear band mode could occur when H/σ fell to a critical value, which satisfied the equilibrium requirement

$$(\mu - 1/2\sigma)n_1^4 + 2(2\mu^* - \mu)n_1^2 n_2^2 + (\mu + 1/2\sigma)n_2^4 = 0, \qquad (4.47)$$

where **n** is the normal to the plane of the band, and for a shear band to exist Eq. (4.47) must have real solutions for **n**. As shown by Hill and Hutchinson (1975), bifurcations from a uniform mode of deformation occur before shear bands are allowed. To explain this we consider symmetric modes of the form

$$\bar{v}_1 = \bar{U}_1(x_1)\cos\left(\frac{m\pi x_2}{L}\right), \quad \bar{v}_2 = \bar{U}_2(x_1)\sin\left(\frac{m\pi x_2}{L}\right), \quad (4.48)$$

where an overbar signifies a bifurcation mode. Here $2L$ and $2a$ are the current total length and width of the crystal, respectively. We let $\zeta = m\pi a/L$ characterize the mode. Hill and Hutchinson (1975) showed that for small ζ the bifurcation stress is

$$\sigma/4\mu^* = 1 + \tfrac{1}{3}\zeta^2 + \tfrac{7}{45}\zeta^4 + O(\zeta^6, \zeta^6\mu^*/\mu). \quad (4.49)$$

Then as long as μ/μ^* is of $O(1)$ or larger, we find that bifurcation into the longest wavelength mode, the diffuse necking mode, occurs shortly after maximum load; maximum load occurs when $\sigma = 4\mu^*$. For very high frequency modes, i.e., as $\zeta \to \infty$, the bifurcation stress was given by Hill and Hutchinson (1975) as

$$\frac{\sigma}{2\mu} = 2\frac{\mu^*}{\mu} + \frac{\sigma}{2\mu}\left(\frac{1 - \sigma/2\mu}{1 + \sigma/2\mu}\right)^{1/2}. \quad (4.50)$$

Peirce et al. (1982) summarized these results as shown in Fig. 51. Here the parameter space is divided into regions where Eq. (4.47) admits 0, 1, or 2 real positive solutions for $(n_1/n_2)^2$; accordingly, the governing field equations are said to be elliptic, parabolic, or hyperbolic. Bifurcation is excluded in the shaded region boarded by line A, which corresponds to maximum load [i.e.,

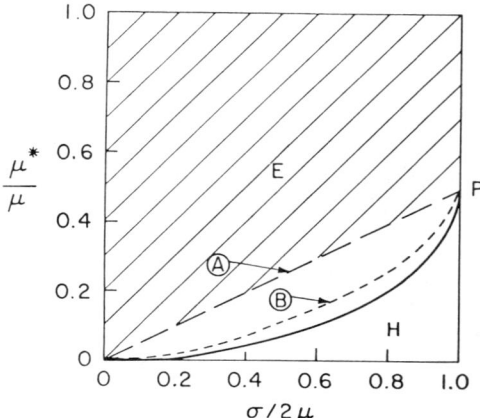

FIG. 51. Elliptic (E), hyperbolic (H), and parabolic (P) regimes in plane strain tension as determined by Eq. (4.47). Here it is assumed that $\mu > 0$. ▨, bifurcation excluded; A, maximum load; B, short wavelength limit.

where H/σ falls to the value $\sin^2 2\phi_0/2(1+q)$ in the rigid plastic limit of Eq. (4.46)]. As pointed out by Peirce et al. (1982), once maximum load is achieved, H/σ is small enough so that $\sigma/2\mu$ may be regarded as a constant equal to $\cos 2\phi_0$, and μ^*/μ as proportional to H/σ. Then, after maximum load, the trajectory in Fig. 51 is approximately vertically downward. Necking may occur after line A is crossed shortly after maximum load, and the short wavelength modes are possible once line B corresponding to Eq. (4.50) is crossed. Finally, as long as $\mu > \sigma/2$, which is easily satisfied after maximum load if $\phi_0 = 30°$ (i.e., if $\phi_0 < 45°$) and $1 \leq q \leq 1.4$, the trajectory crosses the elliptic–hyperbolic interface. This loss in ellipticity corresponds to the admissibility of shear band modes, since then Eq. (4.47) has two real solutions for $(n_1/n_2)^2$. Specifically, if $\mu > \sigma/2$ and $\mu/2\mu^* > 1$, which will certainly be the case for the values of ϕ_0 and q being considered (typically $H/\sigma < 1/5$ after load maximum), shear bands become possible when H/σ satisfies the relation (Asaro, 1979)

$$(H/\sigma)_{\text{crit}} \leq \frac{\cos 2\phi_0 \pm \sin 4\phi_0/\sqrt{2(1+q)}}{2(q + \cos 4\phi_0)} \sin^2 2\phi_0. \tag{4.51}$$

Alternatively, the critical conditions corresponding to satisfying the equilibrium condition of Eq. (4.47) for a shearing mode of the form

$$\bar{v}_i = g_i f(\mathbf{n} \cdot \mathbf{x}), \quad \mathbf{g} \cdot \mathbf{n} = 0, \quad f''(\mathbf{n} \cdot \mathbf{x}) \neq 0 \tag{4.52}$$

can be obtained directly. This mode corresponds to a nonuniform shearing mode if $f''(\mathbf{n} \cdot \mathbf{x}) \neq 0$ in a band of normal $\mathbf{n}$, as sketched in Fig. 52. If θ is the orientation of this band, the critical conditions are (Asaro, 1979)

$$(H/\sigma)_{\text{cr}} = \frac{\cos 2\theta - \cos^2 2\theta/\cos 2\phi_0}{(1-q)\cos^2 2\theta/\cos^2 2\phi_0 + (1+q)\sin^2 2\theta/\sin^2 2\phi_0}. \tag{4.53}$$

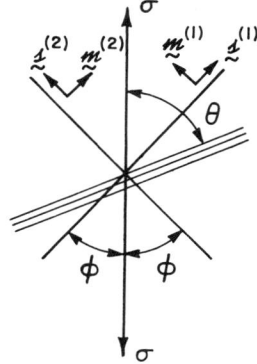

FIG. 52. Illustration of a shear band inclined by the angle θ to the tensile axis.

If $q = 1$, for example, $(H/\sigma)_{cr}$ has a maximum value of 0.05, corresponding to a band orientation $\theta = 37.2°$. In fact, $(H/\sigma)_{cr}$ is positive for any θ in the range $\phi_0 < \theta < 45°$. The bands are thus predicted to be misaligned with respect to the underlying slip systems in the sense of being rotated *away* from the tensile axis. The optimum band is misoriented by about 7.2° from the slip system to which it is nearly parallel. It is also interesting to note that if $\phi \to 45°$, $(H/\sigma)_{cr} \to 0$ (again in the rigid–plastic limit). The effect can be traced directly to the existence of a yield vertex at the current stress point (Asaro, 1979). When viewed in the space of $(\sigma_{22}, \sigma_{12})$, the included angle in the elastic domain is $\pi - 4\phi_0$. Thus "vertex softening" plays an important role in promoting localized shearing.

Two other important kinematic features of the predicted shear band modes are the jumps in *lattice* rotations that must occur across the bands and the ratio of shearing rates on the two active slip systems. The bands that are misaligned with the slip systems must involve slipping on both systems, whereas observations indicate that slip tends to concentrate on that system to which the band is nearly parallel. It is easy to show that for a band nearly aligned with system 1 the ratio of shearing rates in the shearing mode [e.g., Eq. (4.52)] is given by (Peirce et al., 1982)

$$\frac{\Delta\dot{\gamma}^{(2)}}{\Delta\dot{\gamma}^{(1)}} = \frac{\cos 2\phi_0 \sin 2\theta - \sin 2\phi_0 \cos 2\theta}{\cos 2\phi_0 \sin 2\theta + \sin 2\phi_0 \cos 2\theta}, \quad (4.54)$$

where the Δ signifies the difference in a quantity from the prebifurcation state. For the case quoted above, $\phi_0 = 30°$ and $\theta = 37.2°$, this ratio is equal to 0.35, which indicates concentrated shearing on system 1. Furthermore, as will be shown later, shearing modes corresponding to $\theta \neq \phi_0$ also imply *lattice* rotations. In particular, if $\phi_0 < \theta < 45°$, the modes involve lattice rotations *away* from the tensile axis. The important effect of this is to cause an increase in the resolved shear stress on that slip system to which the band is nearly parallel. The band thus becomes geometrically soft on that system on which straining is concentrated. This is what the experimental observations described earlier have shown.

Before continuing with this description of shear bands, it is worthwhile to review some of the constitutive assumptions that underlie the kinematic predictions just outlined. In particular, the trajectory described in connection with Fig. 51 as well as the kinematics of the critical modes predicted from Eq. (4.51) or (4.53) pertain to cases where the inequalities $\mu > \sigma/2$ and $\mu > 2\mu^*$ hold. In the model for a double slipping fcc crystal where $\phi_0 = 30°$ and $\cos 2\phi_0 > 0$, these are likely to hold *after* a strain of a few percent beyond initial yield, since then H/σ is typically less than 1/5 or so. At initial yield, however, H/σ is often larger than unity, and if $q > 1$ it may be that $\mu < \sigma/2$. In this case bifurcation takes place in the parabolic regime of Fig. 51 where a single solution to Eq. (4.47), $(n_2/n_1)^2 = 0$, exists. Although this mode itself involves unloading on one of the systems, a bifurcation mode consisting of a linear combination of the shear band mode and the symmetric double slipping prebifurcation mode can be constructed so that use of the constitutive laws and moduli corresponding to double slipping is justified. On the other hand, as the discussion in Section III,C,1 showed, uniqueness of the slipping mode itself also depends on the relative magnitudes of σ/H and q. In fact, looking back to Eqs. (3.42)–(3.44), we note that Eq. (3.44) is equivalent to $\mu > 0$. Thus predicted bifurcations in the parabolic regime assuming double slip may well occur at conditions where slip mode uniqueness is close to failing. Nonetheless, since the condition $\mu < \sigma/2$ is met at positive values for μ, bifurcation modes of this type can be encountered with active double slipping. Loss of uniqueness in the slipping mode would all but preclude analysis of the crystal deformation mode and in this case a formulation including rate dependence would be required. What would appear to be a minimal requirement for qualitative consistency between a rate-independent analysis and a rate-dependent one is similarity in the slip mode. The lattice and material rotations associated with the shear bands, for example, require double slip, and if there were ambiguity over whether deformation was by single or multiple slip, as could happen in a rate-independent analysis, very different predictions concerning the uniformity of finite deformation would result.

b. Numerical Results

To conclude this discussion some results of the finite elements studies carried out by Peirce et al. (1982) are presented. The calculations were performed using the plane double slip model as in the bifurcation analysis just outlined, with the constitutive laws formulated from Eq. (3.39). This means that the Schmid law was assumed with $\tau^{(\alpha)}$ defined as in Eq. (3.18). As explained by Peirce et al. (1982) and also in Section III,E, there is a significant computational advantage to having the tensor of moduli **C** in Eq. (3.38) symmetric. This in turn requires symmetry of $g_{\alpha\beta}$. To accomplish this Peirce et al. (1982) took the slip plane hardening matrix **h** to have the form

$$h_{\alpha\beta} = H_{\alpha\beta} + 1/2(\boldsymbol{\beta}^{(\beta)}:\mathbf{P}^{(\alpha)} - \boldsymbol{\beta}^{(\alpha)}:\mathbf{P}^{(\beta)}), \tag{3.92}$$

where $H_{\alpha\beta} = H_{\beta\alpha}$ is symmetric and $H_{\alpha\beta}$ was taken as $H_{\alpha\beta} = qH + H(1-q)\delta_{\alpha\beta}$, which makes Eq. (3.92) equivalent to the hardening law assumed in developing Eqs. (3.43)–(3.46), since the two additional terms in Eq. (3.92) cancel when the crystal is in symmetric tension. They then showed that Eq. (3.92) with the elements $H_{\alpha\beta}$ taken from experimental data provided a good description of latent hardening and overshoot; in fact, the added terms in the $\boldsymbol{\beta}$'s led to a degree or so more overshoot when actual crystal data was used to simulate tension tests.

Two finite element calculations will be summarized. In the first the strain-hardening law was directly fitted to Chang and Asaro's (1981) stress–strain data for aluminum–copper alloy crystals, and in the second the hardening law was modeled after a softer but more strongly hardening crystal. The latter case would represent single crystals such as pure copper (see Fig. 49d) or Chang and Asaro's (1982) "overaged" aluminum–copper alloys. We recall that the observations have shown that shear bands tend to be more sharply delineated, with very abrupt changes in lattice orientation in stronger crystals. In softer crystals with high strain-hardening rates the bands tend to be more diffuse and are preceded by much deeper necking. In the calculations necking and shear bands were initiated from small geometrical (thickness) imperfections, described in detail by Peirce et al. (1982).

In the first case the self-hardening on a slip system was described by the law used in the overshoot calculation presented in Section III,E:

$$H_{\alpha\beta} = qH + (1 - q)H\delta_{\alpha\beta}, \tag{4.55}$$

where

$$H(\gamma) = H_0 \operatorname{sech}^2\left(\frac{H_0\gamma}{\tau_s - \tau_0}\right) \tag{4.56}$$

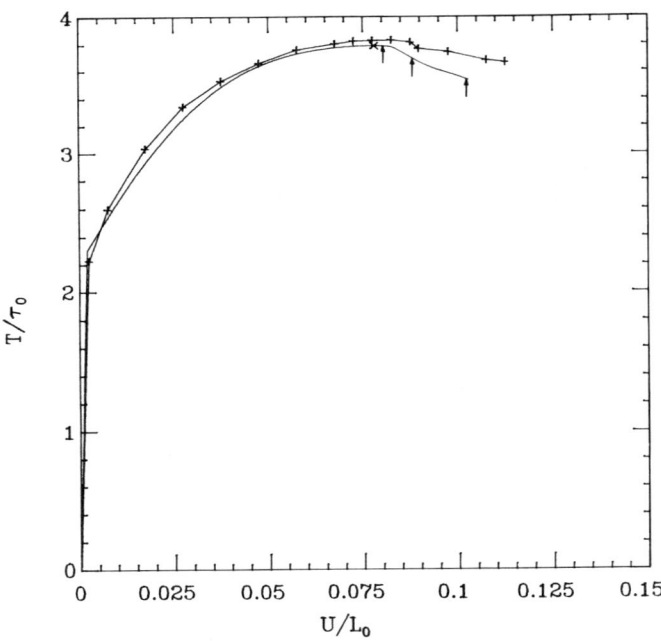

FIG. 53. Load–engineering strain curve for a plane double-slipping crystal with the strain-hardening law (4.56). +, experimental points; ×, maximum load.

In this example $H_0 = 8.9\tau_0$, $\tau_s = 1.8\tau_0$, and $q = 1$; τ_0 is the initial yield strength. Figure 53 shows the computed load (nominal tensile stress) versus end point extension curve, along with the experimentally measured curve. Figure 54 shows the deformed finite element grids at the stages marked by the arrows in Fig. 53. In the calculations, symmetry in the deformation was assumed so that calculations were performed in only one quadrant. Shear bands are evident at or shortly after the abrupt drop in load, as seen, for example, in Fig. 54b. Figure 55 shows data for one quadrant (the crystal's center is in the lower left-hand corner of each figure) for the last stage marked

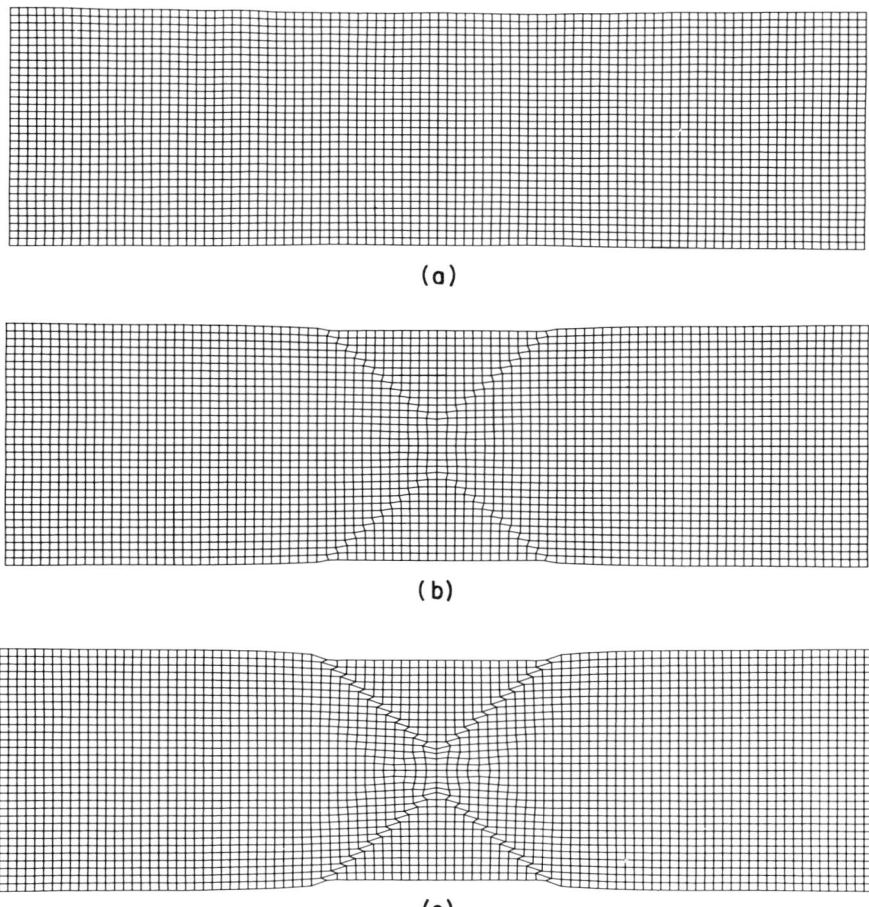

FIG. 54. Deformed meshes for the crystal whose load–strain curve is shown in Fig. 53. The meshes correspond to the stages of deformation indicated by the arrows in Fig. 53.

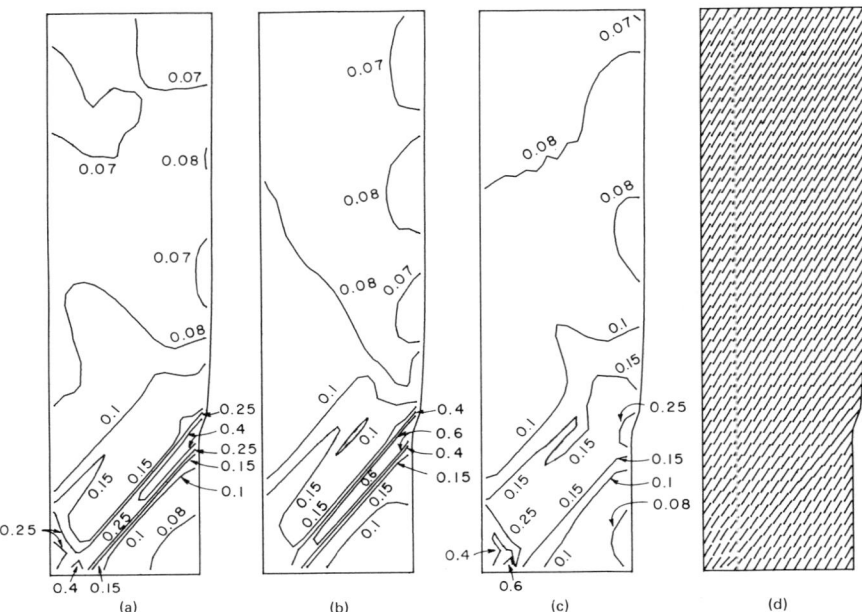

FIG. 55. One quadrant of the deformed crystal shown in Fig. 54c. (a) Contours of constant principal logarithmic strain. (b,c) Contours of constant shear strain on the primary and conjugate slip systems. (d) Plots of the primary slip direction in each mesh. Note the lattice rotation within the shear band.

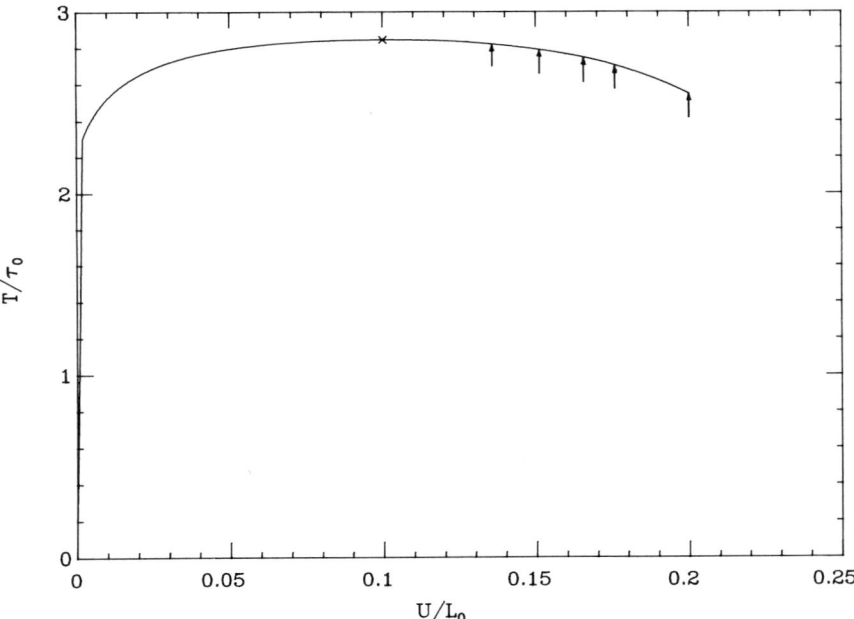

FIG. 56 Load–engineering strain curve for a plane double-slipping crystal with the strain-hardening law (4.58). ×, maximum load.

in Fig. 53. In Fig. 55a contours of maximum principal logarithmic strain are plotted and this clearly shows the intense strains that develop in the band. Figure 55b,c shows contours of glide strain on the primary slip system, to which the shear band is nearly aligned, and on the conjugate system, respectively. In Fig. 55b,c both the intense straining in the band and the transition to a more single-slip mode is evident. In Fig. 55d the primary slip direction is plotted in each of the elements shown and it is seen that in the band the slip direction does indeed tend to rotate away from the tensile axis, as described earlier. This reorientation causes geometrical softening in the band on the intensely slipping system. These results are in close accord with Chang and Asaro's (1981) experiments.

In the second case the hardening law also has the form (4.55), but with H given by

$$H(\gamma) = H_0 \left(\frac{H_0 \gamma}{n \tau_0} + 1 \right)^{n-1}, \qquad (4.57)$$

which is based on the single-slip flow law

$$\tau_c = \tau_0 \left(\frac{H_0 \gamma}{n \tau_0} + 1 \right)^n. \qquad (4.58)$$

This power law form leads to a higher rate of strain hardening at large strains than Eq. (4.56) and to a rather different phenomenology of necking and shear band development. Figure 56 shows the computed load–extension curve, and Fig. 57 the deformed finite element grids at the indicated stages. In this case the load deflection curve goes through a smooth maximum without the abrupt drops typical of the stronger and more weakly strain-hardening materials. In addition, shear bands form much more gradually and are not nearly as sharply delineated as in the former case. The extent of necking is also much larger and this has an interesting effect on the development of nonuniform lattice rotations. To see this we consider the various constant contour plots shown in Figs. 58 and 59. As in the previous example, Fig. 58a shows contours of maximum logarithmic principal strain, whereas Fig. 58b,c shows contours of primary and conjugate glide strain. The shear band forming on the primary slip system is evident from the contours demarcating large strains (e.g., note the 0.7 contour) in Fig. 58b. Figure 58d is different from its counterpart in Fig. 55d in that it shows constant contours of lattice rotation measured in degrees. The sense of rotation is clockwise in the viewing plane, and the angles are measured from the initial orientation of $30°$. As noted earlier, as the lattice and primary slip direction rotates from the initial orientation, where the slip direction is $30°$ from the tensile axis, toward the $45°$ maximum shear stress orientation, the primary system becomes geometrically soft. These nonuniform lattice rotations that produce geometrical softening on one system or the other begin with necking

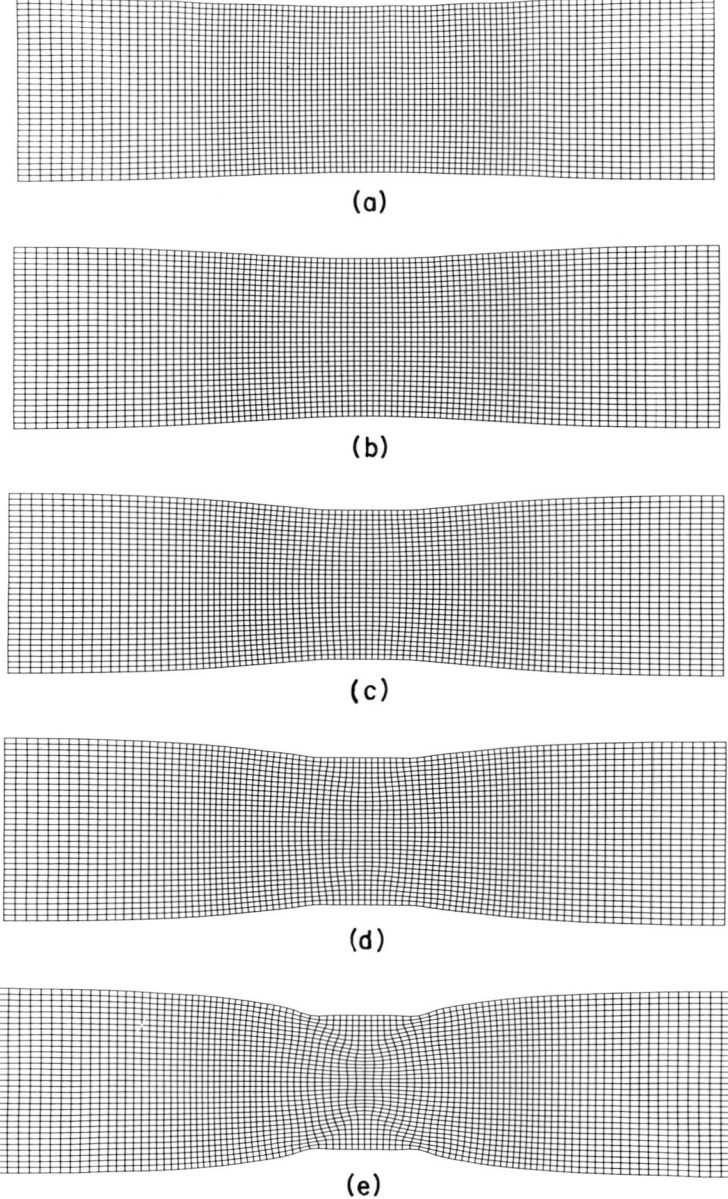

FIG. 57. Deformed meshes for the crystal whose load–strain curve is shown in Fig. 56. The meshes correspond to the stages marked by the arrows in Fig. 56.

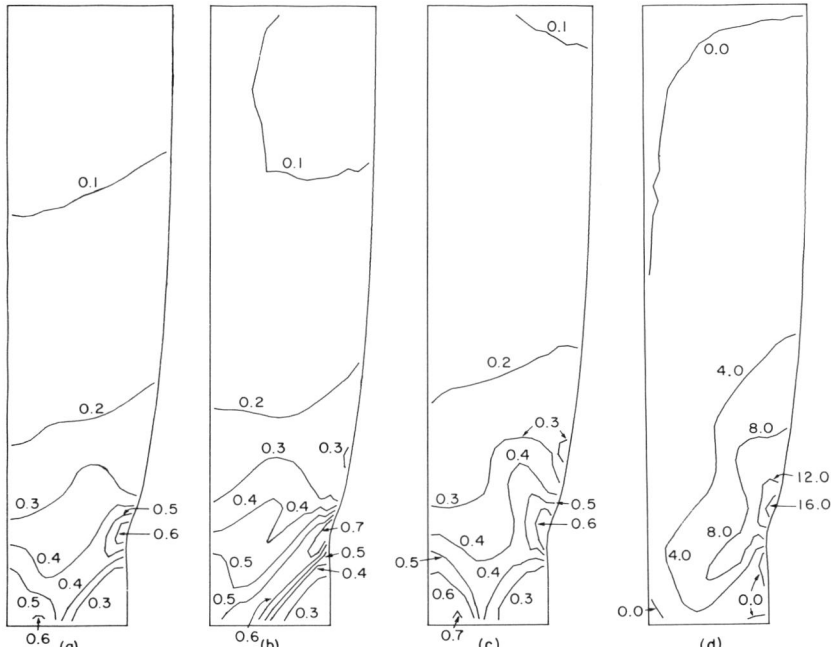

FIG. 58. One quadrant of the deformed crystal shown in Fig. 57e. (a) Contours of constant logarithm principal strain. (b,c) Contours of constant shear strain on the primary and conjugate slip systems. (d) Contours of constant rotation of the primary slip direction away from the tensile axis.

and serve as a kind of crystallographic "orientation imperfection" that triggers localized shearing. Figure 59 shows the set of lattice rotation contours corresponding to the stages of deformation marked by the arrows in Fig. 56. The progression of lattice rotation with necking and the development of localized shearing is evident.

The abrupt change in lattice rotation across shear band boundaries and the development of lattice rotations during necking, both of which cause geometrical softening, clearly emphasize the importance of lattice kinematics in crystal plasticity. Lattice kinematics in turn depend specifically on the *precise* slip mode, so that we again emphasize the importance of slip mode uniqueness. Furthermore, it is likely that analogous patterns of nonuniform lattice orientation will develop within polycrystal grains due to the nonuniform slip modes associated with patchy slip, even if the current overall deformation rates were uniform. Patchy slip is strongly coupled to latent hardening and hence also to the question of slip mode uniqueness. In the calculations of Peirce *et al.* (1982) it was found that when q in Eq. (4.55) was

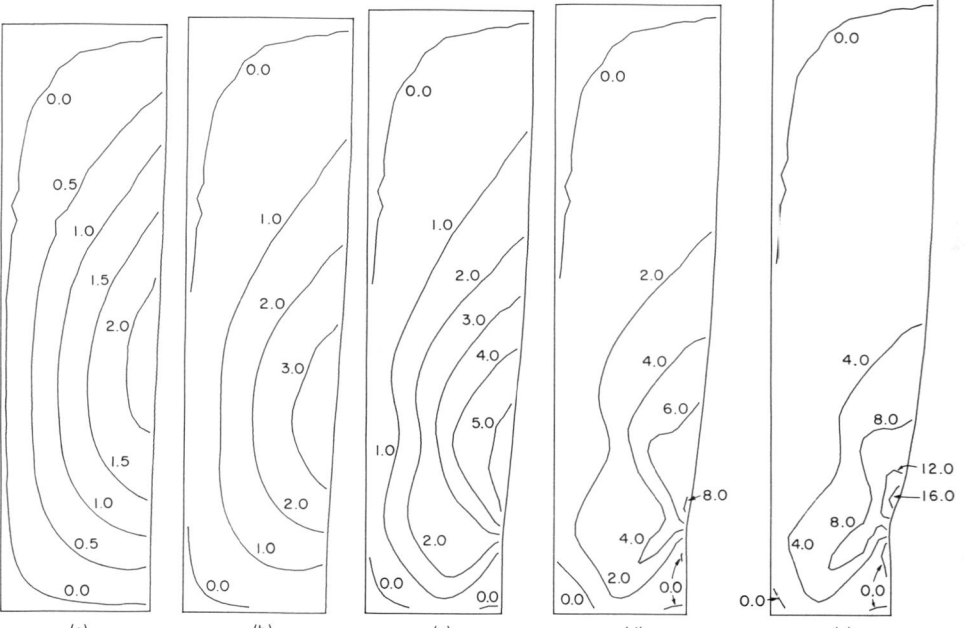

FIG. 59. Contours of constant rotation of the primary slip direction away from the tensile axis. The contour maps correspond to the deformation stages indicated in Fig. 56.

larger than 1.3 or so, a persistent pattern of patchy slip developed throughout the crystal's gauge section. These nonuniform slip modes were initiated by the very slightly nonuniform stress state associated with the small initial thickness variations used to trigger necking and shear bands. Patchy slip, in this case, consists of contiguous regions undergoing single slip on one slip system or the other and regions undergoing double slip. The patches that initially yield in single slip undergo lattice rotations which lead to double slip after only modest strains and thus when the ratio H/σ is still rather large. As indicated by the discussion on pp. 89 and 46 in connection with Eq. (3.44), strong latent hardening as follows with $q > 1$ coupled to the condition that $H/\sigma > 1$ leads to a loss in both slip mode uniqueness and ellipticity of the system's governing equations. In Peirce et al. (1982) the governing equations became parabolic and admitted shear band solutions. With the numerical methods used, this caused the finite element stiffness matrix to become singular and the computational procedure to break down. For these reasons rate-insensitive constitutive laws may not allow a full and satisfactory analysis of finite strain plasticity in multiple-slipping crystals, unless for the case at hand the slip mode expected in a rate-independent model is equivalent with the set of slip systems experiencing appreciable shearing rates in a rate-dependent one. In the model computations outlined in Figs. 53–59 the slip mode was unique and equivalent to that observed experimentally.

Lattice rotations of the type described in Figs. 58d and 59 have been experimentally documented by Saimoto *et al.* (1965) in pure copper and by Lisiecki *et al.* (1981) in pure copper and an aluminum–copper alloy. Figure 49d, shown earlier, is an example taken from Lisiecki *et al.* (1981). The (deformed) grid pattern was obtained by vapor depositing a thin film of carbon on the specimen through a mesh at a stage in the deformation just prior to the appearance of shear bands but after necking had begun. X-Ray diffraction patterns were taken at various points within the neck, and the lattice orientation and values for the Schmid factors determined. Schmid factors are printed at various locations; the circled numbers correspond to one slip system, the uncircled to the other. Since the neck was not perfectly symmetrical, neither was the pattern of lattice rotations. Nonetheless, there is a clear pattern of geometrical softening evidenced by larger Schmid factors on one system in one "quadrant," and on the second system in the adjacent quadrant. Shear bands formed nearly parallel to the softened systems within the appropriate quadrants.

Finally, we again note that it is necessary that the deformation patterns predicted by continuum analyses be linked to the dislocation substructure. In Fig. 49d the crystal axis about which the rotations caused by necking take place is usually the $\langle 111 \rangle$ direction orthogonal to the two active slip directions. In Fig. 49d this axis is also very nearly orthogonal to the plane of the drawing. The very abrupt lattice misorientations that develop across shear band boundaries in aluminum–copper crystals, as documented by Chang and Asaro (1981) and Lisiecki *et al.* (1982), are caused by rotations about the $\langle 110 \rangle$ axis lying along the intersection of primary and conjugate slip planes. Chang and Asaro (1981) have explained that this can result from the formation of dislocation "tilt boundaries" aligned along the primary or conjugate slip planes, as shown in Fig. 60. The dislocations in the boundary are Lomer dislocations produced by the intersection and reaction of primary and conjugate dislocations, as described in Section II,B,2.

V. Some Outstanding Problems

A. Polycrystalline Models

One of the more useful extensions of the analyses outlined previously is the modeling of polycrystalline behavior at finite strain. Finite-strain models should have as a particular goal a description of the development of anisotropy and its influence on constitutive behavior. In polycrystalline metals

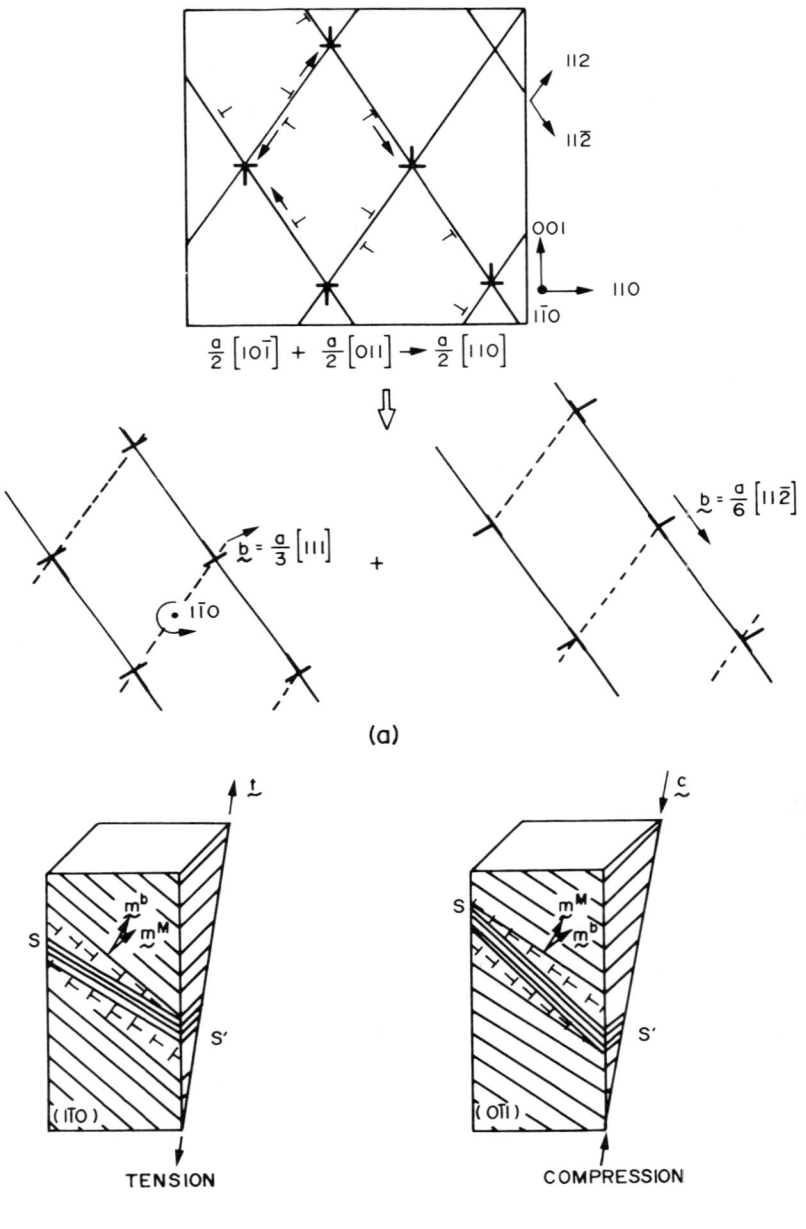

FIG. 60. Chang and Asaro's 1981 model for dislocation tilt boundaries that explain the observed lattice rotations across shear band boundaries. The boundaries are composed of Lomer edge dislocations.

the two main causes of anisotropic plastic response are (1) material anisotropy associated with the nonequiaxed morphology typical of highly deformed grains and the dislocation substructures within them, and (2) crystallographic texture resulting from the rotation of grains into preferred orientations. Both these features of finitely deformed materials impart a strong directionality to the stress–strain response that can play a vital role regarding, among other things, the uniformity and stability of plastic flow. Furthermore, the development of textures by new mechanical processing methods offers a way to enhance strength, as, for example, in the β-titanium alloys studied by Avery and Polan (1974). Theoretical predictions of texture have been in progress for some time (see, for example, Sevillano et al., 1980, and Sowerby et al., 1980), but to date there have been no attempts to incorporate texture into models for large-strain constitutive laws. This represents, then an outstanding problem of considerable importance. This article is concluded with a brief discussion of some of the phenomenological and theoretical aspects involved.

The earliest successful theory for describing polycrystalline stress–strain behavior, and texture development, in terms of single-crystal behavior was contained in Taylor's (1938a,b) rigid–plastic analysis of polycrystalline aggregates. In this Taylor imposed strict compatibility among the grains in the aggregate by subjecting each of them to the same homogeneous displacement (velocity) field. The specific problem he solved was for a polycrystalline fcc metal subjected to uniform symmetric tension. The material behavior was taken to be rate independent and the strain hardening was presumed to be isotropic so that all the $h_{\alpha\beta} = h$.

Taylor's (1938a,b) approach is well known and has been followed since. He argued that five independent slip systems are required to impose an arbitrary incompressible strain and then set out to determine the specific set of systems active in each grain. The grains were assumed to have an initially random distribution of orientations. All slip systems had equal flow stresses $\tau_c^{(\alpha)} = \tau$ and so the incremental work associated with the incremental shears $d\gamma_i^{(\alpha)}$ in the ith grain is

$$\tau \sum_{\alpha=1}^{n} d\gamma_i^{(\alpha)} = \tau\, d\Gamma, \tag{5.1}$$

where Γ is the cumulative shear strain. Taylor chose the particular set of five active systems by minimizing the cumulative shear strain and thus minimizing the work done through it. For the aggregate as a whole containing N grains, this work is

$$\tau \sum_{i=1}^{N} \sum_{\alpha=1}^{n} d\gamma_i^{(\alpha)}. \tag{5.2}$$

The tensile stress can then be computed by equating this to the work done on the aggregate through the tensile strain increment de:

$$\sigma\, de = \frac{\tau}{N} \sum_{i=1}^{N} \sum_{\alpha=1}^{n} d\gamma_i^{(\alpha)}. \tag{5.3}$$

The grain rotations were computed in the following manner. If each grain is subjected to the same displacement field as the aggregate, then each grain undergoes the same rotation. For the case of axisymmetric tension considered by Taylor (1938a,b), this is zero. Then from Eq. (3.8a) it is found that for an increment in time dt

$$\mathbf{\Omega}^P\, dt + \mathbf{\Omega}^*\, dt = \mathbf{\Omega}(\text{aggregate})\, dt = 0, \qquad i = 1, N, \tag{5.4}$$

or

$$\mathbf{\Omega}^*\, dt = -\mathbf{\Omega}_i^P\, dt = -\sum_{\alpha=1}^{n} \mathbf{W}_i^{(\alpha)}\, d\gamma_i^{(\alpha)} \qquad (\text{no sum on } i). \tag{5.5}$$

In Taylor's case $\mathbf{\Omega}_i^*$ represents a rigid lattice spin that reorients the lattice with respect to the principal axes of strain in the material. Figure 61 shows Taylor's (1938a) result for texture development in an aggregate of fcc crystals. The diagram is not a true stereographic projection but "it shows in rectangular coordinates the co-latitude θ and longitude Φ of the specimen axis referred to crystal axes placed with a cubic axis at the pole and a cubic plane as the meridian $\Phi = 0$" (Taylor, 1938a, p. 321). The arrows indicate the sense and extent of rotation from the initial orientation. The smaller letters marked on the arrows correspond to different sets of five slip systems and we note that there are indeed many orientations for individual grains for which there are two distinct choices for the slip mode that gives the same minimum cumulative shear strains. As remarked throughout our discussions of constitutive behavior, such nonuniqueness in slip mode is inherent in rate-independent models for crystalline slip. The rotations and therefore the textures produced by different slip modes can evidently be quite different. Nonetheless, Taylor went on to note that the triangle can be conveniently broken up into the regions marked by the larger letters. For those grains for which the tensile axis is originally in region G, the grains rotate so that a $<111>$ direction tends to become aligned with the tensile axis. For region EC the grains can rotate either toward the $<100>$ direction or across the EC–G boundary, where they subsequently rotate toward the $<111>$ direction. Thus Taylor predicted that the aggregate develops a texture where either $<111>$ or $<100>$, but not $<110>$, crystallographic directions are aligned with the tensile axis. Taylor noted that such textures were experimentally observed. It is important to note that the rotations deduced from Eq. (5.5) are quite different from those calculated in Section III,E in connection with the single-crystal tension test. For axisymmetric compression of

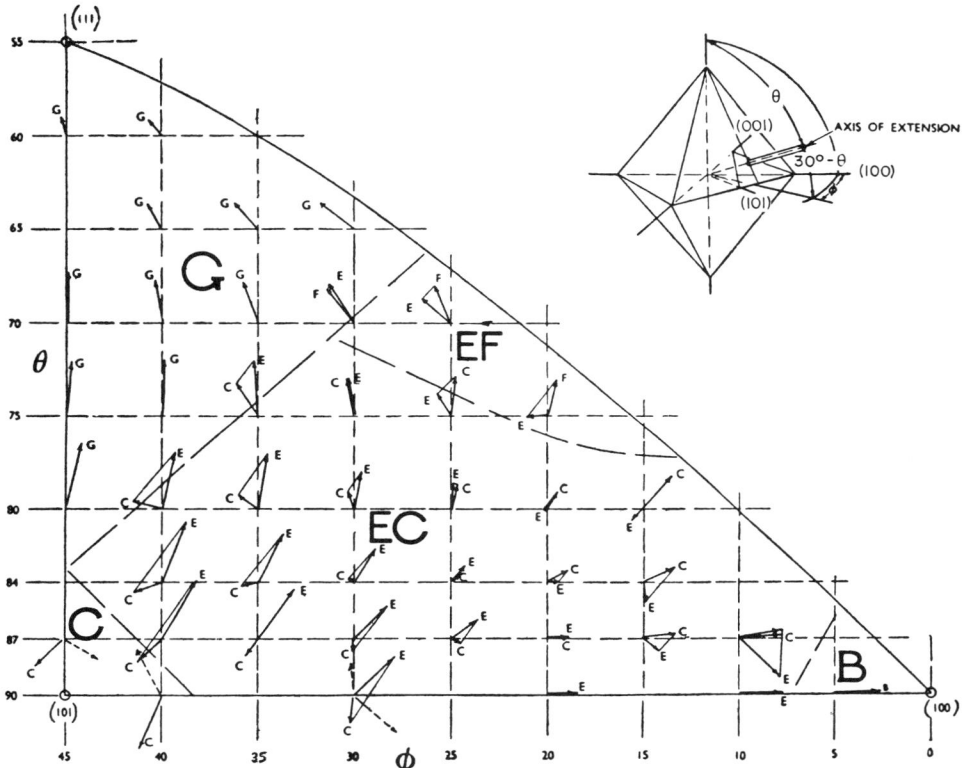

FIG. 61. Taylor's (1938a) plot of lattice rotations and texture development in a fcc polycrystals subjected to axisymmetric tension. The arrows indicate the sense and extent of rotation following an extension of 2.37% for grains whose initial orientations were at the origin of their respective arrows.

a polycrystal the rotations would be equal and opposite to those in tension in the Taylor model. Bishop (1954) in fact determined axisymmetric compression textures for fcc polycrystals and showed that the predominant texture is <100>. He also noted, though, that although experiments on copper, aluminum, and nickel showed close agreement with the Taylor predictions, experiments of α-brass (70% copper–30% zinc and a fcc crystal) did not. The α-brass compression texture possessed a significant <111> component as well. Bishop (1954) noted the studies of Elam (1926), which indicated a strong latent hardening in brass compared to other fcc metals and the inhomogeneity of the deformation in the α-brass grains. He attributed the differences in the copper and α-brass textures to differences in latent hardening.

Figure 61 was constructed for an aggregate extension of 2.37% and it should be noted that even for this modest plastic strain rotations of up to 3–4° are possible. At strains twice or three times this amount rotations of up to 8–10° are expected, and this should be sufficient to cause noticeable anisotropic behavior. To date, though, such anisotropy has not been included in theoretical models for polycrystalline constitutive behavior.

Taylor (1938a,b) also compared experimental stress–strain curves for fcc metals with curves predicted from his analysis ignoring grain rotations. This assumption has been used extensively since. However, rather than describing Taylor's procedure for this, it is perhaps more instructive to call to mind Hutchinson's (1976) more recent discussion of small-strain rate-dependent polycrystalline models. This is particularly worthwhile, since it appears that improved models for texture and finite strain constitutive laws will be based on rate-dependent flow laws, which hopefully lead to unique slip modes. Hutchinson's (1976) calculations in fact incorporate Taylor's rate-independent model.

Hutchinson's (1976) analysis was concerned with steady power-law creep. The flow law he used can be recovered from the more general law (3.80) if the $g^{(\alpha)}$ are constant; $g^{(\alpha)}$ is presumed taken from single-crystal data involving multipe slip. If ε^c represents the instantaneous plastic strain rate of a grain, then the constitutive law for each grain is given by

$$\boldsymbol{\varepsilon}^c = \sum_\alpha \mathbf{P}^{(\alpha)} \dot{\gamma}^{(\alpha)} = \mathbf{M}^c : \boldsymbol{\sigma}^c, \qquad (5.6)$$

where $\mathbf{M}^c$ is the tensor of (creep) compliances given by

$$\mathbf{M}^c = \sum \frac{\dot{a}^{(\alpha)}}{g^{(\alpha)}} \left| \frac{\tau^{(\alpha)}}{g^{(\alpha)}} \right|^{m-1} \mathbf{P}^{(\alpha)} \mathbf{P}^{(\alpha)}. \qquad (5.7)$$

The summations are made over all slip systems. Hutchinson (1976) noted that the $\mathbf{M}^c$ are homogeneous of degree $m - 1$ in the stress, so that

$$\mathbf{M}^c(\lambda \boldsymbol{\sigma}) = \lambda^{m-1} \mathbf{M}^c(\boldsymbol{\sigma}^c). \qquad (5.8)$$

The fact that the Schmid stress $\tau^{(\alpha)}$ is used in Eqs. (3.80) and (5.6) allows for the introduction of two potentials F^c and E^c defined in terms of the dissipation rate as

$$\boldsymbol{\sigma}^c : \boldsymbol{\varepsilon}^c = (m+1) F^c = \frac{m+1}{m} E^c = \sum_\alpha \tau^{(\alpha)} \dot{\gamma}^{(\alpha)}. \qquad (5.9)$$

In terms of these

$$\boldsymbol{\varepsilon}^c = \partial F^c / \partial \boldsymbol{\sigma}^c, \qquad (5.10\text{a})$$

$$\boldsymbol{\sigma}^c = \partial E^c / \partial \boldsymbol{\varepsilon}^c, \qquad (5.10\text{b})$$

and also

$$\mathbf{M}^c = \frac{1}{m} \frac{\partial^2 F^c}{\partial \boldsymbol{\sigma}^c \partial \boldsymbol{\sigma}^c}. \tag{5.11}$$

From Eq. (5.11) it follows that the relation between a change in strain rate and stress is governed by

$$d\boldsymbol{\varepsilon}^c = m\mathbf{M}^c : d\boldsymbol{\sigma}^c. \tag{5.12}$$

The aggregate is subjected to the overall velocity field, prescribed at its boundaries, $v_i = \bar{\varepsilon}_{ij} x_j$, where the $\bar{\varepsilon}_{ij}$ are constant. Since the $\bar{\varepsilon}_{ij}$ are symmetric, this represents a rotationless deformation. This uniform aggregate strain rate is related to the individual grain strain rates $\boldsymbol{\varepsilon}^c$ through the volume average

$$\bar{\boldsymbol{\varepsilon}} \equiv \{\boldsymbol{\varepsilon}^c\} = \frac{1}{v} \int_v \boldsymbol{\varepsilon}_c \, dv. \tag{5.13}$$

The overall stress $\bar{\boldsymbol{\sigma}}$ is related to the stress in the individual grains, $\boldsymbol{\sigma}^c$, through an identical volume average. Then, given Eq. (5.7), $\bar{\boldsymbol{\sigma}}$ is homogeneous of degree $1/m$ in $\bar{\boldsymbol{\varepsilon}}$ and $\bar{\boldsymbol{\varepsilon}}$ is homogeneous of degree m in $\bar{\boldsymbol{\sigma}}$.

Next, using the principle of virtual work, Hutchinson (1976) noted that

$$\bar{\boldsymbol{\sigma}} : \bar{\boldsymbol{\varepsilon}} = \{\boldsymbol{\sigma}^c : \boldsymbol{\varepsilon}^c\}, \tag{5.14a}$$

$$\bar{\boldsymbol{\sigma}} : d\boldsymbol{\varepsilon} = \{\boldsymbol{\sigma}^c : d\boldsymbol{\varepsilon}^c\}, \tag{5.14b}$$

and

$$\bar{\boldsymbol{\varepsilon}} : d\bar{\boldsymbol{\sigma}} = \{\boldsymbol{\varepsilon}^c : d\boldsymbol{\sigma}^c\}. \tag{5.14c}$$

Then, if the overall potentials $F(\bar{\boldsymbol{\sigma}})$ and $E(\bar{\boldsymbol{\varepsilon}})$ are defined as in Eq. (5.9),

$$(m+1)F(\bar{\boldsymbol{\sigma}}) = \frac{m+1}{m} E(\bar{\boldsymbol{\varepsilon}}) = \bar{\boldsymbol{\sigma}} : \bar{\boldsymbol{\varepsilon}}, \tag{5.15}$$

it can be shown that

$$\bar{\boldsymbol{\sigma}} = \partial E / \partial \bar{\boldsymbol{\varepsilon}} \tag{5.16a}$$

and

$$\bar{\boldsymbol{\varepsilon}} = \partial F / \partial \bar{\boldsymbol{\sigma}}. \tag{5.16b}$$

Here $E(\bar{\boldsymbol{\varepsilon}})$ and $F(\bar{\boldsymbol{\sigma}})$ are related to E^c and F^c through volume averages, as in Eq. (5.13). Further, as in Eq. (5.11), an overall compliance is defined as

$$\mathbf{M} = \frac{1}{m} \frac{\partial^2 F}{\partial \bar{\boldsymbol{\sigma}} \partial \bar{\boldsymbol{\sigma}}} \tag{5.17}$$

so that
$$\bar{\varepsilon} = \mathbf{M}:\bar{\sigma} \tag{5.18}$$

and
$$d\bar{\varepsilon} = m\mathbf{M}:d\bar{\sigma}. \tag{5.19}$$

Hutchinson (1976) sought upper bounds for the stress by using Hill's (1965) minimum principle for power-law materials. This states that among all velocity fields $\hat{\mathbf{v}}$ satisfying the imposed velocities on the specimen boundary the actual field minimizes the functional

$$\{E^c(\hat{\varepsilon})\},$$

where $\hat{\varepsilon}$ is the strain rate associated with $\hat{\mathbf{v}}$. The minimum value of this is $E(\bar{\varepsilon})$ and so if $\hat{\varepsilon}$ is taken as $\bar{\varepsilon}$ *in all grains* as in the Taylor model, Hill's (1965) minimum principle states that

$$\{E^c(\bar{\varepsilon})\} \geq E(\bar{\varepsilon}). \tag{5.20}$$

Furthermore, if σ^u denotes the local stress in each grain associated with the strain rates $\bar{\varepsilon}$, then σ^u must satisfy

$$\bar{\varepsilon} = \mathbf{M}^c(\sigma^u):\sigma^u. \tag{5.21}$$

The procedure then is to solve Eq. (5.21) for σ^u for each grain in the aggregate. An interesting feature of this inverse problem noted by Hutchinson (1976) is that for all stresses $\hat{\sigma}$, σ^u maximizes the functional

$$\hat{\sigma}:\bar{\varepsilon} - F^c(\hat{\sigma}) = \hat{\sigma}:\bar{\varepsilon} - \frac{a}{m+1} g \sum_\alpha \left[\frac{\hat{\tau}^{(\alpha)}}{g}\right]^{m+1}, \tag{5.22}$$

where $\hat{\tau}^{(\alpha)} = \mathbf{P}^{(\alpha)}:\hat{\sigma}$ and, in accordance with the Taylor model, we take all the $a^{(\alpha)} = a$ and the $g^{(\alpha)} = g$. Now, as $m \to \infty$, which corresponds to rate-independent rigid–plastic behavior, $F^c \to 0$ if $\hat{\tau}^{(\alpha)} \leq g$. The latter inequality is a statement of the yield criterion that bounds the $\hat{\tau}^{(\alpha)}$. In other words, for all $\hat{\sigma}$ satisfying the yield criterion

$$\hat{\sigma}:\bar{\varepsilon} \leq \sigma^u:\bar{\varepsilon}. \tag{5.23}$$

Equation (5.23) is equivalent to the Bishop–Hill (1951) maximum work principle that allows selection of the active slip systems. This principle is in fact equivalent to Taylor's minimum internal work principle that follows from his minimization of the cumulative shears in isotropically hardening materials.

Hutchinson's (1976) analysis could in principle be followed incrementally to allow the development of finite strains. Grain rotations could then be computed as in the Taylor model. However, for now we will simply use his analysis to reproduce Taylor's (1938a) results for the isotropic fcc tensile curve. If x_3 is the axis of axisymmetric tensile straining, $\bar{\varepsilon}_{33} = -2\bar{\varepsilon}_{11} = -2\bar{\varepsilon}_{22} = \bar{\varepsilon}$, and $\bar{\sigma}_{33} = \bar{\sigma}$, and without loss of generality

$$\sigma^u = (\bar{\varepsilon}/a)^{1/m} g\omega. \tag{5.24}$$

Here ω depends on m and the orientation of the grain. The upper bound statement (5.20) can be written as

$$\{\sigma^u : \bar{\varepsilon}\} \geq \bar{\sigma} : \bar{\varepsilon}. \tag{5.25}$$

If $\bar{\sigma}_0$ is now defined by writing $\bar{\sigma} = (\bar{\varepsilon}/a)^{1/m} \bar{\sigma}_0$, then Eq. (5.25) can be rewritten as

$$\bar{\sigma}_0/g \leq f(m), \tag{5.26}$$

where

$$f(m) = \{\omega : \bar{\varepsilon}\}/\bar{\varepsilon}. \tag{5.27}$$

Volume integrations were replaced by uniformly weighted averages over all possible grain orientations, as described by Hutchinson (1976), and it was found, for example, that $f(1) = \frac{33}{20}$, $f(5) = 2.77$, and $f(\infty) \to 3.06$. The last value of 3.06 corresponding to the rate-independent limit was actually obtained by extrapolating from computed values for $n \leq 10$; this is equal to the value calculated by Taylor (1938) for rate-independent materials.

Figure 62 shows a comparison between experimentally measured stress–strain curves for fcc nickel and stress–strain curves computed using the value of $f(\infty) = 3.06$. Also shown in Fig. 62 is the stress–strain curve computed from the τ versus γ data for a single crystal whose axis was aligned in a $\langle 111 \rangle$ direction. As shown by Kocks (1970), polycrystalline data correlates mostly closely with single-crystal data if the single-crystal data is taken from crystals deforming in multiple slip. Data obtained from $\langle 111 \rangle$ oriented crystals evidently provides an optimum choice of multiple slip. The results shown in Fig. 62 are typical in that they show a reasonably good correspondence between the Taylor model predictions and the experimental data at larger grain sizes. At grain sizes of 2 μm the effect of grain size on both strength and strain hardening is evident. It may be possible to account for some grain size effects by arbitrarily adjusting the single-crystal yield strength values, although it is unlikely that the effects on strain hardening will be properly accounted for as well.

Since Taylor's original treatment of polycrystalline behavior in 1938 various models have been proposed. For example, Lin (1957) extended Taylor's model to include elasticity, but again imposed the restriction of equal total strains in all grains. A notable problem of imposing equal strains is that equilibrium in the stress state is not maintained. Kroner (1961) and

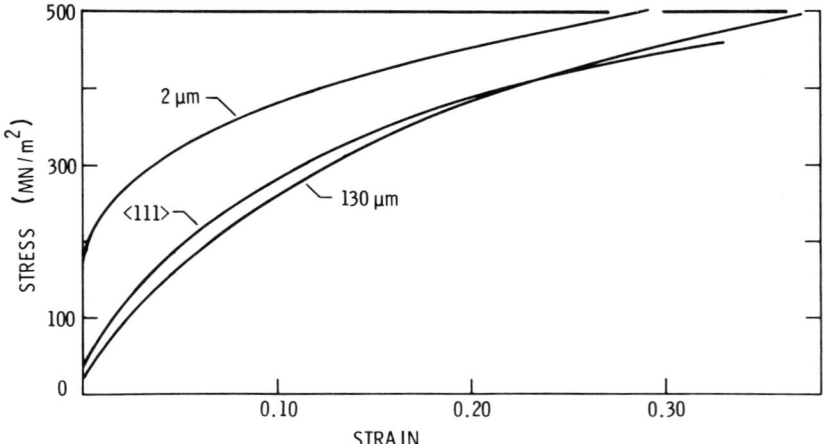

FIG. 62. Polycrystalline stress–strain curves for pure fcc nickel tested in axisymmetric tension. The curves are for polycrystals with grain sizes of 2 and 130 μm. The curve marked ⟨111⟩ was computed from single-crystal τ versus γ data using the Taylor model. This figure is taken from Thompson and Baskes (1975).

Budiansky and Wu (1962) then proposed self-consistent models that, although limited to small strains, gave a more realistic accounting of grain interactions. Hill (1965) developed a self-consistent model that is not limited to small strains in that it evaluates grain interactions by accounting for the overall elastic–plastic response of the aggregate. In essence, the self-consistent method regards the individual grains as misfitting inclusions embedded in a homogeneous matrix, in the spirit of Eshelby's (1957, 1960) well-known "transformation strain" problems. The misfits are related to the differences in the actual strain of the grain and that of the surrounding matrix. The matrix properties are themselves related in a "self-consistent" fashion to those of the grains by suitable volume averages. In the Hill (1965) model the matrix properties correspond to the elastic–plastic state.

The Hill model could in principle, then, be used to carry out the suggested task of determining texture and incorporating the anisotropy caused by texture in constitutive behavior. Hutchinson (1970) has carried out calculations for polycrystals using the Hill (1965) model. His results were also limited to rate-independent materials and small strains and did not account for grain rotations or texture. Nonetheless, his findings illustrated that very strong anisotropic elastic–plastic response accompanied strains on the order of only two to three times the yield strains. The calculations showed that yield vertex structure develops quite early on in the deformation and is manifested in a reduction in various instantaneous moduli governing nonproportional loading. It is likely that crystallographic and material texture will also contribute to such strong directionality in material properties.

Extensions of the Hill (1965) model to include texture could also be carried out using rate-dependent flows laws, as in Eq. (3.80). In fact, Hutchinson (1976) has already done so for steady power-law creep, as described previously. Interestingly enough, the self-consistent results for axisymmetric tensile deformation are not much different from those obtained using the Taylor–Bishop–Hill upper bound. An apparent advantage of the rate-dependent flow laws is that they impart a uniqueness to the slip mode that is characteristically absent from their rate-independent counterparts. However, one aspect of grain deformation that must be faced, and which has been mentioned, is what Peirce et al. (1982) have called patchy slip. As shown in Section II,E, this is a very common feature of grain deformation that, although inherent in single crystals, is very much influenced by grain size and grain interactions. A manifestation of patchy slip is that the actual slip mode could turn out to be quite different from the mode expected assuming that the grain deforms uniformly. It is indeed interesting that in a recent analysis of texture development in rate-independent materials Sowerby et al. (1981) reported that a *random selection* of five independent slip systems gave results in accordance with experiments.

Finally, we note that there seems to be some disagreement in the literature over what part of the rotations or deformation causes texture (Kocks and Chandra, 1981). To provide some guidance Fig. 63 shows some of what is involved. The representative material part in Fig. 63a is meant to be a grain. If the crystallographic orientation of the *material* plane bordered by the heavy lines is followed by aligning an X-ray beam along arrow B, then the orientation of this plane changes as prescribed by $\mathbf{F}^{\text{p}}$. On the other hand, if the orientation with respect to a plane fixed relative to some coordinate frame is determined by aligning the X-ray beam along arrow A, then orientation changes are prescribed by $\mathbf{F}^*$. In the Taylor model the situation would be as depicted in Fig. 63b. Here the orientation of grains with respect to the fixed principal directions of strain are determined and these change as prescribed by $\mathbf{F}^*$.

B. Deformation Shear Bands

The formation of shear bands during tensile straining of single crystals was discussed in Section IV,B. There the vital role of lattice kinematics and geometrical softening in triggering localized plastic flow was illustrated. A similar mode of highly localized deformation, important in metal forming, is very commonly observed in plane strain rolling of both single crystals and

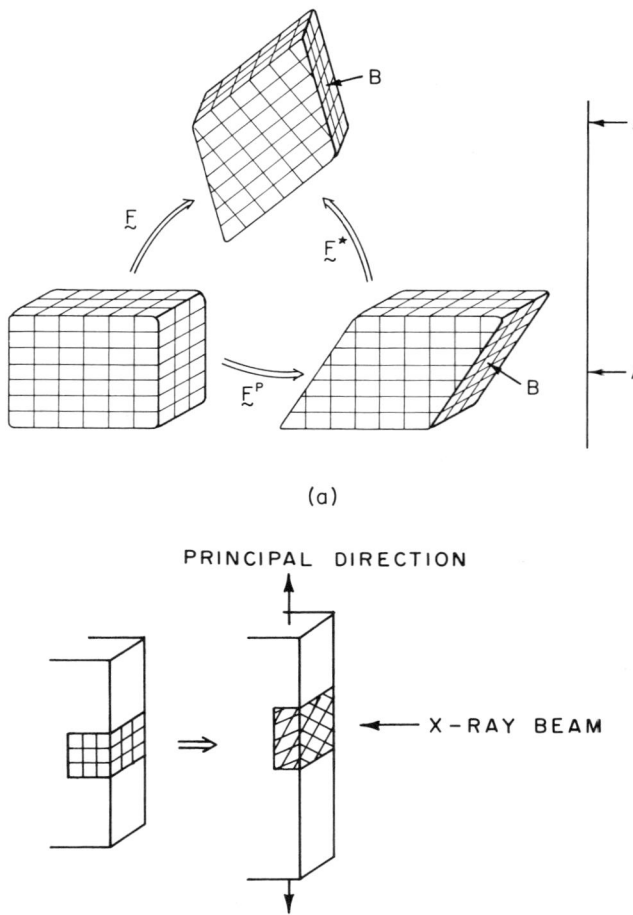

(a)

(b)

FIG. 63. The arrows A and B indicate two possible trajectories for an X-ray beam used to determine lattice orientation and crystallographic texture. Arrow B represents the case where the orientation of a particular material plane is followed. Arrow A represents the case where the lattice orientation relative to a fixed direction is followed.

polycrystals. Figure 64a shows an example of shear bands formed during rolling of a single crystal of copper (Morii and Nakayama, 1981) and Fig. 64b shows a similar example in a polycrystalline aluminum–4.75% copper alloy (Dillamore et al., 1979). In both cases once the shear bands form, they determine the kinematics of subsequent deformation and therefore play a role,

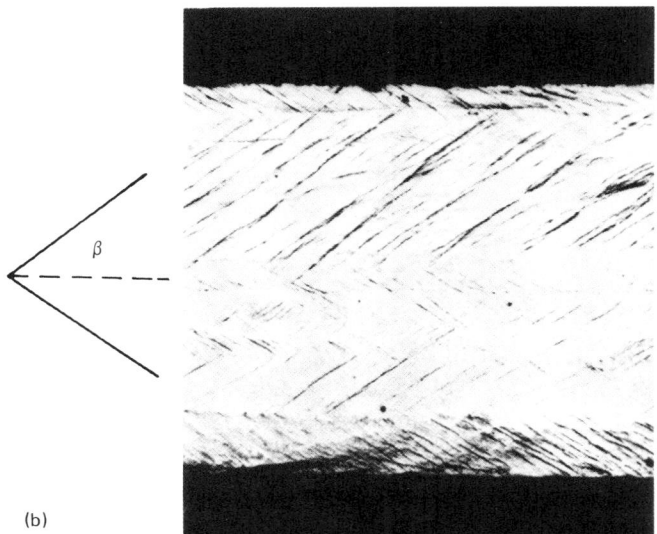

Fig. 64. (a) A shear band formed during the plain strain rolling of a single crystal of copper. This is an electron micrograph that shows the dislocation subgrain structure which develops within the band (Morii and Nakayama, 1981). (b) An optical micrograph of shear bands developed in plane strain rolling of a polycrystalline aluminum–4.75% copper alloy (Dillamore et al., 1979).

as yet not completely documented, on subsequent texture development. Some ideas and a good deal of microstructural characterization of these bands are available (e.g., Morii and Nakayama, 1981; Dillamore et al., 1979; Kriesler and Doherty, 1978) and thus the analysis of these modes within a continuum framework also remains an outstanding problem in the mechanics of crystalline plasticity. In closing, some brief background is provided.

Morii and Nakayama (1981) noted that in single crystals of materials with low stacking fault energy the bands tend to form preferentially within lamellae of mechanical twins, whereas in higher stacking fault energy materials the bands develop without twins. On a macroscopic level the bands are similar in their dimensions and orientation whether they involve twinning or not. For example, in Fig. 64a the bands make an angle of between 30 and 40° to the rolling direction (RD). In the experiment cited here, they documented the micromechanics of the bands in single crystals of pure copper (with a stacking fault energy of the intermediate value 73 ergs/cm^2; Hirth and Lothe, 1968. When the crystals were rolled at a temperature of 77 K, the crystals twinned and the bands indeed formed within the twin lamellae. When the crystals were rolled at room temperature, no twinning was observed and the bands formed at larger strains, but with a similar morphology.

In one interesting experiment the authors prerolled a crystal at 77 K and formed twins without shear bands. Subsequent room-temperature rolling caused shear bands to develop preferentially within the existing twin lamellae in much the same way as at 77 K. Thus the twins can serve as nuclei for shear bands at all temperatures. The mechanism for this apparently involves a reorientation of the lattice into a *geometrically soft* configuration (Morii and Nakayama, 1981). In the case of only higher-temperature rolling they argued that "unequal activation of multiple slip" (Morii and Nakayama, 1981, p. 864) could also lead to lattice reorientations and geometrical softening of a similar type.

Although twinning was not specifically included in the constitutive framework of Section III, the analysis of deformation bands at higher temperatures could proceed along lines very similar to the analysis given in Section IV,C. In this way the essential features of lattice kinematics could be studied. The geometry of the model used, however, should be representative of the special crystal orientations used in the Morii and Nakayama (1981) experiments. Twinning might then be incorporated in the kinematics as an alternative mechanism of inelastic deformation.

For polycrystals, on the other hand, Dillamore et al. (1979) have put forth the very interesting notion that grain rotations can lead to geometrical softening. They in fact described a sort of "texture softening" that depended upon the preexisting texture in the rolled polycrystal. To describe the effect of grain orientation on material stiffness, they introduced a parameter M (sometimes called the Taylor factor) defined as $\varepsilon = \gamma/M$, where ε is the maximum principal strain and γ is the cumulative shear strain of all active slip systems. For a randomly oriented fcc polycrystal, the average value of M in axisymmetric tension is the Taylor value, 3.06. Softening results if M tends to decrease as the grains rotate, since larger principal strains can be produced with smaller slip plane shears and hence with less strain hardening. They found that in polycrystals, textured by plane strain rolling, a shearing mode inclined at angles less than 45° to the rolling direction would lead to grain rotations that forced the grains into "softer" orientations. They further argued that localized shearing began in the individual grains before localized shearing becomes prevalent in the aggregate. This might well be describable using the analysis of Section IV,C. The details of their argument are not reproduced here, but their results and discussion certainly provide a strong impetus for the sort of analysis suggested in the previous section, namely, that of accounting for grain rotation and texture in polycrystalline constitutive laws. In this way a more comprehensive description of large strain deformation of polycrystals could be attained.

ACKNOWLEDGMENTS

This work was supported by the Materials Research Laboratory at Brown University funded by the U.S. National Science Foundation. The many fruitful discussions and collaborations I have had with Professors A. Needleman and J. R. Rice are also gratefully acknowledged.

REFERENCES

Asaro, R. J. (1979). Geometrical effects in the inhomogeneous deformation of ductile single crystals. *Acta Metall.* **27**, 445–453.

Asaro, R. J., and Barnett, D. M. (1976). Applications of the geometrical theorems for dislocations in anisotropic elastic media. *In* "Computer Simulation for Materials Applications" (R. J. Arsenault, J. R. Beeler, Jr., and J. A. Simmons, eds.) NBS, Gaithersburg, Maryland.

Asaro, R. J., and Rice, J. R. (1977). Strain localization in ductile single crystals. *J. Mech. Phys. Solids* **25**, 309–338.

Asaro, R. J., and Hirth, J. P. (1973). Planar dislocation interactions in anisotropic media with applications to nodes. *J. Phys. F* **3**, 1659–1671.

Asaro, R. J., Hirth, J. P., Barnett, D. M., and Lothe, J. (1973). A further synthesis of sextic and integral theories for dislocations and line force s in anisotropic media. *Phys. Status Solidi B* **60**, 261–271.

Ashby, M. F. (1971). The deformation of plastically non-homogeneous alloys. *In* "Strengthening Methods in Crystals" (A. Kelly and R. B. Nicholson, eds.) Halsted, New York.

Avery, D. H., and Polan, N. W. (1974). *Metall. Trans.* **5**, 1159.

Barendreght, J. A., and Sharpe, W. N., Jr. (1973). The effect of biaxial loading on the critical resolved shear stress of zinc single crystals. *J. Mech. Phys. Solids* **21**, 113–123.

Barnett, D. M., and Asaro, R. J. (1972). The fracture mechanics of slit-like cracks in anistropic elastic media: *J. Mech. Phys. Solids* **20**, 353–366.

Barrett, C. S. (1952). "Structure of Metals." McGraw-Hill, New York.

Biot, M. (1965). "Mechanics of Incremental Deformation." Wiley, New York.

Bishop, J. F. W. (1954). A theory of the tensile and compressive textures of face-centered cubic metals. *J. Mech. Phys. Solids* **3**, 130–142.

Bishop, J. F. W., and Hill, R. (1951). A theoretical derivation of the plastic properties of a polycrystalline face-centered metal. *Philos. Mag.* **42**, 414–427.

Boas, W., and Ogilvie, G. J. (1954). The plastic deformation of a crystal in a polycrystalline aggregate. *Acta Metall.* **2**, 655–659.

Bragg, W. H., and Bragg, W. L. (1933). "The Crystalline State." Bell, London.

Brown, K. (1972). Role of deformation and shear banding in the stability of the rolling textures of aluminum and an Al–0.8% Mg alloy. *J. Inst. Metals* **100**, 341–345.

Budiansky, B., and Wu, T. Y. (1962). Theoretical prediction of plastic strains of polycrystals. *Proc. 4th U. S. Nat. Congr. Appl. Mech.*, p. 1175.

Burgers, J. M. (1939). *Proc. Kon. Ned. Akad. Wetenschap.* **42**, 293.

Byrne, J. G., Fine, M. E., and Kelly, A. (1961). Precipitate hardening in an aluminum–copper alloy. *Philos. Mag.* **6**, 1119–1145.

Cahn, R. W. (1951). Slip and polygonization in aluminum. *J. Inst. Metals* **79**, 129–158.

Campbell, J. D., Eteiche A. M., and Tsao M. C. C. (1977). Strength of metals and alloys at high strains and strain rates. *In* "Fundamental Aspects of Structural Alloy Design" (R. I. Saffe and B. A. Wilcox, eds.), pp. 545–563. Plenum, New York.

Chang, Y. W., and Asaro, R. J. (1981). An experimental study of shear localization in aluminum–copper single crystals. *Acta Metall.* **29**, 241–257.

Chiem, C. Y., and Duffy, J. D. (1981) Strain rate history effects and observations of dislocation substructure in aluminum single crystals following dynamic deformation. *Brown Univ.* Report MRL E-137, Providence, Rhode Island.

Dillamore, I. L., Roberts, J. G., and Bush, A. C. (1979). Occurrence of shear bands in heavily rolled cubic metals. *J. Metal Sci.* **13**, 73–77.

Elam, C. F. (1927). Tensile tests on crystals. Part IV.—A copper alloy containing five per cent aluminum. *Proc. R. Soc. London, Ser. A* **115**, 694–702.

Embury, J. D. (1971). Strengthening by dislocation substructures. *In* "Strengthening Methods in Crystals" (A. Kelly, and R. B. Nicholson, eds.) Halsted, New York.

Eshelby, J. D. (1951). The force on an elastic singularity. *Philos. Trans. R. Soc A* **244**, 87–112.

Eshelby, J. D. (1956). The continuum theory of lattice defects. *In* "Solid State Physics" (F. Seitz, and D. Turnbull, eds.), vol. 3, pp. 79–144. Academic Press, New York.

Eshelby, J. D. (1958). The determination of the elastic field of an ellipsoidal inclusion, and related problems. *Proc. R. Soc. London, Ser. A* **241**, 376–396.

Eshelby, J. D. (1961). Elastic inclusions and inhomogeneities. *In* "Progress in Solid Mechanics, vol. II". North-Holland. Publ., Amsterdam.

Ewing, J. A., and Rosenhain, W. (1899). Experiments in micro-metallurgy: effects of strain, preliminary notice. *Proc. R. Soc. London* **65**, 85–90.

Ewing, J. A., and Rosenhain, W. (1900). The crystalline structure of metals. *Philos. Trans. R. Soc. London* **193**, 353–375.

Franciosi, P., Berveiller, M., and Zaoui, A. (1980). Latent hardening in copper and aluminum single crystals. *Acta Metall.* **28**, 273–283.

Frank, F. C., and Read, W. T., Jr. (1950). Multiplication processes for slow moving dislocations. *Phys. Rev.* **79**, 722–724.

Frenkel, J. (1926). Zur theorie der Elastizitätsgrenze und der Festigheit kristallinischer Körper. *Z. Phys.* **37**, 572–609.

Hall, E. O. (1951). The deformation and ageing of mild steel: III discussion of results. *Proc. Phys. Soc. London, Sec. B* **64**, 747–753.

Havner, K. S. and Shalaby, A. H. (1977). A simple mathematical theory of finite distortional latent hardening in single crystals. *Proc. R. Soc. London, Sec. A* **358**, 47–70.

Havner, K. S., Baker, G. S., and Vause, R. F. (1978). Theoretical latent hardening in crystals: I. general equations for tension and compression with applications to F. C. C. crystals in tension. *J. Mech. Phys. Solids* **27**.

Hill, R. (1962). Acceleration waves in solids. *J. Mech. Phys. Solids* **10**, 1–16.

Hill, R. (1965). A self-consistent mechanics of composite materials. *J. Mech. Phys. Solids* **13**, 213–222.

Hill, R. (1966). Generalized constitutive relations for incremental deformation of metals crystals by multislip. *J. Mech. Phys. Solids* **14**, 95–102.

Hill, R., and Havner, K. (1982). Perspectives in the mechanics of elastoplastic crystals". *J. Mech. Phys. Solids*. (to appear).

Hill, R., and Hutchinson, J. W. (1975). Bifurcation phenomena in the plane strain tension test. *J. Mech. Phys. Solids* **23**, 239–264.

Hill, R., and Rice, J. R. (1972). Constitutive analysis of elastic–plastic crystals at arbitrary strain. *J. Mech. Phys. Solids* **20**, 401–413.

Hirth, J. P., and Lothe, J. (1968). "Theory of Dislocations." McGraw-Hill, New York.

Honeycombe, R. W. K. (1950). Inhomogeneities in the plastic deformation of metal crystals: I—occurrence of x-ray asterisms; I—x-ray and optical micrography of aluminum. *J. Inst. Metals* **80**, 45–56.

Hull, A. W. (1919). The positions of atoms in metals. *Proc. Am. Inst. Elec. Eng.* **38**, 1171–1192.

Hutchinson, J. W. (1970). Elastic–plastic behavior of polycrystalline metals and composites. *Proc. R. Soc. London, Sec. A* **319**. 247–272.

Hutchinson, J. W. (1976). Bounds and self–consistent estimates for creep of polycrystalline materials. *Proc. R. Soc. London, Sec. A* **348**, 101–127.

Jackson, P. J., and Basinski, Z. S. (1967). Latent hardening and the flow stress in copper single crystals. *Can. J. Phys.* **45**, 707–735.

Joshi, N. R., and Green, R. E., Jr. (1980). Continuous x-ray diffraction measurement of lattice rotation during tensile deformation of aluminum crystals. *J. Mat. Sci.* **15**, 729–738.

Kelly, A., and Nicholson, R. B. (1971). "Strengthening Methods in Crystals." Halsted, New York.

Klepaczko, J. (1981). The relation of thermally activated flow in BCC metals and ferritic steels to strain rate history and temperature history effects. *Brown University Report DMR-79-23257/2*, Providence, Rhode Island.

Kocks, U. F. (1970). The relation between polycrystal deformation and single-crystal deformation. *Metall. Trans.* **1**, 1121–1142.

Kocks, U. F., and Brown, T. J. (1966). Latent hardening in aluminum. *Acta Metall.* **14**, 87–98.

Kocks, U. F. and Chandra, H. (1982). Slip geometry in partially constrained deformation. *Acta Metall.* **30**, 695–708.

Koehler, J. S. (1952). The nature of work-hardening. *Phys. Rev.* **86**, 52–59.

Krausz, A. S., and Eyring, H. E. (1975). "Deformation Kinetics." Wiley, New York.

Kroner, E. (1961). Zur Plastischen Verformung des Vielkristalls. *Acta Metall.* **9**, 155.

Kuhlman-Wilsdorf, D. (1977). Recent progress in understanding of pure metal and alloy hardening. *In* "Work Hardening in Tension and Fatigue" (A. W. Thompson, ed.). AIME, New York.

Lisiecki, L. L., Nelson, D., and Asaro, R. J. (1982). Lattice rotations, necking and localized deformation in F. C. C. single crystals. *Scripta Met.* (to appear).

Lomer, W. M. (1951). A dislocation reaction in the face–centered cubic lattice. *Philos. Mag.* **42**, 1327–1331.

Love, A. E. H. (1927). "The Mathematical Theory of Elasticity." Cambridge Univ. Press, London and New York.

Mitchell, T. E. (1964). Dislocations and plasticity in single crystals of face–centered cubic metals and alloys. *In* "Progress in Applied Materials Research" (E. G. Sanford, J. H. Fearon, and W. J. McGonnagle, eds.) vol. **6**, pp. 119–237.

Mitchell, T. E. and Thornton, P. R. (1964). The detection of secondary slip during the deformation of copper and α-brass single crystals. *Philos. Mag.* **10**, 315–323.

Miura, S., and Saeki, Y. (1978). Plastic deformation of aluminum bicrystals 100 oriented. *Acta Metall.* **26**, 93–101.

Morii, K., and Nakayama, Y. (1981). Shear bands in rolled copper single crystals. *Trans. Japan Inst. Metals* **22**, 857–864.

Morrison, W. B. (1966). *Trans. ASM* **59**, 825.

Orowan, E. (1934). Zur Kristallplastizität. III. Über den Mechanismus des Gleituorganges. *Z. Phys.* **89**, 634–659.

Orowan, E. (1942). A type of plastic deformation new in metals. *Nature* **149**, 643–644.

Orowan, E. (1963). Dislocations in plasticity. *In* "The Sorby Centennial Symposium on the History of Metallurgy" (C. S. Smisth, ed.). Gordon and Breach, New York.

Pan, J., and Rice, J. R. (1981). Unpublished research, Brown University, Providence, Rhode Island.

Peach, M., and Koehler, J. S. (1950). The forces exerted on dislocations and the stress fields produced by them. *Phys. Rev.* **80**, 436–439.

Peirce, D., Asaro, R. J., and Needleman, A. (1982). An analysis of nonuniform and localized deformation in ductile single crystals *Acta Metall.* (to appear).

Petch, N. J. (1953). The cleavage strength of polycrystals. *J. Iron Steel Inst.* **174**, 25–28.

Piercy, G. R., Cahn, R. W., and Cottrell, A. H. (1955). A study of primary and conjugate slip in crystals of alpha-Brass. *Acta Metall.* **3**, 331–338.

Polanyi, Von, M., (1922). Röntgenographische Bestimmung von Kristallanordnungen. *Naturwissenschaften* **10**, 411–416.

Polanyi, Von, M., (1934). Über eine Art Gitterstorung, die einen Kristall plastisch machen konnte. *Z. Phys.* **89**, 660–664.

Price, R. J., and Kelly, A. (1964). Deformation of age–hardened aluminum alloy crystals—II. fracture". *Acta Metall.* **12**, 979–991.

Ramaswami, B., Kocks, U. F., and Chalmers, B. (1965). Latent hardening in silver and an ag–au alloy. *Trans. AIME* **233**, 927–931.

Reid, C. N., Gilbert, A., and Hahn, G. T. (1966). Twinning slip and catastrophic flow in niobium. *Acta Metall.* **14**, 975–983.

Rice, J. R. (1971). Inelastic constitutive relations for solids: an internal-variable theory and its application to metal plasticity. *J. Mech. Phys. Solids* **19**, 443–455.

Rice, J. R. (1977). The localization of plastic deformation". *In* "Proceedings of 14th International Congress of Theoretical and Applied Mechanics" Vol. I. North-Holland Publ., Amsterdam.

Saiomoto, S., Hasford, W. F., Jr., and Backofen, W. A. (1965). Ductile fracture in copper single crystals. *Philos. Mag.* **12**, 319–333.

Schmid, E. (1924). *Proc, Int. Congr. Appl. Mech.* (*Delft*), 342.

Sevillano, J. G., van Houtle, P., and Aernoudt, E. (1980). Large strain work hardening and textures. *Prog. Mat. Sci.* **25**, 69–411.

Sowerby, R., Dac Viuana, C. S., and Davies, G. J. (1980). The influence of texture on the mechanical response of commercial purity copper sheet in some forming processes. *Mat. Sci. Eng.* **46**, 23–51.

Spitzig, W. A., Sober, R. J., and Richmond, O. (1975). Pressure dependence of yielding and associated volume expansion in tempered martensite. *Acta Metall.* **23**, 885–893.

Swearengen, J. C., and Taggart, R. (1971). Low amplitude cyclic deformation and crack nucleation in copper and copper–aluminum bicrystals. *Acta Metall.* **19**, 543–559.

Taylor, G. I. (1934). The mechanism of plastic deformation of crystals. Part I.—theoretical. *Proc. R. Soc. London, Sec. A* **145**, 362–387.

Taylor, G. I. (1938a). Plastic strain in metals. *J. Inst. Met.* **62**, 307–325.

Taylor, G. I. (1938b). Analysis of plastic strain in a cubic crystal. *In* "Stephen Timoshenko 60th Anniversary Volume" (J. M. Lessels, ed.). Macmillan, New York.

Taylor, G. I., and Elam, C. F. (1923). The distortion of an aluminum crystal during a tensile test. *Proc. R. Soc. London, Sec. A* **102**, 643–667.

Taylor, G. I., and Elam, C. F. (1925). The plastic extension and fracture of aluminum crystals. *Proc. R. Soc. London, Sec. A* **108**, 28–51.

Thompson, A. W. (1975). Polycrystal hardening. *In* "Work Hardening in Tension and Fatigue" (A. W. Thompson, ed.), AIME, New York.

Thompson, A. W., and Baskes, M. I. (1973). The influence of grain size on the work hardening of face-centered-cubic polycrystals". *Philos. Mag.* **28**, 301–308.

Timpe, A. (1905). "Probleme der Spannungsverteilung in ebenen Systemen einfach gelost mit Hilfe der airyschen Funktion". Göttingen Diss., Leipzig.

Volterra, V. (1907). Sur l'équilibre des corps élastiques multiplement convexes. *Ann. Ec. Norm (Ser. 3)* **24**, 401–5 17.

Worthington, P. J., and Smith, E. (1966). The formation of slip bands in polycrystalline 3% Silicon iron in the pre-yield microstrain region. *Acta Metall.* **12**, 1277–1281.

Mechanisms of Deformation and Fracture

M. F. ASHBY

Engineering Department
Cambridge University
Cambridge, England

I.	Introduction	118
	A. Deformation and Fracture Mechanisms	118
	B. Constitutive Laws	118
	C. Deformation and Fracture Maps	121
II.	Mechanisms of General Plasticity and Creep	121
	A. Elastic Collapse	121
	B. Low-Temperature Plasticity by Dislocation Glide	122
	C. Power-Law Creep	125
	D. Diffusional Flow	128
	E. Other Mechanisms	129
III.	Deformation Mechanism Maps	130
	A. Construction of the Maps	130
	B. Deformation Maps for Metals and Ceramics	132
	C. Mechanisms of Deformation and the Selection of Constitutive Laws	146
IV.	Mechanisms of Fracture by General Damage	147
	A. The Ideal Fracture Strength	147
	B. Cleavage	147
	C. Ductile Fracture at Low Temperatures	149
	D. Transgranular Creep Fracture	151
	E. Intergranular Creep-Controlled Fracture	151
	F. Pure Diffusional Fracture	153
	G. Rupture	154
	H. Other Mechanisms of Fracture	154
V.	Fracture Mechanism Maps	155
	A. Construction of the Maps	155
	B. Fracture Maps for Metals and Ceramics	157
	C. Mechanisms of Fracture and Order of Magnitude of K_c and G_c	166
VI.	Conclusions	171
	References	172

I. Introduction

A. Deformation and Fracture Mechanisms

Crystalline solids deform plastically by a number of alternative (and often competing) mechanisms: low-temperature plasticity, twinning, power-law creep, diffusional flow, and so forth. Each has certain characteristics: a rate that depends strongly on temperature, for instance, or on grain size, or which is influenced by a dispersion of a second phase. These characteristics are summarized by the constitutive law for that mechanism, and each has a characteristic regime of dominance, that is, a range of stresses, temperatures, and strain rates over which it is the primary mechanism. The ranges, of course, vary from material to material: Ice creeps at $-10°C$, for instance, while tungsten does not do so below $600°C$. The mechanisms and their ranges are important in engineering design because, for a given application, they determine the proper constitutive law.

Fracture, too, can occur by any one of a number of alternative (and sometimes competing) micromechanisms: cleavage, ductile fracture, rupture intergranular creep fracture, and so forth. Again, each has certain characteristics: negligible ductility, for instance, or a ductility that depends on inclusion density or grain size; and, for given stress states—simple tension, for example—each has a characteristic regime of dominance, that is, a range of stresses and temperatures over which it is the primary mechanism. Naturally, the ranges vary with the material: Nickel fails in a ductile way at room temperature, whereas sodium chloride fails by cleavage. The mechanisms and their ranges are important in engineering design because, for a given application, they determine the ductility of an unnotched component, and the crack advance mechanism in one containing a crack.

This article describes the mechanisms of plasticity and fracture and develops ways of displaying their ranges of dominance and certain of their characteristics.

B. Constitutive Laws

Unless the hydrostatic pressure is very large ($p/K > 10^{-2}$, where K is the bulk modulus), plasticity and creep depend on the deviatoric part of the stress field. We define the equivalent shear stress and strain rate by

$$\sigma_s = \{\tfrac{1}{6}[(\sigma_1 - \sigma_2)^2 + (\sigma_2 - \sigma_3)^2 + (\sigma_3 - \sigma_1)^2]\}^{1/2} \tag{1.1}$$

and
$$\dot{\gamma} = \{\tfrac{2}{3}[(\dot{\varepsilon}_1 - \dot{\varepsilon}_2)^2 + (\dot{\varepsilon}_2 - \dot{\varepsilon}_3)^2 + (\dot{\varepsilon}_3 - \dot{\varepsilon}_1)^2]\}^{1/2}, \tag{1.2}$$

where σ_1, σ_2, and σ_3 and $\dot{\varepsilon}_1$, $\dot{\varepsilon}_2$, and $\dot{\varepsilon}_3$ are the principal stresses and strain rates.

Each mechanism of deformation is described by a rate equation

$$\dot{\gamma} = f(\sigma_s, T, S_i, P_j). \tag{1.3}$$

Here T is the absolute temperature. The set of i quantities S_i are the *state variables* that describe the microstructural state of the material: the dislocation density, the grain size, the cell size, and so forth. Finally, the set of j quantities P_j are the *material properties*: lattice parameter, atomic volume, moduli, diffusion constants, etc.; these can be regarded as constant during deformation.

The state variables S_i may vary over a wide range. A second set of equations is needed to describe their rate of change:

$$dS_i/dt = g_i(\sigma_s, T, S_i, P_j), \tag{1.4}$$

where t is time.

The coupled set of equations (1.3) and (1.4) are the constitutive law for the material. They can be integrated over time to give a complete description of plasticity under any loading history. But although we have satisfactory models that lead to the first set of equations (that for $\dot{\gamma}$), we do not, at present, understand the evolution of structure with strain or time sufficiently well to formulate expressions for the second. To proceed further, we must make simplifying assumptions about the structure.

Two alternative assumptions are used here. The first, and simplest, is the assumption of *constant structure*:

$$S_i = S_i^0. \tag{1.5}$$

Then the rate equation for $\dot{\gamma}$ completely describes plasticity. The second possible assumption is that of *steady state*:

$$dS_i/dt = 0. \tag{1.6}$$

Then the internal variables (dislocation density and arrangement, grain size, etc.) no longer appear explicitly in the rate equations, because they are determined by the external variables of stress and temperature. We can solve for S_1, S_2, etc., in terms of σ_s and T, again obtaining an explicit rate equation for $\dot{\gamma}$.

At high temperatures, deforming materials quickly approach a steady state, and the appropriate rate equations are those for this steady behavior. But at low temperatures a steady state is rarely achieved: Then a constant structure formulation describing flow at a given structure and state of work hardening is the better approximation. The rate equation (1.3) then becomes

$$\dot{\gamma} = f(\sigma_s, T), \qquad (1.7)$$

since both the *P*'s and *S*'s are constant, or can be expressed purely in terms of σ_s and T. It can be applied to problems involving multiaxial stress states by using Eqs. (1.1) and (1.2) together with the associated flow rule. Rate equations of this sort are given further on and are used to construct diagrams that identify the dominant mechanism of deformation and its characteristics.

Fracture, of course, is more complicated. A formulation like that for plasticity is possible and useful. Let the damage within a deforming material be measured by D (it is often possible to think of it as the area fraction of holes, voids, or cracks). It has the value 0 at the start of life, and 1 at the end. Then plasticity and damage accumulation are described by the set of equations

$$\dot{\gamma} = f(\sigma_s, T, S, P, D), \qquad (1.8)$$

$$dS/dt = g(\sigma_s, T, S, P, D), \qquad (1.9)$$

$$dD/dt = h(\sigma_s, \sigma_n, p, T, S, P, D). \qquad (1.10)$$

Here the rate of damage accumulation dD/dt depends on three independent stress components—most conveniently, the shear stress σ_s, the maximum normal stress σ_n, and the hydrostatic pressure p. The assumption of a steady state or a steady structure eliminates Eq. (1.9), and those remaining can then be integrated to give the *time t_f and the strain ε_f to fracture:*

$$t_f = j(\sigma_s, \sigma_n, p, T), \qquad (1.11)$$

$$\varepsilon_f = k(\sigma_s, \sigma_n, p, T). \qquad (1.12)$$

Progress has been made in formulating and integrating such damage rate equations, based on physical models (see, for example, Cocks and Ashby, 1982). But their development is still incomplete, and there exists very little of the data needed to evaluate them. Although they are the obvious starting point for identifying the dominant fracture mechanism (Ashby, 1977), it is not yet a practical route. Instead, we assemble fractographic data and use them to construct empirical diagrams that identify the dominant mechanism of fracture and its characteristics.

C. Deformation Maps and Fracture Maps

The overall plastic behavior, for a given material, can be summarized in a *deformation mechanism map*, a diagram with equivalent stress [Eq. (1.1)] as one axis and temperature or the strain rate [Eq. (1.2)] as the other. The maps are constructed from the constitutive laws that are fitted to the best available data and show the fields of stress, temperature, and strain rate over which each mechanism is dominant (see Sections II and III). They give an overview of the plastic properties of a material and help identify the proper constitutive law for design and the mechanism of flow likely to be dominant in a given application of a material.

In a similar way, the fracture behavior for a given material and stress state can be summarized as a *fracture mechanism map*, a diagram with stress as one axis and temperature as the other, showing the field of dominance of each mechanism. Such maps are constructed by assembling and analyzing fracture observations and data for the material (see Sections IV and V). They give an overview of the micromechanisms by which the material may fail, and help identify the one most likely to be dominant in a given experiment or engineering application. They further help in identifying the mechanism of crack advance likely to be dominant in the monotonic or cyclic loading of cracked materials.

II. Mechanisms of General Plasticity and Creep

When a polycrystalline solid is stressed, it may deform plastically in one of a number of ways. In this section, these deformation mechanisms are described, and rate equations for them are listed. A more detailed discussion was given by Frost and Ashby (1982).

A. Elastic Collapse

The *ideal shear strength* defines a stress above which deformation of a perfect crystal (or of one in which all defects are pinned) ceases to be elastic and becomes catastrophic: At this stress, the crystal structure itself becomes mechanically unstable. Its value for a crystal at 0 K (which we call τ_{TH}) can be calculated from the crystal structure and the interatomic force law by simple statics (Tyson, 1966; Kelly, 1966). Above 0 K the problem becomes a kinetic one, that of calculating the frequency with which dislocation loops nucleate and expand in an initially defect-free crystal. The results of such calculations (Kelly, 1966) show that the temperature dependence of the ideal strength is small, about the same as that of the shear modulus. Plastic flow by collapse of the crystal structure can then be described by

$$\dot{\gamma}_1 = \infty \quad \text{when} \quad \sigma_s \geq \tau_{TH},$$
$$\dot{\gamma}_1 = 0 \quad \text{when} \quad \sigma_s < \tau_{TH}. \quad (2.1)$$

The computations of τ_{TH} lead to values between $\mu/20$ and $\mu/10$, where μ is the shear modulus (the result depends on crystal structure and bonding and on the instability criterion used). For fcc metals we have used $\tau_{TH} = 0.06\mu_0$, taken from the computer calculations of Tyson (1966) based on a Lennard–Jones potential. For bcc metals we have used $\tau_{TH} = 0.1\mu_0$, taken from an analytical calculation of MacKenzie (1959). For all other materials we take $\tau_{TH} = 0.1\mu_0$.

B. Low-Temperature Plasticity by Dislocation Glide

At low temperatures ($T \leq \frac{1}{3}T_M$, where T_M is the melting point) plastic deformation is caused by the glide motion of dislocations. In ductile materials (most metals, for instance) dislocations move easily unless obstacles, deliberately introduced by alloying or work hardening, obstruct them (Fig. 1). The stress required to move a dislocation segment, and thus the shear–yield strength $\hat{\tau}$ at 0 K, increases with increasing concentration of obstacles. Above 0 K, thermal energy allows dislocation segments to penetrate or bypass these obstacles at a slightly lower stress—it makes the obstacles appear to be a little weaker than they really are—and the flow strength falls. Kinetic theory can be used to calculate the shear–strain rate $\dot{\gamma}$ produced by a shear stress σ_s at a temperature T. At an adequate level of approximation

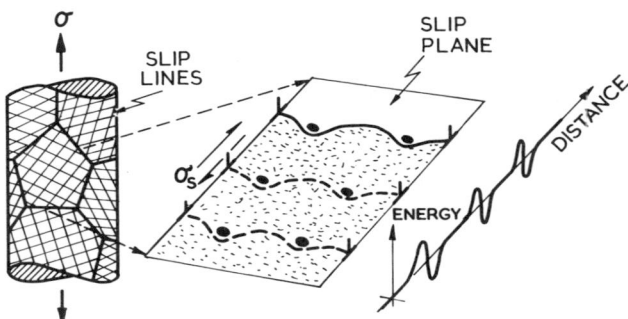

Fig. 1. Low-temperature plasticity limited by discrete obstacles. The strain rate is determined by the kinetics of obstacle cutting.

the rate equation for *obstacle-limited plasticity* (Evans and Rawlings, 1969; de Meester et al., 1973; Kocks et al., 1975) is

$$\dot{\gamma} = \dot{\gamma}_0 \exp -\frac{\Delta F}{kT}\left(1 - \frac{\sigma_s}{\hat{\tau}}\right), \tag{2.2}$$

where $\hat{\tau}$ is the shear strength at 0 K, ΔF is a measure of the strength of obstacles, $\dot{\gamma}_0$ is a kinetic constant (which we take to be 10^6 sec^{-1}; see Frost and Ashby, 1982), and k and T are Boltzmann's constant and the absolute temperature, respectively.

The quantity $\hat{\tau}$ is the "athermal flow strength," the shear strength in the absence of thermal energy. It reflects both the strength and the density (and arrangement) of the obstacles. For a widely spaced, discrete obstacle, $\hat{\tau}$ is proportional to $\mu b/l$, where l is the obstacle spacing, μ is the shear modulus, and b is the Burgers vector of the dislocations. The constant of proportionality depends on their strengths and distributions (Table I): For strong obstacles, such as a carbide dispersion in a steel, it is about 2; for weak obstacles, such as a solid solution of chromium in a steel, it is as small as 0.02.

The temperature dependence of the flow strength and the sensitivity to strain rate are determined by the activation energy ΔF—it characterizes the strength of a single obstacle. It is useful to classify obstacles by their strengths (Table I): Carbides in a steel are strong obstacles ($\Delta F \approx 2\mu b^3$) and usually remain undeformed when the steel is deformed; a solid solution, such as chromium dissolved in a steel, behaves like an array of weak obstacles ($\Delta F \approx 0.2\mu b^3$). Equation (2.2) correctly predicts that weak obstacles give a flow strength that falls steeply with temperature and depends strongly on strain rate.

TABLE I

CHARACTERISTICS OF OBSTACLES

Obstacle strength	ΔF	$\hat{\tau}$	Example
Strong	$>\mu b^3$ ($\sim 2\mu b^3$)	$>\mu b/l$	Dispersions; most precipitates (spacing l)
Medium	$\sim 0.5\mu b^3$	$\sim \mu b/l$	Forest dislocations, radiation damage; small, weak precipitates (spacing l).
Weak	$<0.2\mu b^3$	$<\mu b/l$	Isolated solute atoms (spacing l)

The very high yield strength of ceramics (e.g., oxides, carbides, silicates, ice, and diamond) is not usually caused by obstacles of this sort. The localized, often covalent bonding of these solids imposes a resistance to the motion of the dislocation through them, because bonds must be broken and reformed as it moves (Fig. 2). The kinetic theory of this *lattice resistance-limited plasticity* to plastic flow leads to a rather different law. Again, at an approximate level, it can be written as (Guyot and Dorn, 1967; Kocks et al., 1975)

$$\dot{\gamma} = \dot{\gamma}_p \left(\frac{\sigma_s}{\mu}\right)^2 \exp\left(-\left\{\frac{\Delta F_k}{kT}\left[1 - \left(\frac{\sigma_s}{\hat{\tau}_p}\right)^{3/4}\right]^{4/3}\right\}\right), \quad (2.3)$$

where ΔF_k is the Helmholtz free energy of an isolated pair of kinks and $\hat{\tau}_p$ is, to a sufficient approximation, the flow stress at 0 K. The quantity $\dot{\gamma}_p$ is a kinetic constant that is well approximated as 10^{11} sec^{-1}, and the term in σ_s^2 arises from the variation of the mobile dislocation density with stress.

It should be noted that the values of $\hat{\tau}_p$ (and of $\hat{\tau}$) for single crystals and polycrystals differ. The difference is a Taylor factor: It depends on the crystal structure and on the slip systems activated when the polycrystal deforms. For fcc metals, the appropriate Taylor factor M_s is 1.77; for the bcc metals, it is 1.67 (Kocks, 1970). (These may be more familiar to the reader as the Taylor factors $M = 3.06$ and 2.9, respectively, relating critical resolved shear strength to tensile strength for fcc and bcc metals. Since $\hat{\tau}_p$ and $\hat{\tau}$ are polycrystal shear strengths, the factors we use are smaller by the factor $\sqrt{3}$.) For less symmetrical crystals, the polycrystal shear strength is again a proper average over the individual slip systems that contribute to flow. These often differ markedly and it is then reasonable to identify $\hat{\tau}_p$ with the strength of the hardest of the slip systems (i.e., we take $M_s = 1$, but use $\hat{\tau}_p$ for the hard system). In calculating $\hat{\tau}$ or $\hat{\tau}_p$ from single-crystal data, we have applied the appropriate Taylor factor.

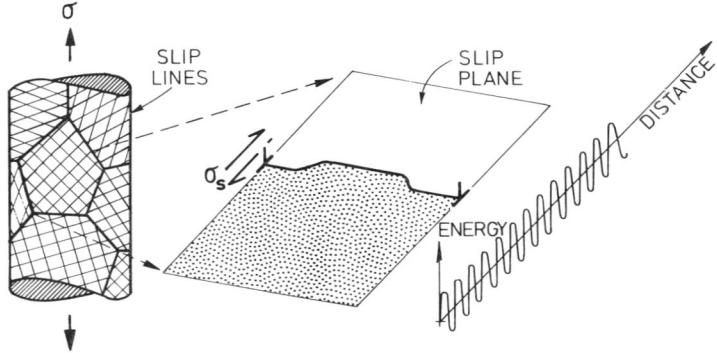

FIG. 2. Low-temperature plasticity limited by a lattice resistance. The strain rate is determined by the kinetics of kink nucleation and propagation.

Under conditions of explosive or shock loading, and in certain metal-forming and machining operations, the strain rate can be large ($<10^3$ sec^{-1}). Then the interaction of a moving dislocation with phonons or electrons can limit its velocity (Klahn et al., 1970; Kumar et al., 1968; Kumar and Kumble, 1969). At still higher rates, dislocation velocities approach that of sound in the solid, and relativistic effects lead to an increasing drag and a steeply rising flow strength (Nabarro, 1967). These strengthening mechanisms are ignored here. Their influence on deformation maps was discussed by Frost and Ashby (1982).

A material that exhibits a lattice resistance, will, in general, also contain obstacles, and these influence the yield strength: Oxides for example, work harden when they are deformed. The two strengthening mechanisms (obstacles and lattice resistance) superimpose. We shall assume that the resulting strain rate is simply the smaller of Eqs. (2.2) and (2.3).

C. Power-Law Creep

At elevated temperatures (about $0.3T_M$ for metals and $0.4T_M$ for most alloys and ceramics) materials creep. The creep rate is found, empirically, to follow a power law, often written as

$$\dot{\gamma} = \dot{\gamma}_0 \left(\frac{\sigma_s}{\sigma_0}\right)^n,$$

with

$$\dot{\gamma}_0 = A \exp\left(-\frac{Q}{RT}\right),$$

where n is the creep exponent and Q is the activation energy for creep. Kinetic models for this *power-law creep* are incomplete, in that they do not explain the observed range of the creep exponent n (typically, $3 < n < 8$). But the models, nonetheless, lead to some understanding of the process by which segments of dislocation move by coupled glide and climb (Fig. 3) and thus permit time-dependent plasticity ("creep") at stresses below that required for glide alone (Weertman, 1956, 1960, 1963; Mukherjee et al., 1969). These models all point to a constitutive law of the form

$$\dot{\gamma}_3 = \frac{A_s \mu b}{kT} \left(\frac{\sigma_s}{\mu}\right)^n D_{\text{eff}},$$

$$D_{\text{eff}} = D_v + 10 \left(\frac{\sigma_s}{\mu}\right)^2 \frac{a_c D_c}{b^2},$$

(2.4)

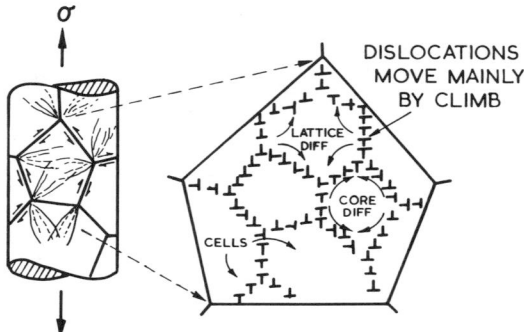

FIG. 3. Power-law creep involving cell formation by climb. Power-law creep limited by glide processes alone is also possible.

where μ is the shear modulus, b is the Burgers vector, D_v is the lattice self-diffusion coefficient, $a_c D_c$ is that for core diffusion times the core area (both vary exponentially with temperature), and A_s is a dimensionless constant. This model-based result has the same form as the empirical equation and is a tolerably good description of steady-state creep data. When D_v is dominant, the creep is lattice diffusion controlled and termed "high-temperature creep"; when the opposite is true, it is termed "low-temperature creep."

At high stresses (above about $10^{-3}\mu$) it is observed that the simple power law breaks down: The measured strain rates are greater than Eq. (2.4) predicts. The process is evidently a transition from climb-controlled to glide-controlled flow (Fig. 4). A number of attempts have been made to describe it in an empirical way (see, for example, Jonas *et al.*, 1969); most lead to a rate equation of the form

$$\dot{\gamma} \propto \exp(\beta\sigma_s),$$

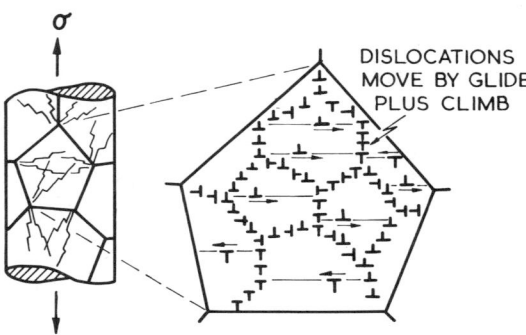

FIG. 4. Power-law breakdown: Glide contributes increasingly to the overall strain rate.

or its generalization (Sellars and Tegart, 1966; Wong and Jonas, 1968)

$$\dot{\gamma} \propto [\sinh(\beta\sigma_s)]^n.$$

In order to obtain exact correspondence of this equation with the power law (2.4), we use the following *rate equation for power-law creep and power-law breakdown*:

$$\dot{\gamma}_6 = \frac{A_s D_{\text{eff}} \mu b}{\alpha^n k T} \left[\sinh\left(\alpha \frac{\sigma_s}{\mu}\right) \right]^n, \qquad (2.5)$$

where α is a dimensionless constant. This equation gives a good design description of high-strain-rate data for a range of metals and alloys (see Frost and Ashby, 1982, for further details).

At very high temperatures ($\geq 0.7 T_M$), power-law creep is often accompanied by repeated waves recrystallization, as shown in Fig. 5 (Hardwick *et al.*, 1961; Nicholls and McCormick, 1970; Hardwick and Tegart, 1961; Stüwe, 1965; Jonas *et al.*, 1969; Luton and Sellars, 1969). Each wave removes or drastically changes the dislocation substructure, allowing a new period of primary creep so that the strain rate (at constant load) oscillates by up to a factor of 10. The phenomenon has been extensively studied in Ni, Cu, Pb, Al, and their alloys (see the previously cited references), usually in torsion tests, but it occurs in any mode of loading. It is known to occur in ceramics such as ice and NaCl, and in both metals and ceramics is most pronounced in very pure samples and least pronounced in heavily alloyed samples containing a dispersion of stable particles.

This *dynamic recrystallization* influences the high-temperature high-stress region of the diagrams shown in Section II,D. When it occurs, the strain rate is higher than that predicted by the steady-state creep equation (2.4), and the

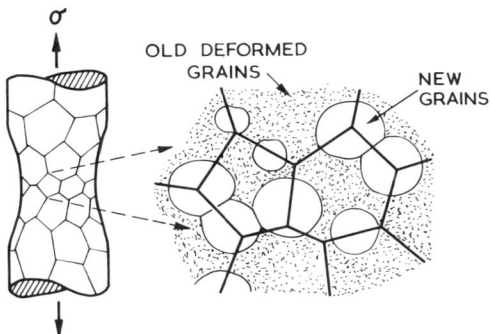

FIG. 5. Dynamic recrystallization replacing deformed by undeformed material, permitting a new wave of primary creep and thus accelerating the creep rate.

apparent activation energy and creep exponent may change also (Jonas et al., 1969). The simplest approach in modeling it is to suppose that repeated waves of primary creep occur with a frequency that depends on the temperature and strain rate, and that each wave follows a primary-creep law with the same activation energy and stress dependence as steady-state creep. But a satisfactory model, even at this level, is not yet available.

In Section II,D we have adopted an empirical approach. The maps show a shaded region at high temperatures, labeled "dynamic recrystallization." It is not based on a rate equation [the contours in this region are derived from Eq. (2.4)], but merely shows the field in which dynamic recrystallization has been observed, or in which (by analogy with very similar materials) it would be expected.

D. Diffusional Flow

At very low stress, dislocation motion stops or becomes so slow that it can be ignored. Creep continues by *diffusional flow*, the diffusive motion of single atoms or ions from sources on grain boundaries that carry a compressive load to sinks on those which carry a tensile one (Fig. 6). Experiments show that the process is diffusion controlled and that the strain rate increases roughly linearly with stress. This mechanism can be modeled well, with the result (Nabarro, 1948; Herring, 1950; Coble, 1963; Lifshitz, 1963; Raj and

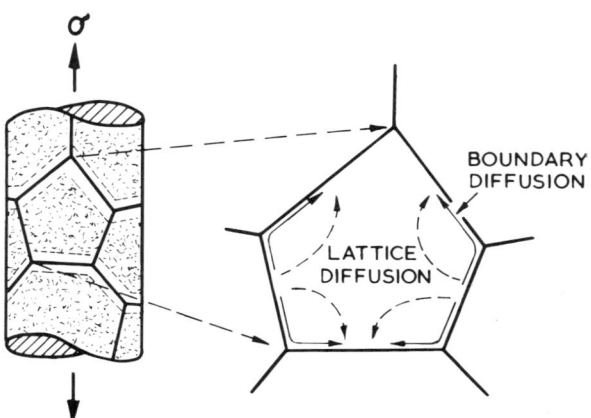

FIG. 6. Diffusional flow by diffusional transport through and around the grains. The strain rate may be limited by the rate of diffusion or by that of an interface reaction.

Ashby, 1971)

$$\dot{\gamma}_4 = \frac{42\sigma_s \Omega}{kT d^2} D_{\text{eff}},$$

where (2.6)

$$D_{\text{eff}} = D_v + \frac{\pi \delta D_b}{d}.$$

Here d is the grain size and δD_b is the grain boundary diffusion coefficient times the boundary thickness. When D_v is dominant, the process is called "lattice diffusion flow" or "Nabarro–Herring Creep"; when the opposite is true, it is called "boundary diffusion flow" or "Coble creep."

This equation is an oversimplification; it neglects the kinetics involved in detaching vacancies from grain boundary sites and reattaching them. This kinetic process can become rate limiting in alloys, particularly those containing a finely dispersed second phase (Ashby, 1969; 1972; Burton, 1972). But pure metals are well described by Eq. (2.6), and it is used, in this form, to construct the maps of Section II,E.

E. Other Mechanisms

Under certain circumstances, other mechanisms of deformation can appear. *Twinning* is an important deformation mechanism at low temperatures in hcp and bcc metals. It is a variety of dislocation glide (Section II,B) involving the motion of partial, instead of complete, dislocations. The kinetics of the process, however, often indicate that nucleation, not propagation, determines the rate of flow, and rate equations describing the process are too uncertain to be of practical value. Twinning is incorporated into the diagrams shown further in an empirical way, by simply indicating where it is observed experimentally.

There is experimental evidence that at sufficiently low stresses and high temperatures materials creep in an anomalous way. The effect was first observed in aluminium by Harper and Dorn (1957) and Harper et al. (1958): They observed linear viscous creep at stresses below $5 \times 10^{-6}\mu$, but at rates much higher than that possible by diffusional flow. This *Harper–Dorn creep* has been observed, too, in lead and tin by Mohamed et al. (1973). The most plausible explanation is that of climb-controlled creep under conditions such that the dislocation density does not change with stress. It can be incorporated into deformation mechanism maps (Frost and Ashby, 1982), but it does not appear on those shown here.

At very high strain rates ($>10^4$ sec^{-1}) *adiabatic heating* can lead to localization of slip and catastrophic local shear. This too can be analyzed and incorporated into the maps, but those shown here do not include it.

III. Deformation Mechanism Maps

A. Construction of the Maps

Figure 7 is a deformation map for commercially pure nickel. It shows, on axes of normalized shear stress and temperature, the fields of dominance of each mechanism of deformation: low-temperature plasticity, power-law

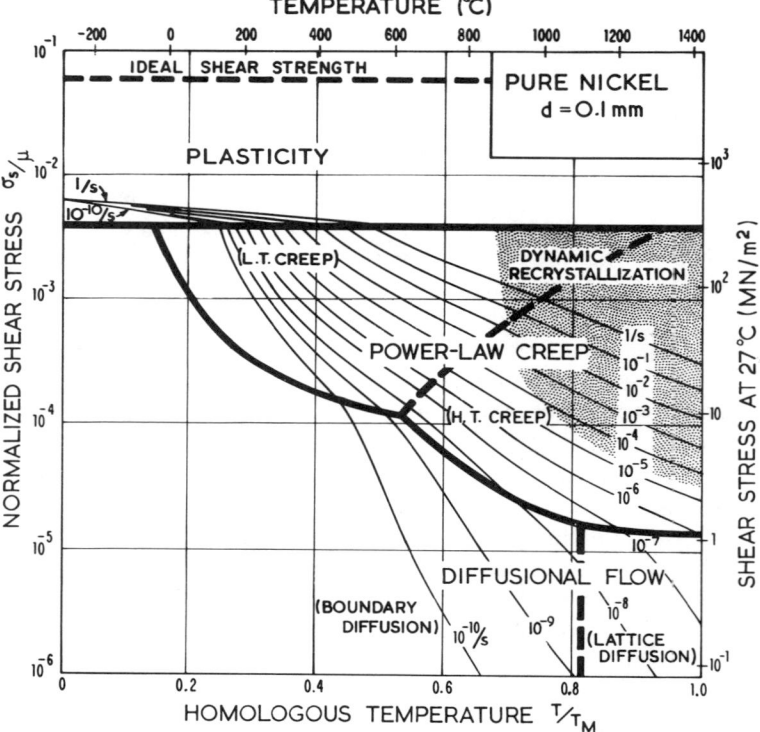

FIG. 7. A deformation mechanism map for nominally pure nickel with a grain size of 0.1 mm. The equations and data used to construct it are described in Section II.

creep, diffusional flow, and so forth. For all but the fcc metals the plasticity field shows two sections, one in which the lattice resistance determines the flow stress, the other in which obstacles of other types become dominant; here the first is absent. The power-law creep field also is subdivided into a section in which lattice diffusion controls the creep rate and one in which core diffusion is more important; the diffusional flow field is similarly subdivided into a section of lattice diffusion control and one of boundary diffusion control.

The maps are based on an exhaustive analysis of previously published work (Ashby, 1972a; Frost, 1973; Ashby and Frost, 1975; Frost and Ashby, 1976; 1982). They are constructed as follows.

First, data for the material properties are gathered: lattice parameter, molecular volume, Burgers vector, moduli and their temperature dependence, and lattice, boundary, and core diffusion coefficients (if they exist). It is often necessary to replot data for moduli and diffusion coefficients in order to make a sensible choice of the material properties, μ_0, $d\mu/dT$, D_{0v}, Q_v, δD_{0b}, Q_b, etc.

Second, data for the hardness, low-temperature yield, and creep parameters are gathered, flow stress as a function of strain rate and temperature, and creep rate as a function of temperature and stress. These data are plotted on the axes used for the maps themselves: $\log_{10}(\sigma_s/\mu)$ and T/T_M. Each datum is plotted as a symbol that identifies its source and is labeled with the value of the third macroscopic variable, $\log_{10} \dot{\gamma}$.

Third, an initial estimate is made of the material properties describing glide (ΔF, $\hat{\tau}$, ΔF_k, $\hat{\tau}_p$, etc.) and creep (n, A, etc.) by fitting Eqs. (2.2)–(2.6) to these data plots. These initial values for the material properties are used to construct a trial map. This is best done by a simple computer program that steps incrementally through the range of σ_s/μ and T/T_M, evaluating and summing the rate equations at each step, and which plots the results in the form shown in Fig. 7. All the maps (regardless of the choice of axes) are divided into *fields*, within each of which a given mechanism is dominant. The *field boundaries* are the loci of points at which two mechanisms contribute equally to the overall strain rate, and are computed by equating pairs (or groups) of rate equations and solving for stress as a function of temperature, as shown in Fig. 8. Superimposed on the fields are *contours of constant strain rate*, obtained by summing the rate equations to give the total strain rate $\dot{\gamma}_{net}$ and plotting the loci of points for which $\dot{\gamma}_{net}$ has given constant values (Fig. 7).

Fourth, the data plots are superimposed on the trial map, allowing the data to be divided into blocks according to the dominant flow mechanism. It is then possible to make a detailed comparison between each block of data and the appropriate rate equation. The material properties are adjusted to

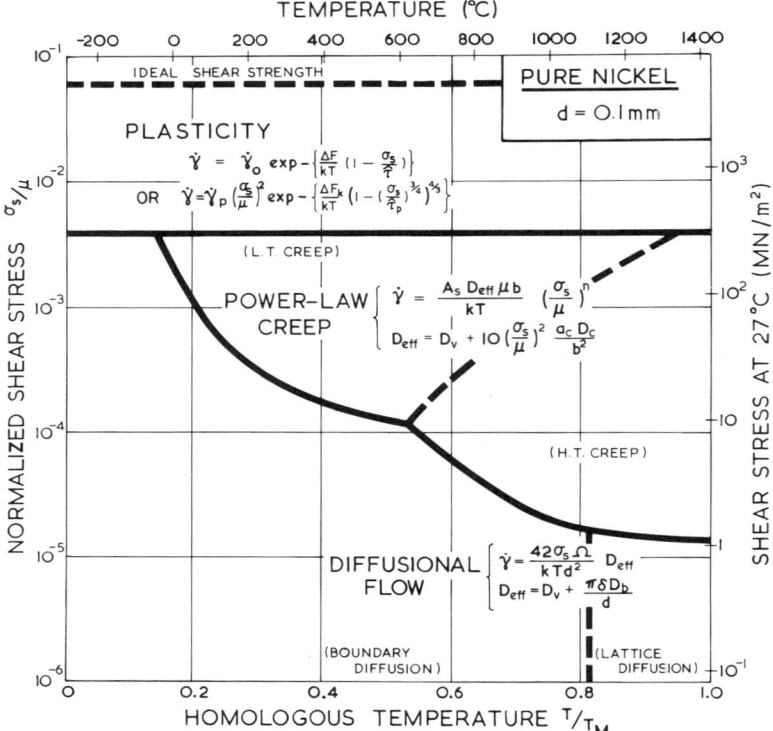

FIG. 8. The construction of a deformation mechanism map. The field boundaries are the loci of points at which two mechanisms (or combination of mechanisms) have equal rates.

give the best fit between theory and experimental data. New maps are now computed, the comparison repeated, and further adjustments made if necessary. It cannot be emphasized too strongly that this detailed comparison with data over the entire range of T, σ_s, and $\dot{\gamma}$ is essential if the final map is to have any real value.

Finally, the adjusted data are tabulated and the maps redrawn, either with data plotted on them or on separate data plots. The results are illustrated by the following subsections.

B. Deformation Maps for Metals and Ceramics

This section describes deformation maps for eight polycrystalline metals and ceramics. Each was chosen because it typifies the class of materials to which it belongs. Further maps and a detailed discussion of the data for each can be found in Frost and Ashby (1982).

1. The fcc Metals: Commercially Pure Nickel and a Nimonic

The fcc metals, above all others, are tough and ductile, even at temperatures close to 0 K. When annealed, they are soft and easily worked, but their capacity for work hardening is sufficiently large so that in the cold-worked state they have useful strength. Their capacity for alloying is also great. This allows the fabrication of a range of materials, such as the aluminium alloys, the beryllium coppers, the stainless steels, and the nickel-based superalloys, which have remarkable yield and creep strengths. The map for nickle (Fig. 7) typifies the commercially pure fcc metals that for a nimonic illustrates some of the effects of alloying.

The fcc metals are remarkable in having an extremely low lattice resistance (certainly less than $10^{-5}\mu$); as a result their yield strength is determined by the density of discrete obstacles or defects they contain.[†] When pure, it is the density and arrangement of dislocations, and thus the state of work hardening, that determines the flow stress. The map shown here (Fig. 7) is for heavily worked nickel. The low-temperature plasticity field is based on data for the tensile strength (Jenkins et al., 1954).

Above about $0.3T_M$, the fcc metals start to creep. Diffusion (which is thought to control creep in these metals) is slower in the fcc structure than in the more open bcc structure; this is reflected in lower creep rates at the same values of σ_s/μ and T/T_M. The creep field is subdivided into a region of low-temperature core-diffusion-controlled creep and a region of high-temperature lattice-diffusion-controlled creep. The power law breaks down for fcc metals near $\sigma_s/\mu = 10^{-3}$, corresponding to a value of α of about 10^3. The power-law creep field of Fig. 7 is based on the references listed on the data plot shown as Fig. 9.

Diffusional flow appears at high temperature and low stress. The field is subdivided into a region in which boundary diffusion controls the creep rate and one in which lattice diffusion is more important. When the grain size is large, this field may be replaced by one of Harper–Dorn creep. The shaded field of dynamic recrystallization is based on the observations reported and reviewed by Ashby et al. (1979).

The nickel-based superalloys (Betteridge and Heslop, 1974) combine a concentrated solid solution (typically of Cr and Co) with a large volume fraction of the precipitate $Ni_3(Al, Ti)$ and a lesser volume fraction of carbides,

---

[†] The refractory fcc metals Ir and Rh are exceptional in exhibiting a large lattice resistance (roughly $10^{-2}\mu$ at 0 K) and failing at low temperature by cleavage on a (100) plane.

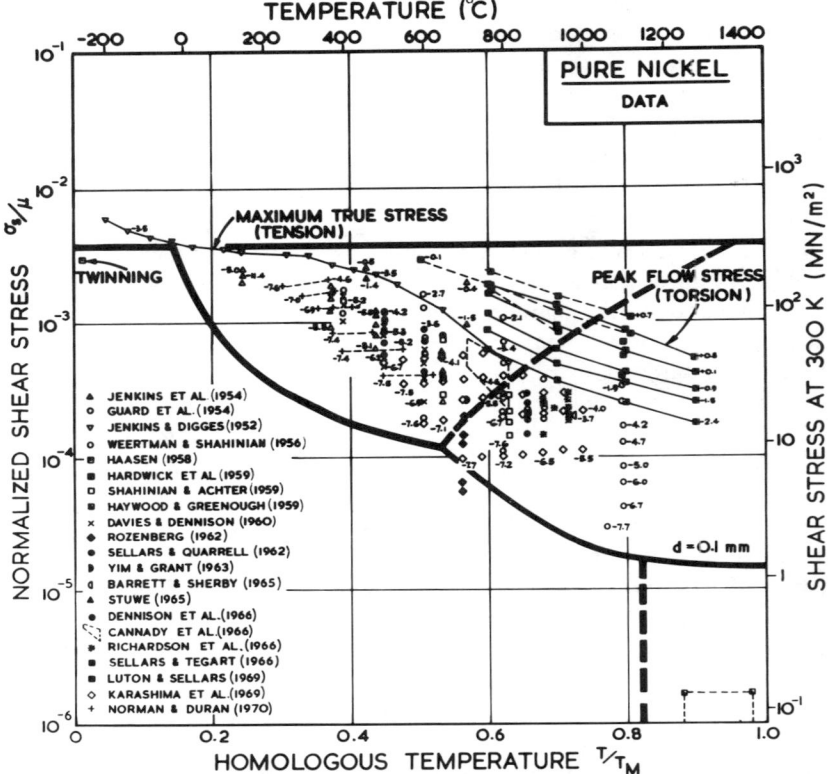

FIG. 9. Data plot for nominally pure nickel. The data are labelled with $\log_{10} \dot{\gamma}$.

much of it in grain boundaries. The map shown in Fig. 10 is largely based on data for MAR-M200 (with the nominal composition given in Table II), but is intended mainly to illustrate the major features of the superalloys generally.

Alloying lowers the melting point substantially (to 1600 K), and raises the shear modulus slightly (to 80 GN/m²) compared with pure nickel. The dispersion of $Ni_3(Al, Ti)$ superimposed on the heavy solid solution strengthening of W and Cr give MAR-M200 (and alloys like it) a yield strength comparable with the ultimate strength of pure nickel, although it is less dependent on temperature (Ver Snyder and Piearcey, 1966). The alloying also greatly reduces the rate of power-law creep, the field of which is based on the data of Webster and Piearcey (1967), Kear and Piearcey (1967), and Leverant and Kear (1970).

These differences can be seen by comparing the map for the superalloy with that for pure nickel. Precipitation strengthening and solution hardening

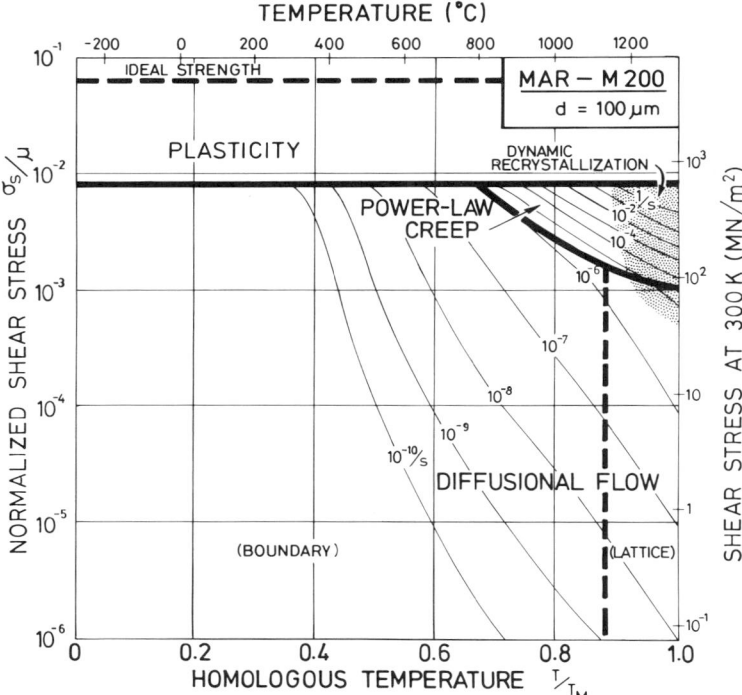

FIG. 10. A deformation map for Mar-M200, a commercial nickel-based superalloy, with a grain-size of 10 mm.

TABLE II

Nominal Composition of MAR-M200 (wt. %)

Al	Ti	W	Cr	Nb	Co	C	B	Zr	Ni
5.0	2.0	12.5	9.0	1.0	10.0	0.15	0.015	0.05	Bal

have raised the yield line and drastically reduced the size of the power-law creep field. They may also change the rate of diffusional flow, though, since there is no experimental evidence for this in MAR-M200, we have made the assumption that it occurs at the same rate as it would in more or less pure nickel.

We are not aware of observations of dynamic recrystallization in MAR-M200, but above 1000°C the γ' phase dissolves, and at a slightly higher temperature the grain boundary carbides do so also ($M_{23}C_6$ at 1040–1095°C,

M_7C_3 at 1095–1150°C; Betteridge and Heslop, 1974). This means that above $0.9T_M$ the alloy is a solid solution, and if data for solid solutions can be used as a guide, we would expect dynamic recrystallization. The shaded field is based on this reasoning.

2. *The bcc Transition Metals: Tungsten*

The refractory bcc metals have high melting points and moduli. Many of their applications are specialized ones that exploit these properties: Tungsten lamp filaments operate at up to 2800°C, and molybdenum furnace windings to 2000°C, for example. They are extensively used as alloying elements in steels and superalloys, raising not only the yield and creep strengths, but the moduli too.

Figure 11 shows a map for tungsten that typifies the bcc metals. Like those for the fcc metals, it shows three principal fields: low-temperature plasticity,

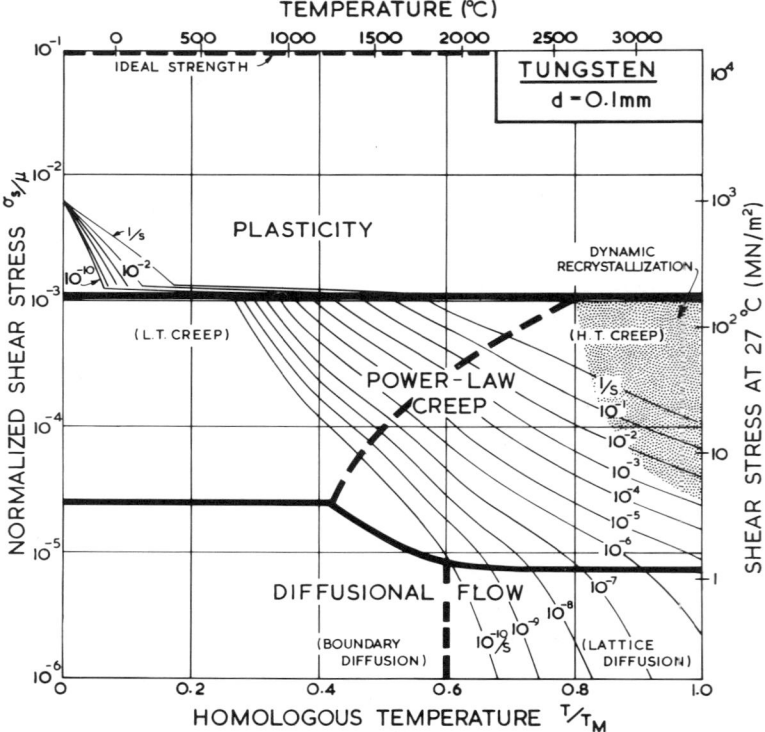

FIG. 11. A deformation map for tungsten with a grain size of 0.1 mm.

power-law creep, and diffusional flow. The principal difference appears at low temperatures (below about $0.15T_M$), where the bcc metals exhibit a yield stress that rises rapidly with decreasing temperature because of a lattice resistance. As a result, the flow stress of bcc metals extrapolates, at 0 K, to a value close to $10^{-2}\mu$, independent of obstacle content for all but extreme states of work hardening. By contrast, the high-temperature strength of the bcc metals is lower than that of the fcc metals because diffusion in the more open bcc lattice is faster than in the close-packed fcc one.

The map is fitted to the data plotted in Fig. 12. The plasticity field for tungsten is based on the polycrystalline yield data of Raffo (1969), which generally agree with the single-crystal critical resolved shear stresses of Koo (1963) and Argon and Maloof (1966), after correction by a Taylor factor.

The high-temperature creep of tungsten has been reviewed by Robinson and Sherby (1969), who demonstrated that most of the available data can be divided into high-temperature creep above about 2000°C and low-temperature creep below. The high-temperature data are from Flagella (1967) (for wrought arc-cast tungsten) and King and Sell (1965). These data show

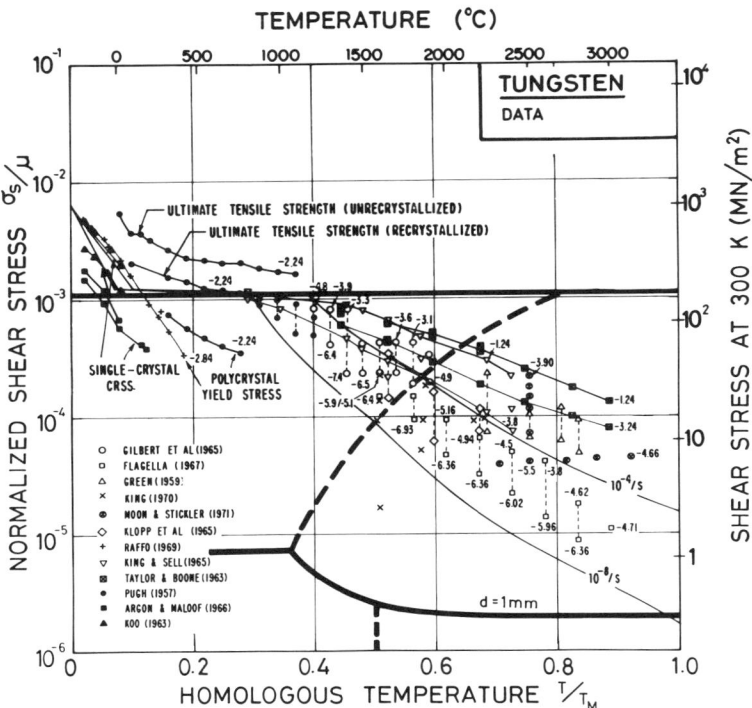

FIG. 12. Data plot for tungsten. Each data point is labeled with $\log_{10} \dot{\gamma}$.

faster creep rates than those of Green (1959) and Flagella (1967) for powder metallurgy tungsten—a common finding with bcc metals. The low-temperature creep region is represented by the data of Gilbert et al. (1965).

The field of dynamic recrystallization is based on the observations of Glasier et al. (1959) and Brodrick and Fritch (1964), who tested tungsten up to $0.998 T_M$.

3. The Hexagonal Metals: Magnesium

The hexagonal metals form the basis of a range of industrial alloys of which those using magnesium, zinc, and titanium are the most familiar. These metals and alloys resemble the fcc metals in many regards, but their structure, though close packed, is of lower symmetry. This necessitates new modes of slip, some of which are opposed by considerable lattice resistance.

The map for magnesium (Fig. 13), which typifies the hcp metals, is very much like that for nickel (Fig. 7). The fields of plasticity, power-law creep, and diffusional flow are of the same general size, though they are displaced to slightly higher normalized stresses. The hcp metals differ from the fcc ones in two regards. First, twinning is much more prevalent, becoming an impor-

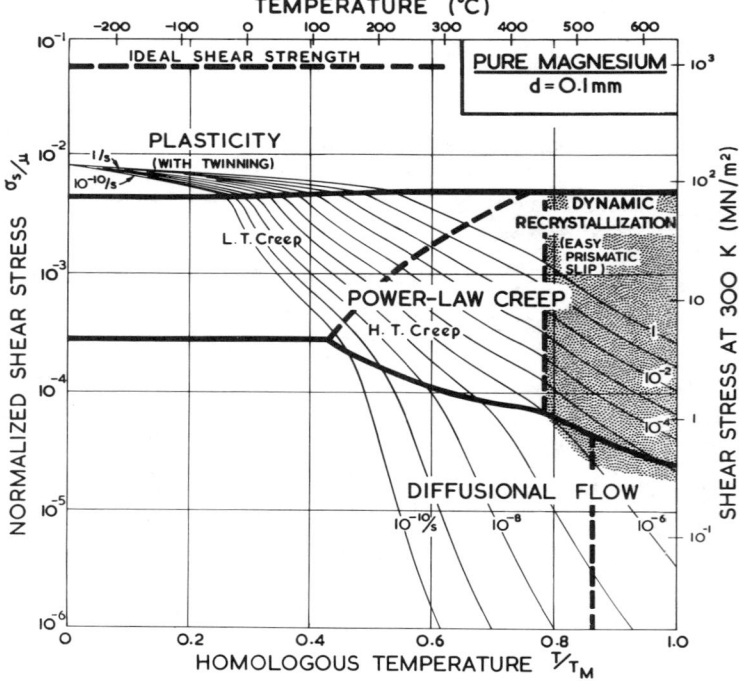

FIG. 13. A deformation map for magnesium with a grain size of 0.1 mm.

tant mode of deformation at low temperatures; second, the low symmetry means that at least two different classes of slip system must operate for polycrystal plasticity. The metals considered here show easy basal slip, with no measurable lattice resistance, but the associated prismatic or pyramidal slip, or twinning, required for polycrystal plasticity is opposed by a large lattice resistance (of general magnitude $5 \times 10^{-3}\mu$). This resistance falls with increasing temperature (like that of the bcc metals) until it is obscured by the obstacle strengthening associated with work hardening. We present here a map for heavily work-hardened magnesium, such that work hardening obscures the lattice resistance at all temperatures.

The map is fitted to the data shown in Fig. 14. The low-temperature plasticity field is based on data of Hauser et al. (1956), Toaz and Ripling (1956), and Flynn et al. (1961).

The field of power-law creep is based on data of Jones and Harris (1963), Roberts (1953), and Tegart (1961). Above $0.8 T_M$ there is evidence for both Mg and Zn for a creep field of anomalously high activation energy, generally explained as resulting from a rapid drop in the resistance to shear of nonbasal systems, though it may simply reflect the onset of dynamic recrystallization.

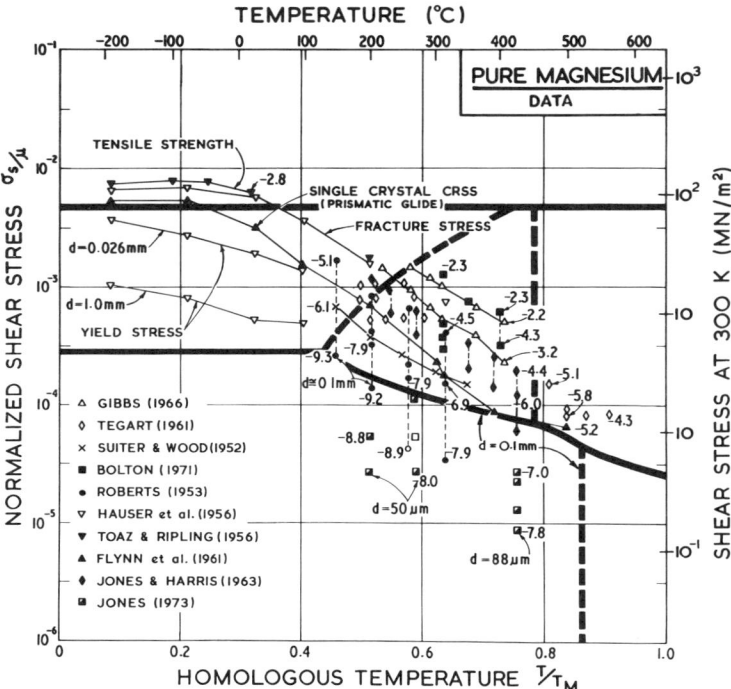

FIG. 14. Data plot for magnesium. Each data point is labeled with $\log_{10} \dot{\gamma}$.

The mechanism is not yet clear, but since it results in important changes in strength, we have included it in an empirical way by fitting separate power laws to data above and below this temperature.

The diffusional flow field is fitted to data of Jones (1973). That of dynamic recrystallization is based on the observations of Gandhi and Ashby (1979).

4. *The Alkali Halides: NaCl (Halite)*

Sodium chloride (Fig. 15) typifies a large group of alkali halides with the rock-salt structure. More than 200 compounds with this structure are known. It occurs most commonly among alkali-metal halides, the oxides of divalent transition metals, the alkaline earth oxides and calcogenides, and the transition metal carbides, nitrides, and hydrides. But at least four mechanically distinguishable subgroups with this structure exist. Two are examined in this section: the alkali halides and the simple oxides (Section B,5). The

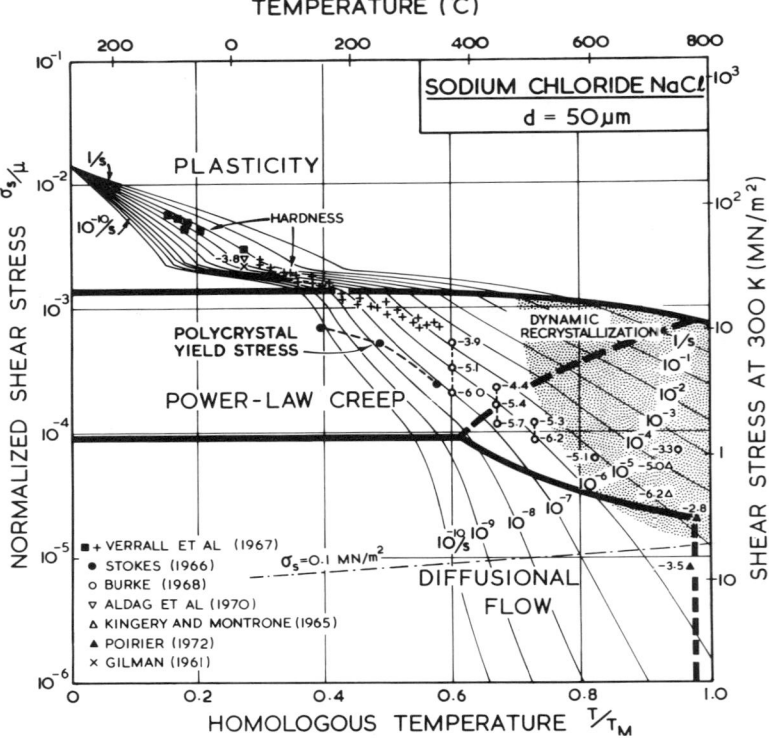

FIG. 15. A deformation map for sodium chloride (NaCl) with a grain size of 50 μm. It is fitted to the data shown.

bonding of each subgroup is different, so that, despite the similarity in structure, the mechanical strengths differ.

Like other polycrystalline materials, their plastic deformation involves a number of distinct mechanisms: low-temperature plasticity, power-law creep, and diffusional flow. All three have been observed in alkali halides, among which LiF and NaCl are the most extensively studied.

Impurities influence the defect structure of ionic solids (and thus the rates of the mechanisms) in ways that are more complicated than in metals (see, for example, Greenwood, 1970). First, divalent impurities alter the concentration of vacancies (and other point defects) and thus affect the diffusion coefficients. An impurity such as calcium reduces the concentration of vacancies on the chlorine sublattice, and thus its rate of diffusion. Since chlorine usually controls the rate of mass transport in the alkali halides, impurities of this sort can greatly slow the rate of power-law creep and diffusional flow. A second effect of impurities is to alter the electrical charge carried by dislocations and grain boundaries (see, for example, Kliewer and Koehler, 1965). This seems to enhance diffusion at or near the core and may explain certain aspects of creep in NaCl. There is a third effect of impurities: Divalent solutes cause rapid solution hardening, raising the yield stress in the dislocation glide regime. Johnston (1962) and Guiu and Langdon (1974) have documented the effect, which is expected in all alkali halides.

For all these reasons, there is a great deal of scatter in published data for alkali halides. We have fitted the equations to the available data, giving the most weight to data from pure polycrystalline specimens.

The plasticity field shows the major influence of the lattice resistance in determining the strength of sodium chloride. It is based on hardness data of Gilman (1961), Verrall *et al.* (1977) and Brown and Ashby (1981). The field of power-law creep is better documented: It is fitted to the data of Stokes (1966), Burke (1968), Poirier (1972), and Heard (1972), who, however, used comparatively impure salt. Evidence for diffusional flow was given by Kingery and Montrone (1965).

Despite the scatter, a consistent picture of the deformation mechanisms of alkali halides emerges from a study of this sort (Frost and Ashby, 1982) and Fig. 15 can be regarded as broadly typifying them.

5. *The Simple Oxides: MgO (Magnesia)*

A number of simple oxides crystallize with the rock-salt structure; they include MgO, CaO, CdO, MnO, FeO, NiO, CoO, SrO, and BaO. Some, like magnesia (Fig. 16), are stable as the stoichiometric oxide MgO. Others, like cobalt monoxide, exist only as the hyper stoichiometric oxide $Co_{1-x}O$,

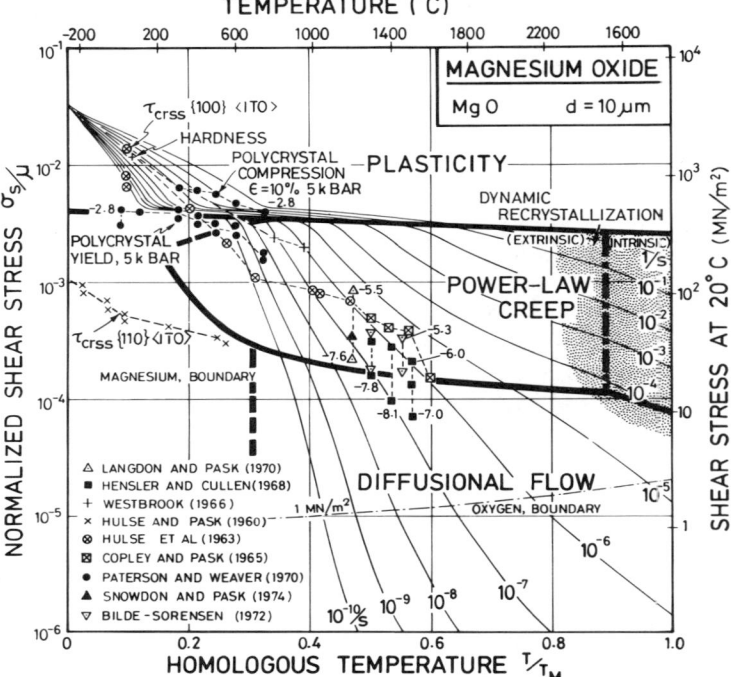

FIG. 16. A deformation map for magnesium oxide (MgO) with a grain size of 10 μm. It is fitted to the data shown.

always oxygen rich because of the presence of Co^{3+} ions. In all, the oxygens are the larger ions. They are packed in a fcc array, with the metal ions occupying the octahedral interstices.

The bonding in these oxides is largely ionic. Like the alkali halides, they slip most easily on $\langle 1\bar{1}0 \rangle$ $\{110\}$, though this provides only two independent slip systems. Polycrystal plasticity is made possible by slip on $\langle 0\bar{1}1 \rangle$ $\{100\}$ (Parker, 1961; Hulse and Pask, 1960; Hulse et al., 1963; Day and Stokes, 1966). But despite this similarity with the alkali halides, the oxides form a distinct mechanical group: Their melting points and moduli are much higher; their normalized strengths at 0 K are a little higher (about $\mu/30$ compared with $\mu/50$); and their low-temperature strength is retained to somewhat higher homologous temperatures. The plasticity field of Fig. 15 is based on the polycrystal yield data of Paterson and Weaver (1970) and the hardness measurements of Westbrook (1966).

Above about $0.4 T_M$ the oxides start to creep. Once they do, their strength falls as fast or faster than that of the alkali halides. When creep of a compound is diffusion controlled, both components—oxygen and metal in these oxides—must move. The creep rate is then controlled by a weighted mean of the diffusivities of the components. But diffusion may be intrinsic or extrinsic, and alternative diffusion paths (lattice and grain boundary paths, for instance) may be available. As a general rule, the creep rate is determined by the fastest path of the slowest species, and this may change with temperature and grain size. The result is a proliferation of different diffusional flow fields—six or more are possible (Stocker and Ashby, 1973). The power-law creep field shown in Fig. 16 is based on the data of Langdon and Pask (1970), Copley and Pask (1965), and Hensler and Cullen (1968). Hensler and Cullen's (1968) measurements give evidence for diffusional flow and have been used to position the field boundary (Frost and Ashby, (1982)).

6. The Refractory Oxides: Al_2O_3 (α-Alumina)

Three common oxides have the α-alumina structure: corundum or sapphire (Al_2O_3), chromium sesquioxide (Cr_2O_3), and hematite (Fe_2O_3). The structure, if described by its smallest unit cell, is rhombohedral, but it is convenient to think of it in terms of a larger hexagonal unit cell. The oxygen ions are packed in a close-packed hexagonal arrangement with metal ions in two-thirds of the octahedral sites. The unoccupied octahedral sites are ordered within each close-packed layer and alternate between layers, repeating every third layer. The hexagonal unit cell contains six oxygen layers.

These oxides are generally harder and more refractory than the rock-salt structured oxides, retaining their strength to higher temperatures. Alumina is used as a structural ceramic, as well as an abrasive, and a coating for cutting tools. Chromium sesquioxide is perhaps most important as a surface layer on stainless steels and nickel-based alloys.

By comparing the map for α-alumina (Fig. 17) with that for magnesia (Fig. 16), one sees that the lattice resistance persists to a much higher temperature ($0.5T_M$ instead of $0.3T_M$) and that creep, at the same values of σ_s/μ and T/T_M, is nearly 10 times slower. Part of this difference is caused by the structure of α-alumina, which imposes restrictions on slip that do not exist in simple hexagonal close packing of ions. Because of the ordered filling of the octahedral sites, certain shears create stacking faults; among these are the simple basal shear and simple twinning, both of which are made much more difficult. The easy slip systems are still those in the basal plane (with slip in the harder prismatic and pyramidal systems appearing at high temperatures), but the overall strength is greater than that of the cubic oxides.

Slip and twinning in alumina were discussed by Kronberg (1957). Because of the complicated packing of aluminium ions, the Burgers vector is large, 2.73 times larger than the oxygen ion spacing. Slip is easiest on the basal plane, but is observed (and required, for polycrystal plasticity) on the prism

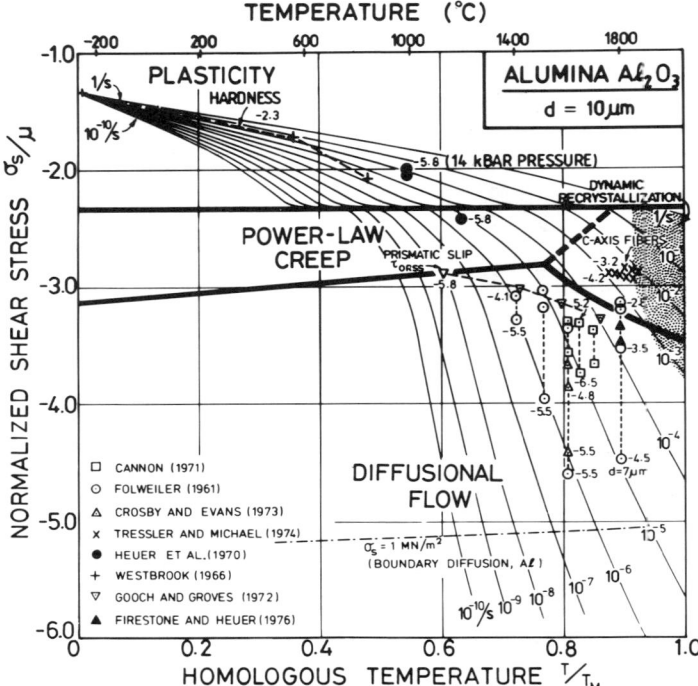

Fig. 17. A deformation map for alumina (Al$_2$O$_3$) with a grain size of 10 μm. It is fitted to the data shown.

and pyramidal planes also. The field of plasticity in Fig. 17 is largely based on the hardness data of Westbrook (1966).

Power-law creep in Al$_2$O$_3$ is much slower than in the rock-salt-structured oxides at the same fraction of the melting point. The map is based on the compression data of Cannon (1971) for large-grained (65 μm) material, taking a creep exponent n of 3. These data are generally consistent with other measurements of the critical resolved shear stress for pyramidal and prismatic slip in single crystals (Tressler and Michael, 1974; Gooch and Groves, 1972).

Clear evidence of diffusional flow is provided by the creep studies of Folweiler (1961) and Crosby and Evans (1973).

7. Ice (H$_2$O)

Ice is a remarkably strong solid. When its normalized strength (σ_s/μ) is compared with that of other solids at the same fraction of their melting point (T/T_M), ice is found to be among the strongest and hardest (Fig. 18).

Mechanisms of Deformation and Fracture 145

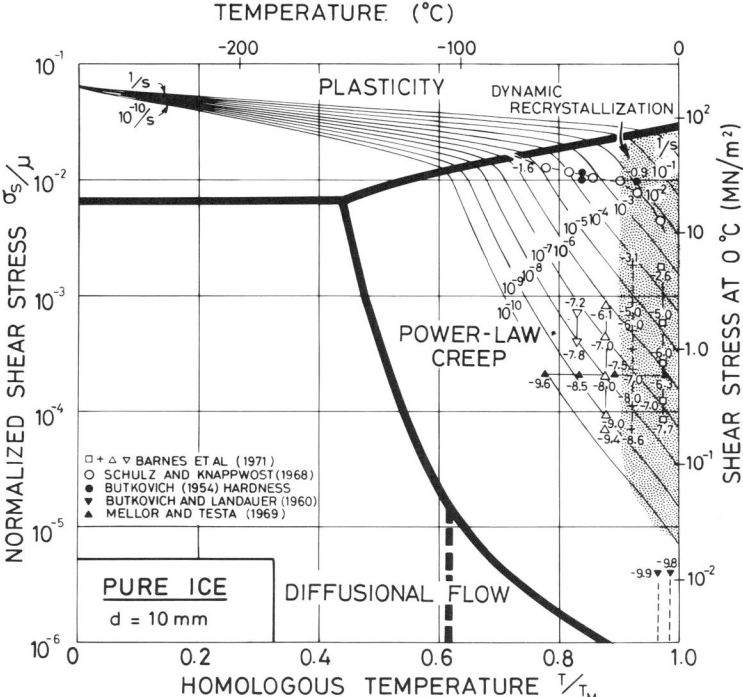

FIG. 18. A deformation map for ice (H_2O) with a grain size of 10 mm. It is fitted to the data shown.

In its mechanical behavior it most closely resembles silicon and germanium, but the unique character if its proton bonds (H bonds) leads to forces opposing dislocation motion that are peculiar to ice.

At atmospheric pressure, ice exists in the hexagonal 1_h form. The motion of dislocations in it is made difficult by a mechanism peculiar to the ice structure (Glen, 1968). The protons that form the bridge between the oxygen ions are asymmetrically placed, occupying one of two possible positions on each bond. The proton arrangement is disordered, even at 0 K, so that the selection of the occupied site on a given bond is random. Shear, with the protons frozen in position on the bonds, creates defects, bonds with no proton on them ("L-type Bjerrum defects") and bonds with two protons ("D-type Bjerrum defects"). With the protons randomly arranged, a defect is, on average, created on every second bond, and the energy F_f of such defects is high—0.64 eV per defective bond (Hobbs, 1974). The result is that the stress required to move a dislocation through ice without rearranging the protons is very large, about $\mu/10$ (Glen, 1968).

At and near absolute zero, it seems probable that dislocations can move only at this high stress—for practical purposes, equal to the ideal strength—and that they create many defects when they do. But at higher temperatures ice deforms plastically at much lower stresses, implying that dislocations can move without creating such large numbers of defects. They can do this if they advance only where the local proton arrangement is favorable, so that no defects (or an acceptably small number of them) are created. The dislocations drift forward with a velocity that depends on the stress and on the kinetics of proton rearrangement.

This drift velocity can be calculated (Whitworth *et al.*, 1976; Frost *et al.*, 1976; Goodman *et al.*, 1981), and from it the strain rate can be estimated. The results of such calculations suggest that near its melting point, ice creeps in a way that approximates a power law. The power law creep field of Fig. 18 is based on the data reviewed by Hobbs (1974), Weertman (1973), and Goodman *et al.* (1981), fitted to a power law (Section II) with $n = 3$.

Observations of the creep of ice in the diffusional flow regime are confused. Bromer and Kingery (1968) found behavior consistent with diffusional flow, but Butkovich and Landauer (1960) and Mellor and Testa (1969) did not. Some of the confusion may be due to the contribution of transient creep (Weertman, 1969): Unless the test is carried to strains exceeding 1%, transients associated with dislocation motion and grain boundary sliding may mask the true diffusional flow; and grains in ice are often large, so that rates of diffusional flow are small. But there seems no reason to suppose that diffusional flow will not occur in ice and that, for the appropriate regime of stress and temperature, it will appear as the dominant mechanism. This predicted regime of dominance is shown on the maps.

C. Mechanisms of Deformation and the Selection of Constitutive Laws

Deformation mechanism maps summarize, in a compact way, information about the plastic properties of a material. They can be used in a number of ways.

They can be used to identify the *mechanism* by which a component or structure deforms in service, thereby identifying the constitutive law or combination of laws that should be used in design. This is of particular importance when materials are used under extreme conditions, where little or no laboratory data are available: It is then common practice to extrapolate the data to the service conditions. If, in doing so, a field boundary is crossed, the extrapolated strain rate may be in error by orders of magnitude.

The maps can give guidance in alloy design and selection. They offer a quick way of determining whether a given metal or alloy is a possible candidate for a given application. And strengthening mechanisms are selective: Alloying, for instance, suppresses power-law creep more than diffusional flow. By identifying the dominant mechanism of flow, the maps help the designer make a rational choice.

Finally, these maps have pedagogical value in allowing a body of detailed information to be presented in a simple way. Each map is a summary of the mechanical behavior of the material and (because of the normalized axes) also gives an approximate description of other related materials with similar bonding, crystal structure, and purity.

IV. Mechanisms of Fracture by General Damage

When a polycrystalline solid is loaded in simple tension, it may fail in one of a number of ways. In this section, these fracture mechanisms are described. Their relevance for crack propagation is considered in Section V. Further details of the method developed here were given by Ashby (1977), Ashby *et al.* (1979), and Gandhi and Ashby (1979).

A. The Ideal Fracture Strength

The ideal fracture strength is the stress that will overcome the interatomic forces in a perfect crystal, causing it to separate on a plane normal to the stress axis. The many calculations (Kelly, 1966; Kelly *et al.*, 1967; Macmillan, 1972) are in general agreement; At an adequate level of accuracy, it is

$$\sigma_{\text{IDEAL}} \approx E/10. \tag{4.1}$$

Fracture occurs when the maximum principal stress σ_n exceeds σ_{IDEAL}. The ideal strength appears as a horizontal line, coincident with the top of the diagram, in further figures.

B. Cleavage

Almost all crystalline solids fail by cleavage if the temperature is sufficiently low; fcc metals and their alloys appear to be the only exceptions (but, see the footnote to Section III,B,1).

Brittle solids—those in which plasticity at low temperatures is limited—generally contain small cracks because of abrasion or corrosion or as growth defects (Fig. 19). Such cracks may propagate at a stress that is lower than that required for slip on any slip system. Fracture then occurs without

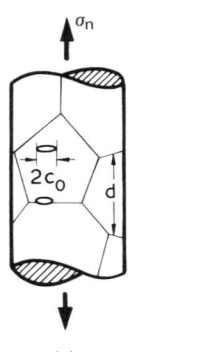
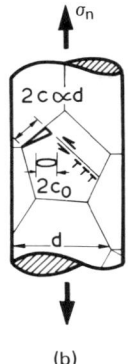

Fig. 19. (a) Brittle solids containing incipient cracks of length $2c_0$. (b) Cracks generated by slip. Their length, $2c$, often scales as the grain size. Either sort of crack may propagate to give a cleavage fracture.

general plasticity at the nominal stress

$$\sigma_f \approx (EG_c/\pi c)^{1/2}, \tag{4.2}$$

where $2c$ is the preexisting crack length, E is Young's modulus, and G_c is the toughness. Since the stress is less than the yield stress of even the softest slip system, no general plasticity is possible (though there may be local plasticity at the crack tip). We call this field *cleavage 1*. Within it, the strength of the solid is determined by the largest crack it contains.

If preexisting cracks are small or absent, then the stress can reach the level required to initiate slip or twinning (Fig. 19). Either can generate internal stresses that can nucleate cracks (McMahon and Cohen, 1965). We call this regime of slip or twin-nucleated cracking *cleavage 2* to distinguish it from cleavage from preexisting flaws (*cleavage 1*). Cracks nucleated in this way generally have a length that is proportional to the grain size d, because this is the wavelength of the internal stresses. If the twinning stress or the flow stress on the easiest slip system (σ_y) exceeds the value σ^* defined by

$$\sigma^* \approx (EG_c/\pi d)^{1/2}, \tag{4.3}$$

a crack propagates as soon as it forms and cleavage fracture occurs at the microyield stress σ_y. But if $\sigma^* > \sigma_y$, then cracks nucleated by slip or twinning will not immediately propagate and the stress will have to be raised further, leading to fracture within the *cleavage 2* field. The field can be regarded as one of cleavage preceded by microplasticity. The fracture is brittle, with negligible ductility ($<1\%$).

As the temperature is raised, the flow stress falls until general plasticity or creep precedes failure—which may, nevertheless, occur by cleavage. We call this regime *cleavage 3* to distinguish it from the regimes of completely brittle fracture (cleavages 1 and 2). Within this field, substantial plastic strain (1–10%) precedes fracture, and this plasticity is sufficient to blunt small preexisting cracks, effectively raising G_c. General plasticity or (often) grain boundary sliding then nucleates a larger grain boundary crack or causes a preexisting crack to grow in a stable manner, until its increased length, coupled with the higher stress caused by work hardening, causes it to propagate unstably as a cleavage crack.

In many metals and ceramics there is a delicate balance between the stress required to cause a crack to propagate by cleavage and that required to cause brittle separation along grain boundaries: Small changes in impurity content, texture, or temperature can cause the crack path to switch from the one to the other; and a mixed transgranular and intergranular fracture is often observed. Both fracture paths are associated with low energy absorption ($G_c = 1$–100 J/m^2). When the fracture path is transgranular (following cleavage planes), we will refer to it as *cleavage (1, 2,* or *3)*; when, instead, it follows the grain boundaries, we shall refer to it as *brittle intergranular fracture (1, 2,* or *3)* or simply *BIF (1, 2,* or *3)*.

C. Ductile Fracture at Low Temperatures

When they do not cleave, polycrystalline solids may fail in a ductile transgranular way. Holes nucleate at inclusions; further plasticity makes them grow; and, when they are large enough, or when the specimen itself becomes mechanically unstable, they coalesce, and the material fractures (Fig. 19).

Consider nucleation first. A hard inclusion (Fig. 20a) disturbs both the elastic and plastic displacement field in a deforming body (Argon *et al.*, 1975; Argon and Im, 1975; Brown and Stobbs, 1971, 1976; Goods and Brown, 1979). The disturbance concentrates stress at the inclusion, and the stress grows as the plastic strain increases, until the inclusion either parts from the matrix or fractures (Goods and Brown, 1979; Le Roy *et al.*, 1980).

Having nucleated in this way, the holes grow until they coalesce to produce a fracture path. Void growth in simple tension is understood, at least approximately (McClintock, 1968; Rice and Tracey, 1969; Le Roy *et al.*, 1980). A spherical void concentrates stress and because of this, it elongates at first at a rate that is about twice that of the specimen itself. As it extends, it becomes ellipsoidal and grows more slowly, until, when very elongated, it extends at the same rate as the specimen itself. Finally, some critical strain

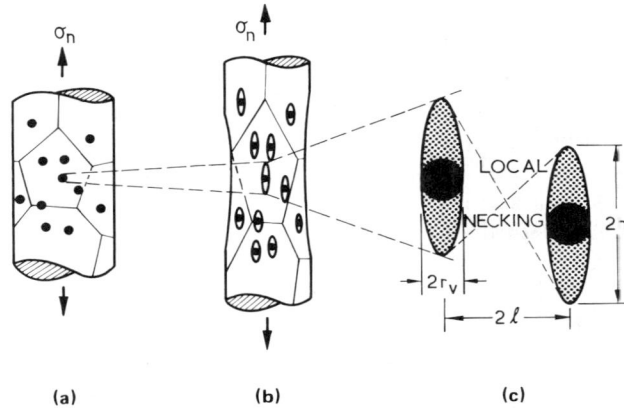

Fig. 20. Ductile and transgranular creep fracture. (a) Ductile fracture and transgranular creep fracture nucleate at inclusions that concentrate stress. (b) The holes elongate as the specimen is extended. (c) They link, causing fracture, when their length $2h$ is about equal to their separation $2l - 2r_v$.

is reached at which plasticity becomes localized, the voids coalesce, and fracture follows with almost no further elongation (Fig. 20b,c).

The physics underlying this localization of flow is not completely clear. Thomason (1968, 1971) and Brown and Embury (1973) both used a critical distance of approach of the growing voids as a criterion for coalescence. Although their models differ in detail, both require that a local slip-line field can be developed between adjacent voids—a condition that is met when the void height $2h$ is about equal to its separation from its neighbors:

$$2h = \alpha(2l - 2r_v),$$

where α is a constant of order unity. If the voids elongate at a rate that, on the average, is faster by a factor of C than the rate of extension of the specimen itself ($1 < C < 2$), then the true strain required for coalescence is simply

$$\varepsilon_\theta = \frac{1}{C} \ln\left[\alpha\left(\frac{2l - 2r_v}{2r_v}\right)\right] \approx \frac{1}{C} \ln\left[\alpha\left(\frac{1}{f_v^{1/2}}\right) - 1\right], \qquad (4.4)$$

where f_v is the volume fraction of inclusions.

It should be recognized that this description fails to include the effects of macroscopic instabilities, necking or the formation of zones of concentrated shear. It is a serious omission, since the onset of necking is known to influence both the nucleation and growth of holes, and localized shear can cause holes to coalesce, giving rise to failure by void sheeting. Yet, as Brown and Embury (1973) and Le Roy et al. (1980) have shown, an equation with the form of Eq. (4.4) reasonably describes data from a number of alloys that fail in a fully ductile way.

Ductile fracture usually follows a transgranular path. But if the density of inclusions or preexisting holes is higher on grain boundaries than it is within the grains, then the fracture path may follow the boundaries, giving rise to a fibrous or ductile intergranular fracture.

D. Transgranular Creep Fracture

At temperatures above $0.3T_M$ metals creep. The flow stress now depends on strain rate, often approximating a power law

$$\dot{\varepsilon} = \dot{\varepsilon}_0(\sigma_s/\sigma_0)^n, \qquad (4.5)$$

where $\dot{\varepsilon}_0$, σ_0, and n are material parameters.

Several new fracture mechanisms now appear. One—the subject of this section—is merely the adaptation to the creep regime of low-temperature ductile fracture: Holes nucleate at inclusions within the grains and grow as the material creeps, until they coalesce to give a transgranular fracture path.

Although the stages of fracture parallel those described for ductile fracture (Fig. 20), their progress is modified in two ways. First, since the material is creeping, the stresses within it tend to be lower than before and nucleation may thereby be postponed to larger strains; second, the strain rate dependence of the flow stress can stabilize flow and thereby postpone the coalescence of voids.

Both ductile fracture and transgranular creep fracture allow comparable large ductilities (20–100% nominal strain to fracture).

E. Intergranular Creep-Controlled Fracture

At lower stresses and longer times to fracture, a transition from a transgranular to an intergranular fracture is observed. Within this new regime, grain boundaries slide, wedge cracks or voids grow on boundaries lying roughly normal to the tensile axis, and a proportionality is sometimes found (Monkman and Grant, 1956) between the time to fracture t_f and the reciprocal of the steady-state creep rate $\dot{\varepsilon}_{ss}$:

$$t_f \dot{\varepsilon}_{ss} = C_{MG}, \qquad (4.6)$$

where C_{MG} is a constant. This suggests that the fracture is still directly controlled by power-law creep, not by some new process (such as boundary diffusion); yet the shapes of the grain boundary voids or cracks imply that local diffusion contributes to their growth.

In certain instances, this behavior can be explained if the nucleation of voids or cracks is controlled by grain boundary sliding, which, in turn, is limited by power-law creep (Crossman and Ashby, 1975). But this explanation is tenable only if the nucleation stage consumes almost all the life of the specimen and the growth stage occupies a small part only of the life. A much more general explanation (Beere and Speight, 1978; Edward and Ashby, 1979; Dyson, 1976, 1978; Cocks and Ashby, 1981) is that the voids, when small, grow by local diffusion, but the rate of diffusion is controlled by power-law creep in the surrounding grains. Void or crack growth is then a result of coupled diffusion and power-law creep, as illustrated in Fig. 21.

Figure 21 shows voids growing by diffusion. But the diffusion field of one cavity extends less than halfway to the next, so that there remains a ligament of material between the voids that must deform if they are to grow. Each void is contained in a cage of creeping material (shaded), the deformation of which controls the rate of cavity growth.

Fracture by one or another sort of coupled diffusion and power-law creep accounts for almost all intergranular creep fractures. But at very low stresses and high temperatures, the diffusion fields of the growing voids overlap, and the containing cage of power-law-creeping material disappears. The resulting regime of pure diffusional growth, which can be regarded as a special limiting case of the mechanism described above, is discussed in Section IV,F.

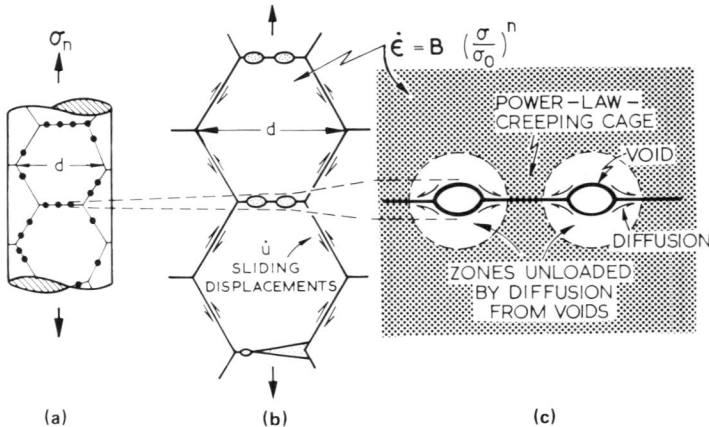

FIG. 21. Intergranular creep-controlled fracture. (a,b) Grain boundary sliding stimulates the nucleation of grain boundary voids. (c) The voids grow by diffusion, but the diffusion fields of neighboring voids do not overlap, so that each void is contained within a cage of power-law-creeping material.

F. Pure Diffusional Fracture

When the temperature is high enough to permit diffusion and the stress so low that power law is negligible, holes on grain boundaries in stressed solids can grow by diffusion along grain boundaries (Hull and Rimmer, 1959; Speight and Harris, 1967; Raj and Ashby, 1975) (Fig. 22). The tensile traction σ_n acting across a grain boundary lowers the chemical potential of atoms or ions there by the amount $\sigma_n \Omega$. If the surface energy of the hole is Γ_s, then a chemical potential difference $\sigma_n \Omega - 2\Gamma_s \Omega / r_h$ exists between sites on the boundary and those on the void surface. Matter flows out of the holes (usually by grain boundary diffusion) and deposits onto the boundary, causing the grains on either side to move apart. Although pure diffusional fracture is encountered only rarely in laboratory tests, there is no reason to doubt that this physical process is a correct description of void growth when power-law creep is absent, as it may often be in ceramics at low stresses.

There are certain complicating factors. Voids do not always maintain their equilibrium near-spherical shape. The faster they are caused to grow, and the larger they become, the greater must be the rate of surface redistribution if equilibrium is to be maintained. The void shape is then determined by the balance between the growth rate and the rate of surface redistribution (Chuang and Rice, 1973; Cocks and Ashby, 1981): voids become flatter and more penny shaped as the stress is raised, until a change of mechanism (to that described in Section IV,D or IV,E) takes place.

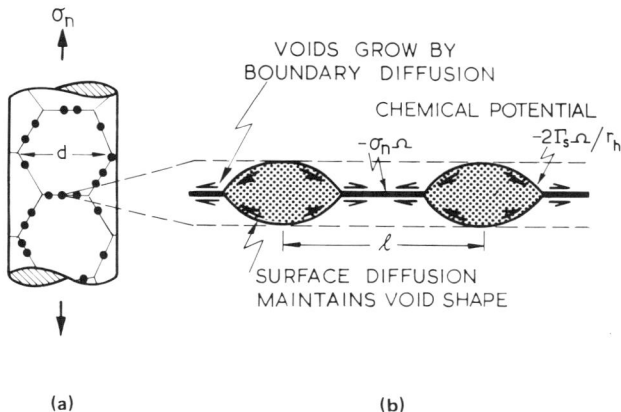

FIG. 22. Diffusional void growth. Voids that lie on boundaries which carry a tensile stress can grow by diffusion. This mechanism is the limiting case of that shown in Fig. 21 when the diffusion fields of the growing voids overlap.

G. Rupture

If no other fracture mechanism intervenes, a material pulled in tension ultimately becomes mechanically unstable. Deformation localizes in a neck or shear band that continues to thin until the cross section has gone to zero, when the material is said to rupture. The strain to rupture depends on the strain at which localization starts and, through this, on the work-hardening characteristics and strain rate sensitivity of the material (Hart, 1967). It also depends on the further strain required to make the initial neck grow until the cross section has become zero; this too depends on the strain rate sensitivity of the material (Burke and Nix, 1975).

Rupture obviously involves large reductions in area. For present purposes, it can be thought of as the tensile failure mode that appears when all other modes are suppressed. It requires either that the nucleation of internal voids be suppressed or (if they nucleate) that they not coalesce. This is commonly observed at high temperatures ($0.7T_M$) when dynamic recovery or recrystallization suppresses void nucleation (Fig. 23).

H. Other Mechanisms of Fracture

Under special circumstances, other mechanisms of fracture can appear. When constrained by a high confining pressure, the mechanisms described above become modified: Ductile fracture (Fig. 20) is replaced by a shear failure, which still involves void growth and linkage, known as "void sheeting"

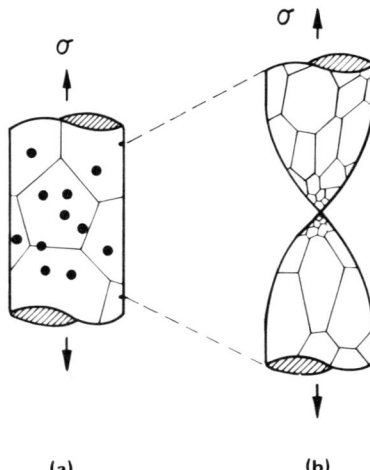

FIG. 23. Rupture at high temperatures. This is generally associated with dynamic recovery or recrystallization.

and cleavage fracture can proceed in a stable manner to produce "cataclastic flow." High strain rates too lead to new or modified modes of fracture: In crystals, adiabatic shear can give rise to catastrophic damage accumulation in a single slip band, and in metallic glasses and polymers, strain-induced dilatation can localize flow and lead to catastrophic failure by local softening or melting. These mechanisms do not appear in the diagrams of Section V.

V. Fracture Mechanism Maps

A. Construction of the Maps

Figure 24 is a fracture map for commercially pure nickel. It shows, on axes of normalized tensile stress and temperature, the fields of dominance of each mechanism of fracture: ductile fracture, transgranular creep fracture,

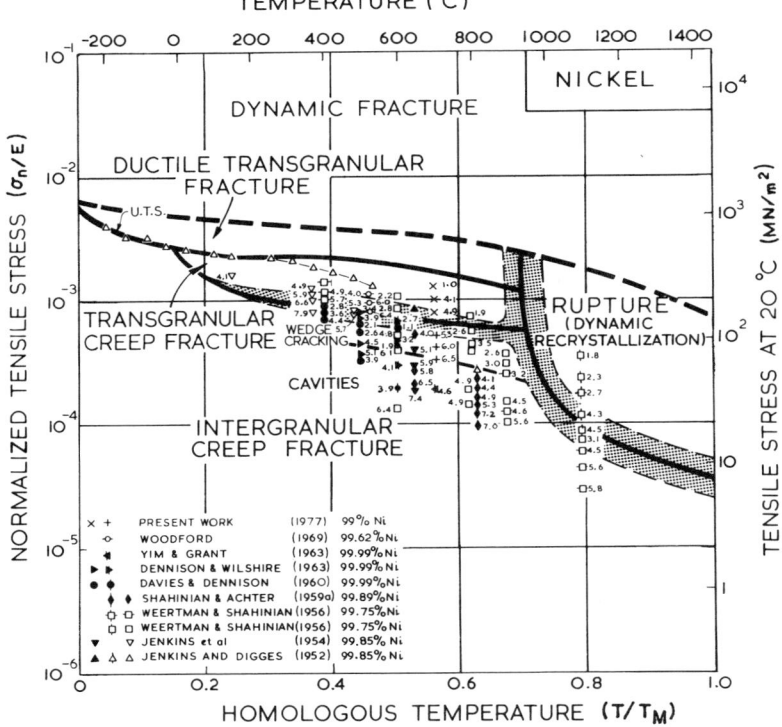

FIG. 24. A fracture map for nickel. Each data point is labeled with $\log_{10} t_f$.

intergranular creep fracture, rupture, and so forth. No distinction is drawn between the two mechanisms of intergranular creep fracture (pure diffusional growth and growth by coupled diffusion and power-law creep) because of the experimental difficulty in distinguishing them; however, a boundary separating the regime of wedge cracking from that of grain boundary growth is shown whenever data make it possible, and this boundary is a rough indicator of the change of mechanism.

The maps are based, in part, on our own experiments and on an exhaustive analysis of previously published work (Ashby, 1977; Ashby et al., 1979; Gandhi and Ashby, 1979). They are constructed as follows.

First, tensile and creep data are assembled for the given material. Then the homologous temperature T/T_M and the normalized tensile stress σ_n/E are tabulated, where σ_n is the nominal stress in a creep or tensile test[†] and E is Young's modulus, adjusted to the temperature of the test, together with the time to fracture t_f, and the strain to fracture, ε_f. The data used to construct the maps shown here are from the sources listed in the figures.

Second, wherever possible, a *mechanism of failure* is assigned to each data point. This is usually based on fractographic observations, though it can sometimes be inferred from sudden changes in the strain or time to fracture. The information is then plotted in the way shown in Fig. 24. The axes are σ_n/E and T/T_M; each data point is plotted as a symbol identifying the investigator and (where possible) the mode of failure, and is labeled with the logarithm of the time to fracture, $\log_{10} t_f$.

Finally, boundaries are drawn separating blocks of data with a given mode of failure. Although the data sometimes conflict and certain points cannot be assigned a mechanism, we have found, in constructing fracture maps for over 30 metals and ceramics, that there is little difficulty in doing this in a way that is consistent with fractographic evidence. The maps are truncated in a stress level corresponding to a strain rate of about 10^4 sec^{-1} (times to failure of about 10^{-4} sec). Above this line, loading and the subsequent deformation are dynamic and involve the propagation of elastic and plastic waves through the material. The region is labeled *dynamic fracture*, although it must be remembered that fast fracture can occur in cracked solids at stress levels that are far lower than this; it is the loading rate which is in the dynamic range.

There are, of course, difficulties and ambiguities in a study of this sort. There is the influence of purity: Strictly, a diagram applies to one purity of metal, or composition of alloy, with one grain size and in one state of heat treatment. Specimen shape is important: Rupture is favored in thin sheet,

---

[†] Creep fracture data refer to tests at constant load; σ_n is the load divided by the initial area of the cross section. Tests at lower temperatures were at constant displacement rate; here σ_n is the tensile strength.

for instance, because the conditions are more nearly those of plane stress. Where possible, we have selected data from round bars, tested in tension, with a ratio of diameter to gauge length of 1:7 or greater. In spite of these difficulties, we have found that the general form of the diagrams is reproducible.

B. Fracture Maps for Metals and Ceramics

This section describes fracture maps for eight polycrystalline metals and ceramics. Each was chosen because it typifies the class of materials to which it belongs. The maps were constructed using data from the sources listed in the figures. Further maps can be found in the articles by Ashby et al. (1979) and Gandhi and Ashby (1979).

1. The fcc Metals: Commercially Pure Nickel and Nimonic

Figure 24 for commercially pure nickel is typical of fracture for many fcc metals and alloys. It shows four mechanism fields. At high stresses and low temperatures, nickel fails by ductile fracture (Section IV,C). As the temperature is raised, the metal starts to creep, and, in the range of temperature and stress indicated in Fig. 24, it fails by transgranular creep fracture (Section IV,D). The fracture mechanism is identical to that of ductile fracture, but the dominant mode of plasticity causing hole growth and linkage has changed: It is power-law creep, not glide plasticity. The boundary between these two fields simply shows where power-law creep becomes the dominant mode of flow.

Below this field lies a field of intergranular creep fracture (Sections IV,E and IV,F). Specimens stressed in this regime fail because holes or wedge cracks nucleate and grow on grain boundaries (often those carrying the largest normal traction) until they link, reducing the cross section of the specimen until plasticity causes the remaining ligaments to fail. Such samples show little or no necking and may fail after very small strains. The transition is a gradual one: Within the shaded band, a mixed mode of fracture—part transgranular, part intergranular—is observed.

As the temperature is raised further, strain-induced grain growth and dynamic recrystallization accompany the creep test (Section IV,G). The result is a broad transition to rupture—necking to a point or chisel edge.

Alloying alters the extent and position of the fields. Stable dispersion suppresses dynamic recrystallization and thereby inhibits rupture. Both dispersion and a solid solution raise the overall strength levels and tend to make the intergranular creep fracture field expand at the expense of the other fields.

This progression is illustrated by the nickel-based superalloys, most of which have evolved from that typified by Nimonic-80 A. Their heat treatment typically consists of a solution treatment at 1080°C followed by aging at 700°C to give a precipitate of γ'-Ni_3(Al, Ti) in a nickel-rich matrix; both phases are solution strengthened by Cr, Co, and often W. In addition, the alloys contain a general dispersion of Ti(C, N) and a grain boundary dispersion of $Cr_{23}C_6$ and other carbides, which help suppress boundary sliding. Minor additions of B, Zr, and Hf improve creep ductility, but for reasons that are not yet completely understood (Holt and Wallace, 1976).

A fracture map for Nimonic-80 A is shown in Fig. 25. Note that all the mechanisms that were identified in pure nickel appear here also, but that they are displaced to much higher stresses and rupture now appears only above the temperature at which the γ' and carbide dispersions have dissolved.

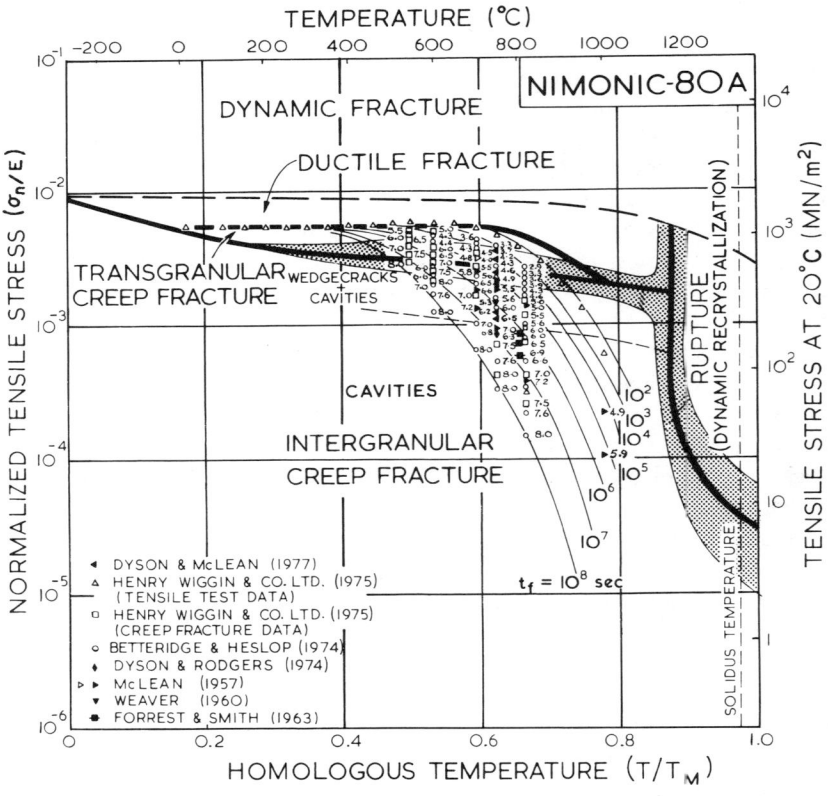

FIG. 25. A fracture map for Nimonic 80A. Each data point is labeled with $\log_{10} t_f$.

2. The bcc Transition Metals: Tungsten

Figure 26, describing commercially pure tungsten, is typical of maps for many bcc refractory metals (Gandhi and Ashby, 1979). If they are precracked or contain flaws, these materials fail at low temperatures by cleavage 1, without any detectable general plasticity (though, of course, there may be plasticity at the crack tip). More usually, twinning or slip nucleates cracks that propagate either by cleavage or along the grain boundaries, leading to cleavage 2 or brittle intergranular fracture 2 (BIF 2).

As the temperature is raised, the flow stress falls rapidly and general yield precedes fracture. The fracture toughness tends to rise and higher stresses are required to initiate an unstable cleavage crack. All bcc metals show a hump in the fracture stress (or UTS) in this temperature range and fail by

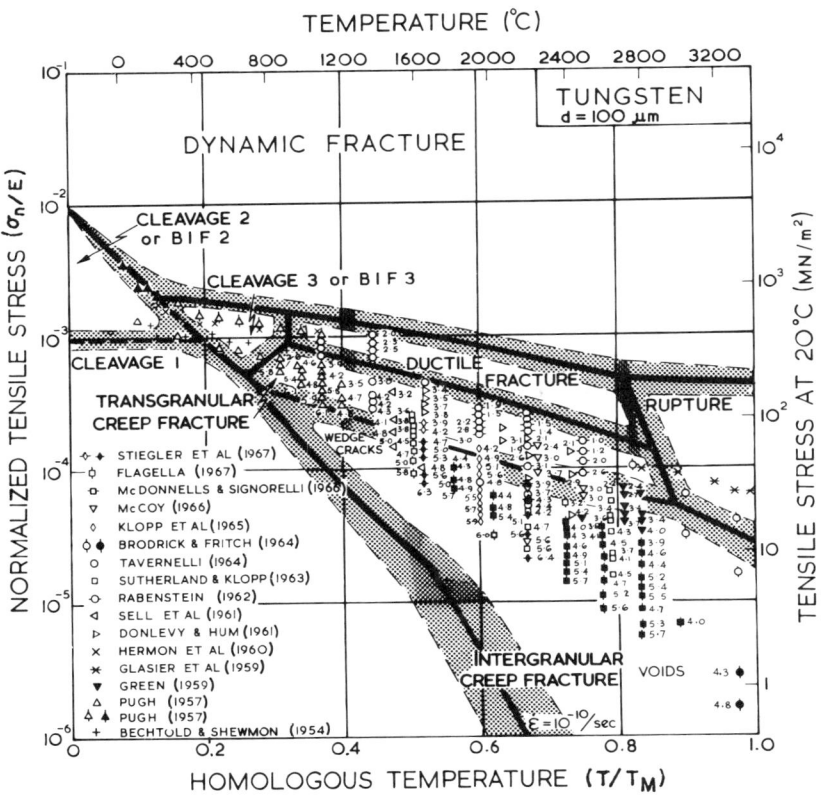

FIG. 26. A fracture map for tungsten. Each data point is labeled with $\log_{10} t_f$.

cleavage 3 or BIF 3, with measurable ductility (1–10%). The extent of this field depends on grain size and the interstitial impurity content.

As the temperature is increased further, there is a transition to a fully ductile fracture. In recrystallized bcc metals, it occurs at about $0.3T_M$; in worked bcc metals, it is displaced to higher temperatures. At slightly higher temperatures still, creep becomes appreciable and at high stresses fracture occurs by transgranular creep fracture.

At low stresses and high temperatures, the fracture mechanism changes to intergranular creep fracture. Wedge cracks or cavities form on the grain boundaries and grow in size and number with creep strain, finally linking to produce a fracture path along the grain boundary. Wedge cracking is typical of the low-temperature ($0.5T_M$) high-stress ($\sigma_n/E \simeq 10^{-4}$) section of the intergranular creep fracture field; cavities form at higher temperatures and lower stresses. Ideally, one would wish to subdivide the field to show this, but the data are rarely complete enough to permit it.

At temperatures above $0.8T_M$, failure occurs by rupture, apparently a consequence of dynamic recovery, recrystallization, and grain growth.

There are slight differences between the bcc refractory metals. Tantalum, for instance, has the least tendency to cleave; chromium has the greatest. On the broad scales of the figures, these differences are slight, and the bcc refractory metals form a well-defined class with remarkably similar maps.

3. The Hexagonal Metals: Magnesium

Figure 27, a map for commercially pure magnesium, typifies metals with hcp structure. If they contain cracks or flaws, these materials fail at low temperatures by cleavage 1. More usually, basal slip or twinning nucleates small cracks that propagate either by cleavage, brittle intergranular fracture or a combination of both, giving a mixed mode of fracture. This field is labeled cleavage 2 or BIF 2. As the temperature is raised, the yield strength falls below the fracture strength and general plasticity precedes fracture, although it is still of the cleavage or brittle intergranular type (cleavage 3 or BIF 3).

All the hexagonal metals we examined (Gandhi and Ashby, 1979) show a transition to a completely ductile mode of fracture, with large ductility, at above $0.3T_M$. At rather higher temperatures (about $0.35T_M$), they start to creep, failing by transgranular creep fracture at high stresses, and by intergranular creep fracture at low stresses. Commercial-purity materials fail by rupture above $0.8T_M$, and the alloys also showed this mode of failure, though at a slightly higher temperature.

At low temperatures, the fracture behavior of hexagonal metals resembles that of ceramics (discussed below). At higher temperatures, more slip systems

Mechanisms of Deformation and Fracture

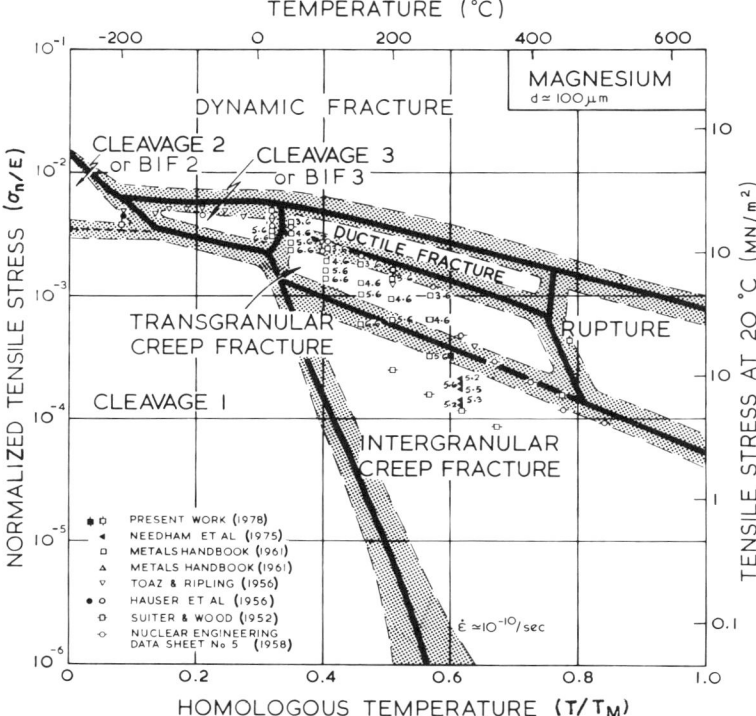

FIG. 27. A fracture map for magnesium. Each data point is labeled with $\log_{10} t_f$.

become available and they behave like fcc metals. On the normalized scales used to plot the maps, the hcp metals are significantly stronger than bcc metals, at both low and high temperatures. We have noted, however, that hcp metals and alloys are particularly prone to intergranular creep fracture, perhaps because the difficult nonbasal slip means that grain boundary sliding is not as readily accommodated by general plasticity as it is in bcc and fcc metals.

4. *The Alkali Halides: NaCl (Halite)*

Figure 28 for polycrystalline sodium chloride typifies the behavior of the alkali halides. At low temperatures ($<0.4T_M$ or 180°C), NaCl clearly shows fracture from preexisting flaws (cleavage 1). When care is taken to avoid them, microslip-nucleated cleavage (cleavage 2) replaces it (Aldag *et al.*, 1970; Stokes, 1966). The lower boundary of this field is drawn at the stress level

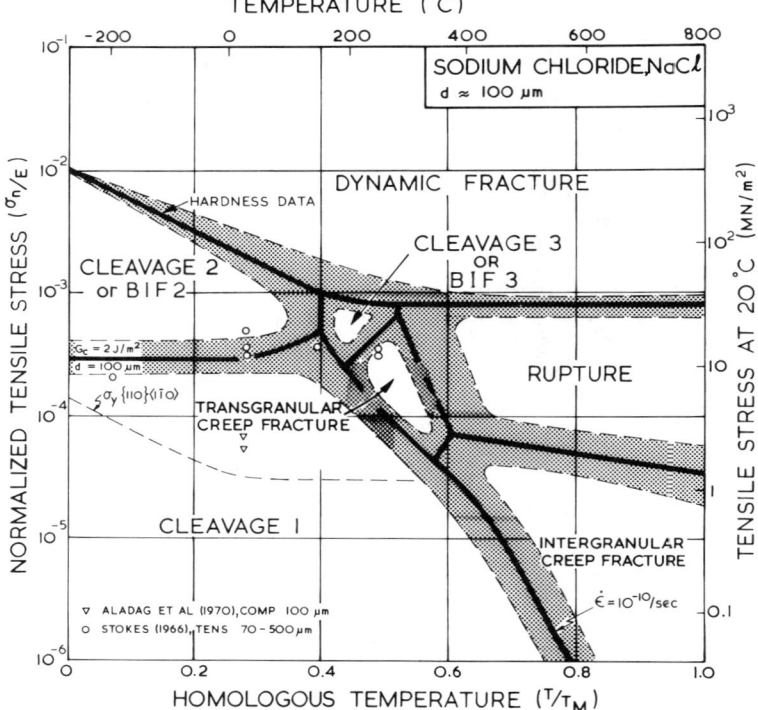

FIG. 28. A fracture map for sodium chloride, NaCl. Each data point is labeled with $\log_{10} t_f$.

required to propagate a crack of length equal to the grain size (100 μm), using a toughness G_c of 2 J/m² (see Section IV,B).

Above $0.4 T_M$, NaCl (and other alkali halides) exhibit general yield before fracture (Stoloff *et al.*, 1963; Stokes, 1966). The fracture, which is by mixed cleavage and brittle intergranular fracture, occurs at a stress that is now higher than the initial yield stress: There is measurable ductility (up to 10%) and some work hardening is necessary before the cleavage crack propagates. This field is labeled "cleavage 3" or "BIF 3."

Above $0.45 T_M$ (200°C) the tensile ductility of polycrystalline NaCl increases by as much as 60% (Stoloff *et al.*, 1963). Pronounced necking is observed in samples tested above $0.47 T_M$, and the mode of failure changes from cleavage 3 to ductile fracture. At still higher temperatures, alkali halides show dynamic recrystallization and fail by rupture (Becher *et al.*, 1975; Koepke *et al.*, 1975).

At low stresses and high temperatures, the alkali halides deform by slow creep. Cavities nucleate and grow on grain boundaries, leading to intergranular creep fracture, just as with metals (Stoloff *et al.*, 1963).

5. The Simple Oxides: MgO (Magnesia)

Magnesia (MgO) is by far the best documented of the oxides with rock-salt structure. A fracture map based on published data is shown in Fig. 29. Below $0.35T_M$ (800°C) MgO fails by type 1 or 2 cleavage, as discussed for the alkali halides. The fracture is completely brittle and occurs at stresses well below the general yield strength (Sweeting and Pask, 1976; Evans et al., 1970; Day and Stokes, 1966; Copley and Pask, 1965; Vasilos et al., 1964; Hulse et al., 1963).

Above $0.4T_M$ (900°C), polycrystalline MgO shows general plasticity and deforms by power-law creep. Over a wide range of stresses and temperatures, failure is by brittle intergranular fracture 3 (Evans et al., 1970; Day and Stokes, 1966; Copley and Pask, 1965). The material has measurable ductility, but plastic flow ultimately generates cracks, one of which propagates to give a fast fracture.

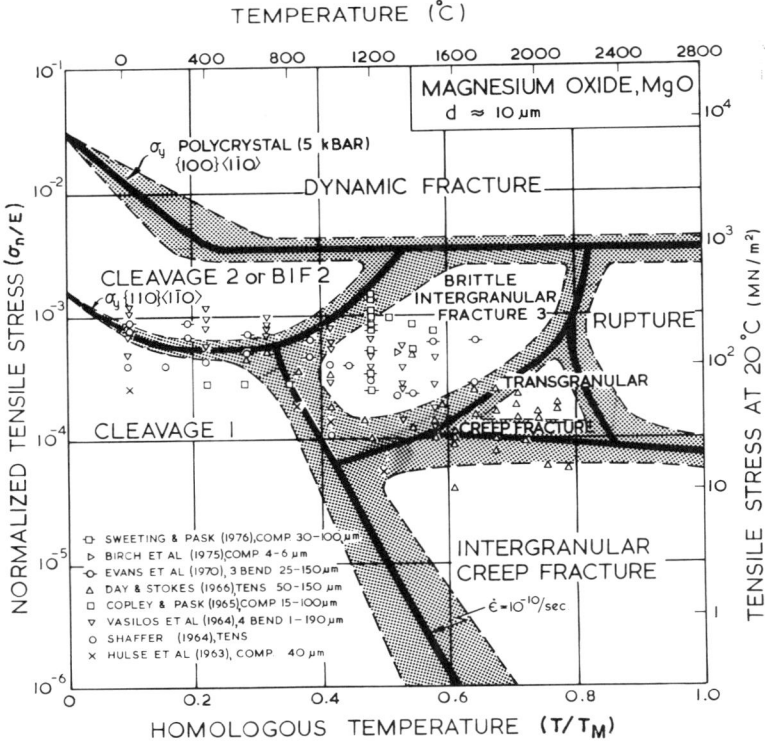

FIG. 29. A fracture map for magnesia, MgO. Each data point is labeled with $\log_{10} t_f$.

At lower stresses there exists a regime of intergranular creep fracture: Slow creep tests above $0.5T_M$ (1200°C) end in this type of failure (Birch et al., 1975; Day and Stokes, 1966). There is evidence for a subdivision of this field for MgO. At the lowest loads, cavities grow on grain boundaries, linking to give an intergranular fracture path. But at higher stresses, at around $0.6T_M$ (1600°C), wedge cracks are observed to an extent that depends on purity (Evans et al., 1970; Day and Stokes, 1966; Copley and Pask, 1965).

Above $0.65T_M$ (1700°C), MgO shows extensive plasticity ($<40\%$), and necking and fracture occur with no sign of grain boundary cavitation (Day and Stokes, 1966). This regime is shown in Fig. 29 as a field of ductile transgranular creep fracture. Single crystals, when strained above 1700°C, neck down and recrystallize in the neck region (Day and Stokes, 1966). This suggests the possibility that at very high temperatures, polycrystalline MgO may also undergo dynamic recrystallization and fail by rupture. No data are available, but a postulated rupture field is shown in Fig. 29.

6. *The Refractory Oxides Alumina: Al_2O_3*

Figure 30 for commercially pure alumina is broadly typical of the maps for the very refractory oxides. Commercial materials are often flawed and fail at low temperatures by cleavage 1. If flaws are avoided, failure occurs when twinning or microslip nucleates cracks; these propagate by cleavage or by grain boundary fracture (cleavage 2 or BIF 2), still with negligible ductility (Lankford, 1978; Simpson and Merrett, 1974; Spriggs et al., 1964; Charles, 1963; Stavrolakis and Norton, 1950).

Additional slip systems appear and general plasticity or creep becomes possible above $0.45T_M$. Over a wide range of temperatures and stresses, failure is by brittle intergranular fracture 3 or, occasionally, cleavage 3 (Teh, 1976; Davies, 1975; Crosby and Evans, 1973; Spriggs et al., 1964). As the temperature is raised further, strain-induced growth and dynamic recrystallization accompany the creep test. This permits added plasticity and caused a broad transition to transgranular creep failure or, in extreme cases, to rupture. At high temperatures and low stresses, the refractory oxides fail by intergranular creep fracture; like the metals, these ceramics show wedge cracking at higher stress levels and failure by void growth at lower stresses (Crosby and Evans, 1973; Davies, 1975).

The boundaries of the fields shown in Fig. 30 have a finite width. Their exact position varies with strain rate, purity, grain size, and method of fabrication, so that a map that describes typical behavior is properly drawn with boundaries of a width which encompasses this variation. In addition, the transition from one mechanism to another can be gradual and can include a regime of mixed-mode fracture. The shading shown on the field boundaries indicates both these effects.

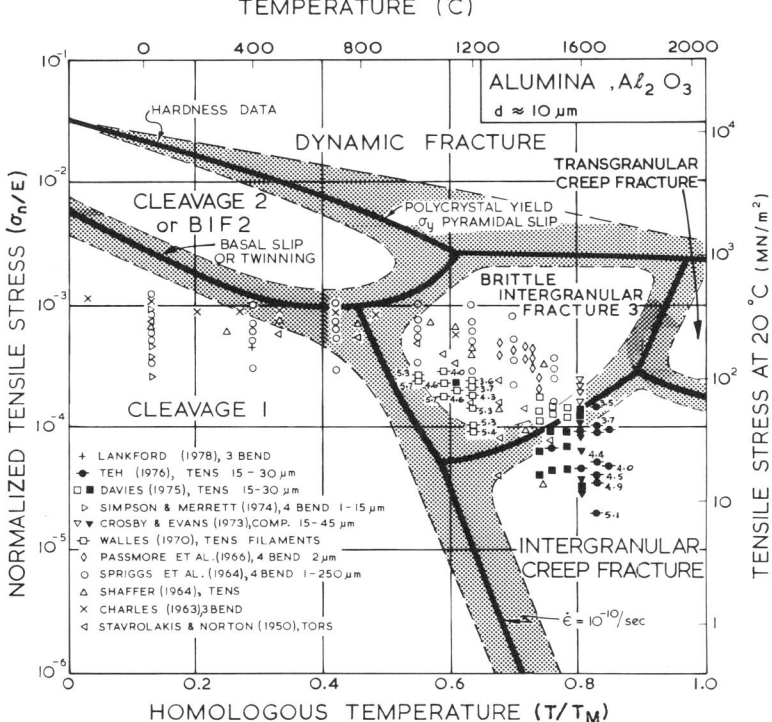

FIG. 30. A fracture map for alumina, Al_2O_3. Each data point is labeled with $\log_{10} t_f$.

These oxides show all the mechanisms that we identified in the bcc and hcp metals. But the totally brittle mechanisms that disappear above $0.1T_M$ in bcc metals and at about $0.2T_M$ in hexagonal metals extend to $0.5T_M$ or above here, and the areas occupied by the other mechanisms (which permit at least limited ductility) are correspondingly smaller.

7. Ice (H_2O)

The fracture strength of ice is important in calculating the forces on off-shore structures in frozen waters and in ice breaking. The regime of dominance of each mechanism in simple tension can be established approximately from published data (Fig. 31). Below about $0.6T_M$ ($-110°C$) ice is completely brittle and fails in the cleavage 1 or cleavage 2 mode. The compression experiments of Parameswaran and Jones (1975), for example, on columnar grained ice showed that cracks first appeared at a low stress (~ 0.3 MN/m^2), probably nucleated by (easy) basal slip; the failure stress was 5 MN/m^2, still well below general yield. Above $0.6T_M$, ice shows some

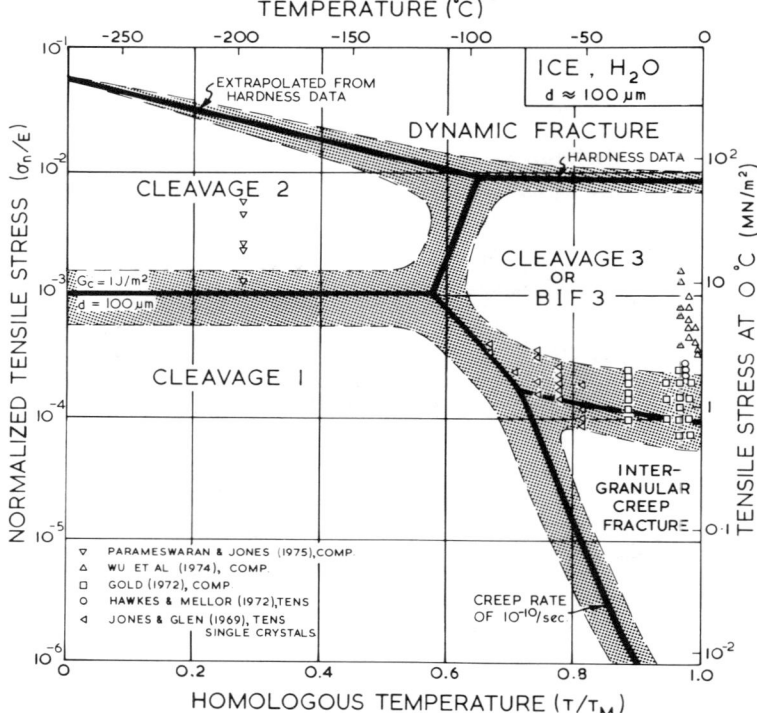

FIG. 31. A fracture map for ice, H_2O. Each data point is labeled with $\log_{10} t_f$.

general plasticity. Most of the fracture data lie between 0.9 and $0.995 T_M$ (Wu et al., 1974; Gold, 1972a,b; Hawkes and Mellor, 1972). The fracture mode changes from cleavage 3 at high stresses to intergranular creep fracture at low stresses. Extensive studies of columnar ice at about $0.9 T_M$ by Gold (1972b) showed fracture by cavitation below 0.6 MN/m^2 and fracture by cleavage above 1 MN/m^2. At higher temperatures, grain growth and recrystallization accompany creep deformation but do not lead to rupture (Frederking and Gold, 1975).

C. Mechanisms of Fracture and Order of Magnitude of K_c and G_c

The diagrams of Section V,B showed the mechanisms by which materials fail in simple tension. At and near a crack tip in a thin sheet, the stress state approximates plane stress and the range of dominance of the mechanisms is almost identical to that corresponding to Figs. 24–31. But in a thick plate, the stress state is a more extreme one, with (in mode 1 loading) a much larger hydrostatic component. This has the effect of expanding the cleavage (or brittle) fields at the expense of those giving large ductilities, but—as far as is known—no new mechanisms of fracture appear. The maps therefore give a qualitative picture of the way in which crack extension mechanisms change with temperature.

We will now examine the approximate magnitudes of the fracture toughness K_c or toughness G_c that we might expect from each mechanism and tabulate the results as a chart of toughness versus mechanism (Fig. 32).

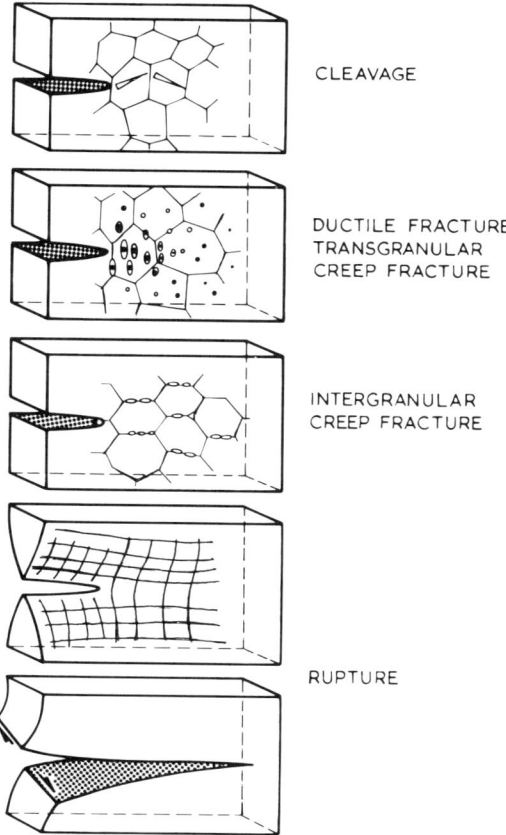

FIG. 32. Crack extension mechanisms and their relation to the fracture mechanisms described in the text.

1. Crack Extension by Cleavage

In the cleavage 1 and 2 regimes cracks extend without general plasticity. In many ceramics, there is little or no crack tip plasticity either: The yield strength is so large that the cohesive strength [Eq. (4.1)] is exceeded before even local slip can take place.

Roughly speaking, this sort of cleavage fracture will occur when the stress at the crack tip is sufficient to pull atoms apart, that is [using Eq. (4.1)], when

$$\sigma(r) = \frac{K}{\sqrt{2\pi r}} > \frac{E}{10}$$

at a distance $r \approx b$ from the tip, where b is the atom size. We thus find

$$K_c \simeq \frac{E}{10}\sqrt{2\pi b}. \tag{5.1}$$

Substituting for E and b, we find the values given in Table III.

In summary, when materials cleave, without plasticity, by the propagation of an atomically sharp crack that simply stretches the atomic bonds at the crack tip until they fail, the fracture toughness is low, from 0.1 to about 2 MPa m$^{1/2}$. Experiments support this: The measured fracture toughness of ice is close to 0.1 MPa m$^{1/2}$; that for alumina is about 3 MPa m$^{1/2}$. Metals are more ductile than this. Although there is negligible general plasticity in the cleavage 1 and cleavage 2 regimes of tungsten, say, or of magnesium, there is appreciable local plasticity and the toughness is larger than the minimum values calculated in Table III.

When general plasticity precedes cleavage failure (cleavage 3), the toughness increases substantially. Experiments in this regime (which straddles room temperature for many bcc and hcp metals) show values of K_c between 2 and 20 MPa m$^{1/2}$ (or of G_c between 30 and 3000 J/m^2).

TABLE III

CLEAVAGE WITH NO PLASTICITY

Material	E (GN/m^2)	b (m)	$K_c \approx (E/10)\sqrt{2\pi b}$ (MPa m$^{1/2}$)	$G_c \approx K_c/E$ (J/m^2)
Tungsten	406		1.8	7.7
Steel	200		0.9	3.8
Magnesium	42	3×10^{-10}	0.2	0.8
Alumina (Al$_2$O$_3$)	390		1.7	7.4
Ice	9.1		0.1	0.2

2. Crack Advance by Ductile Tearing

The ductility of most solids increases with increasing temperature, until it becomes sufficiently large that crack advance by the nucleation, plastic growth, and coalescence of holes (Section IV,C) replaces cleavage. The local criterion for crack advance is that a sufficient plastic strain occurs in a volume element at the crack tip so that the holes in it coalesce.

Consider first a crack tip surrounded by a plastic zone of approximate size

$$r_y = \frac{K}{2\pi\sigma_y^2},$$

beyond which the matrix is elastic. If the solid work hardens such that

$$\sigma = A\varepsilon^n,$$

the strains within the zone are given by

$$\varepsilon(r) = \frac{B}{r^{n/(n+1)}},$$

where A and B are constants. Matching the elastic strains

$$\varepsilon^{el} = \frac{K}{E(2\pi r)^{1/2}}$$

with those in the plastic zone at $r = r_y$, we find within the zone

$$\varepsilon(r) = \left(\frac{K^2}{2\pi\sigma_y^2 r}\right)^{n/n+1} \frac{\sigma_y}{E}. \tag{5.2}$$

If we then require that $\varepsilon > \varepsilon_f$ at a distance $r = r^*$ (which we identify with the inclusion spacing), we find

$$K_c \approx \left[2\pi\sigma_y^2 r^* \left(\frac{E\varepsilon_f}{\sigma_y}\right)^{(n+1)/n}\right]^{1/2}. \tag{5.3}$$

This has the lower limit

$$K_c^{\min} \approx (2\pi\sigma_y r^* E\varepsilon_f)^{1/2} \tag{5.4}$$

for perfect plasticity ($n = \infty$). If $E = 200 \text{ GN/m}^2$, $\sigma_y = 5 \times 10^{-3} E$, $r^* = 10 \,\mu\text{m}$, and $\varepsilon_f = 1$, we find

$$K_c = 100 \text{ MPa m}^{1/2} \quad \text{or} \quad G_c = 5 \times 10^4 \text{ J/m}^2,$$

a very good estimate of the toughness of a tough steel.

Ductile tearing, then, leads to large values of K_c (20–500 MPa m$^{1/2}$) or of G_c (10^4–10^5 J/m^2). Under certain circumstances, the toughness may be

TABLE IV

MAGNITUDES OF G_c AND K_c AND THEIR RELATIONSHIP TO THE MECHANISMS OF CRACK ADVANCE

	Toughness G_c (J/m^2)	Fracture toughness K_c (MPa$\sqrt{\text{m}}$)	G_c/Eb	$K_c/E\sqrt{b}$
Plastic rupture (Section IV,G)	10^6–10^7	500–2000	$\sim 10^5$	>100
Fibrous, by void growth and coalescence (*ductile fracture*, Section IV,C)	10^4–10^6	20–500	10^2–10^4	10–100
Unstable shear: void sheet linkage	10^3–10^5	10–100	$\sim 10^3$	2–20
Mixed fibrous and cleavage	10^3–2×10^4	10–50	$\sim 10^2$	2–10
Cleavage preceded by slip (*cleavage* or *BIF 3*, Section IV, B)	30–300	2–20	1–10	0.5–5
Cleavage without slip (*cleavage* or *BIF 1 and 2*, Section IV,B)	1–100	0.2–5	0.1–1	0.1–2
2 × surface energy, metals	0.5–3	0.4–1	~ 0.1	~ 0.2
2 × surface energy, ceramics	0.5–4	0.2–1	~ 0.1	~ 0.2
2 × surface energy, polymers	0.1–1	0.01–0.05	~ 0.1	~ 0.2

less than this. In certain high-strength alloys, for instance, plasticity at the crack tip becomes localized in bands of intense shear; the crack tip advances by *void sheet linkage*, at a lower value of K. There also exists a regime of mixed fibrous and cleavage fracture, the crack propagating across some grains by cleavage but passing through others by ductile fracture. Again, the toughness is lower than that of a fully ductile fibrous fracture.

3. Crack Advance by Rupture

Thin sheets of pure metals can exhibit crack advance by rupture, local reduction of the section to zero. This requires a local strain of unity in a slice of material with a height about equal to the sheet thickness t. The value of K_c is given by Eq. (5.3) or (5.4), with the inclusion spacing replaced by t:

$$K_c^{min} \approx (2\pi t \sigma_y E)^{1/2}.$$

Taking $t = 1$ mm, $E = 200$ GN/m^2, and $\sigma_y = 5 \times 10^{-3} E$, we have

$$K_c^{min} \approx 1000 \text{ MPa m}^{1/2} \quad \text{or} \quad G_c \approx 5 \times 10^6 \text{ J/m}^2.$$

Rupture, then, leads to the highest toughness of all, which can probably be regarded as an upper limit to the toughness of a material.

The relation between toughness and crack-advance mechanism, as deduced above, is summarized in Table IV.

VI. Conclusions

Complicated phenomena like deformation or fracture, involving many competing mechanisms, can be presented in a simple way by constructing mechanism maps for them. These show the range of the macroscopic variables (stress σ_s and temperature T) over which each mechanism is dominant and the resulting rate of deformation or time to fracture. They can be constructed empirically by tabulating and plotting observations of each mechanism; the fracture maps shown here were made in this way. But if each mechanism can be modeled to give a physically based constitutive law, then a more powerful procedure is to fit the laws to the data and use these to construct the map. This method, which was used to construct the deformation maps shown here, has the merit of displaying the behavior beyond the range of the data from which they were constructed.

When maps for deformation or fracture are plotted on normalized axes of σ_s/μ and T/T_M (where μ is the shear modulus and T_M is the melting temperature), it is found that groups of materials with the same crystal structure, bonding, and purity and in a similar microstructural state (grain size, state of work hardening, etc.) have closely similar maps. The similarities are sufficiently great that one map broadly typifies the fcc metals (of a given purity and in a similar microstructure), one typifies the bcc metals, one the alkali halides, and so forth. This means that a given class of materials (the fcc metals, for instance) all deform and fracture by the same set of mechanisms, operating over roughly the same ranges of σ_s/μ and T/T_M. Another class of materials (the simple oxides, for example) show the same mechanisms, but they operate over a different (and characteristic) range of σ_s/μ and T/T_M.

The maps have utility in summarizing a large body of data in a readily accessible form. They allow quick identification of the dominant mechanism in a given engineering application, and thereby help in selecting the appropriate constitutive law for design.

ACKNOWLEDGMENTS

The figures of this paper first appeared in articles in *Acta Metallurgica* between 1972 and 1981 and from the book "Deformation Mechanism Maps," by H. J. Frost and M. F. Ashby, published by Pergamon Press.

REFERENCES

Aldag, E., Davis, L. A., and Gordon, R. B. (1970). *Phil. Mag.* **21**, 469.
Argon, A., and Im, J. (1975). *Metall. Trans.* **6***A*, 839.
Argon, A., and Maloof, S. R. (1966). *Acta Metall.* **14**, 1449.
Argon, A., Im, J., and Safoglu, R. (1975). *Metall. Trans.* **6***A*, 825.
Ashby, M. F. (1969). *Scripta Metall.* **3**, 837.
Ashby, M. F. (1972a). *Acta Metall.* **20**, 887.
Ashby, M. F. (1972b). *Surface Sci.* **31**, 498.
Ashby, M. F. (1977), in "Fracture 77", D. Taplin, ed. ICF4, Waterloo University Press, Vol. 1, p. 1.
Ashby, M. F., and Frost, H. J. (1975). "Constitutive Equations in Plasticity" (A. Argon, ed., Chap. 4.) MIT Press, Cambridge, Massachusetts.
Ashby, M. F., Gandhi, C., and Taplin, D. M. R. (1979). *Acta Metall.* **27**, 699.
Balluffi, R. W. (1970). *Phys. Stat. Solids.* **42**, 11.
Barrett, C. R., and Sherby, O. D. (1965). *Trans. AIME* **233**, 1116.
Becher, P. F., Rice, R. W., Klein, P. H., and Freiman, S. W. (1975). "Deformation of Ceramic Materials" (R. C. Brandt and R. W. Tressler, eds.), p. 517. Plenum, New York.
Bechtold, J. H., and Shewmon, P. G. (1954). *Trans. ASM* **46**, 397.
Beere, W., and Speight, M. V. (1978). *Metall. Sci.* **4**, 172.
Betteridge, W., and Heslop, J. (1974). "Nimonic Alloys," 2nd ed., Arnold, London.
Bilde-Sörensen, J. B. (1972). *J. Am. Ceram. Soc.* **55**, 606.
Birch, J. M., King, P. J., and Wilshire, B. (1975). *J. Mat. Sci.* **10**, 175.
Bird, J. E., Mukherjee, A. K., and Dorn, J. E. (1969). "Quantitative Relation between Properties and Microstructure," pp. 255–342. Israel Univ. Press, Jerusalem.
Bolling, G. F., and Richman, R. H. (1965). *Acta Metall.* **13**, 709, 723.

Bolton, C. J. (1971). Unpublished work, quoted by I. G. Crossland and R. B. Jones (1972). *Metall. Sci. J.* **6**, 162.
Brodrick, R. F., and Fritch, D. J. (1964). *Proc. ASTM* **64**, 505.
Bromer, D. J. and Kingery, W. D. (1968). *J. Appl. Phys.* **39**, 1688.
Brown, A. M., and Ashby, M. F. (1980). *Scripta Metall.* **14**, 1297.
Brown, A. M., and Ashby, M. F. (1981). *In* "Deformation of Polycrystals: Mechanisms and Microstructures" (T. Leffers, and N. Hanson, eds.) Second Risø Int. Symp., Risø Nat. Lab., Denmark.
Brown, L. M., and Embury, J. D. (1973). "Proceedings of the 3rd International Conference on the Strength of Metals and Alloys. Institute of Metals, London.
Brown, L. M., and Stobbs, W. M. (1971). *Philos. Mag.* **23**, 1201.
Brown, L. M., and Stobbs, W. M. (1976). *Philos. Mag.* **34**, 351.
Burke, M. A., and Nix, W. D. (1975). *Acta Metall.* **23**, 793.
Burke, P. M. (1968). Ph.D. Thesis, Dept. of Stanford University, California. Eng. and Metal.
Burton, B. (1972). *Mat. Sci. Eng.* **10**, 9.
Burton, B. (1973). *Mat. Sci. Eng.* **11**, 337.
Burton, B., and Bastow, B. D. (1973). *Acta Metall.* **21**, 13.
Butkovich, T. R. (1954). Snow, Ice and Permafrost Research Establishment. Research Report 20, p. 1.
Butkovich, T. R., and Landauer, J. K. (1960). Snow, Ice and Permafrost Research Establishment. Research Report 72, p. 1.
Cannaday, J. E., Austin, R. J., and Saxer, R. K. (1966). *Trans. AIME* **236**, 595.
Cannon, R. M. (1971). Ph.D. Thesis, Stanford University.
Charles, R. J. (1963). Studies of the brittle behaviour of ceramic materials. ASD-TR-61-628, p. 467.
Chuang, T., and Rice, J. R. (1973). *Acta Metall.* **21**, 1625.
Coble, R. L. (1963). *J. Appl. Phys.* **34**, 1679.
Cocks, A. C. F., and Ashby, M. F. (1981). *Prog. Mater. Sci.* **27**, p. 189.
Copley, S. M., and Pask, J. A. (1965). *J. Am. Ceram. Soc.* **48**, 139, 636.
Crosby, A., and Evans, P. E. (1973). *J. Mat. Sci.* **8**, 1573.
Crossman, F. W., and Ashby, M. F. (1975). *Acta Metall.* **23**, 425.
Danneberg, W. (1961). *Metall.* **15**, 977.
Davies, C. K. L. (1975). "Physical Metallurgy of Reactor Fuel Elements" (J. Harris, and T. Sykes, eds.), p. 99. Metals Soc., London.
Davies, P. W., and Dennison, J. P. (1960). *J. Inst. Met.* **88**, 471.
Day, R. B., and Stokes, R. J. (1966). *J. Am. Ceram. Soc.* **49**, 345.
de Meester, B., Yin, C., Doner, M., and Conrad, H. (1973). *In* "Rate Processes in Plastic Deformation" (J. C. M. Li, and A. K. Mukherjee, eds.) American Society for Metals, Columbus, Ohio.
Dennison, J. P., and Wilshire, B. (1963). *J. Inst. Met.* **91**, 343.
Dennison, J. P., Llewellyn, R. J., and Wilshire, B. (1966). *J. Inst. Met.* **94**, 130.
Donlevy, A., and Hum, J. K. Y. (1961). *Mach. Design* **11**, 244.
Dyson, B. F. (1976). *Metal Sci.* **10**, 349.
Dyson, B. F. (1978). *Can. Met. Quart.* **18**, 31.
Dyson, B. F., and McLean, D. (1977). *Metal Sci.* **11**, 37.
Dyson, B. F., and Rodgers, M. J. (1974). *Metal Sci.* **8**, 261.
Edward, G. H. and Ashby, M. F. (1979). *Acta Metall.* **27**, 1505.
Evans, A. G., and Rawlings, R. D. (1969). *Phys. Stat. Sol.* **34**, 9.
Evans, A. G., Gilling, D., and Davidge, R. W. (1970). *J. Mat. Sci.* **5**, 187.
Fields, R. J., Weerasooriya, T., and Ashby, M. F. (1980). *Metall. Trans.* **11A**, 333.

Firestone, R. F., and Heuer, A. H. (1976). *J. Am. Ceram. Soc.* **59**, 24.
Flagella, P. N. (1967a). GE-NMPO, GEMP-543, Aug. 31, presented at the Third International Symposium on High Temperature Technology, Asilomar, Calif.
Flagella, P. N. (1967b). *J. AIAA* **5**(2), 281.
Flynn, P. W., Mote, J., and Dorn, J. E. (1961). *Trans. AIME* **221**, 1148.
Folweiler, R. C. (1961). *J. Appl. Phys.* **32**, 773.
Forrest, P. G., and Smith, P. A. (1963). *J. Inst. Met.* **92**, 61.
Frederking, R., and Gold, L. W. (1975). *Can. Geotech. J.* **12**, 456.
Frost, H. J. (1973). Ph.D. Thesis, Harvard University.
Frost, H. J., and Ashby, M. F. (1982). "Deformation Mechanism Maps," Pergamon, Oxford.
Frost, H. J., Goodman, D. J., and Ashby, M. F. (1976). *Philos. Mag.* **33**, 951.
Gandhi, C., and Ashby, M. F. (1979). *Acta Metall.* **27**, 1565.
Gibbs, G. B. (1965). *Mem. Sci. Rev. Met.* **62**, 781.
Gibbs, G. B. (1966). *Philos. Mag.* **13**, 317.
Gilbert, E. R., Flinn, J. E., and Yaggee, F. L. (1965). 4th Symposium on Refractory Metals, Met. Soc. *AIME*, p. 44.
Gilman, J. J. (1961). *Prog. Ceram. Sci.* **1**, 146.
Glasier, L. F., Allen, R. D., and Saldinger, I. L. (1959). Mechanical and physical properties of the refractory metals, *Aerojet*, General Corp., report no. M1826.
Glenn, J. W. (1968). *Phys. kondens. Mater.* **7**, 43.
Gold, L. W. (1972a). National Research Council of Canada, Division of Building Research Technical Paper no. 369.
Gold, L. W. (1972b). *Philos. Mag.* **26**, 311.
Gooch, D. J., and Groves, G. W. (1972). *J. Am. Ceram. Soc.* **55**, 105.
Gooch, D. J., and Groves, G. W. (1973a). *Philos. Mag.* **28**, 623.
Gooch, D. J., and Groves, G. W. (1973b). *J. Mat. Sci.* **8**, 1238.
Goodman, D. J., Frost, H. J., and Ashby, M. F. (1981). *Philos. Mag.* **43**, 665.
Goods, S. H., and Brown, L. M. (1979). *Acta Metall.* **27**, 1.
Green, W. V. (1959). *Trans. AIME* **215**, 1057.
Greenwood, N. N. (1970). "Ionic Crystals, Lattice Defects, and Non-Stoichiometry," Butterworth, London.
Guard, R. W., Keeler, J. H., and Reiter, S. F. (1954). *Trans. AIME* **200**, 226.
Guiu, F., and Langdon, T. G. (1974). *Philos. Mag.* **24**, 145.
Guyot, P., and Dorn, J. E. (1967). *Can. J. Phys.* **45**, 983.
Haasen, P. (1958). *Philos. Mag.* **3**, 384.
"Handbook of Chemistry and Physics," 54th ed. (1973). Chemical Rubber Co., Cleveland, Ohio.
Hardwick, D., and Tegart, W. J. McG. (1961). *J. Inst. Met.* **90**, 17.
Hardwick, D., Sellars, C. M., and Tegart, W. J. McG. (1961). *J. Inst. Met.* **90**, 21.
Harper, J. G., and Dorn, J. E. (1957). *Acta Metall.* **5**, 654.
Harper, J. G., Shepard, L. A., and Dorn, J. E. (1958). *Acta Metall.* **6**, 509.
Hart, E. W. (1957). *Acta Metall.* **5**, 597.
Hart, E. W. (1967). *Acta Metall.* **15**, 351.
Hauser, F. E., Landon, P. R., and Dorn, J. E. (1956). *Trans. AIME* **206**, 589.
Hawkes, I., and Mellor, M. (1972). *J. Glaciology* **11**, 103.
Hayward, E. R., and Greenough, A. P. (1959). *J. Inst. Met.* **88**, 217.
Heard, H. C. (1972). "The Griggs Volume." American Geophysical Union Monograph 16. Menlo Park, California.
Henry Wiggin and Co. Ltd. (1975). Technical Data Sheets, Publications 3663 and 3573A, Birmingham, England.
Hensler, J. H., and Cullen, G. V. (1968). *J. Am. Ceram. Soc.* **51**, 557.

Hermon, E. L., Morgan, R. P., Graves, N. F., and Reinhardt, G. (1960). Investigation of the properties of tungsten and its alloys, Union Carbide, Mat. Res. Lab. Report WADD, TR.60-144, AF(616)5600.
Herring, C. (1950). *J. Appl. Phys.* **21**, 437.
Heuer, A. H., Firestone, R. F., Snow, J. D. and Tullis, J. D. (1970). "Proceedings of the second International Conference on the Strength of Metals and Alloys, Asilomar, California, p. 1165. *ASM*, Columbus, Ohio.
Hobbs, P. V. (1974). "Ice Physics," Oxford, Univ. Press. (Clarendon), London and New York.
Holt, R. T., and Wallace, W. (1976). *Int. Met. Revs.* **21**, 1.
Hull, D., and Rimmer, D. E. (1959). *Philos. Mag.* **4**, 673.
Hulse, C. O., and Pask, J. A. (1960). *J. Am. Ceram. Soc.* **43**, 373.
Hulse, C. O., Copley, S. M., and Pask, J. A. (1963). *J. Am. Ceram. Soc.* **46**, 317.
Jenkins, W. D., and Digges, T. G. (1952). *J. Res. N.B.S.* **48**, 313.
Jenkins, W. D., Digges, T. G., and Johnson, L. R. (1954). *J. Res. N.B.S.* **53**, 329.
Johnston, W. G. (1962). *J. Appl. Phys.* **33**, 2050.
Jonas, J. J., Sellars, C. M., and Tegart, W. J. McG. (1969). *Met. Rev.* **14**, 1.
Jones, R. B. (1973). *J. Shef. Univ. Metall. Soc.* **12**, p. 1.
Jones, R. B. and Harris, J. E. (1963). *From* "Proceedings of the Joint International Conference on Creep," Vol. I, p. 1. Inst. Mech. Eng., London.
Jones, S. J., and Glen, J. W. (1969). *J. Glaciology* **8**, 463.
Karashima, S., Oikawa, H., and Motomiya, T. (1969). *Trans. Jap. Inst. Met.* **10**, 205.
Kear, B. H., and Piearcey, B. J. (1967). *Trans. Metall. Soc. AIME* **239**, 1209.
Kelly, A. (1966). "Strong Solids," Chap. 1. Oxford Univ. Press (Clarendon) London and New York.
Kelly, A., Tyson, W. R., and Cottrell, A. H. (1967). *Philos. Mag.* **15**, 567.
King, G. W. (1970). Westinghouse Report BLR 90284-2, Sept. 14.
King, G. W., and Sell, H. G. (1965). *Trans. AIME* **233**, 1104.
Kingery, W. D., and Montrone, E. B. (1965). *J. Appl. Phys.* **36**, 2412.
Klahn, D., Mukherjee, A. K., and Dorn, J. E. (1970). "Second International Conference on the Strength of Metals and Alloys, Asilomar, California," p. 951. ASM, Columbus, Ohio.
Kliewer, K., and Kockler, J. (1965). *Phys.* (Rev.) *A* **140**, 1226.
Klopp, W. D., Witzke, W. R., and Raffo, P. L. (1965). *Trans. AIME* **233**, 1860.
Kocks, U. F. (1970). *Metall. Trans.* **1**, 1121.
Kocks, U. F., Argon, A. S., and Ashby, M. F. (1975). *Prog. Mat. Sci.* **19**, 1.
Koepke, B. G., Anderson, R. H., and Stokes, R. J. (1975). "Deformation of Ceramic Materials" (R. C. Brandt, and R. E. Tressler, eds.) p. 497. Plenum, New York.
Koo, R. C. (1963). *Acta Metall.* **11**, 1083.
Köster, W. (1948). *Z. Metallk.* **39**, 1.
Kreider, K. G., and Bruggeman, G. (1967). *Trans. AIME* **239**, 1222.
Kronberg, M. L. (1957). *Acta Metall.* **5**, 507.
Kumar, A., and Kumble, R. G. (1969). *J. Appl. Phys.* **40**, 3475.
Kumar, A., Hauser, F. E., and Dorn, J. E. (1968). *Acta Metall.* **16**, 1189.
Langdon, T. G., and Pask, J. A. (1970). *Acta Metall.* **17**, 505.
Lankford, J. (1978). *J. Mat. Sci.* **13**, 351.
Lindler, R., and Parfitt, G. D. (1957). *J. Chem. Phys.* **26**, 182.
Le Roy, G., Embury, J. D., Edward, G., and Ashby, M. F. (1981). *Acta Metall.* **29**, 1509.
Leverant, G. R., and Kear, B. H. (1970). *Metall. Trans.* **1**, 491.
Lifshitz, L. M. (1963). *Sov. Phys. JETP* **17**, 909.
Lundy, T. S., Winslow, F. R., Pawel, R. E., and McHargue, C. J. (1965). *Trans. AIME* **233**, 1533.
Luton, M. J., and Sellars, G. M. (1969). *Acta Metall.* **17**, 1033.

McClintock, F. A. (1968). *J. Appl. Mech.* **35**, 363.
McCoy, H. E. (1966). Creep rupture properties of tungsten and tungsten base alloys, *ORNL*-3992.
McDonnells, D. L., and Signorelli, R. A. (1966). Stress rupture properties of tungsten wire from 1200 to 2500 F, *NASA, TN,* D-3467.
MacKenzie, J. K. (1959). Ph.D. Thesis, Bristol University.
McLean, D. (1957). *J. Inst. Met.* **85**, 468.
McMahon, C. J., and Cohen, M. (1965). *Acta Metall.* **13**, 591.
Macmillan, N. H. (1972). *J. Mat. Sci.* **7**, 239.
Mellor, M., and Testa, R. (1969a). *J. Glaciology* **8**, 131.
Mellor, M., and Testa, R. (1969b). *J. Glaciology* **8**, 147.
"Metals Handbook" (1961), 8th ed. vol. 1. ASM, Columbus, Ohio.
Mohammed, F. A., Murty, K. L., and Morris, J. W., Jr. (1973). *In* "The John E. Dorn Memorial Symposium, Cleveland, Ohio," ASM, Columbus, Ohio.
Monkman, F. C., and Grant, N. J. (1956). *Proc. ASTM* **56**, 593.
Moon, D. M., and Stickler, R. (1971). *Philos. Mag.* **24**, 1087.
Mukherjee, A. K., Bird, J. E., and Dorn, J. E. (1969). *Trans. ASM* **62**, 155.
Nabarro, F. R. N. (1948). Report on a Conference on the Strength of Metals, Phys. Soc. London.
Nabarro, F. R. M. (1967). "Theory of Crystal Dislocations". Oxford Univ. Press, London and New York.
Needam, N. G., Wheatley, J. E., and Greenwood, G. W. (1975). *Acta Metall.* **23**, 23.
Neumann, G. M., and Hirschwald, W. (1966). *Z. Naturforsch.* **21a**, 812.
Nicholls, J. H., and McCormick, P. G. (1970). *Met. Trans.* **1**, 3469.
Norman, E. C., and Duran, S. A. (1970). *Acta Metall.* **18**, 723.
Nuclear Engineering Data Sheet No. 5: Magnesium (1958). *J. Nucl. Eng.,* **3**, 85.
Parameswaran, V. R., and Jones, S. J. (1975). *J. Glaciology* **14**, 305.
Parker, E. R. (1961). "Mechanical Properties of Engineering Ceramics" (W. W. Kriegel and H. Palmour, eds.) p. 61. Wiley, New York.
Passmore, E., Moschetti, A., and Vasilos, T. (1966). *Philos. Mag.* **13**, 1157.
Paterson, M. S., and Weaver, C. W. (1970). *J. Am. Ceram. Soc.* **53**, 463.
Piearcey, B. J., Kear, B. H., and Smashey, R. W. (1967). *Trans. ASM* **60**, 634.
Poirier, J. P. (1972). *Philos. Mag.* **26**, 701.
Pugh, J. W. (1967). *Proc. ASTM* **57**, 906.
Rabenstein, A. S. (1962). Marquardt Corp. contract report no. AF 33 (657)-8706, 281-2Q-3.
Raffo, P. L. (1969). *J. Less-Common Met.* **17**, 133.
Raj, R., and Ashby, M. F. (1971). *Metall. Trans.* **2**, 1113.
Raj, R., and Ashby, M. F. (1975). *Acta Metall.* **23**, 653.
Rice, J. R., and Tracey, D. M. (1969). *J. Mech. Phys. Solids* **17**, 201.
Richardson, G. J., Sellars, C. M., and Tegart, W. J. McG. (1966). *Acta Metall.* **14**, 1225.
Roberts, C. S. (1953). *Trans. AIME* **197**, 1021.
Robinson, S. L., and Sherby, O. D. (1969). *Acta Metall.* **17**, 109.
Rozenberg, V. M. (1962). *Fiz. Metal. Metalloved.* **14**(1), 114.
Sell, H. G., Keith, H. H., Koo, R. C. Schnitzel, R. H., and Corth, R. (1961). Westinghouse Elec. Corp., Report WADD TR 60-37, Part II.
Sellars, C. M., and Quarrell, A. G. (1961). *J. Inst. Met.* **90**, 329.
Sellars, C. M., and Tegart, W. J. McG. (1966). *Mem. Sci. Rev. Met.* **63**, 731.
Shaffer, P. T. B. (1964). "High Temperature Materials, No. 1; Materials Index." Plenum, New York.
Shahinian, P., and Achter, M. R. (1959). *Trans. AIME* **215**, 37.
Simpson, L. A., and Merrett, G. J. (1974). *J. Mat. Sci.* **9**, 685.

Snowden, W. E., and Pask, J. A. (1974). *Philos. Mag.* **29**, 441.
Speight, M. V., and Harris, J. E. (1967). *Met. Sci. J.* **1**, 83.
Spriggs, R. M., Mitchell, J. B., and Vasilos, T. (1964). *J. Am. Ceram. Soc.* **47**, 323.
Stavrolakis, J. A., and Norton, F. H. (1950). *J. Am. Ceram. Soc.* **33**, 263.
Stiegler, J. O., Farrell, K., Loh, B. T. M., and McCoy, H. E. (1967). *Trans. ASM* **60**, 494.
Stocker, R. L., and Ashby, M. F. (1973). *Rev. Geophys. Sp. Phys.* **11**, 391.
Stokes, R. J. (1966). *Proc. Brit. Cer. Soc.* **6**, 189.
Stoloff, N. S., Lezius, D. K., and Johnston, T. L. (1963). *J. Appl. Phys.* **34**, 3315.
Stüwe, H. P. (1965). *Acta Metall.* **13**, 1337.
Suiter, J. W., and Wood, W. A. (1952). *J. Inst. Metals* **81**, 181.
Sutherland, E. C., and Klopp, W. D. (1963). *NASA* TN D-1310.
Sweeting, T. B., and Pask, J. A. (1976). *J. Am. Ceram. Soc.* **59**, 226.
Tavernelli, J. F. (1964). GE report no. 64-Ime-226, Cleveland, Ohio.
Taylor, J. L., and Boone, D. H. (1963). *Trans. ASM* **56**, 643.
Tegart, W. J. McG. (1961). *Acta Metall.* **9**, 614.
Teh, S. K. (1976). Ph.D. Thesis, Materials Department, p. 234. Queen Mary College, University of London.
Thomason, P. F. (1968). *J. Inst. Met.* **96**, 360.
Thomason, P. F. (1971). *Int. J. Fract. Mech.* **7**, 409.
Toaz, M. W., and Ripling, E. J. (1956). *Trans. AIME* **206**, 936.
Tressler, R. E., and Michael, D. J. (1974). *In* "Deformation of Ceramic Materials" (R. C. Bradt, and R. E. Tressler, eds.), p. 195, Plenum, New York.
Tyson, W. R. (1966). *Philos. Mag.* **14**, 925.
Vasilos, T., Mitchell, J. B., and Spriggs, R. L. (1964). *J. Am. Ceram. Soc.* **47**, 606.
Verrall, R. A., Fields, R. J., and Ashby, M. F. (1977). *J. Am. Ceram. Soc.* **60**, 211.
Ver Snyder, F. L., and Piearcey, B. J. (1966). *SAE* June **74**, 36.
Walles, K. F. A. (1970). *Proc. Brit. Cer. Soc.* **15**, 157.
Weaver, C. W. (1960). *J. Inst. Met.* **88**, 296, 462.
Webster, G. A., and Piearcey, B. J. (1967). *Met. Sci. J.* **1**, 97.
Weertman, J. (1956). *J. Mech. Phys. Solids* **4**, 230.
Weertman, J. (1960). *Trans. AIME* **218**, 207.
Weertman, J. (1963). *Trans. AIME* **227**, 1475.
Weertman, J. (1969). *J. Glaciology* **8**, 494.
Weertman, J. (1973). "Physics and Chemistry of Ice" (E. Whalley, S. J. Jones, and L. W. Gold, eds.) p. 362. Ottawa Royal Society of Canada, Ottawa.
Weertman, J., and Shahinian, P. (1956). *Trans. AIME* **206**, 1223.
Westbrook, J. H. (1966). *Rev. Hautes Temper. Refract.* **3**, 47.
Whitworth, R. W., Paren, J. G., and Glen, J. W. (1976). *Philos. Mag.* **33**, 409.
Wong, W. A., and Jones, J. J. (1968). *Trans. AIME* **242**, 2271.
Woodford, D. A. (1969). *Met. Sci.* **3**, 234.
Wu, H. C., Chang, K. J., and Schwarz, R. (1974). *Eng. Fract. Mech.* **20**, 845.
Yim, W. M., and Grant, N. J. (1963). *Trans. AIME* **227**, 868.

Recent Developments Concerning Saint-Venant's Principle

CORNELIUS O. HORGAN

College of Engineering
Michigan State University
East Lansing, Michigan

JAMES K. KNOWLES

Division of Engineering and Applied Science
California Institute of Technology
Pasadena, California

I. Introduction	180
II. A Model Problem	182
A. Flow in a Cylinder	182
B. A Representation for the Exact Solution	185
C. An Energy Inequality	189
D. An Alternative Energy Procedure	192
E. Energy Decay for Other Linear Elliptic Second-Order Problems	195
F. An Upper Bound for the Total Energy $E(0)$	196
G. Higher-Order Energies and Cross-Sectional Estimates	199
H. Pointwise Decay	204
I. A Different Approach: Maximum Principles	210
III. Linear Elastostatic Problems	213
A. Axisymmetric Torsion	213
B. Plane Strain	223
C. Torsionless Axisymmetric Problems	237
D. The Elastic Cylinder and Other Three-Dimensional Problems	239
IV. Principles of Saint-Venant Type in Other Contexts	250
A. The Influence of Nonlinearity	250
B. Slow Stress Diffusion in Shells	255
C. Time-Dependent Problems	257
V. Concluding Comments	261
References	262

I. Introduction

The task of determining, within the framework of the linear theory of elasticity, the stresses and displacements in an elastic cylinder in equilibrium under the action of loads that arise solely from tractions applied to its plane ends has come to be called *Saint-Venant's problem* because of the remarkable investigations of this issue carried out by Barré de Saint-Venant and reported in two classical memoirs published well over a century ago.[†] His approach to the cylinder problem involved judicious anticipation, on physical grounds, of the main features the elastostatic field might be expected to exhibit. In this way he was led to an exact solution of the field equations of the theory that leaves the lateral surface of the cylinder traction-free and for which the *resultant forces and moments* acting on the end faces may be arbitrarily prescribed, subject only to the requirement of overall equilibrium of the cylinder.

Saint-Venant's construction does not permit the arbitrary preassignment of the point-by-point variation of the end tractions giving rise to these forces and moments; indeed, this variation is essentially determined as a consequence of the special assumptions made in connection with his so-called semi-inverse procedure. He himself conjectured, however, that his special solution would be useful as an approximation, valid far enough away from the ends of a sufficiently long cylinder, in cases where the end tractions are statically equivalent to, but not identical with, those for which his solution is rigorously valid. This conjecture, as well as generalizations of it for bodies that are not necessarily long and slender, was studied by Boussinesq (1885). This early work of Saint-Venant and Boussinesq presumably furnished the seeds from which grew a large number of more general assertions, most referring to elastic solids of arbitrary shape and many being rather imprecise, concerning the effect on stresses within the body of replacing the tractions acting over a portion of its surface by statically equivalent ones. Such propositions, especially in textbooks, usually went by the name of *Saint-Venant's principle*, despite the fact that Saint-Venant's original conjecture was intended to apply only to cylinders. Ambiguities in these general statements of the principle led von Mises (1945) to suggest an amended version that was precisely formulated and proved by Sternberg (1954). Saint-Venant's principle as proposed by von Mises and established by Sternberg refers to elastic bodies of very general shape, but the probable existence of a stronger version of the principle appropriate to "thin" bodies—such as the long cylinder with which Saint-Venant was originally concerned—was

[†] Saint-Venant (1856a,b).

remarked upon by both von Mises and Sternberg, as well as by Keller (1965). The latter author discussed the form that such a stronger version might be expected to take.

In the mid-1960s, renewed efforts began in an attempt to establish a principle of the Saint-Venant type that might prove to be applicable to the original cylinder problem as well as to related issues involving thin bodies of other types.[†] Despite the fact that this program has not yet been fully realized, the present article undertakes to survey the progress that has been made since 1965.[‡]

When pursued within the framework of the *linear* theory of elasticity, the task of comparing stress fields arising in, say, a cylinder due to two sets of end tractions with a common resultant force and moment but distinct detailed distributions can be reduced immediately by appealing to superposition to the problem of assessing the stresses in the same cylinder subject to *self-equilibrated*, but otherwise arbitrary, tractions on one end while the remainder of the boundary is traction-free. The principal analytic feature shared by most of the recent approaches to this latter question, as well as to related ones for other geometries, is the attempt to carry out this assessment by first studying the decay of strain energy away from the loaded end of the cylinder (or other body) under consideration. The two papers in which this idea was first exploited in connection with Saint-Venant's principle were those of Toupin (1965a), who treated the cylinder problem, and Knowles (1966), who dealt with similar issues in the theory of plane strain. The notion of examining the distribution of strain energy in an elastic body apparently first appeared in papers concerned with Saint-Venant's principle by Zanaboni (1937a,b,c); Zanaboni did not, however, estimate the rate of decay of energy away from the loaded portion of the boundary, and his results[§] do not appear to be directly related to those of Toupin (1965a) or Knowles (1966).

Questions like those occurring in connection with Saint-Venant's principle in the theory of elasticity may also be posed in other branches of mathematical physics. Many of the ideas and techniques that have been introduced since

---

[†] For a clear discussion of the role played by principles of the Saint-Venant type in the foundations of the theory of thin elastic shells, for example, the reader is referred to the lecture presented by Koiter and Simmonds (1973) at the Thirteenth International Congress of Theoretical and Applied Mechanics, Moscow, 1972.

[‡] For references to earlier work, see, for example, the papers of Sternberg (1954), Toupin (1965a,b) and Knowles (1966), or paragraphs 54–56 of the article by Gurtin (1972).

[§] For a discussion of the work of Zanaboni, see paragraph 10.11 of the book by Fung (1965), as well as the comments by Toupin (1965b). Zanaboni's ideas were developed considerably further by Robinson (1966). Robinson's work yields another version of Saint-Venant's principle but does not, however, furnish information about energy decay rates. We will not discuss Robinson's result here; it is described in paragraph 56a of Gurtin (1972).

1965 find their simplest expression in the context of elementary flow problems governed by Laplace's equation. Section II of this present article contains an extensive exposition of many of these ideas and procedures in this simple setting. Principles of the Saint-Venant type as they have been formulated and analyzed within the linear theory of elasticity proper—as distinguished from approximate theories of shells, for example—are the subject of Section III. Section IV is devoted to work bearing on other aspects of Saint-Venant's principle, such as the influence of nonlinearity, some special phenomena that arise within two-dimensional theories of thin elastic shells, Saint-Venant's principle in diffusion or transient heat conduction, and generalizations to viscoelastic materials. In the course of our discussion, we shall point out a number of open questions, especially in Sections III and IV.

There are several aspects of Saint-Venant's *problem*—as distinguished from Saint-Venant's *principle* as we think of it—that are of current interest but which we will not attempt to review here. One of these deals with minimum-energy (or related) characterizations of Saint-Venant's solutions (see, e.g., Shield and Anderson, 1966; Sternberg and Knowles, 1966; Maisonneuve, 1971; and Ericksen, 1980). A second issue in this category concerns the question raised by Truesdell (1959, 1966, 1978) of the effect on a suitably defined "torsional modulus" of varying the end tractions on a cylinder within the class of tractions that produce a given torque.[†] Finally, although we do review some results dealing with nonlinear effects on stress decay, we do not consider the equivalent of the original Saint-Venant cylinder problem in *finite* elasticity (see Ericksen, 1977a,b, 1979, and Muncaster, 1979).

II. A Model Problem

A. Flow in a Cylinder

The simplest context in which to illustrate principles of the Saint-Venant type is that presented by the Neumann problem for Laplace's equation in a cylindrical region. Let $\mathbb{R}$ be the interior of a cylinder of length l whose cross-section is bounded by one or more piecewise-smooth simple closed curves.[‡] Choose Cartesian coordinates x_1, x_2, and x_3 with the origin in one end of

---

[†] In connection with this question, see Day (1981).
[‡] In this section only cylinders with bounded cross sections are considered.

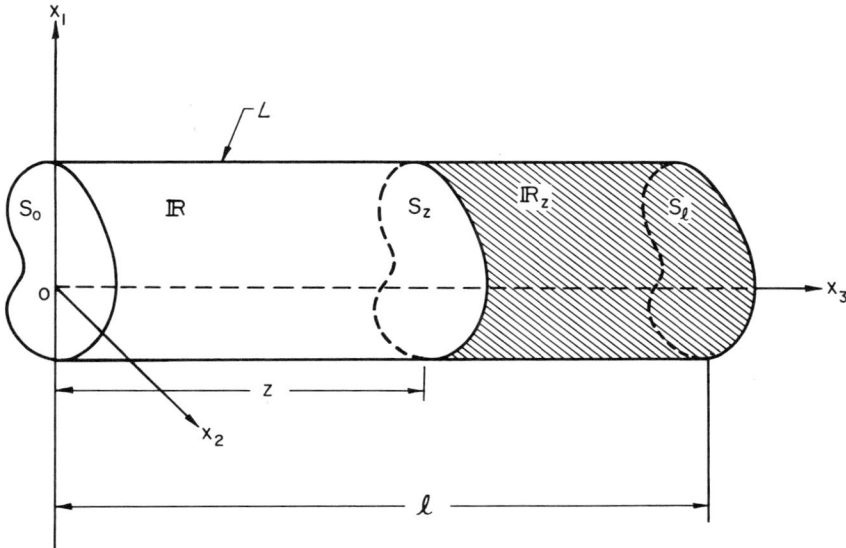

Fig. 1. The cylinder $\mathbb{R}$.

the cylinder and the x_3 axis parallel to the generators (Fig. 1). Suppose that u is twice continuously differentiable on the closure $\bar{\mathbb{R}}$ of $\mathbb{R}$ and satisfies[†]

$$\Delta u \equiv u_{,ii} = 0 \quad \text{on } \bar{\mathbb{R}}. \tag{2.1}$$

Let S_z denote the open cross section of $\mathbb{R}$ for which $x_3 = z$, $0 \le z \le l$. The function u is subject to the boundary conditions

$$u_{,3} = f \quad \text{on } S_0, \qquad u_{,3} = g \quad \text{on } S_l, \tag{2.2}$$

where f and g are given. On the lateral surface L, it is required that

$$\partial u/\partial u = 0 \quad \text{on } L, \tag{2.3}$$

where $\partial/\partial n$ indicates differentiation in the direction of the unit outward normal vector $\mathbf{n}$ on L.

For the existence of a solution u of the prescribed smoothness, it is necessary that the given data f and g be continuously differentiable on $\bar{S}_0$ and $\bar{S}_l$,

---

[†] Latin subscripts have the range 1, 2, 3, while Greek subscripts take only the values 1, 2; when repeated, subscripts are summed over the appropriate range; when preceded by a comma, they indicate differentiation with respect to the corresponding Cartesian coordinate.

respectively, and that they satisfy

$$f_{,\alpha}n_\alpha = 0 \quad \text{on } C_0, \qquad g_{,\alpha}n_\alpha = 0 \quad \text{on } C_1. \tag{2.4}$$

Here C_0 and C_1 are the respective boundaries of S_0 and S_1. Moreover, f and g must also satisfy

$$\int_{S_0} f\, dA = \int_{S_1} g\, dA, \tag{2.5}$$

since the integral over the boundary of $\mathbb{R}$ of the normal derivative of u must vanish.

Although the boundary value problem (2.1)–(2.3) has a number of elementary physical interpretations, it is most convenient for present purposes to regard it as describing the temperature field u associated with a steady flow of heat in a conducting cylinder whose lateral surface is insulated and whose ends are subject to prescribed entry and exit distributions of heat flux characterized by f and g. The common value—call it Q—of the two integrals in Eq. (2.5) represents the resultant heat flow in the positive x_3 direction across either S_0 or S_1. In fact,

$$Q = \int_{S_z} u_{,3}\, dA = \frac{d}{dz} \int_{S_z} u\, dA, \qquad 0 \le z \le l, \tag{2.6}$$

for *any* cross section S_z.

The boundary value problem (2.1)–(2.3) determines u only to within an arbitrary additive constant. To assure the uniqueness of the solution, the normalization adopted here is that which requires the average exit temperature to vanish: Thus u is to satisfy

$$\int_{S_1} u\, dA = 0. \tag{2.7}$$

It follows from Eqs. (2.6) and (2.7) that the average of u over S_z varies linearly with z and is given by

$$\frac{1}{A} \int_{S_z} u\, dA = \frac{Q}{A}(z - l), \qquad 0 \le z \le l, \tag{2.8}$$

where A is the area of S_z.

The case of *uniform* heat flow in the cylinder corresponds to a temperature field u given by

$$u = \frac{Q}{A}(x_3 - l) \qquad \text{on } \mathbb{R}. \tag{2.9}$$

Clearly, u as given in Eq. (2.9) satisfies Eqs. (2.1), (2.3), and (2.7), but the end conditions (2.2) are fulfilled if and only if f and g are constants.

The issue involved in the Saint-Venant principle for the simple problem posed above has to do with the quality of the uniform flow (2.9) as an *approximate* solution of the boundary value problem when the entry and exit distributions f and g are *not* both constants over the end sections.[†] In such cases, significant departures of the exact solution from that given by Eq. (2.9) might be expected to occur only near the ends of the cylinder, and an appropriate principle of the Saint-Venant type would provide a quantitative assessment of the decay of these departures away from the ends. In the present simple situation, such an assessment is easily constructed from the exact solution, as shown in the following section. Since exact solutions are not available for the more complex problems to be discussed later, the remainder of the present section is devoted to a description of methods for providing estimates of the end effects that do *not* depend on representations of the solution. While such indirect methods are not needed for the boundary value problem (2.1)–(2.3), they are illustrated most transparently in this simple setting.

B. A Representation for the Exact Solution

A representation for the solution of the boundary value problem (2.1)–(2.3), (2.7) can be constructed in terms of the eigenvalues and eigenfunctions of the following two-dimensional eigenvalue problem on the cross section of the cylinder:

$$\phi_{,\alpha\alpha} + \lambda^2 \phi = 0 \quad \text{on } S_0, \tag{2.10}$$

$$\partial \phi / \partial n = 0 \quad \text{on } C_0. \tag{2.11}$$

For this problem, $\lambda^2 \equiv \lambda_0^2 = 0$ is an eigenvalue with a corresponding eigenfunction ϕ_0 that is constant on S_0; the remaining eigenvalues are positive and ordered so that $0 < \lambda_1^2 < \lambda_2^2 < \cdots < \lambda_n^2 < \cdots$. The associated eigenfunctions $\{\phi_n\}_{n=0}^{\infty}$ form a complete set and are orthonormal, so that

$$\int_{S_0} \phi_n \phi_m \, dA = \delta_{nm}, \quad n, m = 0, 1, 2, \ldots, \tag{2.12}$$

where δ_{nm} is the Kronecker delta. The *formal* solution to the boundary value problem (2.1)–(2.3), (2.7) may be represented in the form

$$u(x_1, x_2, x_3) = \frac{Q}{A}(x_3 - l) + \sum_{n=1}^{\infty} u_n(x_3) \phi_n(x_1, x_2), \tag{2.13}$$

[†] The uniform flow (2.9) is the analog in the present model problem of Saint-Venant's solution of the *relaxed* problem for the elastic cylinder.

where, for $n = 1, 2, \ldots$, and $0 \leq x_3 \leq l$,

$$u_n(x_3) = -\frac{1}{\lambda_n \sinh(\lambda_n l)} \{f_n \cosh[\lambda_n(l - x_3)] - g_n \cosh(\lambda_n x_3)\}, \quad (2.14)$$

and the constants f_n and g_n are the Fourier coefficients of f and g:

$$f_n = \int_{S_0} f\phi_n \, dA, \quad g_n = \int_{S_l} g\phi_n \, dA, \quad n = 1, 2, \ldots. \quad (2.15)$$

The first term on the right in Eq. (2.13) arises from the zero eigenvalue λ_0^2 and the corresponding constant eigenfunction ϕ_0; it coincides with the special solution (2.9) for uniform flow.

In the presence of sufficient smoothness on the part of the given functions f and g and for a suitably restricted boundary C_0, it may be confirmed that Eq. (2.13) provides the *actual* solution of the underlying boundary value problem. Of greater interest here is the x_3 dependence of the series on the right in Eq. (2.13). From Eq. (2.14) it is easily shown that for $n = 1, 2, \ldots$,

$$|u_n(x_3)| \leq \frac{2}{1 - e^{-2\lambda_1 l}} \left(\frac{|f_n|}{\lambda_n} e^{-\lambda_1 x_3} + \frac{|g_n|}{\lambda_n} e^{-\lambda_1(l - x_3)} \right), \quad 0 \leq x_3 \leq l. \quad (2.16)$$

Since f and g are continuously differentiable on $\bar{S}_0$, their Fourier coefficients satisfy the inequalities[†]

$$\sum_{n=1}^{\infty} \lambda_n^2 f_n^2 \leq \int_{S_0} f_{,\alpha} f_{,\alpha} \, dA, \quad \sum_{n=1}^{\infty} \lambda_n^2 g_n^2 \leq \int_{S_l} g_{,\alpha} g_{,\alpha} \, dA. \quad (2.17)$$

It then follows from Eqs. (2.16) and (2.17) that

$$\left| \sum_{n=1}^{N} u_n(x_3)\phi_n(x_1, x_2) \right| \leq \frac{2}{1 - e^{-2\lambda_1 l}} \left[e^{-\lambda_1 x_3} \left(\sum_{n=1}^{N} \lambda_n^2 f_n^2 \right)^{1/2} \right.$$
$$\left. + e^{-\lambda_1(l-x_3)} \left(\sum_{n=1}^{N} \lambda_n^2 g_n^2 \right)^{1/2} \right] \left(\sum_{n=1}^{N} \frac{\phi_n^2(x_1, x_2)}{\lambda_n^4} \right)^{1/2}$$
$$\leq \frac{2}{1 - e^{-2\lambda_1 l}} (M_0 e^{-\lambda_1 x_3} + M_l e^{-\lambda_1(l - x_3)})$$
$$\times \left(\sum_{n=1}^{N} \frac{\phi_n^2(x_1, x_2)}{\lambda_n^4} \right)^{1/2}, \quad (2.18)$$

where

$$M_0 = \left(\int_{S_0} f_{,\alpha} f_{,\alpha} \, dA \right)^{1/2}, \quad M_l = \left(\int_{S_l} g_{,\alpha} g_{,\alpha} \, dA \right)^{1/2}. \quad (2.19)$$

[†] See paragraph 14.2 of Titchmarsh (1958).

If S_0 has a sufficiently regular boundary, it is possible to prove that the infinite series

$$J = \sum_{n=1}^{\infty} \phi_n^2/\lambda_n^4 \qquad (2.20)$$

converges[†] on the open cross section S_0. If it is further assumed that J is bounded[‡] on S_0, it follows immediately from Eqs. (2.13) and (2.18)–(2.20) that the departure of the solution u from that corresponding to uniform flow satisfies

$$\left| u(x_1, x_2, x_3) - \frac{Q}{A}(x_3 - l) \right| \leq K(M_0 e^{-\lambda_1 x_3} + M_l e^{-\lambda_1(l - x_3)}), \qquad (2.21)$$

where M_0 and M_l are given by Eq. (2.19) and

$$K = \frac{2}{1 - e^{-2\lambda_1 l}} \left(\sup_{S_0} J \right)^{1/2}. \qquad (2.22)$$

The estimate (2.21) makes it clear that the decay of end effects associated with nonuniform entry and exit heat distributions is governed by the smallest positive eigenvalue λ_1^2 of the problem (2.10), (2.11). If $\lambda_1 l \gg 1$, as we shall always assume, the uniform flow (2.9) furnishes a good approximation except in boundary layers near the ends. The extent of these boundary layers is measured by the "characteristic decay length" λ_1^{-1}. For a circular cross section of radius a, an explicit calculation gives $\lambda_1^{-1} = \alpha a$, where $\alpha \doteq 0.543$. For a *hollow* circular cylinder of inner radius b and outer radius a, it turns out that $\lambda_1^{-1} = F(b/a)(a - b)$, where the function F has the property that $F(b/a) \to \pi^{-1}$ as $b/a \to 1$. Thus for a thin-walled tube, the characteristic decay length is of the order of the wall thickness, and hence is much shorter than that for a solid cylinder of the same radius. The solid cylinder is *thin* if $a \ll l$; the *tube* is thin if $a - b \ll l$, regardless of the size of a.

An extensive discussion of the solution (2.13) and its properties as they relate to Saint-Venant's principle may be found in paragraph 2, Chapter II of the thesis of Maisonneuve (1971).

The two-dimensional version of the cylinder problem discussed above involves the construction of a solution $u(x_1, x_2)$ of Laplace's equation

$$\Delta u \equiv u_{,\alpha\alpha} = 0 \quad \text{on } \mathbb{R}, \qquad (2.23)$$

where $\mathbb{R}$ is now the rectangle $0 < x_1 < l$, $0 < x_2 < h$; $u \in C^2(\mathbb{R})$ is required

---

[†] The results required to establish this convergence may be found in paragraphs 17.3 and 18.4 of Titchmarsh (1958).

[‡] If S_0 is, for example, circular or rectangular, the boundedness of J may be verified explicitly.

to satisfy the boundary conditions

$$u_{,1}(0, x_2) = f(x_2), \quad u_{,1}(l, x_2) = g(x_2), \quad 0 \le x_2 \le h, \quad (2.24)$$

$$u_{,2}(x_1, 0) = u_{,2}(x_1, h) = 0, \quad 0 \le x_1 \le l, \quad (2.25)$$

and the normalization condition

$$\int_0^h u(l, x_2)\, dx_2 = 0. \quad (2.26)$$

For this problem to be solvable, it is necessary that f and g satisfy

$$\int_0^h f\, dx_2 = \int_0^h g\, dx_2 \equiv Q, \quad f'(0) = f'(h) = g'(0) = g'(h) = 0, \quad (2.27)$$

where the prime indicates differentiation with respect to the argument.

If f and g are constant on $[0, h]$, the flow is uniform and the corresponding solution of Eqs. (2.23)–(2.26) is

$$u(x_1, x_2) = \frac{Q}{h}(x_1 - l). \quad (2.28)$$

If either f or g fails to be constant, a detailed consideration of the exact solution furnishes the analog of Eq. (2.21):

$$\left| u(x_1, x_2) - \frac{Q}{h}(x_1 - l) \right| \le K(M_0 e^{-\lambda_1 x_1} + M_l e^{-\lambda_1(l - x_1)}), \quad (2.29)$$

where

$$\lambda_1 = \pi/h, \quad (2.30)$$

$$M_0 = \left(\int_0^h (f')^2\, dx_2 \right)^{1/2}, \quad M_l = \left(\int_0^h (g')^2\, dx_2 \right)^{1/2}, \quad (2.31)$$

and

$$K = \left(\frac{h^3}{45} \right)^{1/2} \frac{2}{1 - e^{-2\pi l/h}}. \quad (2.32)$$

[In the two-dimensional case, the analog of J in Eq. (2.20) can be evaluated explicitly, leading to the formula for K given in Eq. (2.32).] As in the three-dimensional problem, we shall always assume that $\pi l/h \gg 1$.

The estimates (2.21) and (2.29) pertain to the solution u itself; the exact solution can, of course, be used to estimate the derivatives of u as well. Indeed, in most contexts it is the *gradient* of u that is of primary physical interest and whose estimation is critical to the establishment of the appropriate principle of the Saint-Venant type.

The implications for Saint-Venant's principle of the exact solution of the two-dimensional problem (2.23)–(2.25) and its axisymmetric analog are discussed by Drake and Chou (1965) for the case in which $g \equiv 0$ and f is "self-equilibrated" in the sense that its integral over $[0, h]$ vanishes.

The main purpose here in considering these simple prototypes of the Saint-Venant problem is to illustrate alternative procedures for obtaining estimates that, like Eq. (2.21) or (2.29), reveal the decay rate λ_1 (or at least a lower bound for it) of the departure from the simple solution (2.9) or (2.28). Since in the more complicated problems to be considered later, exact solutions are rarely available, the alternative techniques about to be illustrated must not rely on representations like Eq. (2.13).

C. An Energy Inequality

All the procedures that have been developed since 1965 for the treatment of Saint-Venant's principle for elasticity problems governed by higher-order equations are based on a consideration of the distribution within the region $\mathbb{R}$ of energy or energy-like quantities.[†] In the present section, the simplest and most natural of the energy decay inequalities is derived for the boundary value problem (2.1)–(2.3), (2.7) and its two-dimensional counterpart (2.23)–(2.26).

Because each of these problems is linear, the essential features of the end effects associated with nonuniform flow can clearly be determined by restricting attention to the special case in which the resultant flow Q is zero, and the exit distribution g vanishes identically. In these circumstances, Eqs. (2.1)–(2.3) and (2.7) reduce to

$$u_{,ii} = 0 \quad \text{on } \mathbb{R}, \tag{2.33}$$

$$u_{,3} = f \quad \text{on } S_0, \qquad u_{,3} = 0 \quad \text{on } S_l, \tag{2.34}$$

$$\partial u/\partial n = 0 \quad \text{on } L, \tag{2.35}$$

$$\int_{S_l} u \, dA = 0, \tag{2.36}$$

where f satisfies the "self-equilibration" condition

$$\int_{S_0} f \, dA = 0. \tag{2.37}$$

[†] For *second-order* problems, maximum principles have also proved useful (see Section II,I).

An immediate consequence of Eqs. (2.33)–(2.35) is the fact that

$$\int_{S_z} u_{,3}\, dA = 0, \qquad 0 \le z \le l; \tag{2.38}$$

from Eq. (2.38) and the normalization condition (2.36), it then follows that for every cross section S_z

$$\int_{S_z} u\, dA = 0, \qquad 0 \le z \le l. \tag{2.39}$$

Because $u_{,3} = 0$ on S_l, one expects [and of course Eqs. (2.21) and (2.19) show] u to decay exponentially away from the end S_0. To show this without recourse to representation (2.13), one may begin by introducing the "energy" distribution

$$E(z) = \int_{\mathbb{R}_z} u_{,i} u_{,i}\, dV, \qquad 0 \le z \le l, \tag{2.40}$$

where $\mathbb{R}_z$ is that portion of the cylinder $\mathbb{R}$ in which $z < x_3 < l$. It is easily shown with the aid of Eqs. (2.33)–(2.35) that

$$E(z) = -\int_{S_z} u_{,3} u\, dA, \qquad 0 \le z \le l; \tag{2.41}$$

moreover, the derivative $E'(z)$ is given by

$$E'(z) = -\int_{S_z} u_{,i} u_{,i}\, dA, \qquad 0 \le z \le l. \tag{2.42}$$

Thus for any constant k,

$$E'(z) + 2kE(z) = -\int_{S_z} (u_{,\alpha} u_{,\alpha} + u_{,3}^2 + 2k u_{,3} u)\, dA, \qquad 0 \le z \le l. \tag{2.43}$$

The objective is to determine a positive value of k (preferably the largest possible one) for which the right-hand side of Eq. (2.43) is nonpositive, thus establishing a differential inequality for $E(z)$ from which its exponentially decaying character will follow. By completing the square appropriately in Eq. (2.43), one finds that

$$E'(z) + 2kE(z) \le -\int_{S_z} (u_{,\alpha} u_{,\alpha} - k^2 u^2)\, dA, \qquad 0 \le z \le l. \tag{2.44}$$

To show that the right-hand side of Eq. (2.44) is nonpositive for a suitable choice of k, one invokes the following integral inequality: If Φ is continuously differentiable on the closure $\bar{S}$ of a bounded plane domain S with a sufficiently regular boundary C, and if

$$\int_S \Phi\, dA = 0, \tag{2.45}$$

then

$$\int_S \Phi_{,\alpha} \Phi_{,\alpha}\, dA \ge \lambda_1^2 \int_S \Phi^2\, dA, \tag{2.46}$$

where λ_1^2 is the smallest positive eigenvalue of the Neumann problem (2.10), (2.11) for the domain[†] S. Choosing $S = S_z$ and $\Phi(x_1, x_2) = u(x_1, x_2, z)$ for fixed z in Eq. (2.46), and observing that Eq. (2.45) is indeed satisfied by virtue of Eq. (2.39), one infers from Eq. (2.44) that

$$E'(z) + 2kE(z) \leq -(\lambda_1^2 - k^2) \int_{S_z} u^2 \, dA, \qquad 0 \leq z \leq l. \qquad (2.47)$$

It is then appropriate to take $k = \lambda_1$ in Eq. (2.47), after which it follows immediately that

$$E(z) \leq E(0) \exp(-2\lambda_1 z), \qquad 0 \leq z \leq l. \qquad (2.48)$$

Thus the energy $E(z)$ contained in $\mathbb{R}_z$ decays exponentially with increasing z with a decay rate at least as great as $2\lambda_1$. It is an easy matter to construct solutions from Eq. (2.13) (with $Q = 0$) whose energy decay rate is precisely $2\lambda_1$, so that in this sense the choice $k = \lambda_1$ supplies the best possible lower bound for the rate of decay.

A result equivalent to Eq. (2.48) but for an n-dimensional cylinder was given by Oleinik (1979a); her argument differs slightly in detail from that used here.

An energy inequality analogous to Eq. (2.48) but for a *Dirichlet* problem for Laplace's equation[‡] in a more general three-dimensional domain was obtained by Weck (1976).

In the two-dimensional problem (2.23)–(2.27) with $g = Q = 0$, one may carry out the above argument without significant change, now making use of the *one-dimensional* version of the inequality (2.46).[§] If $E(z)$ stands for the integral of the square of the gradient of u over the subrectangle $\mathbb{R}_z$ of $\mathbb{R}$ for which $z < x_1 < l$ and $0 < x_2 < h$, it again follows that Eq. (2.48) holds, but now $\lambda_1 = \pi/h$. The two-dimensional version of Eq. (2.48) is a special case of a result of Knowles (1967). For a related *Dirichlet* problem, the energy decay estimate analogous to Eq. (2.48) is a special case of a result obtained by Mieth (1975) (see also Mieth, 1976). In two dimensions, an extremely simple argument of a different kind supporting the idea that the uniform flow (2.28) is a good approximation away from the ends of the rectangle has been given by John (1975).

[†] See Chapter VI of Courant and Hilbert (1953) as well as Chapter V of Mikhlin (1964); λ_1 is proportional to the smallest positive frequency of vibration of a "free" membrane occupying the plane region S.

[‡] We concentrate on the *Neumann* problem because of its analogy with the *traction* boundary value problem in elasticity. The latter problem, in turn, is the one of interest in connection with Saint-Venant's principle.

[§] See p. 402 of Courant and Hilbert (1953).

There remains, however, the problem of obtaining from the energy inequality (2.48) (in two *or* three dimensions) a pointwise estimate for u itself analogous to Eq. (2.29) or (2.21) with $Q = M_l = 0$. Discussion of this technically troublesome matter, as well as the corresponding issue for ∇u, is deferred until Section II,H below.

D. An Alternative Energy Procedure

The simple argument leading to Eq. (2.48) turns out not to generalize readily to the problems involving differential equations of higher order that are taken up later in this article. The present section is devoted to a sketch of a modified version of the preceding analysis that *does* carry over to at least some of these more involved problems.[†] The technique is due to Toupin (1965a) and was used by him in one of the first papers to employ energy decay estimates in connection with Saint-Venant's principle for the elastic cylinder problem.

An important feature of the argument we are about to give is that it makes no use of Eq. (2.39) [or, for that matter, of the normalization (2.7)]; this is essential, because there is no strict analog of the property (2.39) of u in the elastic cylinder problem to be treated in Section III,D,1.

We take Eq. (2.41) as a starting point and first observe that if

$$\bar{u}(x_1, x_2, x_3) = u(x_1, x_2, x_3) - a, \qquad (2.49)$$

where a is independent of x_1, x_2, and x_3 and is as yet undetermined, then by Eq. (2.38)

$$E(z) = -\int_{S_z} u_{,3}\bar{u}\, dA, \qquad 0 \leq z \leq l. \qquad (2.50)$$

We let α be an arbitrary positive number and apply the arithmetic-geometric mean inequality to infer from Eq. (2.50) that

$$E(z) \leq \frac{1}{2}\int_{S_z}\left(\alpha u_{,3}^2 + \frac{1}{\alpha}\bar{u}^2\right)dA, \qquad 0 \leq z \leq l. \qquad (2.51)$$

Supposing z to be fixed in $(0, l)$ and choosing the number δ so that $0 < \delta < l - z$, we may integrate Eq. (2.51) from z to $z + \delta$ to get

$$\int_z^{z+\delta} E(\zeta)\,d\zeta \leq \frac{\alpha}{2}\int_{\mathbb{R}_{z,\delta}} u_{,i}u_{,i}\,dV + \frac{1}{2\alpha}\int_{\mathbb{R}_{z,\delta}} \bar{u}^2\,dV, \qquad (2.52)$$

where $\mathbb{R}_{z,\delta}$ denotes that "slice" of the cylinder $\mathbb{R}$ for which $z < x_3 < z + \delta$. The next step requires an appeal to the three-dimensional version of the

---

[†] For still another argument leading to an energy decay inequality for the boundary value problem (2.33)–(2.36), see Section II,F,2.

inequality (2.46): *If Ψ is a continuously differentiable function on the closure $\bar{D}$ of a bounded three-dimensional domain D with a sufficiently regular boundary $\mathscr{S}$, and if*

$$\int_D \Psi \, dV = 0, \tag{2.53}$$

then

$$\int_D \Psi_{,i} \Psi_{,i} \, dV \geq \omega_1^2 \int_D \Psi^2 \, dV, \tag{2.54}$$

where ω_1^2 is the smallest positive eigenvalue of the three-dimensional (acoustic) eigenvalue problem

$$\psi_{,ii} + \omega^2 \psi = 0 \quad \text{on } D, \qquad \frac{\partial \psi}{\partial n} = 0 \quad \text{on } \mathscr{S}. \tag{2.55}$$

In order to apply Eq. (2.54) to Eq. (2.52) with $D = \mathbb{R}_{z,\delta}$ and $\Psi = \bar{u}$, we must first choose a in Eq. (2.49) so that $\bar{u}$ has a zero average[†] over $\mathbb{R}_{z,\delta}$. Then from Eqs. (2.54) and (2.49),

$$\int_{\mathbb{R}_{z,\delta}} \bar{u}^2 \, dV \leq \frac{1}{\omega_1^2(\delta)} \int_{\mathbb{R}_{z,\delta}} \bar{u}_{,i} \bar{u}_{,i} \, dV = \frac{1}{\omega_1^2(\delta)} \int_{\mathbb{R}_{z,\delta}} u_{,i} u_{,i} \, dV, \tag{2.56}$$

where $\omega_1^2(\delta)$ is the smallest positive eigenvalue[‡] of the problem (2.55) for the slice $\mathbb{R}_{z,\delta}$. Using Eq. (2.56) in Eq. (2.52) yields

$$\int_z^{z+\delta} E(\zeta) \, d\zeta \leq \frac{1}{2}\left(\alpha + \frac{1}{\alpha \omega_1^2(\delta)}\right) \int_{\mathbb{R}_{z,\delta}} u_{,i} u_{,i} \, dV. \tag{2.57}$$

The coefficient in parentheses before the integral sign in Eq. (2.57) is minimized by choosing $\alpha = 1/\omega_1(\delta)$; this choice of α leads to

$$\int_z^{z+\delta} E(\zeta) \, d\zeta \leq \frac{1}{\omega_1(\delta)} \int_{\mathbb{R}_{z,\delta}} u_{,i} u_{,i} \, dV. \tag{2.58}$$

But

$$\int_{\mathbb{R}_{z,\delta}} u_{,i} u_{,i} \, dV = E(z) - E(z+\delta) = -\frac{d}{dz} \int_z^{z+\delta} E(\zeta) \, d\zeta, \tag{2.59}$$

so that Eq. (2.58) may be written as

$$\frac{d}{dz} \int_z^{z+\delta} E(\zeta) \, d\zeta + \omega_1(\delta) \int_z^{z+\delta} E(\zeta) \, d\zeta \leq 0. \tag{2.60}$$

[†] With the normalization (2.7) in force, Eq. (2.39) would show that $a = 0$ and $u = \bar{u}$, but we make no use of this here.

[‡] The notation indicates explicitly that ω_1 depends on the thickness δ of the cylindrical slice $\mathbb{R}_{z,\delta}$; ω_1, of course, also depends on the shape of S_z, but not on z itself.

Integration of this differential inequality from 0 to z furnishes

$$\int_z^{z+\delta} E(\zeta)\,d\zeta \leq \int_0^\delta E(\zeta)\,d\zeta \exp[-\omega_1(\delta)z], \qquad 0 \leq z \leq l - \delta. \tag{2.61}$$

Since $E(z)$ decreases monotonically with increasing z, Eq. (2.61) implies that

$$E(z + \delta) \leq E(0)\exp[-\omega_1(\delta)z], \qquad 0 \leq z \leq l - \delta,$$

or, equivalently,

$$E(z) \leq E(0)\exp[-\omega_1(\delta)(z - \delta)], \qquad 0 < \delta \leq z \leq l. \tag{2.62}$$

Equation (2.62) provides an exponential decay inequality for $E(z)$ over the subcylinder for which $\delta \leq z \leq l$ in terms of the fundamental positive frequency $\omega_1(\delta)$ of free *acoustic* vibrations of the cylindrical slice $\mathbb{R}_{z,\delta}$ whose thickness δ is as yet unspecified. For the acoustic eigenvalue problem (2.55) on the slice $\mathbb{R}_{z,\delta}$, it is not difficult to show that, when δ is chosen so as to maximize $\omega_1(\delta)$, the eigenfunction ψ corresponding to $\omega_1(\delta)$ is independent of x_3; such an eigenfunction, in fact, is a solution of the *two*-dimensional eigenvalue problem (2.10), (2.11), so that $\omega_1(\delta)$ is given by

$$\omega_1(\delta) = \lambda_1, \tag{2.63}$$

and hence is independent[†] of δ. Thus Eq. (2.62) reduces to

$$E(z) \leq E(0)e^{-\lambda_1(z-\delta)}, \qquad 0 < \delta \leq z \leq l. \tag{2.64}$$

Since λ_1, $E(z)$, and $E(0)$ are independent of δ, one may let $\delta \to 0+$ in Eq. (2.64) to get

$$E(z) \leq E(0)e^{-\lambda_1 z}, \qquad 0 \leq z \leq l. \tag{2.65}$$

In comparing the energy inequalities (2.48) and (2.65) obtained by the two different arguments, one observes that the latter underestimates the best possible decay rate by a factor of 2.

It is the analog of Eq. (2.62) [rather than Eq. (2.65)] for the elastic cylinder problem that was obtained by Toupin (1965a) (see Section III,D,1). For the boundary value problem (2.33)–(2.36) for Laplace's equation, the result (2.62) was undoubtedly known to Toupin in 1965.

The inequality (2.65), which in view of Eq. (2.48) is *not* the best possible one, was derived from the exact solution by Maisonneuve (1971).

For the two-dimensional problem (2.23)–(2.27) with $g = Q = 0$, the "slice argument" can be carried out in the same way. The resulting analog of Eq. (2.62) then involves the fundamental frequency of free acoustic vibra-

---

[†] This simplification does not occur when the present argument is adapted to the elastic cylinder problem (see Section III,D,1).

tions for a rectangular slice of thickness δ and height h; this frequency has the value π/h (for $\delta < h$), and the analog of Eq. (2.65) again follows, now with $\lambda_1 = \pi/h$.

Finally, it may be remarked that either of the decay inequalities (2.48) or (2.62) becomes fully explicit once an upper bound is known for the total energy $E(0)$. Such a bound can be obtained with the aid of standard minimum principles from the calculus of variations; we illustrate this briefly in Section II,F.

E. Energy Decay for Other Linear Elliptic Second-Order Problems

The energy argument sketched in Section II,C for the two-dimensional Neumann problem was developed by Knowles (1967) for elliptic equations in "divergence form": $(pu_{,1})_{,1} + (qu_{,2})_{,2} = 0$. The domain is taken to be a rectangle, and the energy estimate is appropriate to the case in which the rectangle is slender.

An exponential decay estimate analogous to Eq. (2.48) was obtained by Mieth (1975) (see also Mieth, 1976) for still more general two-dimensional second-order linear elliptic operators on plane multiply connected domains. He treated boundary conditions of the Dirichlet type.

For domains that are not necessarily thin, exponential decay inequalities for the energy distribution may not be appropriate or may fail altogether, as in the case of certain unbounded domains. Wu (1970) investigated one situation of this kind for Laplace's equation in two dimensions by an argument strictly parallel to that given in Section II,C. His results include, for example, the case of a ring sector domain $\mathbb{R}$ bounded by circular arcs $r = a$ and $r = b$, $0 < a < b \leq \infty$, and radial lines $\theta = 0$ and $\theta = \alpha$, $0 < \alpha \leq 2\pi$. For a harmonic function on $\mathbb{R}$ whose normal derivative vanishes at $\theta = 0$, $\theta = \alpha$, and $r = b$, Wu showed that the energy $E(r)$ stored beyond the radius r satisfies $E(r) \leq E(a)(a/r)^{2\pi/\alpha}$.

Results for a broad range of mixed boundary value problems for general second-order linear elliptic equations in n dimensions have been obtained by Oleinik and Yosifian (1977a). They established an energy decay principle of the Saint-Venant type for weak solutions and for a general class of not necessarily thin, not necessarily bounded domains. Again, the decay need not be exponential. For example, in unbounded *conical* domains, they obtained a power-law decay inequality. Oleinik and Yosifian used their energy results as a priori estimates to establish uniqueness and existence theorems in unbounded domains. The former are generalizations of the Phragmén–Lindelöf theorem for harmonic functions.

Berdichevskii (1978) also discussed energy decay inequalities and associated cross-sectional eigenvalue problems for a general class of second-order elliptic operators.

F. An Upper Bound for the Total Energy $E(0)$

1. An Upper Bound for $E(0)$

We sketch here a procedure for constructing an upper bound for the total energy $E(0)$ appearing in the energy decay inequalities (2.48) and (2.62). The scheme described here is based on a variational principle and makes no use of representations (2.13) and (2.14).

If u is the solution of the Neumann problem (2.33)–(2.36), a standard minimum principle of the calculus of variations[†] asserts that for any vector field $\mathbf{v}$ that is continuously differentiable on $\mathbb{R}$ and satisfies

$$v_{i,i} = 0 \quad \text{on } \mathbb{R}, \tag{2.66}$$

as well as

$$v_3 = f \quad \text{on } S_0, \qquad v_3 = 0 \quad \text{on } S_l, \qquad \mathbf{v} \cdot \mathbf{n} = 0 \quad \text{on } L, \tag{2.67}$$

one has

$$\int_\mathbb{R} |\mathbf{v}|^2 \, dV \geq \int_\mathbb{R} |\nabla u|^2 \, dV \equiv E(0). \tag{2.68}$$

It is a simple matter to verify directly that Eqs. (2.66) and (2.67) imply Eq. (2.68).

Let us choose for $\mathbf{v}$ a vector field with components of the form

$$v_\alpha = -\xi_\alpha(x_1, x_2)\zeta'(x_3), \qquad v_3 = f(x_1, x_2)\zeta(x_3), \tag{2.69}$$

where f is the given heat input in Eq. (2.34) and ξ_1, ξ_2, and ζ are (as yet) undetermined functions. Assume that ξ_1 and ξ_2 are continuously differentiable on $\bar{S}_0$ and satisfy

$$\xi_{\alpha,\alpha} = f \quad \text{on } \bar{S}_0, \qquad \xi_\alpha n_\alpha = 0 \quad \text{on } C_0. \tag{2.70}$$

Assume further that ζ is twice continuously differentiable on $[0, l]$ and such that

$$\zeta(0) = 1, \qquad \zeta(l) = 0. \tag{2.71}$$

Then $\mathbf{v}$ will satisfy Eqs. (2.66) and (2.67), so that by Eq. (2.68)

$$E(0) \leq \int_\mathbb{R} |\mathbf{v}|^2 \, dV = I \int_0^l [\zeta'(x_3)]^2 \, dx_3 + H \int_0^l [\zeta(x_3)]^2 \, dx_3, \tag{2.72}$$

[†] See paragraph 53, Chapter VII of Mikhlin (1964).

where

$$I = \int_{S_0} \xi_\alpha \xi_\alpha \, dA > 0, \qquad H = \int_{S_0} f^2 \, dA > 0. \qquad (2.73)$$

One now applies the procedures of the calculus of variations so as to determine ξ_1, ξ_2, and ζ such that the right member of Eq. (2.72) is minimized, subject to the constraint $(2.70)_1$ and the boundary conditions $(2.70)_2$, (2.71). When this is done, one finds that ξ_1 and ξ_2 are given by

$$\xi_\alpha = V_{,\alpha} \qquad \text{on } \bar{S}_0, \qquad (2.74)$$

where the function V satisfies

$$V_{,\alpha\alpha} = f \quad \text{on } \bar{S}_0, \qquad \partial V/\partial n = 0 \quad \text{on } C_0; \qquad (2.75)$$

without loss of generality, it may be assumed that

$$\int_{S_0} V \, dA = 0. \qquad (2.76)$$

Moreover, the optimum ζ is given by

$$\zeta(x_3) = \frac{\sinh[\mu(l - x_3)]}{\sinh(\mu l)}, \qquad (2.77)$$

where

$$\mu = \sqrt{H/I}. \qquad (2.78)$$

Substituting from Eq. (2.77) into Eq. (2.72) and carrying out the integration, one obtains

$$E(0) \leq \frac{H}{\mu} \coth(\mu l). \qquad (2.79)$$

In Eqs. (2.78) and (2.79) H is given in terms of the datum f by the second part of Eq. (2.73), and from the first part of Eq. (2.73) and Eq. (2.74),

$$I = \int_{S_0} |\nabla V|^2 \, dA. \qquad (2.80)$$

Applying the divergence theorem to Eq. (2.80) and making use of Eq. (2.75), one obtains

$$I = -\int_{S_0} fV \, dA. \qquad (2.81)$$

By employing the Schwarz inequality and inequality (2.46) with $\Phi = V$ [see Eq. (2.76)], one finds from Eq. (2.81) that

$$I \leq \frac{1}{\lambda_1} \left(\int_{S_0} f^2 \, dA \int_{S_0} |\nabla V|^2 \, dA \right)^{1/2}. \qquad (2.82)$$

In view of the second part of Eq. (2.73) and Eqs. (2.80) and (2.78), inequality (2.82) may be written as

$$\mu \geq \lambda_1. \tag{2.83}$$

Since $\mu^{-1} \coth(\mu l)$ decreases monotonically in μ, Eqs. (2.83) and (2.79) and the second part of Eq. (2.73) imply the final estimate

$$E(0) \leq \frac{\coth(\lambda_1 l)}{\lambda_1} \int_{S_0} f^2 \, dA. \tag{2.84}$$

From the exact solution (2.13), (2.14) (recall that $g \equiv 0$ under present circumstances), one can show that if $f = \phi_1$, equality holds in Eq. (2.84). In this sense, Eq. (2.84) is a best possible inequality. It may be derived readily from representation (2.13), as was done by Maisonneuve (1971).

When Eq. (2.84) is utilized in either Eq. (2.48) or (2.62), there follows a fully determinate upper bound for the distribution of energy in terms of given data.

For the self-equilibrated version of the two-dimensional Neumann problem (2.23)–(2.27) ($g = Q = 0$), a similar argument shows that inequality (2.84) holds with $\lambda_1 = \pi/h$, where S_0 now stands for the end segment $x_1 = 0$, $0 < x_2 < h$, of the rectangle $\mathbb{R}$.

2. An Associated Decay Inequality

We digress briefly here to show how the upper bound (2.84) for $E(0)$ can *itself* be used to establish an energy decay estimate different from those already discussed.

For the boundary value problem (2.33)–(2.36), the entry distribution f is self-equilibrated [see Eq. (2.37)]; thus $Q = 0$, so Eq. (2.6) reads

$$\int_{S_z} u_{,3} \, dA = 0, \qquad 0 \leq z \leq l, \tag{2.85}$$

for every cross section S_z. As a result, one may immediately apply Eq. (2.84) to the subcylinder $\mathbb{R}_z$ to obtain

$$E(z) \leq \frac{\coth[\lambda_1(l-z)]}{\lambda_1} \int_{S_z} u_{,3}^2 \, dA, \qquad 0 \leq z < l. \tag{2.86}$$

It is possible to show by a direct calculation based only on Eqs. (2.33) and (2.35) that

$$\int_{S_z} (u_{,3}^2 - u_{,\alpha} u_{,\alpha}) \, dA = c = \text{const}, \qquad 0 \leq z \leq l, \tag{2.87}$$

i.e., the integral in Eq. (2.87) is independent[†] of z. Using the boundary condition of the second part of Eq. (2.34), one concludes from Eq. (2.87) that

$$c = -\int_{S_l} u_{,\alpha} u_{,\alpha} dA \leq 0. \tag{2.88}$$

It follows from Eqs. (2.87), (2.42), and (2.88) that

$$\int_{S_z} u_{,3}^2 dA = \frac{1}{2}\int_{S_z} u_{,3}^2 dA + \frac{1}{2}\int_{S_z} u_{,\alpha} u_{,\alpha} dA + \frac{c}{2}$$

$$= -\frac{1}{2} E'(z) + \frac{c}{2} \leq -\frac{1}{2} E'(z), \qquad 0 \leq z \leq l. \tag{2.89}$$

Making use of Eq. (2.89) in Eq. (2.86) gives

$$E(z) \leq -\frac{\coth[\lambda_1(l-z)]}{2\lambda_1} E'(z), \qquad 0 \leq z < l. \tag{2.90}$$

Integration of this differential inequality on the interval $[0, z]$ immediately furnishes the decay estimate

$$E(z) \leq E(0)\left(\frac{\cosh[\lambda_1(l-z)]}{\cosh(\lambda_1 l)}\right)^2, \qquad 0 \leq z \leq l. \tag{2.91}$$

One can show that Eq. (2.48) implies both Eqs. (2.65) and (2.91), so that Eq. (2.48) is always the strongest of the three energy decay inequalities. It can also be shown that if $\lambda_1 l \geq \log\sqrt{3}$, Eq. (2.91) implies Eq. (2.65); otherwise the reverse is true. Thus for sufficiently long cylinders, Eq. (2.91) is a stronger decay estimate than Eq. (2.65).

For the two-dimensional problem, Eq. (2.91) holds with $\lambda_1 = \pi/h$, and the immediately preceding discussion applies in this case as well.

G. HIGHER-ORDER ENERGIES AND CROSS-SECTIONAL ESTIMATES

1. *Higher-Order Energies*

The energy decay inequalities derived in the preceding sections may be regarded as providing *mean-square* estimates on $\mathbb{R}_z$ for the first derivatives of the solution of the Neumann problem. While it is possible to obtain from

---

[†] The "conservation law" (2.87) is a consequence of the fact that a variational principle equivalent to the boundary value problem under consideration has a certain invariance property with respect to translations of the axial coordinate. Such conservation laws are intimately connected with Noether's theorem (see, e.g., p. 262 of Courant and Hilbert, 1953); they occur in the theory of elasticity as well, where they have given rise to a large literature (see, e.g., Knowles and Sternberg, 1972). The relevance to Saint-Venant's cylinder problem of the analog in elasticity of Eq. (2.87) has been discussed by Ericksen (1977a,b, 1979) and Muncaster (1979).

these estimates *pointwise* bounds for u and ∇u that are valid in the interior[†] of $\mathbb{R}$, pointwise estimates which remain valid up to the boundary of $\mathbb{R}$ require mean-square estimates of higher derivatives of u. In order to construct these, it is helpful to first introduce the notion of "higher-order energies."

The argument of Section II,C shows that

$$E_1(z) \equiv E(z) = \int_{\mathbb{R}_z} |\nabla u|^2 \, dV, \qquad 0 \le z \le l, \tag{2.92}$$

satisfies an inequality of the form

$$E_1(z) \le E_1(0) e^{-2kz}, \qquad 0 \le z \le l, \tag{2.93}$$

where $k = \lambda_1$. Since u is assumed to be twice continuously differentiable on $\overline{\mathbb{R}}$, and since it is then infinitely differentiable on $\mathbb{R}$ itself, we may set

$$v = u_{,3} \qquad \text{on } \overline{\mathbb{R}} \tag{2.94}$$

and observe that v satisfies

$$v_{,ii} = 0 \qquad \text{on } \mathbb{R}, \tag{2.95}$$

$$\begin{aligned} v &= f \qquad \text{on } S_0, \\ v &= 0 \qquad \text{on } S_l, \end{aligned} \tag{2.96}$$

$$\partial v / \partial n = 0 \qquad \text{on } L. \tag{2.97}$$

If one then defines a *second-order energy* $E_2(z)$ by

$$E_2(z) = \int_{\mathbb{R}_z} |\nabla v|^2 \, dV, \qquad 0 \le z \le l, \tag{2.98}$$

it is easy to show by an argument virtually identical to that of Section II,C that

$$E_2(z) \le E_2(0) e^{-2kz}, \qquad 0 \le z \le l, \tag{2.99}$$

where the value of k is the same as that entering Eq. (2.93). In view of Eqs. (2.98) and (2.94), inequality (2.99) furnishes mean-square estimates on $\mathbb{R}_z$ of the second derivatives $u_{,3i}$, and, by the differential equation (2.33), of the "cross-sectional Laplacian" $u_{,\alpha\alpha}$ as well.

By requiring still greater smoothness[‡] on the part of u (and hence f), we may introduce

$$w = v_{,3} = u_{,33} \qquad \text{on } \overline{\mathbb{R}} \tag{2.100}$$

[†] This is carried out in Section II,H.
[‡] $u \in C^3(\overline{\mathbb{R}})$.

and observe that w is a solution of a Neumann problem of the same type as that which determines u:

$$w_{,ii} = 0 \quad \text{on } \bar{\mathbb{R}}, \tag{2.101}$$

$$\begin{aligned} w_{,3} &= -f_{,\alpha\alpha} \quad \text{on } S_0, \\ w_{,3} &= 0 \quad \text{on } S_l, \end{aligned} \tag{2.102}$$

$$\partial w/\partial n = 0 \quad \text{on } L. \tag{2.103}$$

[Note that $-f_{,\alpha\alpha}$ satisfies the self-equilibration condition analogous to Eq. (2.37) because of the first part of (2.4).] It is thus clear that the *third-order energy*

$$E_3(z) = \int_{\mathbb{R}_z} |\nabla w|^2 \, dV, \quad 0 \leq z \leq l, \tag{2.104}$$

will satisfy

$$E_3(z) \leq E_3(0)e^{-2kz}, \quad 0 \leq z \leq l. \tag{2.105}$$

From Eq. (2.105) one obtains mean-square estimates on $\mathbb{R}_z$ of the third derivatives $u_{,33i}$ as well as of $u_{,\alpha\alpha i}$.

As in Section II,F,1, one can show on the basis of a minimum principle that the total second-order energy $E_2(0)$ satisfies

$$E_2(0) \leq \frac{\coth(\lambda_1 l)}{\lambda_1} \int_{S_0} |\nabla f|^2 \, dA. \tag{2.106}$$

The analogy between the w problem (2.101)–(2.103) and the u problem (2.33)–(2.37), together with Eq. (2.84), immediately furnishes an upper bound for $E_3(0)$:

$$E_3(0) \leq \frac{\coth(\lambda_1 l)}{\lambda_1} \int_{S_0} (\Delta f)^2 \, dA, \tag{2.107}$$

where Δ denotes the two-dimensional Laplacian. When Eqs. (2.106) and (2.107) are used in Eqs. (2.99) and (2.105), respectively, there follow fully explicit decay estimates for the two higher-order energies $E_2(z)$ and $E_3(z)$.

While the preceding discussion has been based on the analysis of Section II,C, an entirely similar program could be carried out using the "slice argument" of Section II,D.

2. Cross-Sectional Estimates

In order to obtain pointwise estimates valid up to the boundary of $\mathbb{R}$, as we shall do in Section II,H,2, it is necessary first to construct mean-square estimates *on the cross section* S_z (rather than on $\mathbb{R}_z$). We sketch here one procedure for doing so. Let $\psi \in C^2(\mathbb{R})$ be an *arbitrary* function, and let its associated first- and second-order energies F_1 and F_2 be defined by

$$F_1(z) = \int_{\mathbb{R}_z} |\nabla \psi|^2 \, dV, \quad F_2(z) = \int_{\mathbb{R}_z} |\nabla \psi_{,3}|^2 \, dV, \quad 0 \le z \le l. \quad (2.108)$$

Set

$$\Psi_i(z) = \int_{S_z} \psi_{,i}^2 \, dA, \quad 0 \le z \le l, i = 1, 2, 3, \quad (2.109)$$

and observe that

$$\Psi_i(z) = \Psi_i(l) - \int_z^l \Psi_i'(x_3) \, dx_3$$

$$= \Psi_i(l) - 2 \int_{\mathbb{R}_z} \psi_{,i} \psi_{,i3} \, dV, \quad 0 \le z \le l, \text{ no sum on } i. \quad (2.110)$$

This identity, together with Eq. (2.108), immediately yields

$$\Psi_i(z) \le \Psi_i(l) + 2[F_1(z) F_2(z)]^{1/2}, \quad 0 \le z \le l, i = 1, 2, 3. \quad (2.111)$$

In some of the applications that we shall make of Eq. (2.111) $\Psi_i(l)$ is known, but for others it is not. When the latter is the case, we estimate $\Psi_i(l)$ in terms of F_1 and F_2 in the following way. Suppose that δ is momentarily arbitrary, except that $0 < \delta < l$. Then

$$\Psi_i(l) = \int_{l-\delta}^l \frac{d}{dz}\left(\frac{\delta - (l-z)}{\delta} \Psi_i(z)\right) dz, \quad i = 1, 2, 3 \quad (2.112)$$

or

$$\Psi_i(l) = \int_{l-\delta}^l \left[-\frac{1}{\delta} \Psi_i(z) + \left(1 - \frac{l-z}{\delta}\right) \Psi_i'(z)\right] dz. \quad (2.113)$$

Thus by Eqs. (2.109) and (2.113),

$$\Psi_i(l) = -\frac{1}{\delta} \int_{\mathbb{R}_{l-\delta}} \psi_{,i}^2 \, dV + 2 \int_{\mathbb{R}_{l-\delta}} \left(1 - \frac{l-z}{\delta}\right) \psi_{,i} \psi_{,i3} \, dV, \quad \text{no sum on } i, \quad (2.114)$$

so that bearing Eq. (2.108) in mind, assuming[†] $kl > 1$, and choosing $\delta = 1/2k$, we obtain

$$\Psi_i(l) \le 2k F_1\left(l - \frac{1}{2k}\right) + 2\left[F_1\left(l - \frac{1}{2k}\right) F_2\left(l - \frac{1}{2k}\right)\right]^{1/2}. \quad (2.115)$$

Inequalities (2.111) and (2.115) together furnish an estimate of the cross-sectional mean square (2.109) in terms of the two energies F_1 and F_2.

[†] See the discussion following Eq. (2.22).

In our *first* application of these results, we choose $\psi = u$, so that $F_1 = E_1$ and $F_2 = E_2$ [see Eqs. (2.92) and (2.98)]; it then follows from Eqs. (2.111) and (2.115) (with $i = 1$ or 2) and the decay inequalities (2.93) and (2.99) that

$$\int_{S_z} u_{,\alpha}^2 \, dA \leq 2e\{kE_1(0) + [E_1(0)E_2(0)]^{1/2}\}e^{-2kl}$$
$$+ 2[E_1(0)E_2(0)]^{1/2} e^{-2kz}, \qquad 0 \leq z \leq l, \alpha = 1, 2. \quad (2.116)$$

When $\psi = u$ and $i = 3$ in Eq. (2.111), we may capitalize on the fact that $u_{,3} = 0$ on S_l, so that $\Psi_3(l) = 0$, in order to avoid Eq. (2.115) altogether and obtain directly from Eqs. (2.111), (2.93), and (2.99) the simpler estimate

$$\int_{S_z} u_{,3}^2 \, dA \leq 2[E_1(0)E_2(0)]^{1/2} e^{-2kz}, \qquad 0 \leq z \leq l. \quad (2.117)$$

Next we take $\psi = v = u_{,3}$, so that $F_1 = E_2$ and $F_2 = E_3$. Bearing the second part of Eq. (2.96) in mind, we now find from Eqs. (2.111), (2.115), (2.94), (2.99), and (2.105) that

$$\int_{S_z} v_{,\alpha}^2 \, dA = \int_{S_z} u_{,3\alpha}^2 \, dA \leq 2[E_2(0)E_3(0)]^{1/2} e^{-2kz}, \qquad 0 \leq z \leq l, \quad (2.118)$$

and

$$\int_{S_z} v_{,3}^2 \, dA = \int_{S_z} u_{,33}^2 \, dA \leq 2e\{kE_2(0) + [E_2(0)E_3(0)]^{1/2}\}e^{-2kl}$$
$$+ 2[E_2(0)E_3(0)]^{1/2} e^{-2kz}, \qquad 0 \leq z \leq l. \quad (2.119)$$

In view of Eq. (2.39), we may choose $\Phi = u$ and $S = S_z$ in Eq. (2.46) and use the resulting inequality, together with Eq. (2.116), to obtain a mean-square estimate over S_z of u itself:

$$\int_{S_z} u^2 \, dA \leq \frac{2e}{\lambda_1^2} \{kE_1(0) + [E_1(0)E_2(0)]^{1/2}\}e^{-2kl}$$
$$+ \frac{2}{\lambda_1^2} [E_1(0)E_2(0)]^{1/2} e^{-2kz}, \qquad 0 \leq z \leq l. \quad (2.120)$$

Note that Eq. (2.119), in view of Eq. (2.33), supplies a mean-square estimate over S_z of the "cross-sectional Laplacian" $u_{,\alpha\alpha}$.

Again, results of the kind obtained above may also be derived on the basis of Eq. (2.62) rather than Eq. (2.48).

The higher-order energy decay inequalities (2.99) and (2.105), the total energy bounds (2.106) and (2.107), and the cross-sectional estimates (2.116)–(2.120) also hold for the corresponding *two-dimensional* Neumann problem. In this case, the associated higher order energies are defined in strict analogy with Eqs. (2.94) and (2.98) and (2.100) and (2.104); however, in Eqs. (2.116) and (2.118), α takes only the value 2, and x_3 derivatives are to be replaced by x_1 derivatives in Eqs. (2.117)–(2.119). In the two-dimensional case, the analogs of (2.116)–(2.119), together with the differential equation $u_{,\alpha\alpha} = 0$, furnish cross-sectional estimates of *all* first and second derivatives of u. In contrast, Eqs. (2.118) and (2.119) provide mean-square estimates of only *certain* second derivatives in the three-dimensional problem.

When coupled with the total energy bounds (2.84), (2.106), and (2.107), the preceding cross-sectional estimates are fully determined. Corresponding results hold in two dimensions.

Higher-order energies in the two-dimensional case were introduced (although along slightly different lines) by Knowles (1967), who considered a more general class of second-order linear elliptic operators. In the same article he obtained the two-dimensional analogs of Eqs. (2.116), (2.117), and (2.120)—also for more general equations—by essentially the arguments given here. Upper bounds on first- and second-order total energies were also constructed in the same article.

An analogous notion of higher-order energies had been employed earlier by Shield (1965) in connection with problems of the dynamic stability of continuous linear systems. In both Knowles (1967) and Shield (1965), the higher-order energies ultimately serve to furnish pointwise estimates. In the latter work, *time* plays the role of the axial coordinate in the preceding analysis. Indeed, there would seem to be a striking parallel between the energy arguments employed in connection with principles of Saint-Venant type and those used in discussions of dynamic stability (especially in stability arguments of the Lyapunov type). The apparent analogy between the spatial variable in the elastic cylinder problem and time in dynamical problems was in fact perceived by Saint-Venant himself[†] (see also the remarks on p. 74 of Muncaster, 1979).

H. Pointwise Decay

1. *Interior Estimates*

The discussion in the present section is concerned with the construction of pointwise estimates, valid only in the interior of $\mathbb{R}$, for the solution u and its gradient in the three-dimensional Neumann problem (2.33)–(2.36) with a "self-equilibrated" entry distribution f [see Eq. (2.37)]. These estimates

---

[†] In discussing the role played by his special solutions of the elastic cylinder problem, Saint-Venant (1856a), p. 314, commented as follows: "Il s'établit ici, dans l'espace, une sorte d'état permanent semblable à celui qui est produit, dans le temps, par l'action continue de causes constantes qui finissent par effacer l'effet des causes initiales d'un grand nombre de phénomènes."

are based on the first-order energy inequality (2.48)† and the mean value theorem for harmonic functions.

Let $\mathbb{R}^{(d)}$ be an open cylinder whose closure $\overline{\mathbb{R}^{(d)}}$ is contained in the open cylinder $\mathbb{R}$, whose generators are parallel to those of $\mathbb{R}$, and for which the minimum distance from $\mathbb{R}^{(d)}$ to the boundary of $\mathbb{R}$ is $2d$. Let $\mathbf{x} = (x_1, x_2, z)$ be a point in $\overline{\mathbb{R}^{(d)}}$ and denote by B the interior of the sphere of radius d centered at $\mathbf{x}$. According to the mean value theorem for harmonic functions,

$$u(\mathbf{x}) = \frac{3}{4\pi d^3} \int_B u \, dV, \tag{2.121}$$

because B lies entirely in $\mathbb{R}$. Thus

$$u^2(\mathbf{x}) \leq \frac{3}{4\pi d^3} \int_B u^2 \, dV \leq \frac{3}{4\pi d^3} \int_{\mathbb{R}_{z-d}} u^2 \, dV. \tag{2.122}$$

Since u satisfies Eq. (2.39), inequality (2.46) (with $\Phi = u$ and $S = S_z$), together with Eq. (2.122), furnishes

$$u^2(\mathbf{x}) \leq \frac{3}{4\pi d^3 \lambda_1^2} \int_{\mathbb{R}_{z-d}} u_{,\alpha} u_{,\alpha} \, dV$$

$$\leq \frac{3}{4\pi d^3 \lambda_1^2} E_1(z - d), \quad \mathbf{x} \in \overline{\mathbb{R}^{(d)}}. \tag{2.123}$$

Combining Eq. (2.123) with Eq. (2.48) yields an *interior estimate*:

$$|u(x_1, x_2, z)| \leq \left(\frac{3}{4\pi d^3 \lambda_1^2} E_1(0)\right)^{1/2} e^{-\lambda_1(z-d)}, \quad \mathbf{x} \in \overline{\mathbb{R}^{(d)}}. \tag{2.124}$$

Since ∇u is also harmonic, Eq. (2.121) holds with u replaced by $u_{,i}$, $i = 1, 2, 3$. It follows immediately that

$$|u_{,i}(x_1, x_2, z)| \leq \left(\frac{3}{4\pi d^3} E_1(0)\right)^{1/2} e^{-\lambda_1(z-d)}, \quad \mathbf{x} \in \overline{\mathbb{R}^{(d)}}. \tag{2.125}$$

Inequalities (2.124) and (2.125) establish the exponential decay of u and $u_{,i}$ on the subcylinder $\mathbb{R}^{(d)}$. It is clear, however, that these estimates deteriorate if $\mathbb{R}^{(d)}$ is allowed to expand to the boundary of $\mathbb{R}$, since then $d \to 0$. One would, in fact, not expect a pointwise estimate such as Eq. (2.124) or (2.125) to remain in force at the boundary‡ of $\mathbb{R}$, based as they are on essentially local considerations and ignoring as they do the details of the boundary of S_z. An estimate that does remain satisfactory near this boundary must

---

† They could equally well be based on Eq. (2.62).
‡ See the comments of Toupin (1965a, p. 84), paraphrasing a remark attributed to Saint-Venant himself, concerning pointwise estimates in the elastic cylinder problem.

rest on more stringent assumptions concerning the smoothness of u (and hence the smoothness of f and the regularity of C_0).

Results entirely analogous to Eqs. (2.124) and (2.125) hold for the two-dimensional problem. They have been derived for a more general (not necessarily simply connected) plane domain by Mieth (1975, 1976) by essentially the argument given here. Mieth (1975) also derived—by other methods—an interior estimate for a *Dirichlet* problem involving a more general two-dimensional linear elliptic operator.

2. Estimates Valid up to the Boundary

If the solution u of the Neumann problem is smooth enough to tolerate the analysis of Section II,G pertaining to higher-order energies and cross-sectional mean squares, it is possible to construct pointwise estimates of u and its first derivatives, which, in contrast to Eqs. (2.124) and (2.125), remain valid up to the boundary. We shall sketch here the details of these constructions for the *two-dimensional* Neumann problem; afterward we shall comment briefly on the corresponding issues in the three-dimensional case.

Let $\mathbb{R}$ be the open rectangle $0 < x_1 < l$, $0 < x_2 < h$, and let $u \in C^3(\overline{\mathbb{R}})$ be the solution of the two-dimensional problem (2.23)–(2.27) with $g \equiv 0$ and thus with f self-equilibrated. Let $x_1 = z$ be fixed, with $0 \le z \le l$, and write the differential equation (2.23) and the boundary conditions (2.25) at $x_2 = 0$ and $x_2 = h$ as follows:

$$\left.\begin{array}{ll} u_{,22} = -u_{,11}, & 0 \le x_2 \le h \\ u_{,2} = 0, & x_2 = 0 \text{ and } x_2 = h \end{array}\right\} x_1 = z. \qquad (2.126)$$

We treat Eq. (2.126) as a one-dimensional boundary value problem for $u(z, x_2)$ on the interval $0 \le x_2 \le h$, proceeding as if the right-hand side of the differential equation in Eq. (2.126) were known. One needs to observe that, as in the three-dimensional problem [see Eq. (2.39)],

$$\int_0^h u(z, x_2)\, dx_2 = 0, \qquad 0 \le z \le l. \qquad (2.127)$$

It follows from Eqs. (2.126) and (2.127) that u may be "represented" as

$$u(z, x_2) = \int_0^h N(x_2, \eta) u_{,11}(z, \eta)\, d\eta, \qquad 0 \le x_2 \le h, \qquad (2.128)$$

where the "one-dimensional Neumann kernel" N is given by

$$N(x_2, \eta) = \begin{cases} (h - \eta)^2/2h, & 0 \le x_2 \le \eta \le h, \\ (h^2 + \eta^2 - 2hx_2)/2h, & 0 \le \eta \le x_2 \le h. \end{cases} \qquad (2.129)$$

[Note that $N(x_2, \eta)$ has a zero average with respect to x_2 on $[0, h]$ for each η; moreover, $N(x_2, \eta)$ is *not* symmetric in (x_2, η).]

Using the differential equation in Eq. (2.126), followed by an integration by parts and an appeal to the boundary conditions in Eq. (2.126), we obtain

$$u(z, x_2) = \int_0^h \frac{\partial N}{\partial \eta}(x_2, \eta) u_{,2}(z, \eta) \, d\eta. \tag{2.130}$$

Since

$$\left|\frac{\partial N}{\partial \eta}(x_2, \eta)\right| \leq 1, \qquad 0 \leq \eta \leq h, 0 \leq x_2 \leq h, \tag{2.131}$$

we find from Eq. (2.130) that

$$u^2(z, x_2) \leq h \int_0^h u_{,2}^2(z, \eta) \, d\eta, \, 0 \leq x_2 \leq h, 0 \leq z \leq l. \tag{2.132}$$

From the two-dimensional analog of Eq. (2.116), we conclude that

$$u^2(z, x_2) \leq 2he\{kE_1(0) + [E_1(0)E_2(0)]^{1/2}\}e^{-2kl}$$
$$+ 2h[E_1(0)E_2(0)]^{1/2}e^{-2kz}, \qquad 0 \leq z \leq l, 0 \leq x_2 \leq h. \tag{2.133}$$

[Recall that $kl > 1$; this assumption underlies Eq. (2.115) and hence Eqs. (2.116) and (2.133) as well.] Here E_1 and E_2 are the appropriate (two-dimensional) first- and second-order energies

$$E_1(z) = \int_{R_z} u_{,\alpha} u_{,\alpha} \, dA, \qquad E_2(z) = \int_{R_z} u_{,\alpha 1} u_{,\alpha 1} \, dA, \tag{2.134}$$

and the decay constant k is given by

$$k = \pi/h. \tag{2.135}$$

If one now uses the two-dimensional versions of the total energy bounds (2.84) and (2.106), one obtains from Eqs. (2.133) and (2.135) a pointwise decay estimate of u in terms of the integrals of f^2 and $(f')^2$. If the one-dimensional version of inequality (2.46) is used to estimate the former of these integrals in terms of the latter, there results an upper bound for u^2 that may be directly compared with that obtained from Eq. (2.29) (with $Q = M_l = 0$). For a thin rectangle ($l \gg h$), each of these upper bounds is of the form

$$u^2(z, x_2) \leq ch^3 \left(\int_0^h (f')^2 \, dx_2\right) e^{-2\pi z/h}, \qquad 0 \leq z \leq l, 0 \leq x_2 \leq h. \tag{2.136}$$

For the bound obtained from Eq. (2.133), $c \doteq 1.3$. The corresponding bound based on Eq. (2.29) (and hence on a representation for the exact solution) has $c \doteq 0.1$.

To estimate the derivative $u_{,1}$, we return to Eq. (2.128) and write

$$u_{,1}(z, x_2) = \int_0^h N(x_2, \eta) u_{,111}(z, \eta) \, d\eta. \tag{2.137}$$

As in the argument leading to Eq. (2.132), we find

$$u_{,1}^2(z, x_2) \leq h \int_0^h u_{,12}^2(z, \eta) \, d\eta. \tag{2.138}$$

From Eq. (2.138) and the two-dimensional version of Eq. (2.118) (with $\alpha = 2$) we obtain a pointwise bound for $u_{,1}^2$ analogous to the estimate (2.133) for u^2:

$$u_{,1}^2(z, x_2) \leq 2h[E_2(0)E_3(0)]^{1/2} e^{-2\pi z/h}, \qquad 0 \leq z \leq l, 0 \leq x_2 \leq h. \tag{2.139}$$

To illustrate the fully explicit pointwise estimate, we record the result of using the total energy bounds—for two dimensions—from Eqs. (2.106) and (2.107) in Eq. (2.139):

$$u_{,1}^2(z, x_2) \leq \frac{2h^2}{\pi} \coth\left(\frac{\pi l}{h}\right) \left(\int_0^h (f')^2 \, dx_2 \int_0^h (f'')^2 \, dx_2 \right)^{1/2} e^{-2\tau z/h}, \tag{2.140}$$

$$0 \leq z \leq l, 0 \leq x_2 \leq h.$$

A similar argument supplies a pointwise estimate for $u_{,2}$; we omit it here.

For the three-dimensional problem (2.33)–(2.36), an approach to pointwise estimates that is parallel to that outlined previously depends on the theory of the *two-dimensional* Neumann problem for Poisson's equation. As before, one writes the differential equation (2.33) and the boundary condition (2.35) as

$$u_{,\alpha\alpha} = -u_{,33} \qquad \text{on } S_z, \tag{2.141}$$

$$\frac{\partial u}{\partial n} = 0 \qquad \text{on } C_z, \tag{2.142}$$

where C_z is the translation of C_0 to S_z. Suppose that a *Neumann kernel* (or Green's function of the second kind) $N(x_1, x_2; \xi_1, \xi_2)$ exists for the Laplace operator on the plane domain S_z; let N be normalized by the requirement that

$$\int_{S_z} N(x_1, x_2; \xi_1, \xi_2) \, dA_x = 0. \tag{2.143}$$

Then because $u_{,33}$ in Eq. (2.141) satisfies

$$\int_{S_z} u_{,33} \, dA = 0, \tag{2.144}$$

we have from Eqs. (2.141) and (2.142) that[†]

$$u(x_1, x_2, z) = \int_{S_z} N(x_1, x_2; \xi_1, \xi_2) u_{,33}(\xi_1, \xi_2, z) \, dA_\xi. \tag{2.145}$$

[†] See Titchmarsh (1958, paragraph 14.18), or Bergman and Schiffer (1953, p. 263). We follow the former author's conventions as to N, in particular as regards the normalization (2.143). As a result, N is not symmetric here. Bergman and Schiffer employ a different normalization that assures that their Neumann kernel *is* symmetric.

Let

$$\bar{N}(x_1, x_2) = \frac{1}{A} \int_{S_z} N(x_1, x_2; \xi_1, \xi_2) \, dA_\xi, \qquad (x_1, x_2) \in S_z, \qquad (2.146)$$

where A is the area of S_z. In view of Eq. (2.144), we may write Eq. (2.145) as

$$u(x_1, x_2, z) = \int_{S_z} [N(x_1, x_2; \xi_1, \xi_2) - \bar{N}(x_1, x_2)] u_{,33}(\xi_1, \xi_2, z) \, dA_\xi. \qquad (2.147)$$

Although N has a logarithmic singularity when $x_\alpha = \xi_\alpha$, it is square integrable on S_z, so that we may infer from Eq. (2.147) that

$$u^2(x_1, x_2, z) \leq \int_{S_z} \tilde{N}^2(x_1, x_2; \xi_1, \xi_2) \, dA_\xi \int_{S_z} u_{,33}^2(\xi_1, \xi_2, z) \, dA_\xi, \qquad (2.148)$$

where we have written

$$\tilde{N}(x_1, x_2; \xi_1, \xi_2) = N(x_1, x_2; \xi_1, \xi_2) - \bar{N}(x_1, x_2). \qquad (2.149)$$

One can show that when the normalization (2.143) is in force,

$$\int_{S_z} \tilde{N}^2(x_1, x_2; \xi_1, \xi_2) \, dA_\xi = J(x_1, x_2), \qquad (x_1, x_2) \in S_z, \qquad (2.150)$$

where J is given by Eq. (2.20). If J is bounded[†] on S_z, it follows from Eqs. (2.148) and (2.150) that

$$|u(x_1, x_2, z)| \leq \left(\sup_{S_z} J \right)^{1/2} \left(\int_{S_z} u_{,33}^2 \, dA \right)^{1/2}, \qquad (x_1, x_2) \in S_z, \, 0 \leq z \leq l. \qquad (2.151)$$

When combined with Eqs. (2.119), (2.106), and (2.107), inequality (2.151) furnishes a fully explicit pointwise estimate of u that is comparable in form to Eq. (2.21), where, in the latter, $Q = M_l = 0$.

If Eq. (2.145) is used as a starting point, the pointwise estimation of ∇u presents technical difficulties, since the gradient of N is not square integrable[‡] on S_z. These can be overcome by considering still higher-order energies, but we shall not pursue this issue here.

The procedure described previously is parallel to—and was suggested by—the technique used by Shield (1965) to obtain pointwise bounds in his analysis of dynamic stability.

In two dimensions, a scheme for the pointwise estimation of u was developed by Knowles (1967) for more general second-order elliptic operators. The program followed there does not depend on the theory (or, in fact, the

---

[†] Recall that if S_z is, for example, circular or rectangular, the boundedness of J may be verified explicitly.

[‡] See Shield (1965, p. 662), where similar difficulties arise.

existence) of the Neumann kernel, and is based on arguments akin to those used to establish inequalities of the Sobolev type.[†]

Making use of higher-order energies in much the same way as in Knowles (1967), Wu (1970) obtained pointwise estimates for the Neumann problem for Laplace's equation in plane domains that are not necessarily thin.

Pointwise estimates valid *at* and *near* the boundary were constructed for Dirichlet problems for Laplace's equation, as well as for more general elliptic operators, in plane, not necessarily simply connected domains by Mieth (1975) on the basis of entirely different procedures.

I. A Different Approach: Maximum Principles

It has been observed by Wheeler *et al.* (1975) (see also Wheeler and Horgan, 1976) that pointwise decay estimates for the derivatives of solutions of the two-dimensional Neumann problem[‡] (2.23)–(2.27) in the self-equilibrated case ($g = Q = 0$) can be derived with the aid of maximum principles for second-order elliptic equations. As pointed out by these authors, an argument based on the maximum principle—when it is available—leads to explicit, although not necessarily best possible, results more readily than does an analysis based on energy integrals and the pointwise estimates associated with them. This ease of application, however, is achieved at a substantial cost in generality: Extensions to higher-order equations—such as arise in most situations in elastostatics—appear to be out of the question, since appropriate maximum principles are not available. Neither linearity nor ellipticity, however, is essential (see Sections IV,A,1 and IV,C,1).

To apply arguments based on maximum principles (or comparison principles) one converts the Neumann problem (2.23)–(2.27) (with $g = Q = 0$) to a Dirichlet problem for a function v, conjugate to u, that satisfies

$$u_{,1} = v_{,2}, \quad u_{,2} = -v_{,1} \quad \text{on } \bar{\mathbb{R}}. \tag{2.152}$$

(Since u is twice continuously differentiable on $\bar{\mathbb{R}}$, so is v.) Then

$$v_{,\alpha\alpha} = 0 \quad \text{on } \bar{\mathbb{R}}, \tag{2.153}$$

$$v(0, x_2) = \int_0^{x_2} f(\eta)\, d\eta \equiv F(x_2), \quad v(l, x_2) = 0, \; 0 \le x_2 \le h, \tag{2.154}$$

$$v(x_1, 0) = v(x_1, h) = 0, \quad 0 \le x_1 \le l, \tag{2.155}$$

where, by virtue of Eq. (2.27) with $Q = 0$, the function F satisfies $F(h) = 0$.

[†] See the discussion in Section III,D,3.

[‡] The domain $\mathbb{R}$ on which the Neumann problem was considered by Wheeler *et al.* (1975) is actually more general than the rectangle treated here.

The pointwise estimate

$$|v(x_1, x_2)| \le \frac{h}{2}\left(\max_{[0,h]} |f(x_2)|\right) \sin\left(\frac{\pi}{h} x_2\right) e^{-(\pi/h)x_1}, \qquad 0 \le x_1 \le l, 0 \le x_2 \le h, \tag{2.156}$$

can then be established. The decay constant in Eq. (2.156) is the best possible.

To prove Eq. (2.156), set

$$w(x_1, x_2) = M \sin\left(\frac{\pi}{h} x_2\right) e^{-(\pi/h)x_1} \qquad \text{on } \mathbb{R}, \tag{2.157}$$

where the constant M is defined by

$$M = \max_{0 \le x_2 \le h} \left(\frac{|F(x_2)|}{\sin[(\pi/h)x_2]}\right) \ge 0; \tag{2.158}$$

M is finite, since $F(x_2)$ is continuously differentiable for $0 \le x_2 \le h$ and $F(0) = F(h) = 0$. Now observe that w is nonnegative, satisfies Laplace's equation, and

$$w(x_1, 0) = v(x_1, 0) = w(x_1, h) = v(x_1, h) = 0, \qquad 0 \le x_1 \le l. \tag{2.159}$$

Furthermore,

$$w(l, x_2) \ge 0 = v(l, x_2), \qquad 0 \le x_2 \le h, \tag{2.160}$$

and

$$w(0, x_2) = M \sin\left(\frac{\pi x_2}{h}\right) \ge v(0, x_2), \qquad 0 \le x_2 \le h, \tag{2.161}$$

the last step following by virtue of the definition of M in Eq. (2.158). It follows from the maximum principle for harmonic functions (see, e.g., Protter and Weinberger, 1967) that

$$v \le w \qquad \text{on } \mathbb{R}. \tag{2.162}$$

A similar argument shows that

$$-w \le v \qquad \text{on } \mathbb{R}, \tag{2.163}$$

so that one obtains

$$|v| \le w \qquad \text{on } \mathbb{R}. \tag{2.164}$$

It can be shown (Wheeler et al., 1975; Wheeler and Horgan, 1976) that

$$M \le \frac{h}{2} \max_{[0,h]} |f(x_2)|. \tag{2.165}$$

Estimate (2.156) now follows from Eqs. (2.157), (2.164), and (2.165).

An estimate for the derivative $v_{,2}$ [and so, from the first part of Eq. (2.152), for $u_{,1}$] on the long sides of the rectangle is now easily established. Thus, since v and w both vanish at $x_2 = 0$ and $x_2 = h$ for each x_1 in $0 \le x_1 \le l$, it follows from Eq. (2.164) that

$$|v_{,2}(x_1,0)| \leq w_{,2}(x_1,0), \quad |v_{,2}(x_1,h)| \leq -w_{,2}(x_1,h), \quad 0 \leq x_1 \leq l, \quad (2.166)$$

and so

$$|v_{,2}(x_1,0)| = |u_{,1}(x_1,0)| \leq M \frac{\pi}{h} e^{-(\pi/h)x_1}, \quad 0 \leq x_1 \leq l, \quad (2.167)$$

with an identical result for $|v_{,2}(x_1,h)| = |u_{,1}(x_1,h)|$. In view of Eq. (2.165), we may write Eq. (2.167) as

$$|u_{,1}(x_1,0)| \leq \frac{\pi}{2} \max_{[0,h]} |f(x_2)| e^{-(\pi/h)x_1}, \quad 0 \leq x_1 \leq l, \quad (2.168)$$

with an identical estimate holding for $|u_{,1}(x_1,h)|$. Since $u_{,2}$ vanishes on the long sides of the rectangle, Eq. (2.168) and its counterpart at $x_2 = h$ provide explicit estimates for the *gradient* of u on these portions of the boundary, exhibiting the desired exponential decay. The decay rate is the best possible.

The establishment of estimates for the derivatives of v (and thereby for the derivatives of u) that are valid on the entire closed rectangle requires considerable effort of a technical nature. Using the maximum principle for gradients of harmonic functions,[†] Wheeler et al. (1975) obtained a decay estimate for the gradient of the solution of the Neumann problem for a class of curvilinear "strip" domains. For the special case of the rectangle $\mathbb{R}$ considered here, their final result has the form

$$u_{,\alpha}(x_1,x_2)u_{,\alpha}(x_1,x_2) \leq A e^{-\pi x_1/2h} \quad \text{on } \mathbb{R}, \quad (2.169)$$

where the constant A is given by

$$A = [1 + 4(1 - e^{-\pi l/h})^{-2}] \left\{ \max\left[\left(1 + \frac{h^2}{l^2}\right)^{1/2} \bar{f} + \frac{h}{2}\bar{f}', \frac{\pi}{2}\bar{f} \right] \right\}. \quad (2.170)$$

Here

$$\bar{f} = \max_{[0,h]} |f(x_2)|, \quad \bar{f}' = \max_{[0,h]} |f'(x_2)|, \quad (2.171)$$

where f is the given function in Eq. (2.24). It may be observed that the decay rate associated with the right-hand side of Eq. (2.169) underestimates the best possible value by a factor of 4. It was pointed out by Wheeler et al. (1975) that the decay rate in Eq. (2.169) can be improved (without major changes in their argument) at the cost of enlarging the constant A. This improvement, however, cannot be made to yield the best possible decay rate.

The foregoing results were generalized by Wheeler and Horgan (1976) to uniformly elliptic equations of the form $(pu_{,1})_{,1} + (qu_{,2})_{,2} = 0$ on rectangular domains.

[†] See, e.g., Protter and Weinberger (1967).

III. Linear Elastostatic Problems

A. Axisymmetric Torsion

1. Decay Estimates

One of the simplest elastostatic problems to which an analysis of the type described in Section II can be applied is that of the pure axisymmetric torsion of a body of revolution according to the linear equilibrium theory of homogeneous isotropic elastic materials.[†] A detailed account of the formulation and analysis of this problem, as well as of most of the results summarized in this section, may be found in Knowles and Sternberg (1966).

Consider a homogeneous isotropic elastic body of revolution whose interior occupies the domain $\mathbb{R}$ and whose meridional cross section may be either that of a *hollow* body or that of a *solid*, as shown in Fig. 2. It is assumed for simplicity that $\mathbb{R}$ has plane (circular or annular) ends S_0 and S_l; the distance between them is l. Let x_1, x_2, and x_3 be Cartesian coordinates with origin 0 at the center of S_0 and let the x_3 axis be along the axis of revolution of $\mathbb{R}$, while r, θ, and z are cylindrical coordinates with origin 0 with the z axis coinciding with the x_3 axis. If $\mathbb{R}$ is hollow, it is described by

$$\mathbb{R} = \{(r,\theta,z) | h_1(z) < r < h_2(z), 0 \le \theta < 2\pi, 0 < z < l\}, \tag{3.1}$$

while for the solid body

$$\mathbb{R} = \{(r,\theta,z) | 0 < r < h(z), 0 \le \theta < 2\pi, 0 < z < l\}. \tag{3.2}$$

The functions h_α or h that describe the shape of $\mathbb{R}$ are assumed to be continuously differentiable on $[0,l]$. The problem of axisymmetric torsion requires the determination of a stress tensor field τ and a displacement vector field $\mathbf{u}$ that, in the absence of body forces, satisfy the basic field equations of linear elastostatics for a homogeneous isotropic material with shear modulus $\mu > 0$ and Poisson's ratio v, $-1 < v < \frac{1}{2}$. The lateral surface of $\mathbb{R}$ is to be free of traction, and on the plane ends

$$\begin{aligned}\tau_{zr} = \tau_{zz} = 0 \quad &\text{on } S_0 \text{ and } S_l, \\ \tau_{z\theta} = f(r) \quad \text{on } S_0, \quad \tau_{z\theta} &= g(r) \quad \text{on } S_l,\end{aligned} \tag{3.3}$$

where τ_{zr}, τ_{zz}, and $\tau_{z\theta}$ are cylindrical components of τ and f and g are the given axisymmetric (torsional) shear tractions. For overall equilibrium it is necessary that f and g satisfy

$$\int_{S_0} rf(r)\,dA = \int_{S_l} rg(r)\,dA = T, \tag{3.4}$$

where T is the applied scalar torque.

[†] This theory is presented, for example, in Sokolnikoff (1956) and Gurtin (1972).

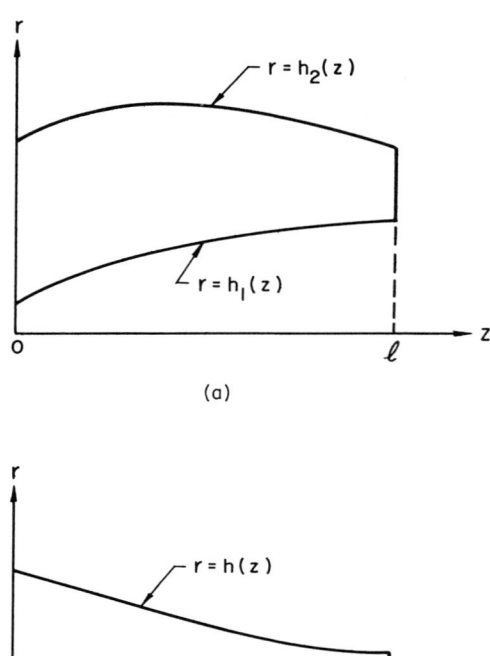

FIG. 2. Meridional cross section of a body of revolution. (a) Hollow body. (b) Solid body.

In the context of axisymmetric torsion, a Saint-Venant principle must provide a quantitative estimate of the effect upon the stored energy, stress, and deformation of replacing the given torsional loading by tractions that are statically equivalent[†] to those specified in Eq. (3.3) and which are again purely axisymmetric and torsional. Because the problem is linear, it is clearly sufficient to consider only the case for which a self-equilibrated load[‡] $f(r)$ is applied to S_0 while S_l is traction free, so that $g(r) \equiv 0$.

It is possible to reduce the foregoing elastostatic boundary value problem to one of the Neumann type for a second-order elliptic differential equation satisfied by the "angular twist" $\phi(r, z) = r^{-1} u_\theta(r, z)$, where u_θ is the circumferential displacement (the only nonvanishing cylindrical component of

[†] Thus the replacement loading results in the same torque T.
[‡] Thus $T = 0$.

displacement present in the problem). Alternately, the final second-order boundary value problem may be formulated as one of the Dirichlet type in terms of a stress function $\psi(r, z)$ for which

$$\tau_{r\theta} = -\frac{\mu}{r^2}\frac{\partial \psi}{\partial z}, \qquad \tau_{z\theta} = \frac{\mu}{r^2}\frac{\partial \psi}{\partial r}, \qquad (3.5)$$

while

$$\tau_{zz} = \tau_{\theta\theta} = \tau_{rr} = \tau_{rz} = 0 \qquad \text{on } \mathbb{R}. \qquad (3.6)$$

The differential equation for ψ is

$$\frac{\partial}{\partial r}\left(\frac{1}{r^3}\frac{\partial \psi}{\partial r}\right) + \frac{\partial}{\partial z}\left(\frac{1}{r^3}\frac{\partial \psi}{\partial z}\right) = 0 \qquad \text{on } \mathbb{R}. \qquad (3.7)$$

For the case of a self-equilibrated load at $z = 0$ and a traction-free end at $z = l$, the boundary conditions for the *hollow* body are

$$\psi(r, 0) = F(r) \equiv \frac{1}{\mu}\int_{h_1(0)}^{r} \rho^2 f(\rho)\,d\rho, \qquad h_1(0) \leq r \leq h_2(0),^\dagger \qquad (3.8)$$

$$\psi(r, l) = 0, \qquad h_1(l) \leq r \leq h_2(l), \qquad (3.9)$$

$$\psi[h_1(z), z] = \psi[h_2(z), z] = 0, \qquad 0 \leq z \leq l. \qquad (3.10)$$

For the *solid* body, Eqs. (3.8)–(3.10) are replaced by

$$\psi(r, 0) = F(r) \equiv \frac{1}{\mu}\int_0^r \rho^2 f(\rho)\,d\rho, \qquad 0 \leq r \leq h(0), \qquad (3.11)$$

$$\psi(r, l) = 0, \qquad 0 \leq r \leq h(l), \qquad (3.12)$$

$$\psi[h(z), z] = 0, \qquad 0 \leq z \leq l, \qquad (3.13)$$

$$\psi(0, z) = 0, \qquad 0 \leq z \leq l. \qquad (3.14)$$

[Actually, one must have $\partial\psi/\partial r = O(r^3)$ and $\partial\psi/\partial z = O(r^3)$ as $r \to 0$ in the case of the solid body.]

The stored energy per unit volume W associated with a stress field τ in an isotropic homogeneous elastic solid is given by[‡]

$$W = \frac{1}{4\mu}\left(\tau_{ij}\tau_{ij} - \frac{v}{1+v}\tau_{ii}\tau_{jj}\right), \qquad (3.15)$$

where τ_{ij} are the Cartesian components of τ. Because of Eqs. (3.5) and (3.6), W in the present circumstances reduces to

$$W = \frac{1}{2\mu}(\tau_{r\theta}^2 + \tau_{z\theta}^2) = \frac{\mu}{2r^4}\left[\left(\frac{\partial\psi}{\partial r}\right)^2 + \left(\frac{\partial\psi}{\partial z}\right)^2\right]. \qquad (3.16)$$

[†] Self-equilibration requires that F vanish at $r = h_2(0)$ (hollow body) or $r = h(0)$ (solid body).
[‡] See p. 85 of Sokolnikoff (1956).

Let $\mathbb{R}_z$ stand for the portion of $\mathbb{R}$ for which $x_3 > z$ and set

$$E(z) = \int_{R_z} W \, dV, \qquad 0 \le z \le l. \tag{3.17}$$

It is shown by an argument differing only in detail from that used in Section II,C that

$$E(z) \le E(0) \exp\left(-2 \int_0^z \lambda_1(\zeta) \, d\zeta\right), \qquad 0 \le z \le l, \tag{3.18}$$

where in the case of the *hollow* body $\lambda_1(z)$ is the smallest positive root of the equation

$$Y_2[\lambda h_1(z)] J_2[\lambda h_2(z)] - J_2[\lambda h_1(z)] Y_2[\lambda h_2(z)] = 0, \tag{3.19}$$

whereas in the *solid* case λ_1 is the smallest positive root of

$$J_2[\lambda h(z)] = 0. \tag{3.20}$$

Here J_2 and Y_2 are the second-order Bessel functions of the first and second kinds, respectively. For the solid body, Eqs. (3.18) and (3.20) give

$$E(z) \le E(0) \exp\left(-2\alpha \int_0^z \frac{d\zeta}{h(\zeta)}\right), \qquad 0 \le z \le l, \tag{3.21}$$

where $\alpha \doteq 5.135$ is the smallest positive zero of J_2.

In general, the rate of decay associated with Eq. (3.18) [or Eq. (3.21)] is not constant because of the variable radii of the cross sections of $\mathbb{R}$. In the special case of a solid circular cylinder of radius h, Eq. (3.21) shows that the energy decay rate is at least $2\alpha/h$. In this instance, the boundary value problem (3.7), (3.11)–(3.14) can, of course, be solved explicitly, and it may be confirmed that the decay rate $2\alpha/h$ is the best possible one. A similar result holds for the hollow circular cylinder; for a *thin* shell, this yields a characteristic decay length proportional to the shell thickness.

Pointwise interior estimates for the shear stresses $\tau_{z\theta}$ and $\tau_{r\theta}$ can be obtained from the energy inequality (3.18) by means of a procedure parallel to that described in Section II,H,1. The principal tool employed is a mean value theorem due to Diaz and Payne (1958) for isotropic homogeneous elastic solids: If $\mathbf{x}$ is a point in $\mathbb{R}$ and B is a sphere centered at $\mathbf{x}$, contained in $\mathbb{R}$, and of radius d, then

$$\tau_{ij}(\mathbf{x}) = \frac{15}{8\pi d^5} \int_B \tau_{ik}(\mathbf{y})(5\eta_j \eta_k - \eta_m \eta_m \delta_{kj}) \, dV, \tag{3.22}$$

where

$$\eta_i = y_i - x_i. \tag{3.23}$$

An application of Schwarz's inequality to Eq. (3.22) leads to the inequality

$$|\tau_{ij}(\mathbf{x})| \le \frac{15}{\sqrt{\pi}} d^{-3/2} \left(\int_B \tau_{km} \tau_{km} \, dV\right)^{1/2}. \tag{3.24}$$

Finally, $\tau_{km}\tau_{km}$ can be estimated in terms of the strain energy density $W(\tau)$ by the inequality[†]

$$\tau_{km}\tau_{km} \leq \frac{4\mu}{\beta(v)} W(\tau), \qquad (3.25)$$

valid for homogeneous isotropic materials, where

$$\beta(v) = \min\left(1, \frac{1-2v}{1+v}\right), \qquad -1 < v < \frac{1}{2}. \qquad (3.26)$$

Combining Eqs. (3.24), (3.25), and (3.17), one obtains

$$|\tau(r,z)| \leq \frac{30}{\sqrt{\pi}} \left(\frac{\mu}{\beta(v)d^3}\right)^{1/2} \sqrt{E(z-d)}, \qquad (3.27)$$

where τ stands for either $\tau_{r\theta}$ or $\tau_{z\theta}$. Together Eqs. (3.27) and (3.18) furnish the interior decay estimate

$$|\tau(r,z)| \leq \frac{30}{\sqrt{\pi}} \left(\frac{\mu}{\beta(v)d^3}\right)^{1/2} \sqrt{E(0)} \exp\left(-\int_0^{z-d} \lambda_1(\zeta)\,d\zeta\right). \qquad (3.28)$$

The *mean twist* $\Theta(z)$ defined by

$$\Theta(z) = \frac{1}{A(z)} \int_{S_z} \frac{\partial}{\partial z}\left(\frac{1}{r} u_\theta(r,z)\right) dA, \qquad 0 \leq z \leq l, \qquad (3.29)$$

where $A(z)$ is the area of the cross section S_z, can be estimated for the *hollow body*[‡] in terms of $E(z)$ as follows:

$$|\Theta(z)| \leq \left(\frac{32\pi[h_2(z) - h_1(z)](l-z)E(z)}{\mu A^2(z) h_1^3(z)}\right)^{1/2}, \qquad 0 \leq z \leq l. \qquad (3.30)$$

This result, in conjunction with Eq. (3.18), provides a decay inequality for $\Theta(z)$.

Bounds for the total energy $E(0)$ are constructed in paragraph 3 of Knowles and Sternberg (1966) by means of a minimum principle associated with the boundary value problem for ψ, as in Section II,F,1.

2. Error Estimates for Thin Shells in Torsion

One of the fundamental problems concerning the foundations of the theory of thin elastic shells is that of estimating the error committed in approximating the solution of a shell problem—considered three-dimensional—by

---

[†] See Knowles and Sternberg (1966) or Horgan (1973).
[‡] See paragraph 5 of Knowles and Sternberg (1966) for details; the analogous estimate for the solid body presents difficulties and has not been obtained.

means of results from a two-dimensional shell theory.[†] The simplest class of problems for curved shells—as distinguished from flat plates—for which one might hope to investigate this issue is that of axisymmetric torsion by end loads of shells of revolution. Such a study was undertaken by Ho and Knowles (1970) using energy arguments. The problem was reconsidered by Horgan and Wheeler (1976a, 1977a) using comparison principles.

In order to summarize these results, it is necessary to consider a hollow body slightly different from that described in Section III,A,1 and shown in Fig. 2. In terms of the cylindrical coordinates r, θ, and z, let C be a curve lying in the plane $\theta = 0$ and described by the parametric equations

$$r = r_0(\xi), \quad z = z_0(\xi), \quad 0 \leq \xi \leq \bar{l}, \tag{3.31}$$

where r_0 and z_0 are sufficiently smooth functions of the arc length ξ on C and $\bar{l}$ is the length of C. The closed meridional cross section of the shell of revolution of thickness h to be considered consists of all points in the plane $\theta = 0$ whose distance from the midcurve C do not exceed $h/2$ (Fig. 3). The shell itself is obtained by rotating this cross section about the z axis.

It is convenient to reformulate the boundary value problem for the axisymmetric torsion of such a shell in terms of an orthogonal curvilinear coordinate

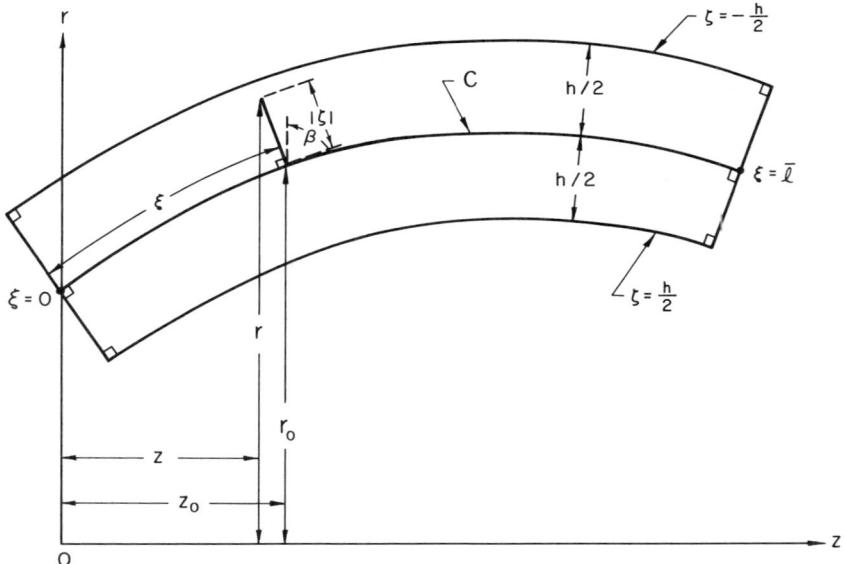

FIG. 3. Meridional cross section of a thin shell of revolution.

[†] See the extensive discussion of this matter in Koiter and Simmonds (1973).

system (ξ, θ, ζ), where ζ is the (signed) perpendicular distance from a typical point in the shell to the midsurface; ξ and ζ are related to r and z by

$$r = r_0(\xi) - \zeta \sin[\beta(\xi)], \qquad z = z_0(\xi) + \zeta \cos[\beta(\xi)], \qquad (3.32)$$

where

$$\tan[\beta(\xi)] = z'_0(\xi)/r'_0(\xi). \qquad (3.33)$$

For pure axisymmetric torsion, the displacement components u_ξ and u_ζ vanish, and the components of stress are given by

$$\tau_{\xi\theta} = \frac{\mu r}{1 + \zeta/R_\xi} \frac{\partial \phi}{\partial \xi}, \qquad \tau_{\zeta\theta} = \mu r \frac{\partial \phi}{\partial \zeta}, \qquad (3.34)$$

$$\tau_{\xi\xi} = \tau_{\theta\theta} = \tau_{\zeta\zeta} = \tau_{\xi\zeta} \equiv 0, \qquad (3.35)$$

where

$$\phi(\xi, \zeta) = \frac{1}{r(\xi, \zeta)} u_\theta(\xi, \zeta) \qquad (3.36)$$

is the angle of twist and

$$\frac{1}{R_\xi} = -\beta'(\xi), \qquad \frac{1}{R_\theta} = -\frac{\sin[\beta(\xi)]}{r_0(\xi)} \qquad (3.37)$$

are the principal midsurface curvatures. The equations of equilibrium, together with Eqs. (3.34) and (3.35), reduce to

$$\frac{\partial}{\partial \xi}\left[r^3\left(1 + \frac{\zeta}{R_\xi}\right)^{-1} \frac{\partial \phi}{\partial \xi}\right] + \frac{\partial}{\partial \zeta}\left[r^3\left(1 + \frac{\zeta}{R_\xi}\right) \frac{\partial \phi}{\partial \zeta}\right] = 0. \qquad (3.38)$$

[An alternate approach would lead to the differential equation (3.7) for the stress function ψ, but in terms of the coordinates ξ and ζ.] For axisymmetric torsional end loading, the boundary conditions accompanying Eq. (3.38) are

$$\zeta = \pm \frac{h}{2}, \qquad \frac{\partial \phi}{\partial \zeta} = 0, \qquad (3.39)$$

$$\xi = 0, \qquad \frac{\partial \phi}{\partial \xi} = \frac{1 + \zeta/R_\xi}{\mu r} \bar{f}(\zeta),$$

$$\xi = \bar{l}, \qquad \frac{\partial \phi}{\partial \xi} = \frac{1 + \zeta/R_\xi}{\mu r} \bar{g}(\zeta); \qquad (3.40)$$

here $\bar{f}$ and $\bar{g}$ are the given torsional shear tractions applied to the ends of the shell. They must satisfy

$$\int_{-h/2}^{h/2} r^2(0,\zeta)\bar{f}(\zeta)\,d\zeta = \int_{-h/2}^{h/2} r^2(\bar{l},\zeta)\bar{g}(\zeta)\,d\zeta = T/2\pi \tag{3.41}$$

where T is the torque.

To facilitate the discussion of a "thin" shell it is convenient to nondimensionalize the boundary value problem with the help of the following parameters: Let

$$R = \min\left[\min_{0 \le \xi \le \bar{l}} |R_\xi(\xi)|,\; \min_{0 \le \xi \le \bar{l}} r_0(\xi)\right], \tag{3.42}$$

and put

$$L = \min(R, \bar{l}), \quad \varepsilon = h/L, \quad x = \xi/L, \quad y = \zeta/h, \quad l = \bar{l}/L. \tag{3.43}$$

The shell is *thin* if $\varepsilon \ll 1$; the new independent variables x and y range over the rectangle $0 \le x \le l$, $-\frac{1}{2} \le y \le \frac{1}{2}$. The boundary value problem (3.38)–(3.40) can be written as follows:

$$\frac{\partial}{\partial x}\left(p\frac{\partial\phi}{\partial x}\right) + \frac{\partial}{\partial y}\left(q\frac{\partial\phi}{\partial y}\right) = 0, \quad 0 \le x \le l,\; -\frac{1}{2} \le y \le \frac{1}{2}, \tag{3.44}$$

$$y = \pm\frac{1}{2}, \quad \frac{\partial\phi}{\partial y} = 0, \tag{3.45}$$

$$x = 0, \quad p\frac{\partial\phi}{\partial x} = f, \qquad x = l, \quad p\frac{\partial\phi}{\partial x} = g, \tag{3.46}$$

where

$$p = p(x,y;\varepsilon) = \frac{\varepsilon^2 r^3}{1 + \varepsilon Ly/R_\xi}, \quad q = q(x,y;\varepsilon) = r^3\left(\frac{1 + \varepsilon Ly}{R_\xi}\right), \tag{3.47}$$

r is given by Eq. (3.32), and

$$f(y;\varepsilon) = \varepsilon^2 \frac{L}{\mu} r^2 \bar{f}(hy), \qquad g(y;\varepsilon) = \frac{\varepsilon^2 L}{\mu} r^2 \bar{g}(hy). \tag{3.48}$$

The fact that the torque on any cross section $\xi = \text{const}$ is T may be expressed as

$$\int_{-1/2}^{1/2} p\frac{\partial\phi}{\partial x}\,dy = \frac{\varepsilon T}{2\pi\mu}, \quad 0 \le x \le l. \tag{3.49}$$

The unknown angle of twist ϕ is regarded as a function of x, y, and ε, and its behavior for small ε is of particular interest.

A simple approximate solution $\tilde{\phi}(x, y; \varepsilon)$ of the problem (3.44)–(3.46) can be constructed for small ε by noting that for fixed x and y, p is small compared to q, suggesting that the first term in the differential equation be neglected. If this is done, the free surface conditions (3.45) and the remaining part of the differential equation show that $\tilde{\phi}$ is independent of y. The end conditions (3.46) thus cannot be imposed on $\tilde{\phi}$, but the torque requirement (3.49) can be used to find[†]

$$\tilde{\phi}(x; \varepsilon) = \frac{\varepsilon T}{2\pi\mu} \int_0^x \left(\int_{-1/2}^{1/2} p(x', y; \varepsilon) \, dy \right)^{-1} dx'. \quad (3.50)$$

The approximations $\tilde{\tau}_{z\theta}$ and $\tilde{\tau}_{\xi\theta}$ to the stresses that follow when Eq. (3.50) is substituted into Eq. (3.34) are, in physical variables,

$$\tilde{\tau}_{\zeta\theta} \equiv 0, \qquad \tilde{\tau}_{\xi\theta} = \frac{T}{2\pi} \frac{r(\xi,\zeta)}{1 + \zeta/R_\xi} \left(\int_{-h/2}^{h/2} \frac{r^3(\xi,\zeta) \, d\zeta}{1 + \zeta/R_\xi} \right)^{-1}. \quad (3.51)$$

The object is to obtain error estimates for the approximations $\phi \sim \tilde{\phi}$ and $\tau \sim \tilde{\tau}$ associated with Eqs. (3.50) and (3.51), keeping in mind that ε is small. Let

$$\hat{\phi} = \phi - \tilde{\phi}. \quad (3.52)$$

From Eqs. (3.50) and (3.44) one can easily show that $\hat{\phi}$ is a solution of the following boundary value problem:

$$\frac{\partial}{\partial x}\left(p \frac{\partial \hat{\phi}}{\partial x}\right) + \frac{\partial}{\partial y}\left(q \frac{\partial \hat{\phi}}{\partial y}\right) = H, \qquad 0 \le x \le l, \ -\tfrac{1}{2} \le y \le \tfrac{1}{2}, \quad (3.53)$$

$$y = \pm\tfrac{1}{2}, \quad \frac{\partial \hat{\phi}}{\partial y} = 0,$$

$$x = 0, \quad p\frac{\partial \hat{\phi}}{\partial x} = \hat{f}, \qquad x = l, \quad p\frac{\partial \hat{\phi}}{\partial x} = \hat{g}, \quad (3.54)$$

where

$$\hat{f}(y, \varepsilon) = \frac{\varepsilon^2 L}{\mu} r^2(0, hy) f(hy) - p(0, y; \varepsilon) \frac{\partial \tilde{\phi}}{\partial x}(0; \varepsilon), \quad (3.55)$$

$$\hat{g}(y, \varepsilon) = \frac{\varepsilon^2 L}{\mu} r^2(l, hy) g(hy) - p(l, y; \varepsilon) \frac{\partial \tilde{\phi}}{\partial x}(l; \varepsilon), \quad (3.56)$$

and H is the known function given by

$$H = -\frac{\partial}{\partial x}\left(p \frac{\partial \tilde{\phi}}{\partial x}\right). \quad (3.57)$$

[†] If in Eq. (3.50) one replaces p by its first approximation for small ε [see Eq. (3.47)], one can obtain an even simpler result due apparently to Love (1944, p. 567).

It is readily verified that $\hat{f}$, $\hat{g}$, and H satisfy

$$\int_{-1/2}^{1/2} \hat{f}\, dy = \int_{-1/2}^{1/2} \hat{g}\, dy = 0, \tag{3.58}$$

$$\int_{-1/2}^{1/2} H(x, y; \varepsilon)\, dy = 0, \qquad 0 \le x \le l. \tag{3.59}$$

The analysis given by Ho and Knowles (1970) provides *pointwise* estimates of the error $\hat{\phi}$ and its first derivatives of the form

$$|\hat{\phi}(x, y; \varepsilon)| \le K_1(\varepsilon) + K_2(\varepsilon)(e^{-kx/\varepsilon} + e^{-k(l-x)/\varepsilon}), \qquad 0 \le x \le l, \ -\tfrac{1}{2} \le y \le \tfrac{1}{2}, \tag{3.60}$$

where K_1 and K_2 are fully determined functions of ε that also depend on the given tractions.† The decay constant k in Eq. (3.60) is given by

$$k = \frac{\pi}{8}\left(\frac{\min[r_0(\xi)]}{\max[r_0(\xi)]}\right)^{3/2}. \tag{3.61}$$

Using Eq. (3.60) and the corresponding bounds for the derivatives of $\hat{\phi}$, one may show that the errors in the stresses associated with the approximation (3.50), (3.51) satisfy

$$\tau_{\zeta\theta} - \tilde{\tau}_{\zeta\theta} = O(\varepsilon^{3/2}), \qquad \tau_{\xi\theta} - \tilde{\tau}_{\xi\theta} = O(\varepsilon^{3/2}) \tag{3.62}$$

as $\varepsilon \to 0$, uniformly for $0 < \delta \le x \le l - \delta$, $-\tfrac{1}{2} \le y \le \tfrac{1}{2}$, where δ is fixed, small enough, and positive.‡

The argument employed by Ho and Knowles (1970) to establish these results is based on decay inequalities for first-, second-, and third-order energies and associated pointwise estimates. There are, however, substantial differences in detail between the analysis in Sections II,G and II,H and that of Ho and Knowles (1970). Some of these arise from the presence of the nonhomogeneous term H in the differential equation (3.53).

An alternative argument based on comparison principles is provided by Horgan and Wheeler (1976a, 1977a). As in Section II,I, boundary estimates for the stress errors are readily obtained. For the component of stress that does not vanish on the faces $\zeta = \pm h/2$ ($y = \pm 1/2$), such a boundary estimate yields

$$\tau_{\xi\theta} - \tilde{\tau}_{\xi\theta} = O(\varepsilon^2), \qquad y = \pm\tfrac{1}{2}, \tag{3.63}$$

as $\varepsilon \to 0$, uniformly for $0 < \delta \le x \le l - \delta$. This is slightly *stronger* than the result (3.62) furnished by energy arguments. On the other hand, bounds

---

† The estimate (3.60) presupposes a suitable determination of the arbitrary constant that may be added to $\hat{\phi}$.

‡ Thus the error estimates (3.62) apply outside the boundary layers at the ends of the shell.

valid on the *entire* closed region $0 < \delta \leq x \leq l - \delta$, $-1/2 \leq y \leq 1/2$, are more difficult to obtain and result in stress-error estimates slightly weaker than those of Eq. (3.62).

Illustrative examples for circular cylindrical, spherical, and conical shells were also given by Horgan and Wheeler (1977a).

B. Plane Strain

1. *Isotropic Materials*

a. An Energy Decay Inequality

With only a few exceptions, the boundary value problems of primary interest in the theory of elasticity are governed by differential equations— or systems of such equations—whose order is greater than 2. The extension to higher-order problems of the arguments employed in the foregoing sections to establish principles of the Saint-Venant type proves to be a matter of some difficulty.

The simplest of these higher-order boundary value problems is the two-dimensional one associated with plane strain (or generalized plane stress) for a homogeneous isotropic material. In the present section, we discuss a version of Saint-Venant's principle appropriate to this setting as given by Knowles (1966), improved by Flavin (1974), and generalized on the basis of a different argument by Oleinik and Yosifian (1978a,b). We will also briefly describe some related results.

Let $\mathbb{R}$ be the interior of a fixed cross section of a homogeneous isotropic elastic cylinder in a state of plane strain in the absence of body forces.[†] Except where explicitly stated otherwise, we assume that $\mathbb{R}$ is *simply connected* and bounded by a piecewise-smooth simple closed curve C. Let $\mathbf{t}(s)$ be the (two-dimensional) traction vector acting at a point on C located by its arc length s measured positively counterclockwise along C from a fixed reference point. Suppose that on a subarc C_1 of C, $\mathbf{t}$ is replaced by a new distribution of traction $\hat{\mathbf{t}}$, giving rise to the same resultant force and moment on C as are produced by $\mathbf{t}$. A Saint-Venant principle should yield a quantitative estimate of the effect on the stress field in $\mathbb{R}$ of such a replacement; the estimate should presumably show that at points remote enough from C_1 the effect is small. In view of the linearity of the underlying theory, it is sufficient to consider only the case in which the given traction $\mathbf{t}$ vanishes on $C_0 = C - C_1$ and thence to estimate the stresses in this circumstance. Of

---

[†] For a discussion of plane strain and generalized plane stress, see Section D-VII of Gurtin (1972).

course **t** is now necessarily self-equilibrated on C_1:

$$\int_{C_1} \mathbf{t}\, ds = \int_{C_1} \mathbf{x} \times \mathbf{t}\, ds = \mathbf{0}, \tag{3.64}$$

where **x** is the position vector to points on C_1.

Let x_1 and x_2 be Cartesian coordinates in the plane of $\mathbb{R}$ arranged in such a way (Fig. 4) that the x_2 axis passes through the end points of the loaded arc C_1; it is then assumed that $x_1 \leq 0$ on C_1, and $x_1 \geq 0$ on C_0. Finally, the origin is chosen so that the minimum value of x_2 on C_0 is zero.

The stresses $\tau_{\alpha\beta}$ are given on $\overline{\mathbb{R}} = \mathbb{R} + C$ in terms of the Airy stress function ϕ by

$$\tau_{\alpha\beta} = \varepsilon_{\alpha\lambda}\varepsilon_{\beta\mu}\phi_{,\lambda\mu}, \tag{3.65}$$

where $\varepsilon_{11} = \varepsilon_{22} = 0$, $\varepsilon_{12} = -\varepsilon_{21} = 1$, and ϕ satisfies the biharmonic equation

$$\Delta\Delta\phi \equiv \phi_{,\alpha\alpha\beta\beta} = 0 \qquad \text{on } \mathbb{R}. \tag{3.66}$$

The requirement that C_0 be traction-free leads to the conclusion[†] that

$$\phi = \phi_{,1} = \phi_{,2} = 0 \qquad \text{on } C_0. \tag{3.67}$$

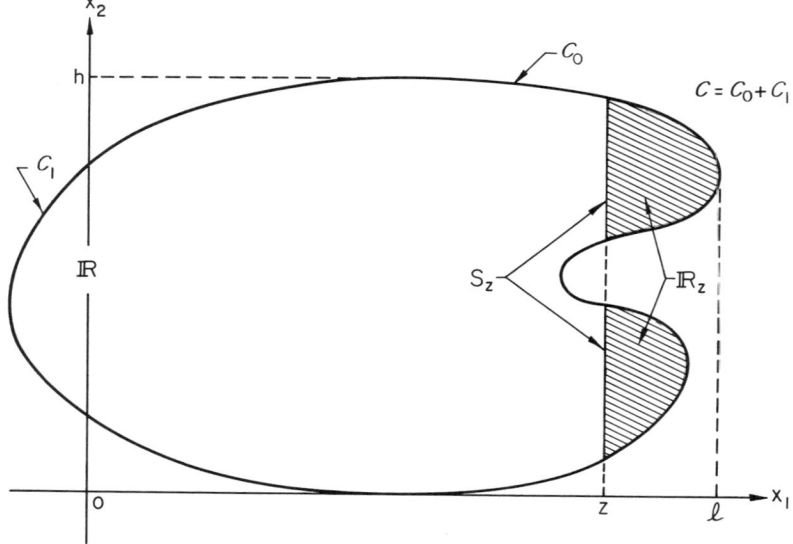

FIG. 4. Plane domain $\mathbb{R}$ and subdomain $\mathbb{R}_z$.

[†] If $\mathbb{R}$ is multiply connected, then the simple *homogeneous* Dirichlet boundary conditions (3.67) need not hold on traction-free interior components of the boundary (see Gurtin, 1972, pp. 158–159).

Recent Developments Concerning Saint-Venant's Principle   225

We assume that ϕ is continuously differentiable twice on $\bar{\mathbb{R}}$ and four times on $\mathbb{R}$ itself.

Let

$$l = \max_{C_0} x_1, \qquad h = \max_{C_0} x_2 \qquad (3.68)$$

(see Fig. 4). Denote by $\mathbb{R}_z$ the set of points (x_1, x_2) in $\mathbb{R}$ for which $x_1 > z$ and $0 \leq z < l$, and let

$$S_z = \{(x_1, x_2) \in \bar{\mathbb{R}} \mid x_1 = z\}, \qquad 0 \leq z \leq l. \qquad (3.69)$$

Note that S_z need not be connected.

A natural quadratic integral associated with the biharmonic operator that is also simply related to the strain energy stored in $\mathbb{R}_z$ is given by[†]

$$E(z) = \int_{\mathbb{R}_z} \phi_{,\alpha\beta} \phi_{,\alpha\beta} \, dA, \qquad 0 \leq z < l. \qquad (3.70)$$

The energy inequality obtained by Knowles (1966) asserts that

$$E(z) \leq 2E(0)e^{-2kz}, \qquad 0 \leq z \leq l, \qquad (3.71)$$

where

$$k = \frac{\pi}{h}\left(\frac{\sqrt{2}-1}{2}\right)^{1/2} \doteq \frac{1.4}{h}, \qquad (3.72)$$

and h is given in Eq. (3.68).

By modifying the proof given by Knowles (1966), Flavin (1974) succeeded in showing that Eq. (3.71) holds with a larger value of k:

$$k = \frac{\pi}{h}\frac{1}{\sqrt{2}} \doteq \frac{2.2}{h}. \qquad (3.73)$$

In contrast to the results for the Laplace equation, Eq. (3.73) does not furnish a best possible estimate of the decay rate. We shall later discuss this aspect of the results further.

We outline the proof of Eqs. (3.71) and (3.72) with only minor rearrangements of the argument as given by Knowles (1966). Let k be an arbitrary fixed positive constant and introduce

$$G(z) = E(z) + 2k \int_z^l E(\zeta) \, d\zeta, \qquad 0 \leq z \leq l. \qquad (3.74)$$

[†] When Poisson's ratio v for the material at hand is zero, the right-hand side of Eq. (3.70) is proportional to the strain energy stored in $\mathbb{R}_z$. For a simply connected domain $\mathbb{R}$, the stresses associated with the traction boundary value problem under consideration are independent of all elastic constants (see Gurtin, 1972, pp. 155–156). Thus the results are not limited in relevance to the case $v = 0$.

With the help of the two-dimensional divergence theorem, it is possible to represent $E(z)$, its derivative $E'(z)$, and the integral occurring in Eq. (3.74) as line integrals[†] over S_z and then to establish the identity

$$G'(z) + 2kG(z) = -\left(\int_{S_z} (\phi_{,11}^2 + \phi_{,22}^2 + 2\phi_{,12}^2)\, dx_2 \right.$$

$$\left. - 4k^2 \int_{S_z} (\phi_{,1}^2 + \phi_{,2}^2 - \phi\phi_{,11})\, dx_2 \right), \qquad 0 \le z \le l. \quad (3.75)$$

Suppose momentarily that there is a value of $k > 0$ such that the right-hand side of Eq. (3.75) is nonpositive. It then follows that for this k

$$G'(z) + 2kG(z) \le 0, \qquad 0 \le z \le l, \quad (3.76)$$

so that

$$G(z) \le G(0)e^{-2kz}, \qquad 0 \le z \le l. \quad (3.77)$$

According to Eq. (3.74), $G(z) \ge E(z)$, so Eq. (3.77) implies that

$$E(z) \le G(0)e^{-2kz}, \qquad 0 \le z \le l. \quad (3.78)$$

It is not difficult to show[‡] that

$$G(0) \le 2E(0); \quad (3.79)$$

Eqs. (3.78) and (3.79) then furnish the decay inequality (3.71).

It remains to show that there *is* a suitable value of k. To this end we prove the following.

Lemma. *For any function $\psi \in C^2(\mathbb{R}_0)$ such that $\psi = \psi_{,1} = \psi_{,2} = 0$ on C_0, the inequality*

$$\int_{S_z} (\psi_{,11}^2 + \psi_{,22}^2 + 2\psi_{,12}^2)\, dx_2$$

$$\ge 4\kappa^2 \left| \int_{S_z} (\psi_{,1}^2 + \psi_{,2}^2 - \psi\psi_{,11})\, dx_2 \right|, \qquad 0 \le z < l, \quad (3.80)$$

holds, with

$$\kappa = \left(\frac{\sqrt{2} - 1}{2} \right)^{1/2} \frac{\pi}{h}. \quad (3.81)$$

[†] See Eqs. (3.6), (3.10), and (3.9) in Knowles (1966).
[‡] For details, see Knowles (1966, pp. 10–11).

Recent Developments Concerning Saint-Venant's Principle 227

Proof. Let I stand for the integral on the left-hand side of Eq. (3.80). Then for any constants α and β with $0 \leq \alpha \leq 1$ and $\beta > 0$,

$$I = \int_{S_z} [(|\psi_{,11}| - \beta|\psi|)^2 + 2\psi_{,12}^2 + \alpha\psi_{,22}^2$$
$$+ (1-\alpha)\psi_{,22}^2 + 2\beta|\psi\psi_{,11}| - \beta^2\psi^2]\,dx_2. \tag{3.82}$$

We now make use of the following inequality of the Wirtinger type: *If Φ is any continuously differentiable function on S_z for which $\Phi = 0$ at the end points of the constituent line segments of S_z, then*

$$\int_{S_z} \Phi_{,2}^2 \, dx_2 \geq \frac{\pi^2}{h^2} \int_{S_z} \Phi^2 \, dx_2.^\dagger \tag{3.83}$$

We first apply Eq. (3.83) in Eq. (3.82) by choosing $\Phi = \psi_{,1}$ and $\Phi = \psi_{,2}$ in turn; it follows that

$$I \geq \int_{S_z} \left(2\frac{\pi^2}{h^2}\psi_{,1}^2 + \alpha\frac{\pi^2}{h^2}\psi_{,2}^2 + 2\beta|\psi\psi_{,11}|\right)dx_2$$
$$+ \int_{S_z} \left((1-\alpha)\frac{\pi^2}{h^2}\psi_{,2}^2 - \beta^2\psi^2\right)dx_2. \tag{3.84}$$

We now estimate the *second* integral in Eq. (3.84) by applying Eq. (3.83) again, now with $\Phi = \psi$. Thus

$$I \geq \int_{S_z} \left(\frac{2\pi^2}{h^2}\psi_{,1}^2 + \alpha\frac{\pi^2}{h^2}\psi_{,2}^2 + 2\beta|\psi\psi_{,11}|\right)dx_2$$
$$+ \left((1-\alpha)\frac{\pi^4}{h^4} - \beta^2\right)\int_{S_z}\psi^2\,dx_2. \tag{3.85}$$

Next let α and β be chosen so that $\beta = \sqrt{1 - \alpha\pi^2/h^2}$ and $2\beta = \alpha\pi^2/h^2$; this gives $\alpha = 2(\sqrt{2} - 1) < 1$ and $\beta = (\sqrt{2} - 1)\pi^2/h^2$, so that Eq. (3.85) yields

$$I \geq \frac{\pi^2}{h^2} \int_{S_z} (2\psi_{,1}^2 + \alpha\psi_{,2}^2 + \alpha|\psi\psi_{,11}|)\,dx_2$$
$$\geq \alpha\frac{\pi^2}{h^2}\int_{S_z}(\psi_{,1}^2 + \psi_{,2}^2 + |\psi\psi_{,11}|)\,dx_2$$
$$\geq \alpha\frac{\pi^2}{h^2}\left|\int_{S_z}(\psi_{,1}^2 + \psi_{,2}^2 - \psi\psi_{,11})\,dx_2\right|. \tag{3.86}$$

In view of the definition of I and the value of α, the lemma is proved.

---

† See p. 185 of Hardy, *et al.* (1967); Eq. (3.83) reflects the fact that π^2/h^2 is the lowest eigenvalue of the problem of the clamped vibrating string.

We complete the proof of Eqs. (3.71) and (3.72) by noting that the lemma may be applied in Eq. (3.75) with $\psi = \phi$ to show that Eq. (3.76) holds with

$$k = \kappa, \tag{3.87}$$

which, by Eq. (3.81), is equivalent to Eq. (3.72).

To obtain the larger value of the decay rate found by Flavin (1974), one shows that the lemma remains valid when the larger value of κ given by

$$\kappa = \frac{1}{\sqrt{2}} \frac{\pi}{h} \tag{3.88}$$

is used in inequality (3.80). The essential feature of Flavin's proof of the sharper version of the lemma is the use, in addition to Eq. (3.83), of the inequality[†]

$$\int_{S_z} \Psi_{,22}^2 \, dx_2 \geq \frac{4\pi^2}{h^2} \int_{S_z} \Psi_{,2}^2 \, dx_2, \tag{3.89}$$

valid for any $\Psi \in C^2(S_z)$ such that $\Psi = \Psi_{,2} = 0$ at the end points of the subsegments of S_z. This inequality, with $\Psi = \psi$, is used in the step that proceeds from Eq. (3.82) to Eq. (3.84) to estimate $\psi_{,22}$ in mean square in terms of $\psi_{,2}$, wherever the former occurs. Subsequent details are similar to those given above.

An improvement of another kind in the energy inequality was made by Mieth (1975), who showed that Eq. (3.71) can be replaced by

$$E(z) \leq 2E(0)e^{-2kz}(1 + e^{-4kz})^{-1}, \qquad 0 \leq z \leq l, \tag{3.90}$$

with k given by Eq. (3.72), thus repairing an obvious shortcoming in Eq. (3.71) at $z = 0$. We omit the details.[‡]

b. An Alternative Argument

Oleinik and Yosifian (1978a,b) have given an energy decay inequality for the plane strain problem just considered that generalizes Eq. (3.71) and is derived by a different argument.[§] We will state their result and sketch their procedure here, again with slight alterations of detail. They proved that

$$E(z) \leq \frac{E(0)}{r(z)}, \qquad 0 \leq z \leq l, \tag{3.91}$$

[†] See Mikhlin (1964, p. 252). Equation (3.89) reflects the fact that $4\pi^2/h^2$ is the lowest eigenvalue of the buckling problem for a clamped beam.
[‡] See Mieth (1975, pp. 36–38).
[§] See also Oleinik (1979a,b).

where $r(z)$ is the solution of the initial value problem

$$r''(z) - m(z)r(z) = 0, \quad 0 \leq z \leq l,$$
$$r(0) = 1, \quad r'(0) = 0, \tag{3.92}$$

and $m(z)$ is any continuous function on $[0, l]$ for which

$$0 < m(z) \leq \inf_{\psi} \left\{ \frac{\int_{S_z} \psi_{,\alpha\beta} \psi_{,\alpha\beta} \, dx_2}{\left| \int_{S_z} (\psi_{,\alpha} \psi_{,\alpha} - \psi \psi_{,11}) \, dx_2 \right|} \right\}, \quad 0 \leq z \leq l. \tag{3.93}$$

In Eq. (3.93), the infimum is taken over all $\psi \in C^2(\mathbb{R}_0)$ for which $\psi = \psi_{,1} = \psi_{,2} = 0$ on C_0. The lemma established above [see Eq. (3.80)] shows that the infimum in Eq. (3.93) exists and is positive for each z, and that one *may* (but need not) choose

$$m(z) = 4\kappa^2 = \text{const}, \tag{3.94}$$

where κ is given by Eq. (3.81), or by Eq. (3.88) if Flavin's improved version of the lemma is invoked. The decay inequality (3.91)—through the choice of the function $m(z)$—makes it possible to account for the "variable width" of the domain $\mathbb{R}_0$ in the x_2 direction.[†]

We now outline the proof of Eq. (3.91). One can show with the help of Eqs. (3.66) and (3.67) and the divergence theorem that for *any* function $R(x_1, z)$ that is continuously differentiable for $0 \leq x_1 \leq l$ and $0 \leq z \leq l$ and has a piecewise continuous second derivative $R_{,11}$ there, one has the identity

$$\int_{\mathbb{R}_0} R(x_1, z) \phi_{,\alpha\beta}(x_1, x_2) \phi_{,\alpha\beta}(x_1, x_2) \, dA$$
$$= \int_{\mathbb{R}_0} R_{,11}(x_1, z)(\phi_{,\alpha} \phi_{,\alpha} - \phi \phi_{,11}) \, dA$$
$$+ \int_{S_0} R_{,1}(0, z)(\phi_{,\alpha} \phi_{,\alpha} - \phi \phi_{,11}) \, dx_2$$
$$+ \int_{S_0} R(0, z)(\phi \phi_{,111} - 2\phi_{,2} \phi_{,21} - \phi_{,1} \phi_{,11}) \, dx_2. \tag{3.95}$$

Now choose

$$R(x_1, z) = \begin{cases} r(x_1), & 0 \leq x_1 \leq z \\ r'(z)x_1 + r(z) - zr'(z), & z \leq x_1 < l \end{cases} \quad 0 \leq z \leq l, \tag{3.96}$$

where r satisfies Eq. (3.92). Thus R coincides with r for $0 \leq x_1 \leq z$ and is continued linearly in x_1 for $z \leq x_1 \leq l$; R has the required smoothness. By Eqs. (3.96) and (3.92) the identity (3.95) reduces to

$$\int_{\mathbb{R}_0} R(x_1, z) \phi_{,\alpha\beta} \phi_{,\alpha\beta} \, dA = \int_{\mathbb{R}_0 - \mathbb{R}_z} r''(x_1)(\phi_{,\alpha} \phi_{,\alpha} - \phi \phi_{,11}) \, dA$$
$$+ \int_{S_0} (\phi \phi_{,111} - 2\phi_{,2} \phi_{,21} - \phi_{,1} \phi_{,11}) \, dx_2. \tag{3.97}$$

[†] The analysis leading to Eq. (3.71) can be modified so as to account for this feature also.

Making use of definition (3.70) as well as of Eqs. (3.66) and (3.67) and the divergence theorem, one can show[†] that the integral over S_0 in Eq. (3.97) is exactly $E(0)$. Bearing in mind the differential equation in Eq. (3.92), we may thus write Eq. (3.97) as

$$\int_{\mathbb{R}_0} R(x_1, z)\phi_{,\alpha\beta}\phi_{,\alpha\beta}\, dA = E(0) + \int_{\mathbb{R}_0 - \mathbb{R}_z} m(x_1)r(x_1)(\phi_{,\alpha}\phi_{,\alpha} - \phi\phi_{,11})\, dA$$

$$= E(0) + \int_0^z m(x_1)r(x_1)$$

$$\times \left(\int_{S_{x_1}} (\phi_{,\alpha}\phi_{,\alpha} - \phi\phi_{,11})\, dx_2\right) dx_1. \tag{3.98}$$

It is not difficult to show from Eq. (3.92) that $r'(z) > 0$, and hence $r(z) > 0$, for $0 \le z \le l$. It then follows from Eqs. (3.98) and (3.93) that

$$\int_{\mathbb{R}_0} R(x_1, z)\phi_{,\alpha\beta}\phi_{,\alpha\beta}\, dA \le E(0) + \int_{\mathbb{R}_0 - \mathbb{R}_z} r(x_1)\phi_{,\alpha\beta}\phi_{,\alpha\beta}\, dA. \tag{3.99}$$

Since by Eq. (3.96), $R(x_1, z) = r(x_1)$ for $0 \le x_1 \le z$, we may conclude from Eq. (3.99) that

$$\int_{\mathbb{R}_z} R(x_1, z)\phi_{,\alpha\beta}\phi_{,\alpha\beta}\, dA \le E(0). \tag{3.100}$$

But by Eq. (3.96) and the fact that $r'(z) > 0$, we have $R(x_1, z) \ge r(z)$ for $z \le x_1 \le l$, so Eqs. (3.100) and (3.70) yield

$$r(z)E(z) \le E(0), \tag{3.101}$$

which is the desired result, Eq. (3.91).

If one chooses $m(z)$ to be constant, as in Eq. (3.94), then the solution $r(z)$ of the initial value problem (3.92) can be explicitly determined, and Eq. (3.91) becomes

$$E(z) \le \frac{E(0)}{\cosh(2\kappa z)} = \frac{2E(0)e^{-2\kappa z}}{1 + e^{-4\kappa z}}. \tag{3.102}$$

If one uses the value of κ given by Eq. (3.88), then Eq. (3.102) reduces exactly to Mieth's result, Eq. (3.90), except that the decay rate is now that of Flavin (1974), given by Eq. (3.73). Moreover, Eq. (3.102) of course implies Flavin's version (3.71), (3.73) of the energy decay inequality, as remarked by Oleinik and Yosifian (1978a,b).

Other choices of $m(z)$ may be appropriate or necessary. Thus Oleinik and Yosifian (1978b) chose $m(z)$ so that for a domain that "expands" like a sector, a power law of energy decay results. On the other hand, they showed that for a domain that "contracts" as one departs from the loaded arc, the energy decay is faster than exponential.

[†] See Eq. (3.6) of Knowles (1966).

Oleinik and Yosifian also used energy decay inequalities to derive theorems of the Phragmén–Lindelöf type for the biharmonic equation in unbounded domains, as well as to investigate solutions near nonregular boundary points.

c. Discussion and Further Results

The simplest instance of the plane strain problem considered previously is that for which $\mathbb{R}$ is the rectangle $0 < x_1 < l$, $0 < x_2 < h$ carrying a self-equilibrated traction on the end $x_1 = 0$ and free of traction on the remaining three sides. The corresponding boundary value problem for the biharmonic equation is much more difficult to deal with explicitly than is its analog for Laplace's equation, because separation of variables leads to a non-self-adjoint eigenvalue problem on the interval $0 \leq x_2 \leq h$. Nevertheless, the rectangle problem—or its counterpart for the semi-infinite strip[†]—offers an opportunity to assess the quality of the estimated decay rates discussed previously. The eigenvalue problems involved have been extensively studied, beginning with the work of Fadle (1940) and Papkovich (1940). One of the first to recognize the relationship between these problems and Saint-Venant's principle was Timoshenko (1934) (see also Horvay, 1957a,b,c). More recently, Gregory (1980) has established the completeness of the relevant eigenfunctions.[‡]

We consider the semi-infinite strip $0 \leq x_1 < \infty$, $0 \leq x_2 \leq h$, and let

$$\phi(x_1, x_2; \lambda) = \left\{\left(\frac{x_2}{h} - 1\right)\sin\left(\frac{\lambda x_2}{h}\right) \pm \frac{x_2}{h}\sin\left[\lambda\left(\frac{x_2}{h} - 1\right)\right]\right\}e^{-\lambda x_1/h}, \quad (3.103)$$

where λ satisfies

$$\lambda \pm \sin \lambda = 0, \quad \operatorname{Re} \lambda > 0.\text{[§]} \quad (3.104)$$

All roots of Eq. (3.104) are complex, as are the functions ϕ themselves. The real (or imaginary) part of ϕ is biharmonic, vanishes together with its normal derivative at $x_2 = 0$ and $x_2 = h$, and has finite energy in the strip as well. The roots λ of Eq. (3.104) of *smallest* real part are such that

$$\operatorname{Re} \lambda \doteq 4.2 \quad (3.105)$$

(Timoshenko and Goodier, 1970, p. 62). Thus the rate of decay of the energy $E(z)$ associated with the real (or imaginary) part of the function ϕ for which Eq. (3.105) holds is $8.4/h$. From Gregory's completeness theorem, one finds

---

[†] It is not difficult to extend the arguments used previously to show that the energy decay inequalities remain valid for the case in which $\mathbb{R}$ is a semi-infinite strip.

[‡] The literature on the traction problem for the rectangle or strip is too vast to survey here; many references are given by Bogy (1975) and Johnson and Little (1965). The remarks on pp. 61 and 62 of Timoshenko and Goodier (1970) are especially relevant to our purposes.

[§] The ambiguous signs in Eqs. (3.103) and (3.104) are to be chosen in the same way.

that "any" solution of the biharmonic equation in the strip that vanishes together with its normal derivative on the long sides and has finite total energy is a superposition of the eigenfunctions (3.103) over the complex spectrum determined by Eq. (3.104). Thus, in general, the exact decay rate of the energy distribution is $8.4/h$. Comparing this with the estimate $2k \doteq 4.4/h$ associated with Eq. (3.73), one sees that the energy arguments here produce results that are well short of the best possible ones.[†]

We return now to the general problem. One can construct upper bounds for the total energy $E(0)$ contained in $\mathbb{R}_0$ appearing in the estimates (3.71) and (3.91) with the help of variational principles in much the same way as in Section II,F,1. This was done for a certain class of domains by Knowles (1966) and also by Mieth (1975). The details are complicated and will not be given here.

There is a mean value theorem for the biharmonic equation analogous to that for harmonic functions.[‡] This permits the derivation of *interior* pointwise estimates of the second derivatives of ϕ—and hence, by Eq. (3.65), of the stresses—from the energy decay inequalities in much the same way as in Section II,H,1. Such procedures were carried out by Knowles (1966) and Mieth (1975).

A theory of higher-order energies parallel to that described in Section II,G for Laplace's equation has not been developed for the biharmonic problem. Nevertheless, pointwise stress estimates valid *near* the boundary have been obtained by Mieth (1975) by entirely different methods. Roseman (1967) has proposed a method for obtaining pointwise bounds on the stresses that do not deteriorate near the boundary in the case where $\mathbb{R}$ is a rectangle. His arguments were based on mean-square estimates and Sobolev's inequality; the results, however, are rather inexplicit. In general, the techniques for obtaining pointwise information valid on $\mathbb{R}_0$ are elaborate.

Various generalizations of the issues discussed previously have been considered. Thus Mieth (1975) permitted $\mathbb{R}$ to be multiply connected, but he employed the boundary conditions (3.67) on the internal boundaries. His results in this regard are therefore not directly relevant to the plane strain problem of elasticity in the case of a multiply connected region containing traction-free internal boundaries.[§]

Finally, we note that Breuer and Roseman (1977a) have shown how to describe energy decay inequalities for domains that are "thin" but for which

---

[†] Whether energy arguments can produce "best possible" estimates of the decay rate for the biharmonic problem is an open question.

[‡] See Diaz and Payne (1958).

[§] If $\mathbb{R}$ is multiply connected, then the simple *homogeneous* Dirichlet boundary conditions (3.67) need not hold on traction-free interior components of the boundary (see Gurtin, 1972, pp. 158–159).

the ratio h/l is *not* the appropriate parameter with which to characterize slenderness.

2. The Anisotropic Case

Horgan (1972a) has obtained a generalization of the energy decay inequality (3.71) that is valid for those anisotropic elastic materials admitting a state of plane strain. The elastic cylinder with simply connected cross section $\mathbb{R}$ considered in the preceding section is now assumed to be composed of a homogeneous material with a plane of elastic symmetry, and the generators of this cylinder are taken to be perpendicular to this plane. A state of plane strain is then possible in the cylinder,[†] so that the components u_i of the displacement vector field $\mathbf{u}$ are such that $u_\alpha = u_\alpha(x_1, x_2)$ and $u_3 \equiv 0$. The constitutive law for the anisotropic material considered then relates the non-vanishing components of infinitesimal strain $e_{\alpha\beta} = \frac{1}{2}(u_{\alpha,\beta} + u_{\beta,\alpha})$ to the in-plane stresses $\tau_{\alpha\beta}$ as follows:

$$e_{11} = \beta_{11}\tau_{11} + \beta_{12}\tau_{22} + \beta_{16}\tau_{12},$$
$$e_{22} = \beta_{21}\tau_{11} + \beta_{22}\tau_{22} + \beta_{26}\tau_{12}, \qquad (3.106)$$
$$2e_{12} = \beta_{61}\tau_{11} + \beta_{62}\tau_{22} + \beta_{66}\tau_{12},$$

where the constants $\beta_{pq} = \beta_{qp}$ ($p, q = 1, 2, 6$) are the elastic compliances of the material in plane strain, and the notation is that of Lekhnitskii (1963). For the isotropic special case,

$$\beta_{11} = \beta_{22} = \frac{1-\nu}{2\mu}, \qquad \beta_{12} = \beta_{21} = -\frac{\nu}{2\mu}, \qquad \beta_{66} = \frac{1}{\mu},$$
$$\beta_{16} = \beta_{61} = \beta_{26} = \beta_{62} = 0, \qquad (3.107)$$

where ν is Poisson's ratio and μ is the shear modulus.

As in the isotropic case, the problem of determining the stresses $\tau_{\alpha\beta}$ arising in $\mathbb{R}$ from a self-equilibrated load acting on the arc C_1 (Fig. 4) and leaving the remainder C_0 of the boundary traction-free is reducible to a boundary value problem for the Airy stress function ϕ.[‡] The boundary conditions (3.67) for ϕ on C_0 remain the same, but the biharmonic differential equation (3.66) is to be replaced by[§]

$$\beta_{22}\phi_{,1111} - 2\beta_{26}\phi_{,1112} + (2\beta_{12} + \beta_{66})\phi_{,1122} - 2\beta_{16}\phi_{,1222}$$
$$+ \beta_{11}\phi_{,2222} = 0 \quad \text{on } \mathbb{R}. \qquad (3.108)$$

[†] See Lekhnitskii (1963, paragraph 21).
[‡] Equation (3.65) continues to hold.
[§] See paragraph 22 of Lekhnitskii (1963).

The strain energy density W for the material under consideration is given by

$$W = \tfrac{1}{2}\tau_{\alpha\beta}e_{\alpha\beta} = \tfrac{1}{2}(\beta_{11}\tau_{11}^2 + \beta_{22}\tau_{22}^2 + \beta_{66}\tau_{12}^2 + 2\beta_{12}\tau_{11}\tau_{22}$$
$$+ 2\beta_{16}\tau_{11}\tau_{12} + 2\beta_{26}\tau_{22}\tau_{12}). \tag{3.109}$$

It is assumed that W is positive definite,[†] so that the matrix

$$\mathbf{B} = \begin{vmatrix} \beta_{11} & \beta_{12} & \beta_{16}/\sqrt{2} \\ \beta_{12} & \beta_{22} & \beta_{26}/\sqrt{2} \\ \beta_{16}/\sqrt{2} & \beta_{26}/\sqrt{2} & \beta_{66}/2 \end{vmatrix} \tag{3.110}$$

has the same property. The energy stored in the subregion $\mathbb{R}_z$ of $\mathbb{R}$ is then given from Eqs. (3.109) and (3.65) as

$$E(z) = \tfrac{1}{2}\int_{\mathbb{R}_z} (\beta_{11}\phi_{,22}^2 + \beta_{22}\phi_{,11}^2 + \beta_{66}\phi_{,12}^2 + 2\beta_{12}\phi_{,11}\phi_{,22}$$
$$- 2\beta_{16}\phi_{,22}\phi_{,12} - 2\beta_{26}\phi_{,11}\phi_{,12})\,dA. \tag{3.111}$$

When one specializes to the isotropic case with the help of Eq. (3.107) and then sets $v = 0$, $E(z)$ of Eq. (3.111) becomes proportional to the energy functional (3.70).

Horgan (1972a) showed that $E(z)$ satisfies

$$E(z) \leq 2E(0)e^{-2kz}, \qquad 0 \leq z \leq l, \tag{3.112}$$

with

$$k = \frac{\pi}{\sqrt{2}h}\left(\frac{sb}{\beta_{22}}\right)^{1/2}, \tag{3.113}$$

where $b > 0$ is the smallest eigenvalue of the matrix $\mathbf{B}$ in Eq. (3.110) and s is the smallest positive root of the equation

$$s^4 + \left(2 + \frac{\beta_{66}}{\beta_{22}}\right)s^3 + \left(2\frac{\beta_{66}}{\beta_{22}} - 3 - 4\frac{\beta_{26}^2}{\beta_{22}^2}\right)s^2$$
$$- 2\left(1 + \frac{\beta_{66}}{\beta_{22}}\right)s + 2 = 0. \tag{3.114}$$

The estimated decay rate (3.113) is thus dependent on the values of the elastic constants β_{pq}, in contrast to the corresponding result [see Eq. (3.72) or (3.73)] for isotropic materials, as one might expect. When Eq. (3.113) is specialized to the isotropic case and v is set equal to zero, Eq. (3.113) reduces to Eq. (3.72).

[†] This assures that the differential equation (3.108) is elliptic.

It has been shown by Horgan (1972a) that for a material that is *not* isotropic, the value of k in Eq. (3.113) is always *smaller* than the value given in Eq. (3.72) for the isotropic case. Bearing in mind that these estimated values are merely *lower bounds* for the *actual* rates of decay, this result does not preclude the possibility that the relationship between the actual rates is reversed, but it does suggest that the "characteristic decay length" associated with Saint-Venant's principle is least for isotropic materials.

Pointwise interior estimates for stresses in the anisotropic case were also considered by Horgan (1972a) on the basis of a mean value theorem.

3. Application to Composite Materials

If the anisotropic material considered in the preceding subsection is *transversely isotropic* with its preferred direction parallel to the x_1 axis,[†] one has

$$\beta_{11} = \frac{1}{E_L}\left(1 - v_{LT}^2 \frac{E_T}{E_L}\right), \qquad \beta_{12} = \frac{-v_{LT}(1 + v_{TT})}{E_L}, \qquad \beta_{22} = \frac{1 - v_{TT}^2}{E_T},$$

$$\beta_{16} = \beta_{26} = 0, \qquad \beta_{66} = \frac{1}{\mu_{LT}}. \qquad (3.115)$$

Here L and T (which are, of course, exempt from the summation convention) refer to the longitudinal (or x_1) and transverse directions, respectively, while v, E, and μ with subscripts stand for Poisson's ratios, Young's moduli, and a shear modulus. For this class of materials, the differential equation (3.108) can be written in the form

$$\left(\frac{\partial^2}{\partial x_1^2} + \varepsilon_t^2 \frac{\partial^2}{\partial x_2^2}\right)\left(\varepsilon_c^2 \frac{\partial^2}{\partial x_1^2} + \frac{\partial^2}{\partial x_2^2}\right)\phi = 0 \qquad \text{on } \mathbb{R}, \qquad (3.116)$$

where $1/\varepsilon_t^2$ and ε_c^2 are, respectively, the larger and smaller roots ξ of the equation

$$\beta_{11}\xi^2 - (2\beta_{12} + \beta_{66})\xi + \beta_{22} = 0. \qquad (3.117)$$

In discussing idealized models for the mechanical response of a fiber-reinforced composite material, Everstine and Pipkin (1971)[‡] employed the theory of transversely isotropic elastic materials, to which they adjoined the additional assumptions of small extensibility in the fiber direction[§] and small bulk compressibility. With these assumptions, the parameters ε_t and

---
[†] See Horgan (1972b).
[‡] See also Spencer (1972) and the recent survey article by Pipkin (1979).
[§] In the present setting, the fibers are parallel to the x_1 axis.

ε_c are given approximately by

$$\varepsilon_t \sim (\mu_{LT}/E_L)^{1/2}, \qquad \varepsilon_c \sim (\mu_{LT}/K)^{1/2}, \tag{3.118}$$

where K is a bulk modulus.[†] These are now treated as small parameters, tending to zero in a fixed ratio.

If a self-equilibrated traction is applied to the arc C_1 (Fig. 4), the energy distribution $E(z)$ decays in accordance with Eqs. (3.112), (3.113), and (3.114). Horgan (1972b) has shown that in the limit as ε_t and ε_c tend to zero, the estimated decay rate k of Eq. (3.113) satisfies

$$k \sim \frac{\pi}{\sqrt{2h}} \varepsilon_t, \tag{3.119}$$

suggesting a very slow decay rate in the fiber direction. His analysis can in fact be used to show that

$$k = O(\varepsilon_t/h) \qquad \text{as } \varepsilon_t \to 0 \tag{3.120}$$

continues to hold, *whether or not ε_c is small*. Thus the expectation of slow decay in the fiber direction persists in the presence of small fiber extensibility, even though the bulk compressibility is not small.[‡]

It is of interest to note that Everstine and Pipkin (1971) have constructed special solutions (exact within the framework of their model) for certain half-space problems in which the *actual* rate of decay in the fiber direction satisfies Eq. (3.120).

Slow decay of end effects in the fiber direction might be anticipated from another viewpoint (Horgan, 1982). In the limit $\varepsilon_t \to 0$, the type of the differential equation (3.116) changes from elliptic to parabolic, the lines $x_2 = \text{const}$ becoming characteristics. Thus, in this limit, Saint-Venant's principle ceases to hold, and end effects are transmitted without attenuation along the characteristics, i.e., in the direction of the fibers.

Slow stress decay for highly anisotropic materials has been observed experimentally. In the course of conducting torsional pendulum tests designed to measure the longitudinal shear modulus of a polymeric composite, Folkes and Arridge (1975) encountered difficulties (because of end effects) in obtaining values of this modulus that were independent of specimen aspect ratios. Meaningful results were obtained only for samples whose aspect ratios exceeded 100. The data of Folkes and Arridge (1975) for polystyrene fibers in a matrix of polybutadiene indicated that the value of the parameter ε_t for this composite material is about 0.06. Equation (3.120) then suggests the possibility of a "characteristic decay length" for end

---

[†] See, for example, Pipkin (1979, p. 19).
[‡] The *first* part of Eq. (3.118) continues to hold, although the second does not.

effects that is of the order of several specimen widths for this material. Further tests were described by Arridge et al. (1976) and Arridge and Folkes (1976).

The effect of severe anisotropy has been further examined by Choi and Horgan (1977). They studied a transversely isotropic rectangular strip, undertaking an *exact* analysis parallel to that leading to Eqs. (3.103) and (3.104) for isotropic materials. For the *exact* decay rate, they obtained the asymptotic estimate

$$k \sim 2\pi\varepsilon_t/h \qquad \text{as } \varepsilon_t \to 0, \tag{3.121}$$

making explicit the order estimate (3.120) furnished by the energy approach. Numerical results have been obtained that show that for a graphite–epoxy composite with $\varepsilon_t \doteq 0.17$, the exact decay rate is almost four times smaller than that predicted by Eq. (3.105) for an isotropic material.

Choi and Horgan (1978) also studied exact decay rates for plane problems involving sandwich strips (with isotropic layers) subject to self-equilibrated end loads. For the case of a relatively soft inner core, they found a slow rate of decay of end effects, in qualitative agreement with earlier photoelastic experimental studies carried out by Alwar (1970). Related finite element calculations have been reported by Rao and Valsarajan (1980) and Dong and Goetschel (1982).

C. Torsionless Axisymmetric Problems

1. *Isotropic Solid Circular Cylinders*

The problem of estimating the rate of decay of energy and stresses in an isotropic solid circular cylinder subject on one end to a torsionless, *axisymmetric*, self-equilibrated traction was considered by Knowles and Horgan (1969). The analysis closely parallels that of Knowles (1966) for plane strain and furnishes as its principal result an energy decay inequality of the following form:

$$E(z) \leq 2\left(\frac{1 - v + |v|}{1 - v - |v|}\right) E(0) e^{-2kz}, \qquad 0 \leq z \leq l, \tag{3.122}$$

where $E(z)$ is the strain energy stored in the cylinder beyond a distance z from the loaded end. Here l is the length of the cylinder, v is Poisson's ratio, and

$$k = \frac{1}{a}\min\left[\frac{q}{\sqrt{2}}, \left(\frac{(q^4 + p^2q^2)^{1/2} - q^2}{2}\right)^{1/2}\right], \tag{3.123}$$

where a is the radius, while p and $q = q(v)$ are the smallest positive roots of

$$J_1(p) = 0, \qquad qJ_0(q) - (1-v)J_1(q) = 0; \tag{3.124}$$

J_0 and J_1 are the usual Bessel functions. Numerical results furnished by Knowles and Horgan (1969) show that for $v = 3/10$, the value of k in Eq. (3.123) is $k \doteq 1.4/a$. Comparison with numerical calculations of the exact decay rate (see, e.g., Little and Childs, 1967) indicates that this value of k underestimates the exact decay rate by a factor of about 2; a similar comparison applies for other values of v with $0 \leq v \leq 1/2$.

It may be that the estimated decay rate (3.123) as given by Knowles and Horgan (1969) can be improved—as in the plane case—by following a procedure analogous to that used by Flavin (1974).

2. The Hollow Circular Cylinder

An analysis of the kind carried out by Knowles and Horgan (1969) can be adapted to the case of a *hollow* circular cylinder of isotropic material subject on one end to an axisymmetric self-equilibrated traction. The result is an energy inequality like Eq. (3.122), but in which a now represents the midsurface radius of the tube, and the decay constant k depends on v and the ratio h/a of the wall thickness h to the midsurface radius: $k = (1/a)\hat{k}(v, h/a)$. For a *thin* shell ($h/a \ll 1$), it turns out that $\hat{k}(v, h/a) = O(1)$ as $h/a \to 0$, so that the characteristic decay length furnished by the energy estimate is of the order of the midsurface radius.

On the basis of systematically constructed theories of thin shells,[†] it is to be anticipated that the actual decay length for a thin circular cylindrical shell under a self-equilibrated axisymmetric end load is of the order of $\sqrt{ah}$, rather than a, so that the result of the energy argument just described is far too conservative. Moreover, one would further anticipate on the basis of shell theory that an even *shorter* decay length—of the order of h itself—prevails if the end traction, in addition to being self-equilibrated, produces zero resultant radial force and bending moment per unit circumferential length of midsurface. Whether energy arguments can be refined to the extent necessary to confirm these decay lengths as suggested by shell theory is an open question.

3. Transversely Isotropic Circular Cylinders

Horgan (1974) has generalized the approach of Knowles and Horgan (1969) to cover the case of a solid cylinder of transversely isotropic material subject on one end to a torsionless axisymmetric self-equilibrated load. The axis of

---

[†] See, for example, Johnson and Reissner (1959).

elastic symmetry is parallel to the axis of the cylinder. An energy decay inequality similar in form to Eq. (3.122) is obtained in which the decay constant k depends on the elastic constants of the material. When this decay constant—which, of course, is only a lower bound on the actual rate of decay—is compared with the numerically determined values of the exact decay rate for various special materials as reported by Warren et al. (1967), k is found to be conservative by a factor that varies from roughly 2 to about 2.5, depending on the material.[†] Horgan (1974) also examined the limiting behavior of k for a highly anisotropic material. If $\varepsilon = E_T/E_L$, where E_T and E_L are the transverse and longitudinal Young's moduli of the material and a is the radius of the cylinder, one finds that $k = O(\varepsilon^{1/2}/a)$ as $\varepsilon \to 0$, again suggesting a very slow decay of end effects in the "stiff" direction for this extreme case.

D. THE ELASTIC CYLINDER AND OTHER THREE-DIMENSIONAL PROBLEMS

1. Energy Decay by the "Slice Argument"

The problems considered up to now in Section III have all been two-dimensional in the sense that the governing partial differential equations involve only two independent variables. The question of comparing stress distributions produced by statically equivalent loads, however, first arose in connection with the genuinely three-dimensional Saint-Venant problem of the deformation of an elastic cylinder by prescribed surface tractions over its plane ends. As before, it is sufficient because of linearity to consider only the case in which the boundary of the cylinder is traction-free, except for one end, which carries a given self-equilibrated traction. It is this problem that was treated by Toupin (1965a) in the first paper to approach the matter of Saint-Venant's principle on the basis of energy decay inequalities. We will now describe and discuss Toupin's result.

In Fig. 1, let $\mathbb{R}$ now represent the interior of the region[‡] occupied by a cylinder composed of homogeneous but not necessarily isotropic elastic material with strain energy density

$$W(\mathbf{e}) = \tfrac{1}{2} c_{ijkl} e_{ij} e_{kl}, \qquad c_{ijkl} = c_{jikl} = c_{klij}. \qquad (3.125)$$

In this quadratic form, the e_{ij} are the components of the strain tensor $\mathbf{e}$, while the constants c_{ijkl} are the components of the fourth-order elasticity tensor of the material. For such a material, the stresses τ_{ij} are related to the

---

[†] See Horgan (1982) for additional numerical results.

[‡] The coordinate system as well as the notation used here for cross sections and subregions of $\mathbb{R}$ is that introduced in Section II,A.

infinitesimal strains e_{ij} through

$$\tau_{ij} = c_{ijkl} e_{kl} \qquad \text{on } \mathbb{R}, \tag{3.126}$$

and the e_{ij} are in turn connected to the displacements u_i by

$$e_{ij} = \tfrac{1}{2}(u_{i,j} + u_{j,i}) \qquad \text{on } \mathbb{R}. \tag{3.127}$$

In the absence of body forces, the local equilibrium equations satisfied by the stresses are

$$\tau_{ij,j} = 0 \qquad \text{on } \mathbb{R}. \tag{3.128}$$

If **n** stands for the unit outward normal vector on the boundary of $\mathbb{R}$, the traction vector **t** at points of this boundary has components

$$t_i = \tau_{ij} n_j. \tag{3.129}$$

The governing boundary value problem[†] then consists of equations (3.126)–(3.128) together with the boundary conditions

$$t_i = 0 \qquad \text{on } L + S_l, \tag{3.130}$$

$$t_i = t_i^* \qquad \text{on } S_0, \tag{3.131}$$

where $\mathbf{t}^*$ is the given end traction on S_0; $\mathbf{t}^*$ must necessarily be self-equilibrated:

$$\int_{S_0} \mathbf{t}^* \, dA = \int_{S_0} \mathbf{x} \times \mathbf{t}^* \, dA = \mathbf{0}. \tag{3.132}$$

It is assumed that $W(\mathbf{e})$ is positive definite, so that there are positive constants μ_M and μ_m—the maximum and minimum elastic moduli, respectively, of the material—such that

$$\tfrac{1}{2}\mu_m e_{ij} e_{ij} \leq W(\mathbf{e}) \leq \tfrac{1}{2}\mu_M e_{ij} e_{ij}. \tag{3.133}$$

The energy stored beyond a distance z from the loaded end of the cylinder is

$$E(z) = \int_{\mathbb{R}_z} W \, dV, \qquad 0 \leq z \leq l; \tag{3.134}$$

with the help of Eqs. (3.125)–(3.130) and the divergence theorem, one can easily show that $E(z)$ is also given by

$$E(z) = -\frac{1}{2} \int_{S_z} \tau_{3i} u_i \, dA, \qquad 0 \leq z \leq l. \tag{3.135}$$

The definition (3.134) immediately yields

$$E'(z) = -\int_{S_z} W \, dA, \qquad 0 \leq z \leq l. \tag{3.136}$$

[†] For a full discussion of the field equations and boundary value problems of linear elasticity, the reader is referred to the article by Gurtin (1972).

Equations (3.134)–(3.136) are the respective analogs of Eqs. (2.40)–(2.42) in our analysis of the model flow problem for Laplace's equation in Section II. By arguments precisely parallel to those used for the latter problem in Section II,D and leading to Eq. (2.62), it can be shown that in the present instance

$$E(z) \leq E(0)\exp[-2k(\delta)(z-\delta)], \quad 0 < \delta \leq z \leq l, \quad (3.137)$$

where

$$k(\delta) = \tfrac{1}{2}\omega_1(\delta)\sqrt{\rho\mu_m}/\mu_M, \quad (3.138)$$

ρ is the mass density of the material, and $\omega_1(\delta)$ is the smallest positive frequency of *free* elastic vibrations[†] of a slice of thickness δ taken perpendicular to the generators of the cylinder.[‡] This was the result obtained by Toupin (1965a) (see also Toupin, 1965b).

In the proof of Eqs. (3.137) and (3.138), the analog of the additive constant a in Eq. (2.49) is now an additive *rigid body displacement field*, adjusted so as to make applicable the analog for *elastic* vibrations of the inequality (2.54) associated with the *acoustic* eigenvalue problem (2.55). The reader is referred to paragraph 3 of Toupin (1965a) for details.

For the case of an *isotropic* material, Toupin obtained a pointwise estimate of the strains e_{ij}, valid away from the boundary of $\mathbb{R}$, with the help of a mean value theorem. He also discussed the extension to noncylindrical bodies of the argument leading to Eq. (3.137).

One presumably wishes to choose δ in Eq. (3.138) in such a way as to make $\omega_1(\delta)$, and therefore the decay constant k, as large as possible. Because of the complexity of the totality of the modes of vibration of an elastic cylinder, however, the dependence of $\omega_1(\delta)$ on δ is not readily determined. As a result, it is not clear how an optimal disposition of δ is to be made. On the one hand, it is clear that a small δ is desirable, since Eq. (3.137) provides an estimate only for $z \geq \delta$. On the other hand, it is to be expected that choosing δ too small will result in a small value of $\omega_1(\delta)$, owing to the presence of low-frequency flexural vibrations in the spectrum of a thin elastic plate. Similarly, a value of δ that is large compared to the maximum diameter of the cross section would result in a small frequency $\omega_1(\delta)$ because of the low-frequency modes associated with the bending vibrations of a rod. Thus, although it is clear that there *is* an optimum choice of δ, it is rather inaccessible; consequently, it is difficult to turn Eq. (3.138) into an *explicit* estimate for the decay rate, even for, say, a solid circular cylinder.

[†] For a discussion of the eigenvalue problem for free elastic vibrations, see paragraphs 76–78 of Gurtin (1972).

[‡] Since $\omega_1^2(\delta)\rho$ is independent of ρ, the decay constant k of Eq. (3.138) does not actually depend on ρ, as, of course, it should not.

The complicated behavior of $\omega_1(\delta)$ stands in marked contrast to that of its counterpart in the model flow problem for Laplace's equation: There, the analog of $\omega_1(\delta)$ is in fact independent of δ [see Eq. (2.63)]. In the present situation, one cannot take the step analogous to that leading from Eq. (2.62) to Eq. (2.64), and from there to Eq. (2.65).

An analog of Toupin's energy decay inequality (3.137) has been given by Fichera (1977, 1978) for the case in which *each* end of an isotropic cylinder carries a self-equilibrated traction.

2. A Decay Estimate in Terms of Cross-Sectional Properties

Berdichevskii (1974) has constructed an energy decay inequality for a semi-infinite cylinder of arbitrary cross section in which the estimated rate of decay is expressed solely in terms of material constants and parameters pertaining to the geometry of the cross section. We shall present only the broad outlines of his argument, which is too intricate to be fully reproduced here. Although his analysis was carried out for anisotropic materials and for regions $\mathbb{R}$ more general than cylinders, we will confine our attention here to the case of an isotropic cylinder.

We continue to use the notation associated with Fig. 1, except that now $l = \infty$, so that the cylinder $\mathbb{R}$ is semi-infinite. As before, the lateral surface L is traction free, the end S_0 carries a (sufficiently smooth) self-equilibrated traction, and the elastostatic field is assumed to decay rapidly enough as $x_3 \to \infty$ to keep the total energy in the cylinder finite.

In order to state Berdichevskii's result and describe his procedure, we shall need to make use of the fundamental inequality (2.46) for real-valued functions Φ with zero average on a bounded plane domain S, as well as two additional inequalities of a general character. The first of these is *Korn's inequality*[†]: Let S be a bounded plane domain with a sufficiently regular boundary. There is a smallest positive constant K (Korn's constant), depending only on S, such that

$$\int_S v_{\alpha,\beta} v_{\alpha,\beta} \, dA \leq K \int_S \tfrac{1}{4}(v_{\alpha,\beta} + v_{\beta,\alpha})(v_{\alpha,\beta} + v_{\beta,\alpha}) \, dA, \qquad (3.139)$$

for all two-dimensional vector fields $\mathbf{v}(x_1, x_2)$ *that are continuously differentiable on the closure* $\bar{S}$ *of* S *and satisfy the constraint*

$$\int_S (v_{\alpha,\beta} - v_{\beta,\alpha}) \, dA = 0. \qquad (3.140)$$

[†] For a proof of Korn's inequality, as well as references to Korn's original work, see Friedrichs (1947), who treated the n-dimensional case; see also Bernstein and Toupin (1960) and Payne and Weinberger (1961).

In order to state the second inequality that we require, we consider a three-dimensional vector field $\hat{\mathbf{u}}(x_1, x_2, x_3)$ that is continuously differentiable on the closure $\bar{D}$ of a bounded three-dimensional domain D with a sufficiently regular boundary. Let

$$\hat{e}_{ij} = \tfrac{1}{2}(\hat{u}_{i,j} + \hat{u}_{j,i}) \qquad \text{on } \bar{D} \tag{3.141}$$

and set

$$W(\hat{\mathbf{e}}) = \mu\left(\hat{e}_{ij}\hat{e}_{ij} + \frac{v}{1-2v}\hat{e}_{ii}\hat{e}_{jj}\right), \tag{3.142}$$

where μ and v are constants, $\mu > 0$, and $-1 < v < 1/2$. (If $\hat{\mathbf{u}}$ were a displacement field in an isotropic elastic solid occupying the region D, $\hat{\mathbf{e}}$ would be the strain tensor and $W(\hat{\mathbf{e}})$ would be the strain energy density.[†]) Suppose that the boundary of D contains a plane domain S (with a sufficiently regular boundary) that lies in a plane perpendicular to the x_3 axis. Then there is a largest positive constant c, depending only on S, D, and v, such that

$$c \int_S \hat{u}_3^2 \, dA \leq \int_D \frac{1}{\mu} W(\hat{\mathbf{e}}) \, dV \tag{3.143}$$

for all three-dimensional vector fields $\hat{\mathbf{u}}$ that are continuously differentiable on the closure $\bar{D}$ of D and satisfy the three constraints

$$\int_D \hat{u}_3 \, dV = 0, \tag{3.144}$$

$$\int_D (\hat{u}_{3,\alpha} - \hat{u}_{\alpha,3}) \, dV = 0.\text{[‡]} \tag{3.145}$$

We are now in a position to describe Berdichevskii's main result, specialized to the semi-infinite isotropic cylinder $\mathbb{R}$. Let $\mathbf{u}$ be the displacement field in $\mathbb{R}$, $\mathbf{e}$ the associated strain tensor, and $W(\mathbf{e})$ the energy density as given in Eq. (3.142), but with $\hat{\mathbf{e}}$ replaced by $\mathbf{e}$. The energy $E(z)$ stored in the subcylinder $\mathbb{R}_z$ for which $z < x_3 < \infty$ is then

$$E(z) = \int_{\mathbb{R}_z} W(\mathbf{e}) \, dV. \tag{3.146}$$

Berdichevskii *first* showed that

$$E(z) \leq E(0)e^{-2kz}, \qquad 0 \leq z < \infty, \tag{3.147}$$

[†] W as given in Eq. (3.142) is the special case of Eq. (3.125) appropriate to an isotropic elastic material with shear modulus μ and Poisson's ratio v (see p. 85 of Gurtin, 1972).

[‡] The symbol c in Eq. (3.143) is denoted by b_n in Berdichevskii (1974); the dimension of c is reciprocal length. See paragraph 3 of Berdichevskii (1974) for an indication of the proof of Eq. (3.143); the result depends on inequality (2.54), the *three-dimensional* version of Korn's inequality, and the *trace inequality*. For a discussion of the latter, see Lions and Magenes (1972) and Horgan (1979).

where

$$k = k[c(\delta)] = 2\left(\frac{1-2v}{1-v}\right)c(\delta)\left\{1 + \left[1 + 2\sqrt{2}\left(\frac{1-2v}{1-v}\right)\frac{\sqrt{K}}{\lambda_1}c(\delta)\right]^{1/2}\right\}^{-2}. \quad (3.148)$$

Here v is Poisson's ratio, K is Korn's constant for the cross section S_0 of the cylinder, λ_1^2 is the smallest positive eigenvalue of the Neumann problem (2.10), (2.11) for S_0 [see Eq. (2.46)], and $c(\delta)$ is the constant associated with inequality (3.143) *for the special choices* $D = \mathbb{R}_{z,\delta}$ *and* $S = S_z$. As in Section II,D, $\mathbb{R}_{z,\delta}$ is the cylindrical slice of $\mathbb{R}$ lying between the cross sections S_z and $S_{z+\delta}$. The constant $c(\delta)$ depends on the cross section S_0 and on v, as well as on δ, but it is clearly independent of z. It must be emphasized that Eq. (3.147) holds, with $k[c(\delta)]$ given by Eq. (3.148), *for every positive* δ.

The appearance in Eq. (3.148) of the constant $c(\delta)$ that depends on the slice thickness δ is reminiscent of the occurrence in Toupin's estimated decay rate (3.138) of the vibration frequency $\omega_1(\delta)$. The crucial difference, however, lies in the fact that the δ dependence of $c(\delta)$ is much more accessible than that of $\omega_1(\delta)$. After establishing Eqs. (3.147) and (3.148),[†] Berdichevskii investigated $c(\delta)$ along the following lines. Noting that $k(c)$ as given in Eq. (3.148) is an increasing function of c, so that a lower bound for c supplies through Eq. (3.148) a lower bound for $k(c)$, he proved that

$$c(\delta) \geq \frac{1}{2}\left(\frac{13}{70}\delta + \frac{6}{5}\frac{I}{J}\frac{5K + \lambda_1\delta^2}{\lambda_1^4\delta^3}\right)^{-1}, \quad \delta > 0, \quad (3.149)$$

where I and J are the polar moment of inertia and the (geometric) torsional rigidity of the cross section S_0, respectively. If the origin of coordinates is taken at the centroid of S_0,

$$I = \int_{S_0} x_\alpha x_\alpha \, dA; \quad (3.150)$$

moreover,

$$J = I - \int_{S_0} \phi_{,\alpha}\phi_{,\alpha} \, dA, \quad (3.151)$$

where ϕ is the warping function of the theory of torsion[‡] and is determined by the Neumann problem

$$\phi_{,\alpha\alpha} = 0 \quad \text{on } S_0, \quad (3.152)$$

$$\frac{\partial \phi}{\partial n} = -\varepsilon_{\alpha\beta} x_\alpha n_\beta \quad \text{on } C_0. \quad (3.153)$$

[†] Actually, Berdichevskii obtained a result more general than Eq. (3.147) that is valid for anisotropic materials, as well as for certain noncylindrical regions $\mathbb{R}$.

[‡] See Sokolnikoff 1956, paragraph 47); one has $0 < J \leq I$.

The proof of Eq. (3.149)—or, rather, of its counterpart for anisotropic materials†—involves an ingenious argument based on an inequality, derived in an earlier paper by Berdichevskii (1973), that compares the three-dimensional elastic energy functional for $\mathbb{R}_{z,\delta}$ with the two-dimensional energy functional for S_z associated with the refined plate-bending theory of Reissner (1945).

The last step consists in observing that the right-hand side of Eq. (3.149) tends to zero both as $\delta \to 0$ and as $\delta \to \infty$; it therefore has a maximum for an optimally chosen δ, say $\delta = \delta^*$. When δ^* is determined explicitly and inserted into Eq. (3.149), the latter becomes

$$c(\delta^*) \geq c^* \equiv \frac{5\sqrt{546}}{104} \lambda_1 \left(\frac{J}{I}\right)^{1/2} \frac{\{1 + [1 + (65/7)(J/I)K]^{1/2}\}^{1/2}}{2 + [1 + (65/7)(J/I)K]^{1/2}}. \quad (3.154)$$

[This equation was not given explicitly in Berdichevskii (1974); see, however, his Eq. (6.18).] The final energy decay inequality is then obtained by choosing $\delta = \delta^*$ in Eq. (3.148), and then replacing $c(\delta^*)$ by c^*. Thus by Eq. (*3.147*)

$$E(z) \leq E(0)e^{-2k^*z}, \quad 0 \leq z < \infty, \quad (3.155)$$

where

$$k^* = k(c^*), \quad (3.156)$$

with c^* as given in Eq. (*3.154*) and the function $k(c)$ as defined by Eq. (*3.148*). In this remarkable result, the estimated decay rate k^* is expressed solely in terms of Poisson's ratio v and the geometrical constants λ_1, K, and J/I pertaining to the cross section S_0. Of the latter three constants, K and J/I depend only on the *shape* (but not the size) of S_0; the characteristic cross-sectional *length* entering the estimated decay rate is thus again λ_1^{-1}, as was the case in the model problem for Laplace's equation treated in Section II. Note that the cross section S_0 need not be simply connected.

It may be remarked that to obtain a lower bound for k^* for a particular cross section, it would suffice to have a *lower* bound for λ_1 and *upper* bounds for I/J and K. Techniques for the construction of such bounds, insofar as λ_1 and I/J are concerned, have been the subject of extensive study (see, for example, Payne and Weinberger, 1960, and Diaz, 1960). The corresponding issue for Korn's constant K has been less widely studied (see, however, Bernstein and Toupin, 1960; Payne and Weinberger, 1961; Dafermos, 1968; Horgan and Knowles, 1971; and Horgan, 1975a,b).

---

† The counterpart of Eq. (3.149) is Eq. (6.17) in Berdichevskii (1974); the symbol B_1 appearing in Eq. (6.17) is a misprint and should be replaced by A_1. Moreover, in place of the ratio I/J, Berdichevskii's Eq. (6.17) contains a factor $(1 - \kappa)^{-1}$, where κ is defined by his Eq. (6.7). With the help of known properties of the torsional rigidity (see Diaz, 1960), one can show that $\kappa = 1 - J/I$.

The task of assessing the quality of Berdichevskii's estimated decay rate in comparison with "exact" results is more troublesome than is the corresponding issue for either plane strain (Section III,B,1,c) or axisymmetric deformations (Section III,C,1). For a *general* cross section S_0, Synge (1945) has stated the two-dimensional, non-self-adjoint eigenvalue problem to which one is led upon seeking solutions of the field equations (for an isotropic material) whose x_3 dependence is exponential. For the special case of a solid circular cross section, Synge showed that his results furnish the same equation for the determination of the relevant (complex) eigenvalues as had been found earlier by Dougall (1913a,b). A recent numerical investigation of these eigenvalues by Stanley and Bleustein (1972)[†] indicated that the *longest* characteristic decay length (or *slowest* actual decay rate) is *not* associated with axisymmetric deformations. Their results show that the actual decay rate is *not greater* than $2.2/a$ when $v = 0$, or $2.1/a$ when $v = 0.3$, where a is the radius of the cylinder.

For a *circular* cross section S_0, the torsional rigidity J and the polar moment of inertia I coincide[‡]; moreover, $\lambda_1 \doteq 1.841 a^{-1}$. Payne and Weinberger (1961) have shown that $K = 4$ for a circle. Thus Berdichevskii's estimated decay rate k^* of Eqs. (3.156) and (3.154) can be evaluated explicitly in this case. It is, of course, proportional to a^{-1}. Berdichevskii (1974) reported the following values:

v	ak^*
0	0.18
1/4	0.14
1/3	0.11

In view of the remarks just made pertaining to the *actual* decay rate, the value of k^* is thus clearly very conservative.

For a *hollow* circular cross section of thickness h and midsurface radius a, one can evaluate k^* asymptotically as $h/a \to 0$ in order to investigate the estimated decay rate for a thin shell as predicted by Berdichevskii's result. [Such a calculation makes use of the known values of λ_1 and J/I for a circular ring, as well as the value of Korn's constant for a circular annulus as determined by Dafermos (1968).] One finds that $k^*a = O(1)$ in this limit, again falling substantially short of the result $ka = O(\sqrt{a/h})$ predicted by shell theory, as does the estimate based on an analysis like that of Knowles and Horgan (1969) (see Section III,C,2).

[†] See also Klemm and Little (1970).
[‡] Note from Eqs. (3.152) and (3.153) that $\phi \equiv \text{const}$ in this circumstance, so that Eq. (3.151) reduces to $J = I$.

Returning to the case of a general cross section, we note from Eqs. (3.148), (3.154), and (3.156) two circumstances in which the estimated decay rate k^* is small. First, it can be shown that for slender cross sections with little resistance to torsion (a thin rectangle or ellipse, for example), c^* will be small because J/I is small [see Eq. (3.154)], and so k^* itself is small. Second, Eqs. (3.148) and (3.156) show that $k^* \to 0$ as $v \to 1/2$, leading to a vanishing *estimated* decay rate in the limiting case of an incompressible material.[†]

The fact that Berdichevskii's estimated decay rate is conservative in the various circumstances described here does not detract from the remarkable character of his analysis nor of the result itself. His arguments are distinguished not only by their ingenuity, but by scrupulous attention to the issue of characterizing all constants that arise as explicitly as possible and by a determined effort to use "best possible" inequalities at each stage.

3. *Total Energy and Pointwise Estimates*

All of the energy decay inequalities for elastic cylinders appearing in Sections III,C and III,D involve the total energy $E(0)$ associated with the deformation. These inequalities become fully determinate once an upper bound for $E(0)$ has been constructed. In the case of the cylinder problem of Section III,D, an upper bound for $E(0)$ has been obtained by Dou (1966) for an isotropic cylinder of finite length with square cross section.[‡] The result is expressed in terms of mean squares over S_0 of the given traction and its first derivatives. The analysis is based on the principle of minimum complementary energy, which is the analog in linear elasticity of the minimum principle used in Section II,F,1 to bound the total energy in the model problem.

Total energy estimates, valid for not necessarily cylindrical domains and applicable to finite as well as linear elastostatics, have been given by Breuer and Roseman (1980).

The problem of constructing pointwise estimates for the stresses in an isotropic cylinder carrying a self-equilibrated load on the end S_0 (and otherwise traction-free) presents no difficulties if only interior estimates are desired. These can be constructed with the help of the mean value theorem (3.22); in fact, Eqs. (3.24) and (3.25) remain valid for the general cylinder problem. Together they furnish the following estimate for the stresses:

$$|\tau_{ij}(x_1, x_2, z)| \le \frac{30}{\sqrt{\pi}} \left(\frac{\mu}{\beta(v)}\right)^{1/2} d^{-3/2}[E(z-d) - E(z+d)]^{1/2}, \quad (3.157)$$

[†] The numerical results of Stanley and Bleustein (1972) indicate a positive actual decay rate when $v = 1/2$.

[‡] See also Gutiérrez (1976).

where $\beta(v)$ is given by Eq. (3.26), μ is the shear modulus, d is the distance from the point (x_1, x_2, z) to the boundary of the cylinder, and $E(z)$ is the energy stored beyond a distance z from the loaded end S_0. Inequality (3.157), when used in conjunction with either Toupin's energy inequality (3.137) or Berdichevskii's result (3.155), furnishes an interior decay inequality for the stresses.

The obvious deterioration of the bound (3.157) as $d \to 0$ makes it clear that an alternative procedure is needed if one demands a stress estimate that is to retain its utility at or near the boundary of the cylinder. As remarked by Toupin (1965a), the nature of the boundary of the cross section—which does not directly enter the interior estimate (3.157)—must play a major role in the construction of any estimate that is to remain valid up to that boundary. This is easily seen by considering as an example a cross section S_z that is circular except for a radial crack entering S_z from its periphery. At least *some* self-equilibrated loads when applied to the end S_0 will produce singular stress concentrations at the tip of the crack. Although the stresses will decay exponentially along a line L interior to the cylinder and parallel to its generators, the "amplitude" of this exponential will increase without bound as the line L is allowed to approach the generator through the crack tips of the cross sections.

For an isotropic cylinder with a cross section that is bounded by a closed curve (or curves) whose curvature is a twice continuously differentiable function of arc length, and which satisfies certain additional hypotheses, a pointwise estimate—valid on the closure $\bar{\mathbb{R}}$ of the cylinder—for the stresses has been given by Roseman (1966). His result may be stated as follows:

$$|\tau_{ij}(x_1, x_2, z)| \leq M \left(\frac{\mu}{\beta(v)}\right)^{1/2} \rho^{-3/2} [E(z - \rho) - E(z + \rho)]^{1/2}, \quad (3.158)$$

where M and ρ are positive constants that depend on the detailed behavior of the boundary of the cross section, but not on the load, the point $\mathbf{x} = (x_1, x_2, z)$, the material constants μ and v, or the length l of the cylinder. The significant distinction between Eq. (3.158) and the interior estimate (3.157) is, of course, the fact that d in the latter depends on $\mathbf{x}$ and tends to zero as $\mathbf{x}$ approaches the boundary, whereas ρ in Eq. (3.158) is independent of $\mathbf{x}$.

Roseman's proof of Eq. (3.158) is based on an analysis that is not identifiable as directly related to higher-order energy considerations of the kind described in Sections II,G and II,H for the Neumann problem for Laplace's equation. It does, however, make use of mean-square estimates of second derivatives of stresses, as well as of inequalities of the Sobolev type. The details are elaborate and will not be reproduced here.

Using different methods, Fichera (1977, 1978) established an interior estimate for the strains e_{ij} whose general form is the same as that given by Eq. (3.157) for the stresses τ_{ij}. He also obtained a pointwise estimate for $\mathbf{e}$

Recent Developments Concerning Saint-Venant's Principle   249

valid *near* a regular point of the boundary under smoothness assumptions on L that are less restrictive than those needed by Roseman (1966).

4. Further Results

Various modifications, extensions, and generalizations of the energy arguments described previously have been proposed in the literature.

Palamà (1976) has derived an energy decay estimate for the general three-dimensional isotropic cylinder problem (as well as for certain noncylindrical domains) by making use of an inequality[†] similar in character to Eq. (3.143), which is central to the analysis given by Berdichevskii (1974). Palamà's estimated decay rate, however, is not expressed solely in terms of material properties and geometrical characteristics of the cross section of the cylinder.

Weck (1976) reconsidered energy decay inequalities for regions that need not be cylindrical, but we were unable to follow his presentation.

Batra (1978) has provided the analog of Toupin's result (see Section III,D,1) for a helical spring composed of an anisotropic elastic material, and Berglund (1977) has generalized Toupin's analysis so that it applies to materials with couple stresses.

Energy decay estimates for the three-dimensional elasticity operator for anisotropic, not necessarily homogeneous materials were obtained by Oleinik and Yosifian (1977b,c) for certain problems involving noncylindrical domains in which the displacement vector is prescribed to vanish over part or all of the boundary. These results were applied by Oleinik and Yosifian to the study of uniqueness issues and questions pertaining to the behavior of solutions of the elasticity equations near irregular points of the boundary.

Flavin (1978) has applied Berdichevskii's technique to an isotropic cylinder problem in which the lateral surface is *fixed* (rather than traction-free). His decay estimate, which is simpler than Berdichevskii's because of the more cooperative boundary condition, may be relevant to the assessment of end effects in the three-dimensional "plane problem" for which the associated plane strain problem involves prescribed displacements.[‡]

Biollay (1980) has treated a problem for the semi-infinite isotropic elastic cylinder in which displacements are specified on S_0 and are required to vanish on the lateral surface. He obtained a decay estimate for the cross-sectional mean square of the displacement vector; the estimated decay rate involves only Poisson's ratio and the smallest eigenvalue for the differential equation (2.10) and the boundary condition $\phi = 0$ (i.e., the fundamental eigenvalue for the "clamped membrane").

[†] See Eq. (3.10) of Palamà (1976).

[‡] For a discussion of the three-dimensional plane problem, see Gurtin (1972, paragraphs 45 and 46).

Finally, Berdichevskii (1978) provided a general discussion of energy decay inequalities as they relate to principles of the Saint-Venant type. He also considered the cross-sectional eigenvalue problems associated with representations of the exact solutions of the underlying boundary value problems.

IV. Principles of Saint-Venant Type in Other Contexts

A. THE INFLUENCE OF NONLINEARITY

1. *Elastostatics*

In the treatment of Saint-Venant's principle for a cylindrical body within the framework of infinitesimal linear elasticity theory, it is sufficient because of superposition to consider only the case in which a self-equilibrated traction acts on one end of the cylinder while the remainder of the boundary is traction-free. As we have seen in Section III, the elastostatic field in such a problem decays exponentially with distance from the loaded end, and the rate of decay depends only on the geometry of the cross section of the cylinder and on the elastic moduli of the material.

When the issues underlying Saint-Venant's principle are examined in the context of the theory of *finite* elasticity,[†] new features of importance must be taken into account. First, since the nonlinearity of the theory rules out superposition, it is no longer possible to reduce the problem to the case of *self-equilibrated* end loads. Second, even if buckling or other instabilities do not arise, there is the possibility that the difference between corresponding field quantities associated with two statically equivalent loadings might not be exponentially small away from the ends of the cylinder when the loads are not small. Even if exponential decay does prevail, the decay rate might now depend on detailed features of the loading, as well as on geometrical or material characteristics.

Few attempts have been made to investigate Saint-Venant's principle in nonlinear elasticity.[‡] Roseman (1967) considered a plane strain problem for

---

[†] A Saint-Venant principle is assumed in connection with the solutions of a number of special problems in the finite theory, such as that for torsion of an incompressible circular cylinder.

[‡] Although they do not bear directly on Saint-Venant's principle, some results concerning spatial decay for nonlinear elliptic partial differential equations have been obtained by Roseman (1973, 1974), Knowles (1977a), Horgan and Wheeler (1977b), and Horgan and Olmstead (1979). The formulation and certain features of Saint-Venant's problem for the cylinder in finite elasticity have been discussed by Ericksen (1977a,b, 1979) and Muncaster (1979).

a homogeneous isotropic elastic material; the domain of interest is a rectangle in the undeformed state. The deformation images of the long sides of the rectangle are taken to be traction-free, while the short sides carry separately *self-equilibrated* but otherwise arbitrary smooth loadings. Roseman showed that at sufficiently large distances from the ends of the rectangle, the Lagrangian finite strain components are bounded in absolute value by a function of position that decays exponentially with distance from the short sides, provided that the deformation throughout the rectangle satisfies certain conditions which limit its severity and appear to be difficult to translate into restrictions on the end loads. Moreover, the rate of decay in Roseman's estimate is characterized in such a way that little can be said about its magnitude, even in comparison with the known lower bound for the corresponding decay rate for the problem when formulated on the basis of infinitesimal theory. The arguments used by Roseman (1967) involve mean-square integral estimates and Sobolev inequalities. A generalization of these results to the three-dimensional problem for the cylinder was carried out by Breuer and Roseman (1977b).

Recently the present authors (Horgan and Knowles, 1981) have established a version of Saint-Venant's principle in the simplest possible setting within the exact theory of finite elasticity: *finite antiplane shear* of an infinitely long cylinder[†] composed of homogeneous isotropic *incompressible* material. The constitutive law is assumed to belong to a special class of such laws that admit nontrivial states of finite antiplane shear.[‡] The undeformed cross section of the cylinder is a semi-infinite strip, the long sides of which remain traction-free, while the short side carries a prescribed shear traction directed parallel to the generators. This given traction is not necessarily uniformly distributed, and the associated average shear stress τ (and hence the resultant shear force) need not vanish. At infinity in the strip, it is assumed that the displacement is that of a simple shear (parallel to the generators) with shear stress τ.

The mathematical problem of concern may be written as a boundary value problem of the Neumann type for a single quasi-linear second-order elliptic partial differential equation on the semi-infinite strip $\mathbb{R} = \{(x_1, x_2) | x_1 > 0, 0 < x_2 < h\}$ comprising the cross section of the undeformed cylinder. One seeks a function $u(x_1, x_2)$—the out-of-plane displacement field—that is continuously differentiable on $\bar{\mathbb{R}}$, twice continuously differentiable on $\mathbb{R}$, and which satisfies the differential equation

$$[M(|\nabla u|)u_{,\alpha}]_{,\alpha} = 0 \quad \text{on } \mathbb{R}, \tag{4.1}$$

[†] In such a deformation, the displacement vector is parallel to the generators of the cylinder and is independent of the coordinates parallel to the generators.

[‡] See Knowles (1976). For a general discussion of finite antiplane shear, see Knowles (1977b) and Gurtin (1981).

subject to the boundary conditions

$$u_{,2}(x_1, 0) = u_{,2}(x_1, h) = 0, \qquad 0 \leq x_1 < \infty, \tag{4.2}$$

$$M(|\nabla u(0, x_2)|) u_{,1}(0, x_2) = f(x_2), \qquad 0 \leq x_2 \leq h. \tag{4.3}$$

Here f is the given shear traction, and $M(\gamma) \in C^1(-\infty, \infty)$ is a given even positive function called the *modulus of shear at shear strain* γ; M is related to the strain energy density of the incompressible material at hand.[†] To complete the statement of the boundary value problem, we adjoin to Eqs. (4.1)–(4.3) a condition at infinity that requires that

$$u(x_1, x_2) \sim \gamma x_1 \quad \text{as } x_1 \to \infty, \qquad 0 \leq x_2 \leq h, \tag{4.4}$$

where the constant γ—the shear strain at infinity—is such that

$$\tau = M(\gamma)\gamma, \qquad -\infty < \gamma < \infty, \tag{4.5}$$

while

$$\tau = \frac{1}{h} \int_0^h f(x_2)\, dx_2 \tag{4.6}$$

is the *average shear traction*, and τh the resultant force per unit axial length, acting on the end of the strip at $x_1 = 0$; τ is not necessarily positive.

A displacement field on $\mathbb{R}$ for which $u \equiv \gamma x_1$ would satisfy Eqs. (4.1), (4.2), and (4.4) exactly, but would not fulfill Eq. (4.3) unless f were constant. Such a displacement field is called a *simple shear* with shear strain γ; the associated shear stress is then precisely τ, where τ is related to γ by the stress–strain relation (4.5). It can be shown that if the differential equation is uniformly elliptic, the odd function $M(\gamma)\gamma$ is monotonically increasing and tends to $+\infty$ as $\gamma \to +\infty$, so that Eq. (4.5) supplies a unique γ for each prescribed τ.

The principle of the Saint-Venant type established by Horgan and Knowles (1981) compares the state of simple shear—with shear stress τ—with the field arising from a nonuniform end traction f whose average is also τ. It is shown that along either long side of the strip, the values of the nonvanishing component of shear stress τ_{31} satisfy the inequality[‡]

$$|\tau_{31} - \tau| \leq q e^{-kx_1}, \qquad x_2 = 0 \text{ or } x_2 = h, 0 \leq x_1 < \infty. \tag{4.7}$$

Here

$$q = \frac{\pi}{h} \max_{0 \leq x_2 \leq h} \left(\frac{\left| \int_0^{x_2} f(\eta)\, d\eta - \tau x_2 \right|}{\sin(\pi x_2/h)} \right) \geq 0; \tag{4.8}$$

[†] For small γ, $M \sim \mu$, where μ is the infinitesimal shear modulus, so that upon linearization Eq. (4.1) becomes Laplace's equation.

[‡] An estimate of the shear stresses valid on the *entire* region could presumably be obtained, but only with considerable technical effort (see Section II,I).

q is a measure of the departure of f from uniformity; $q = 0$ if and only if f is constant. The estimated decay rate k in Eq. (4.7) is of the form

$$k = \frac{\pi}{h} v(\tau, q), \qquad (4.9)$$

where the function $v(\tau, q)$ depends on the strain energy density of the material, as well as on the average end stress τ and the nonuniformity parameter q.

For a so-called *neo-Hookean* material—sometimes invoked as a crude model for the mechanical behavior of rubber—$M(\gamma)$ is constant, Eq. (4.1) reduces exactly to Laplace's equation, and it follows from results described in Section II that $v(\tau, q) = 1$ for *all* τ and q. For such a material the stress–strain relation (4.5) is exactly linear at all strains.

Horgan and Knowles (1981) treated materials that are either *hardening* $[M'(\gamma) > 0$ for all $\gamma \geq 0]$ or *softening* $[M'(\gamma) < 0]$ in shear. For each of these classes of materials, they gave explicit characterizations of $v(\tau, q)$ in terms of the strain energy density of the material. In the special case of a self-equilibrated end traction ($\tau = 0$), their results for hardening materials yield $v(0, q) = 1$ for all $q > 0$, so that the *actual* rate of decay is at least as great for such materials as that given by infinitesimal theory. For softening materials, $v(0, q) < 1$ for all q, so the *actual* decay rate in the self-equilibrated case *may* be *less* than that associated with infinitesimal theory.

For any given $\tau \neq 0$, the value of $v(\tau, q)$ exceeds unity for *softening* materials if q is small enough, thus assuring that the actual decay rate in the nonlinear problem exceeds that of infinitesimal theory at all load levels τ in these circumstances. For *both* hardening and softening materials, $v(\tau, q)$ tends, as $q \to 0$, to the *actual* rate of decay associated with the boundary value problem obtained upon linearization of Eq. (4.1) *about the state of simple shear with shear stress τ*.

For a subclass of the materials considered (the so-called "power-law" materials), Horgan and Knowles (1981) showed that the estimated decay rate decreases with increasing q in both the hardening and softening cases, whatever the value of τ. On the other hand, it decreases or increases with increasing τ (for fixed q) according to whether the power-law material is hardening or softening.

A more general comparison is also made of the shear stresses arising from two different end tractions that have the same average value, *neither* of which need be uniformly distributed.

The foregoing results were obtained by Horgan and Knowles using a technique based on a comparison principle for second-order quasilinear elliptic operators.[†]

[†] See Gilbarg and Trudinger (1977) for a discussion of comparison principles for such operators; see Section II,I for a related approach to *linear* second-order problems.

2. Viscous Flows in Pipes and Channels

A principle of the Saint-Venant type involving nonlinear effects also arises in connection with incompressible viscous flows in pipes and channels. Of concern is the classical entry problem of laminar flow theory involving the development of velocity profiles in the inlet region. Horgan and Wheeler (1978) presented an analysis of this issue within the general framework of the Navier–Stokes equations governing the steady laminar flow of an incompressible viscous fluid in a cylindrical pipe of arbitrary cross section. The end effect studied involves a comparison between two distinct solutions of the Navier–Stokes equations, namely, a base flow with arbitrary entrance profile and the corresponding fully developed solution (e.g., Hagen–Poiseuille flow for the case of a circular pipe). The flow development is analyzed by consideration of the spatial evolution of the *difference* between these flows. The velocity field associated with this difference satisfies a condition of zero net inflow, corresponding to the "self-equilibration" condition arising in elasticity. Using generalizations of the energy decay arguments of Section III,B,1, Horgan and Wheeler (1978) established an exponential decay estimate for the energy dissipation function associated with this difference velocity field for a small enough Reynolds number. The estimated decay rate is characterized in terms of the Reynolds number, the prescribed entry profile of the base flow, and cross-sectional properties of the pipe. These results yield upper bounds for the "entrance length" through which the flow develops.

The two-dimensional version of the foregoing problem, concerning flow development to the (parabolic) Poiseuille flow in a semi-infinite parallel-plate channel, was investigated by Horgan (1978). In this case a more explicit treatment is possible; moreover, the analogy with considerations of Saint-Venant's principle in elasticity becomes more apparent. Introduction of a stream function $\psi(x_1, x_2)$ for the *difference* velocity field leads to a boundary value problem for a single fourth-order nonlinear elliptic equation. In nondimensional form, this equation reads

$$\frac{1}{R}\Delta\Delta\psi + \varepsilon_{\alpha\beta}\psi_{,\alpha}\Delta\psi_{,\beta} - \frac{3}{2}(1-x_2^2)\Delta\psi_{,1} - 3\psi_{,1} = 0, \qquad 0 \leq x_1 < \infty, \; -1 \leq x_2 \leq 1,$$

(4.10)

where Δ is the two-dimensional Laplacian and R is the Reynolds number (based on channel half-width). On the long sides of the channel, one requires ψ and $\psi_{,2}$ to vanish, ψ and $\psi_{,1}$ are prescribed at the entrance $x_1 = 0$, $-1 \leq x_2 \leq 1$, and suitable conditions are imposed at infinity.

(In the formal limiting case of vanishing Reynolds number, one obtains a boundary value problem for the biharmonic equation governing the development of Stokes flows. This problem is formally equivalent to that discussed in Section III,B,1 in connection with Saint-Venant's principle in linear isotropic plane strain.† Thus the results described there may be immediately applied to yield estimates for the two-dimensional Stokes flow problem.)

Using generalizations of the energy decay arguments described in Section III,B,1, Horgan (1978) established an exponential decay estimate for the energy dissipation associated with solutions ψ of Eq. (4.10), provided that the Reynolds number R is small enough. The estimated rate of decay is given in terms of the Reynolds number and the prescribed entry profile of the base flow.

Issues related to those just described have been considered by Yosifian (1978, 1979) and Amick (1977, 1978) in the course of their general studies of solutions of the stationary Navier–Stokes equations in domains with boundaries extending to infinity. Yosifian (1978)‡ established an energy decay estimate for weak solutions of Dirichlet problems for the Stokes system in three-dimensional (not necessarily cylindrical) domains, again with boundaries that extend to infinity. He used this result as an a priori estimate to establish a uniqueness theorem for velocity fields with possibly unbounded energies. Similar results for the Navier–Stokes equations, as well as a result concerning the behavior of solutions near irregular boundary points, were announced by Yosifian (1979). The work of Amick (1977) is concerned with steady flows in two- and three-dimensional domains that are cylindrical outside some bounded set. Methods of functional analysis are used to establish existence, for a small enough Reynolds number, of weak solutions of the Navier–Stokes equations with finite energies that tend to appropriate Poiseuille flows at infinity. Estimates for the rate of approach, shown to be exponential, were provided by Amick (1978).

B. Slow Stress Diffusion in Shells

Stress decay phenomena of a more complicated character can occur in connection with solid structures described by more elaborate elliptic systems of partial differential equations. As an illustration, we shall describe briefly the "slow stress diffusion" that occurs in certain kinds of thin elastic shells subject to self-equilibrated loads.§

---

† The problem of Section III,B,1 is, of course, now to be considered on a semi-infinite strip.
‡ See also Oleinik (1979b).
§ The principles of the Saint-Venant type implicitly associated with the discussion to follow refer to comparisons, *within the two-dimensional framework of shell theory itself*, of stress fields arising from self-equilibrated loads, rather than to the relationship between shell theory and *three-dimensional* elastostatics. The implications of slow stress diffusion in thin-walled aeronautical structures were pointed out by Hoff (1945).

Wan (1975) furnished an example of such slow stress diffusion in a thin *shallow* isotropic elastic shell whose undeformed shape is that of a semi-infinite strip that has been axially "pretwisted." He studied stress and deflection fields that decay suitably at infinity and which leave the long sides of the strip free of stress resultants and couples. The field equations of shallow shell theory can be reduced in this case to a coupled pair of fourth-order linear partial differential equations for two unknown functions $w(x_1, x_2)$ (the *deflection* in the strip) and $\phi(x_1, x_2)$ (a stress function). In an appropriate nondimensional form, these are

$$\left.\begin{array}{l}\Delta\Delta w + \Lambda\phi_{,12} = 0 \\ \Delta\Delta\phi - \Lambda w_{,12} = 0\end{array}\right\} \quad 0 \leq x_1 < \infty, -1 \leq x_2 \leq 1, \quad (4.11)$$

where Δ is the two-dimensional Laplacian and the parameter Λ is given by

$$\Lambda = \frac{4\beta b^2}{h}[3(1-v^2)]^{1/2}. \quad (4.12)$$

In Eq. (4.12), the constants β, b, h, and v stand, respectively, for the pretwist per unit axial length, the half-width of the plate, the plate thickness, and Poisson's ratio. On the long edges, one requires

$$\phi_{,11} = \phi_{,12} = w_{,22} + vw_{,11} = (2-v)w_{,112} + w_{,222} = 0 \quad \text{at } x_2 = \pm 1, 0 \leq x_1 < \infty, \quad (4.13)$$

corresponding to vanishing stress resultants and bending couple.

For given values of the width $2b$ and the pretwist β, Λ is large compared to unity if the thickness h is small enough. It is this case that was treated by Wan (1975), who used a perturbation procedure based on the assumption $\Lambda \gg 1$ to study solutions of Eqs. (4.11) and (4.13) which decay exponentially as $x_1 \to \infty$. He found two classes of such solutions.[†] In the first of these, the characteristic decay length (referred now to *physical* variables, rather than dimensionless ones) is proportional to $\Lambda^{-1/3}b$ and is thus *short* compared to that for an *untwisted* plate, which is of the order of the width of the strip. In the second class, however, the decay length is proportional to Λb and is therefore many times the strip width. Thus even for a strip whose pretwist is *small* in the sense that $\beta b \ll 1$, the decay length associated with self-equilibrated loads may be very large compared to that for an *untwisted* plate, which is of the order of the width of the strip.

Another instance of slow stress diffusion in shell theory was described by Simmonds (1977)[‡], who used perturbation methods to treat semi-infinite shells of revolution (cylinders or cones, for example) subject to self-

---

[†] All decaying solutions, of course, necessarily correspond to self-equilibrated loads at $x_1 = 0$.
[‡] The paper was actually presented in 1973.

Recent Developments Concerning Saint-Venant's Principle 257

equilibrated but not necessarily axisymmetric end loads. For a cylindrical shell, he of course found the slowly decaying stress states with characteristic decay length of the order of $\sqrt{ah}$, where h is the thickness and a the midsurface radius. This decay length is large compared to the shell thickness. For conical shells, the corresponding stress decay follows a power law rather than an exponential one.

As far as we know, energy decay arguments have not been successfully applied to problems of the type discussed by Wan (1975) and Simmonds (1977).

C. Time-Dependent Problems

1. *Parabolic Equations*

All of the results discussed so far refer to problems in which the governing system of differential equations is elliptic. It is natural to ask whether principles of the Saint-Venant type apply to parabolic differential equations as well.[†]

Boley (1958) was apparently the first to address this question; he considered the heat equation, and his viewpoint was that associated with the von Mises–Sternberg version of Saint-Venant's principle. Comments more directly related to the issue of spatial decay for transient heat conduction appear in Boley (1960a).

To describe for the heat equation a principle of the Saint-Venant type of the kind discussed in the foregoing sections of this review, we return to the cylinder $\mathbb{R}$ of Fig. 1, but we now regard it as occupied by a homogeneous isotropic heat conductor of constant thermal diffusivity α whose lateral surface is insulated. If $u(\mathbf{x}, t) = u(x_1, x_2, x_3, t)$ is the time-dependent temperature field[‡] in $\mathbb{R}$, we require that it satisfy the heat equation

$$u_{,ii} - \frac{1}{\alpha}\dot{u} = 0 \qquad \text{on } \mathbb{R} \times [0, \infty), \tag{4.14}$$

as well as the boundary condition

$$\frac{\partial u}{\partial n} = 0 \qquad \text{on } L \times [0, \infty). \tag{4.15}$$

We consider only temperature fields that are initially uniform and hence we

---

[†] For remarks on the corresponding question for *hyperbolic* equations, see Section IV,C,3.
[‡] The function u is assumed to be continuous, together with its derivatives $\dot{u}$, $u_{,i}$, and $u_{,ij}$, on $\mathbb{R} \times [0, \infty)$; a superposed dot indicates time differentiation.

impose the initial condition

$$u(\mathbf{x}, 0) = 0, \qquad \mathbf{x} \in \mathbb{R}. \tag{4.16}$$

To complete the statement of a boundary–initial value problem for the cylinder, we prescribe the entry and exit heat flows:

$$u_{,3} = f \qquad \text{on } S_0 \times [0, \infty), \tag{4.17}$$

$$u_{,3} = g \qquad \text{on } S_l \times [0, \infty), \tag{4.18}$$

where $f(x_1, x_2, t)$ and $g(x_1, x_2, t)$ are given functions. We suppose that f and g are such as to permit the existence of a solution u of the assumed smoothness to the boundary–initial value problem (4.14)–(4.18); the latter problem is the transient analog of the *steady* heat flow problem (2.1)–(2.3) discussed in Section II. For the present problem, one of course need *not* impose the equilibrium condition (2.5).

Let

$$Q_0(t) = \int_{S_0} f\, dA, \qquad Q_l(t) = \int_{S_l} g\, dA \tag{4.19}$$

be the instantaneous *total* fluxes across S_0 and S_l. Suppose now that the functions $f(x_1, x_2, t)$ and $g(x_1, x_2, t)$ in Eqs. (4.17) and (4.18) were *replaced* by $Q_0(t)/A$ and $Q_l(t)/A$, respectively, where A is the area of a cross section of $\mathbb{R}$. Let $\bar{u}$ be the solution of the boundary value problem (4.14)–(4.18) with f and g so replaced; clearly $\bar{u}$ is independent of x_1 and x_2: $\bar{u} = \bar{u}(x_3, t)$. It is the solution of the following *one-dimensional* boundary–initial value problem[†]:

$$\bar{u}_{,33} = \frac{1}{\alpha} \dot{\bar{u}}, \qquad 0 \le x_3 \le l,\ t \ge 0, \tag{4.20}$$

$$\bar{u}_{,3}(0, t) = Q_0(t)/A, \qquad t \ge 0, \tag{4.21}$$

$$\bar{u}_{,3}(l, t) = Q_l(t)/A, \qquad t \ge 0, \tag{4.22}$$

$$\bar{u}(x_3, 0) = 0, \qquad 0 \le x_3 \le l. \tag{4.23}$$

One sees easily that

$$\bar{u}(z, t) = \frac{1}{A} \int_{S_z} u(x_1, x_2, x_3, t)\, dA, \tag{4.24}$$

so that $\bar{u}(z, t)$ is indeed the *average* over the cross section S_z of the solution u of the *original* boundary–initial value problem (4.14)–(4.18). Moreover, $\bar{u}(x_3, t)$ is *also* the analog in the present setting of the simple uniform flow (2.9) in the *steady* case. The "Saint-Venant question" is now roughly posed

---

[†] An explicit representation for $\bar{u}(x_3, t)$ can be given using elementary methods, but we do not need it here.

as follows: *Is $|u - \bar{u}|$—or perhaps $|\nabla u - \nabla \bar{u}|$—in some sense small at each time instant except near the ends?*

To study this question, it is clearly sufficient because of superposition to examine the *original* problem (4.14)–(4.18) for the special case in which $g \equiv 0$ and f is "self-equilibrated" at each instant:

$$\int_{S_0} f(x_1, x_2, t) \, dA = 0, \qquad 0 \le t < \infty. \tag{4.25}$$

Thus we now wish to estimate the size of the solution $u(\mathbf{x}, t)$ of the following boundary–initial value problem:

$$u_{,ii} - \frac{1}{\alpha} \dot{u} = 0 \qquad \text{on } \mathbb{R} \times [0, \infty), \tag{4.26}$$

$$\frac{\partial u}{\partial n} = 0 \qquad \text{on } L \times [0, \infty), \tag{4.27}$$

$$u(\mathbf{x}, 0) = 0, \qquad \mathbf{x} \in \bar{\mathbb{R}}, \tag{4.28}$$

$$u_{,3} = f \quad \text{on } S_0 \times [0, \infty), \qquad u_{,3} = 0 \quad \text{on } S_l \times [0, \infty), \tag{4.29}$$

where f is self-equilibrated,

$$\int_{S_0} f \, dA = 0, \qquad t \ge 0. \tag{4.30}$$

Let

$$E(z, t) = \alpha \int_0^t \int_{\mathbb{R}_z} u_{,i} u_{,i} \, dV \, d\tau + \tfrac{1}{2} \int_{\mathbb{R}_z} u^2 \, dV, \qquad 0 \le z \le l, t \ge 0 \tag{4.31}$$

where u satisfies Eqs. (4.26)–(4.29). By using arguments that are similar to those employed in Section II,C and which make essential use of Eq. (4.30), one can show (Knowles, 1971) that

$$E(z, t) \le E(0, t) e^{-2\lambda_1 z}, \qquad 0 \le z \le l, t \ge 0, \tag{4.32}$$

where λ_1 is the smallest positive eigenvalue of the Neumann problem (2.10), (2.11). Thus the estimated decay rate $2\lambda_1$ in Eq. (4.32) is identical with that for the energy $E(z)$ associated with the *steady* flow problem [see Eq. (2.48)]; end effects for transient heat conduction therefore decay spatially at least as rapidly as their counterparts in the steady case.† This is perhaps not surprising, since some solutions of the heat equation (4.26) tend to solutions of Laplace's equation as $t \to \infty$.

---

† This was observed by Boley (1960a) for a more general domain on the basis of a different argument.

Prior to the work of Knowles (1971), a spatial decay estimate for the problem (4.26)–(4.29) had been given by Edelstein (1969). He employed an energy-like function different from $E(z,t)$ in Eq. (4.31) above, and his estimated decay rate is time-dependent, tending to zero[†] as $t \to \infty$. Sigillito (1970) extended Edelstein's result to cover other boundary conditions.

It is possible to obtain pointwise estimates from the decay inequality (4.32) or from its counterpart in Edelstein (1969); the latter reference gives details.

When the Neumann boundary conditions (4.27) and (4.29) are replaced by conditions of the Dirichlet type, pointwise exponential decay estimates may be obtained by arguments based on comparison principles (see Horgan and Wheeler, 1975a, 1976b).

Energy arguments of the kind used by Edelstein (1969) have been extended by Sigillito (1974) to a certain linear third-order "pseudoparabolic" equation of physical interest. The same equation has been treated by Horgan and Wheeler (1975b) by generalizing the analysis of Knowles (1971).

Results for a broad range of mixed boundary–initial value problems for rather general second-order linear parabolic equations were derived by Oleinik and Yosifian (1976). These authors established an energy decay inequality for weak solutions and for a fairly general class of domains, and they used the result to obtain uniqueness theorems, as well as information pertaining to the large-time behavior of solutions.

Certain nonlinear parabolic problems have also been investigated by Edelstein (1971), Nunziato (1974), and Horgan and Wheeler (1975a, 1976b).

2. Viscoelasticity

The von Mises–Sternberg formulation of Saint-Venant's principle was extended to the linear theory of isothermal quasi-static deformations of isotropic *viscoelastic* materials by Sternberg and Al-Khozaie (1964). Generalizations of the work of Toupin (1965a) to certain classes of viscoelastic materials were carried out by Edelstein (1970) and Neapolitan and Edelstein (1973).

3. Elastodynamics

Insofar as we are aware, there are no results for a hyperbolic system of the kind describing elastic wave propagation that are similar in character to those reviewed in the present article in connection with principles of the

---

[†] Edelstein (1969) considered a slightly more general parabolic equation and treated n-dimensional noncylindrical regions as well. There are significant inadequacies of a *dimensional* nature with Edelstein's analysis; when these are corrected, his result is altered, but its qualitative features remain as described above.

Saint-Venant type for elliptic and parabolic problems.[†] In view of the fact that high-frequency effects may propagate with little spatial attenuation, one would not *expect* to find unqualified decay estimates of the kind discussed here in problems involving elastic wave propagation, even if the end loads are self-equilibrated at each instant (see Boley, 1955, 1960b).

V. Concluding Comments

The model problem for Laplace's equation considered in Section II furnishes a standard for principles of the Saint-Venant type. In this simple setting, the nonrepresentational arguments leading to energy decay inequalities, cross-sectional estimates, and pointwise bounds seem to work naturally, leading to a theory that is essentially complete and which fulfills all the requirements one might wish to impose.

For problems of fourth and higher orders of interest in elasticity, the present state of the theory of Saint-Venant's principle is quite different. Even in the simplest problem of plane strain—that of a semi-infinite isotropic strip subject to a self-equilibrated end load—the energy arguments seem more contrived than those for the Neumann problem associated with Laplace's equation, and best possible estimates of the decay rate have not been obtained. This state of affairs is unsatisfactory, not only because of the quantitative inadequacy of the resulting bounds on physical quantities, but because it may reflect a weakness of the methods as well.

Perhaps this deficiency manifests itself most apparently in connection with the problem of the torsionless axisymmetric self-equilibrated loading of a hollow isotropic circular cylinder described in Section III,C,2. Here an energy argument, which is patterned directly after that used in the plane strain problem, leads to an estimated characteristic decay length that for a *thin* shell is proportional to the midsurface radius. From shell theory, however, it is known that the correct decay length in this case is proportional to the geometric mean of the midsurface radius and shell thickness. Thus the estimated decay length obtained on the basis of an energy argument overestimates the proper result by a factor that is very large for a thin shell. Indeed, for thin shells, energy decay arguments have so far failed to provide estimates of appropriate strength in all but the simplest of problems, that of axisymmetric torsion.

For Saint-Venant's original problem of the elastic cylinder, the best result presently available is that of Berdichevskii (1974). In qualitative terms, his

---

[†] There has been, however, some discussion of the issue: See, for example, Boley (1955, 1960b) and Grandin and Little (1974).

estimated decay rate is satisfactory in that it depends only on the cross-sectional geometry and material properties. Quantitatively, it would appear to be very conservative, and it suffers from the same inadequacy for thin shells as does the result for the torsionless axisymmetric problem just described. In view of Berdichevskii's effort to keep all the inequalities arising in his analysis as sharp as possible, it would seem unlikely that minor modifications of his procedure will lead to a significant improvement of his result.

Despite the evident progress toward the clarification of Saint-Venant's principle that has taken place in the past, important open questions remain, and opportunities for new contributions are plentiful.

ACKNOWLEDGMENTS

The authors are grateful to Eli Sternberg, who read the manuscript and made a number of helpful suggestions.

One of the authors (C. O. H.) acknowledges the support of the National Science Foundation under Grant MEA 78-26071. This article was completed during the summer of 1981, while this author was a Visiting Associate in Applied Mechanics at the California Institute of Technology.

REFERENCES

Alwar, R. S. (1970). Experimental verification of Saint-Venant's principle in a sandwich beam. *AIAA J.* **8**, 160–162.

Amick, C. J. (1977). Steady solutions of the Navier–Stokes equations in unbounded channels and pipes. *Ann. Scuola Norm. Sup. Pisa Cl. Sci., Serie IV*, **4**, 473–513.

Amick, C. J. (1978). Properties of steady Navier–Stokes solutions for certain unbounded channels and pipes. *Nonlinear Anal.* **2**, 689–720.

Arridge, R. G. C., and Folkes, M. J. (1976). Effect of sample geometry on the measurement of mechanical properties of anisotropic materials. *Polymer* **17**, 495–500.

Arridge, R. G. C., Barham, P. J., Farrell, C. J., and Keller, A. (1976). The importance of end effects in the measurement of moduli of highly anisotropic materials. *J. Materials Sci.* **11**, 788–790.

Batra, R. C. (1978). Saint-Venant's principle for a helical spring. *J. Appl. Mech. (Trans. ASME, Ser. E)* **45**, 297–301.

Berdichevskii, V. L. (1973). An energy inequality in the theory of plate bending. *Prikl. Mat. Mekh.* **37**, 940–944 [*J. Appl. Math. Mech.* **37**, 891–896].

Berdichevskii, V. L. (1974). On the proof of the Saint-Venant principle for bodies of arbitrary shape. *Prikl. Mat. Mekh.* **38**, 851–864 [*J. Appl. Math. Mech.* **38**, 799–813 (1975)].

Berdichevskii, V. L. (1978). Energy methods in certain problems of damping of solutions. *Prikl. Mat. Mech.* **42**, 136–151 [*J. Appl. Math. Mech.* **37**, 140–156.]

Berglund, K. (1977). Generalization of Saint-Venant's principle to micropolar continua. *Arch. Ration. Mech. Anal.* **64**, 317–326.

Bergman, S., and Schiffer, M. (1953). "Kernel Functions and Elliptic Differential Equations in Mathematical Physics." Academic Press, New York.

Bernstein, B., and Toupin, R. A. (1960). Korn inequalities for the sphere and circle. *Arch. Ration. Mech. Anal.* **6**, 51–64.

Biollay, Y. (1980). First boundary value problem in elasticity: Bounds for the displacements and Saint-Venant's principle. *Z. Angew. Math. Phys.* **31**, 556–567.

Bogy, D. B. (1975). Solution of the plane end problem for a semi-infinite elastic strip. *Z. Angew. Math. Phys.* **26**, 749–769.

Boley, B. A. (1955). Application of Saint-Venant's principle in dynamical problems. *J. Appl. Mech. (Trans. ASME)* **22**, 204–206.

Boley, B. A. (1958). Some observations on Saint-Venant's principle. *In* "Proc. 3rd U.S. Nat. Cong. Appl. Mech.," pp. 259–264. ASME, New York.

Boley, B. A. (1960a). Upper bounds and Saint-Venant's principle in transient heat conduction. *Quart. Appl. Math.* **18**, 205–207.

Boley, B. A. (1960b). On a dynamical Saint-Venant principle. *J. Appl. Mech. (Trans. ASME)* **27**, 74–78.

Boussinesq, J. (1885). "Application des potentiels à l'étude de l'équilibre et des mouvements des solides élastiques." Gauthier-Villars, Paris.

Breuer, S., and Roseman, J. J. (1977a). Saint-Venant's principle in linear two-dimensional elasticity for non-striplike domains. *Arch. Ration. Mech. Anal.* **66**, 19–29.

Breuer, S., and Roseman, J. J. (1977b). On Saint-Venant's principle in three-dimensional nonlinear elasticity. *Arch. Ration. Mech. Anal.* **63**, 191–203.

Breuer, S., and Roseman, J. J. (1980). A bound on the strain energy for the traction problem in finite elasticity with localized nonzero surface data. *J. Elasticity* **10**, 11–22.

Choi, I., and Horgan, C. O. (1977). Saint-Venant's principle and end effects in anisotropic elasticity. *J. Appl. Mech. (Trans. ASME, Ser. E)* **44**, 424–430.

Choi, I., and Horgan, C. O. (1978). Saint-Venant end effects for plane deformation of sandwich strips. *Int. J. Solids Struct.* **14**, 187–195.

Courant, R., and Hilbert, D. (1953). "Methods of Mathematical Physics," Vol. 1. John Wiley, New York.

Dafermos, C. M. (1968). Some remarks on Korn's inequality. *Z. Angew. Math. Phys.* **19**, 913–920.

Day, W. A. (1981). Generalized torsion: the solution of a problem of Truesdell's. *Arch. Ration. Mech. Anal.*, **76**, 283–288.

Diaz, J. B. (1960). Upper and lower bounds for quadratic integrals, and at a point, for solutions of linear boundary value problems. *In* "Boundary Value Problems in Differential Equations" (R. E. Langer, ed.), pp. 47–83. Univ. Wisconsin Press, Madison.

Diaz, J. B., and Payne, L. E. (1958). Mean value theorems in the theory of elasticity. *In* "Proc. 3rd U.S. Nat. Cong. Appl. Mech.," pp. 293–303. ASME, New York.

Dong, S. B., and Goetschel, D. B. (1982). Edge effects in laminated composite plates. *J. Appl. Mech. (Trans. ASME)* **49**, 129–135.

Dou, A. (1966). Upper estimate of the potential elastic energy of a cylinder. *Comm. Pure and Appl. Math.* **19**, 83–93.

Dougall, J. (1913a). An analytical theory of the equilibrium of an isotropic elastic rod of circular section. *Trans. R. Soc. Edinburgh* **49**, 895–978.

Dougall, J. (1913b). The method of permanent and transitory modes of equilibrium in the theory of thin elastic bodies. *In* "Proceedings of the Fifth International Congress of Mathematicians" (E. W. Hobson and A. E. H. Love, eds.), vol. 2, pp. 328–340. Cambridge Univ. Press, London and New York.

Drake, R. L., and Chou, P. C. (1965). Upper bounds and Saint-Venant's principle for incompressible potential-flow fields. *J. Appl. Mech. (Trans. ASME)* **32**, 661–664.

Edelstein, W. S. (1969). A spatial decay estimate for the heat equation. *Z. Angew. Math. Phys.* **20**, 900–905.

Edelstein, W. S. (1970). On Saint-Venant's principle in linear viscoelasticity. *Arch. Ration. Mech. Anal.* **36**, 366–380.

Edelstein, W. S. (1971). Further study of spatial decay estimates for semilinear parabolic equations. *J. Math. Anal. Appl.* **35**, 577–590.

Ericksen, J. L. (1977a). On the formulation of St. Venant's problem. In "Nonlinear Analysis and Mechanics: Heriot-Watt Symposium" (R. J. Knops, ed.), vol. 1, pp. 158–186. Pitman, London.

Ericksen, J. L. (1977b). Special topics in elastostatics. In "Advances in Applied Mechanics" (C.-S. Yih, ed.), vol. 17, pp. 189–244. Academic Press, New York.

Ericksen, J. L. (1979). On Saint-Venant's problem for thin-walled tubes. *Arch. Ration. Mech. Anal.* **70**, 7–12.

Ericksen, J. L. (1980). On the status of Saint-Venant's solutions as minimizers of energy. *Int. J. Solids Struct.* **16**, 195–198.

Everstine, G. C., and Pipkin, A. C. (1971). Stress channelling in transversely isotropic elastic composites. *Z. Angew. Math. Phys.* **22**, 825–834.

Fadle, J. (1940). Die Selbstspannungs-Eigenwertfunktionen der quadratischen Scheibe. *Ingenieur-Archiv* **11**, 125–149.

Fichera, G. (1977). Il principio di Saint-Venant: Intuizione dell'ingegnere e rigore del matematico. *Rend. di Mat. Serie VI* **10**, 1–24.

Fichera, G. (1978). Remarks on Saint-Venant's principle. In "Complex Analysis and Its Applications" (I. N. Vekua 70th Birthday volume), pp. 543–554. Nauka, Moscow. Reprinted in *Rend. di Mat. Serie VI* **12** (1979), 181–200.

Flavin, J. N. (1974). On Knowles' version of Saint-Venant's principle in two-dimensional elastostatics. *Arch. Ration. Mech. Anal.* **53**, 366–375.

Flavin, J. N. (1978). Another aspect of Saint-Venant's principle in elasticity. *Z. Angew. Math. Phys.* **29**, 328–332.

Folkes, M. J., and Arridge, R. G. C. (1975). The measurement of shear modulus in highly anisotropic materials: The validity of St. Venant's principle. *J. Phys. D: Appl. Phys.* **8**, 1053–1064.

Friedrichs, K. O. (1947). On the boundary-value problems of the theory of elasticity and Korn's inequality. *Ann. Math.* **48**, 441–471.

Fung, Y.-C. (1965). "Foundations of Solid Mechanics." Prentice-Hall, Englewood Cliffs, New Jersey.

Gilbarg, D., and Trudinger, N. S. (1977). "Elliptic Partial Differential Equations of Second order." Springer-Verlag, Berlin and New York.

Grandin, H. T., and Little, R. W. (1974). Dynamic Saint-Venant region in a semi-infinite elastic strip. *J. Elasticity* **4**, 131–146.

Gregory, R. D. (1980). The traction boundary-value problem for the elastostatic semi-infinite strip; existence of solution and completeness of the Papkovich–Fadle eigenfunctions. *J. Elasticity* **10**, 295–327.

Gurtin, M. E. (1972). The linear theory of elasticity. In "Handbuch der Physik" (S. Flügge, ed.), vol. VI a/2, pp. 1–295. Springer-Verlag, Berlin and New York.

Gurtin, M. E. (1981). Topics in finite elasticity, CBMS-NSF Regional Conference Series in Applied Mathematics, No. 35. SIAM, Philadelphia.

Gutiérrez, A. (1976). Acotacion de la energia potencial elastica de un cilindro de seccion cuadrada. *Rev. Real Acad. Ci. Exact. Fis. Natur. Madrid* **70**, 549–573.

Hardy, G. H., Littlewood, J. E., and Pólya, G. (1967). "Inequalities," 2nd ed. Cambridge Univer. Press, London and New York.

Ho, C.-L., and Knowles, J. K. (1970). Energy inequalities and error estimates for torsion of elastic shells of revolution. *Z. Angew. Math. Phys.* **21**, 352–377.

Hoff, N. J. (1945). The applicability of Saint-Venant's principle to airplane structures. *J. Aero. Sciences* **12**, 455–460.

Horgan, C. O. (1972a). On Saint-Venant's principle in plane anisotropic elasticity. *J. Elasticity* **2**, 169–180.

Horgan, C. O. (1972b). Some remarks on Saint-Venant's principle for transversely isotropic composites. *J. Elasticity* **2**, 335–339.
Horgan, C. O. (1973). On the strain energy density in linear elasticity. *J. Eng. Math.* **7**, 231–234.
Horgan, C. O. (1974). The axisymmetric end problem for transversely isotropic circular cylinders. *Int. J. Solids. Struct.* **10**, 837–852.
Horgan, C. O. (1975a). Inequalities of Korn and Friedrichs in elasticity and potential theory. *Z. Angew. Math. Phys.* **26**, 155–164.
Horgan, C. O. (1975b). On Korn's inequality for incompressible media. *SIAM J. Appl. Math.* **28**, 419–430.
Horgan, C. O. (1978). Plane entry flows and energy estimates for the Navier-Stokes equations. *Arch. Ration. Mech. Anal.* **68**, 359–381.
Horgan, C. O. (1979). Eigenvalue estimates and the trace theorem. *J. Math. Anal. Appl.* **69**, 231–242.
Horgan, C. O. (1982). Saint-Venant's principle in anisotropic elasticity theory. *In* "Mechanical Behavior of Anisotropic Solids" (J. P. Boehler, ed.), Proc. of Euromech. Colloquium 115, Villard de Lans, 1979, pp. 853–868 Editions Scientifiques du Centre National de la Recherche Scientifique, Paris.
Horgan, C. O., and Knowles, J. K. (1971). Eigenvalue problems associated with Korn's inequalities. *Arch. Ration. Mech. Anal.* **40**, 384–402.
Horgan, C. O., and Knowles, J. K. (1981). The effect of nonlinearity on a principle of Saint-Venant type. *J. Elasticity* **11**, 271–291.
Horgan, C. O., and Olmstead, W. E. (1979). Exponential decay estimates for a class of non-linear Dirichlet problems. *Arch. Ration. Mech. Anal.* **71**, 221–235.
Horgan, C. O., and Wheeler, L. T. (1975a). On maximum principles and spatial decay estimates for heat conduction. *In* "Proceedings of the Twelfth Annual Meeting, Society of Engineering Science, University of Texas, Austin," pp. 331–339, Univ. of Texas Press, Austin.
Horgan, C. O., and Wheeler, L. T. (1975b). A spatial decay estimate for pseudoparabolic equations. *Lett. Appl. Eng. Sci.* **3**, 237–243.
Horgan, C. O., and Wheeler, L. T. (1976a). Saint-Venant's principle and the torsion of thin shells of revolution. *J. Appl. Mech. (Trans. ASME, Ser. E)* **43**, 663–667.
Horgan, C. O., and Wheeler, L. T. (1976b). Spatial decay estimates for the heat equation via the maximum principle. *Z. Angew. Math. Phys.* **27**, 371–376.
Horgan, C. O., and Wheeler, L. T. (1977a). Maximum principles and pointwise error estimates for torsion of shells of revolution. *J. Elasticity* **7**, 387–410.
Horgan, C. O., and Wheeler, L. T. (1977b). Exponential decay estimates for second-order quasilinear elliptic equations. *J. Math. Anal. Appl.* **59**, 267–277.
Horgan, C. O., and Wheeler, L. T. (1978). Spatial decay estimates for the Navier–Stokes equations with application to the problem of entry flow. *SIAM J. Appl. Math.* **35**, 97–116.
Horvay, G. (1957a). Biharmonic eigenvalue problem of the semi-infinite strip. *Q. Appl. Math.* **15**, 65–81.
Horvay, G. (1957b). Saint-Venant's principle: a biharmonic eigenvalue problem. *J. Appl. Mech. (Trans. ASME)* **24**, 381–386.
Horvay, G. (1957c). Some aspects of Saint-Venant's principle. *J. Mech. Phys. Solids* **5**, 77–94.
John, F. (1975). A priori estimates, geometric effects, and asymptotic behavior. *Bull. Am. Math. Soc.* **81**, 1013–1023.
Johnson, M. W., and Little, R. W. (1965). The semi-infinite strip. *Quart. Appl. Math.* **22**, 335–344.
Johnson, M. W., and Reissner, E. (1959). On the foundations of the theory of thin elastic shells. *J. Math. and Phys.* **37**, 371–392.
Keller, H. B. (1965). Saint-Venant's procedure and Saint-Venant's principle. *Q. Appl. Math.* **22**, 293–304.

Klemm, J. L., and Little, R. W. (1970). The semi-infinite elastic cylinder under self-equilibrated end loading. *SIAM J. Appl. Math.* **19**, 712–729.

Knowles, J. K. (1966). On Saint-Venant's principle in the two-dimensional linear theory of elasticity. *Arch. Ration. Mech. Anal.* **21**, 1–22.

Knowles, J. K. (1967). A Saint-Venant principle for a class of second-order elliptic boundary value problems. *Z. Angew. Math. Phys.* **18**, 473–490.

Knowles, J. K. (1971). On the spatial decay of solutions of the heat equation. *Z. Angew. Math. Phys.* **22**, 1050–1056.

Knowles, J. K. (1976). On finite anti-plane shear for incompressible elastic materials. *J. Aust. Math. Soc.* **19** (ser. B), 400–415.

Knowles, J. K. (1977a). A note on the spatial decay of a minimal surface over a semi-infinite strip. *J. Math. Anal. Appl.* **59**, 29–32.

Knowles, J. K. (1977b). The finite anti-plane shear field near the tip of a crack for a class of incompressible elastic solids. *Int. J. Fracture* **13**, 611–639.

Knowles, J. K., and Horgan, C. O. (1969). On the exponential decay of stresses in circular elastic cylinders subject to axisymmetric self-equilibrated end loads. *Int. J. Solids Struct.* **5**, 33–50.

Knowles, J. K., and Sternberg, E. (1966). On Saint-Venant's principle and the torsion of solids of revolution. *Arch. Ration. Mech. Anal.* **22**, 100–120.

Knowles, J. K., and Sternberg, E. (1972). On a class of conservation laws in linearized and finite elastostatics. *Arch. Ration. Mech. Anal.* **44**, 187–211.

Koiter, W. T., and Simmonds, J. G. (1973). Foundations of shell theory. *In* "Applied Mechanics: Proceedings of the 13th International Congress of Theoretical and Applied Mechanics, Moscow, 1972" (E. Becker and G. K. Mikhailov, eds.), pp. 150–176. Springer-Verlag, Berlin and New York.

Lekhnitskii, S. G. (1963). "Theory of Elasticity of an Anisotropic Elastic Body" (P. Fern, trans.). Holden-Day, San Francisco, California.

Lions, J. L., and Magenes, E. (1972). "Non-Homogeneous Boundary Value Problems and Applications," rev. ed., vol. 1 (P. Kenneth, trans.). Springer-Verlag, Berlin and New York.

Little, R. W., and Childs, S. B. (1967) Elastostatic boundary region problem in solid cylinders. *Quart. Appl. Math.* **25**, 261–274.

Love, A.E.H. (1944). "A Treatise on the Mathematical Theory of Elasticity," 4th ed. Dover, New York.

Maisonneuve, O. (1971). Sur le principe de Saint-Venant. Thése, Université de Poiters.

Mieth, H.-J. (1975). Über abklingende Lösungen elliptischer Randwertprobleme (Prinzip von Saint-Venant). Dissertation, Technische Hochschule Darmstadt.

Mieth, H. J. (1976). Ein Saint-Venantsches Prinzip für eine Klasse von modifizierten Dirichlet-Problemen. *Z. Angew Math. Mech.* **56**, T258–T260.

Mikhlin, S. G. (1964). "Variational Methods in Mathematical Physics." (T. Boddington, trans.). Macmillan, New York.

Mises, R. von (1945). On Saint-Venant's principle. *Bull. Amer. Math. Soc.* **51**, 555–562.

Muncaster, R. G. (1979). Saint-Venant's problem in nonlinear elasticity: A study of cross-sections. *In* "Nonlinear Analysis and Mechanics: Heriot-Watt Symposium" (R. J. Knops, ed.), vol. IV, pp. 17–75. Pitman, London.

Neapolitan, R. E., and Edelstein, W. S. (1973) Further study of Saint-Venant's principle in linear viscoelasticity. *Z. Angew. Math. Phys.* **24**, 823–837.

Nunziato, J. W. (1974). On the spatial decay of solutions in the nonlinear theory of heat conduction. *J. Math. Anal. Appl.* **48**, 687–698.

Oleinik, O. A. (1979a). Energetic estimates analogous to the Saint-Venant principle and their applications. *In* "Equadiff IV" (J. Fabera, ed.), Lecture Notes in Mathematics, vol. 703 pp. 328–339. Springer-Verlag, Berlin and New York.

Oleinik, O. A. (1979b). Applications of the energy estimates analogous to Saint-Venant's principle to problems of elasticity and hydrodynamics. *In* "Lecture Notes in Physics," vol. 90, pp. 422–432. Springer-Verlag, Berlin and New York.

Oleinik, O. A., and Yosifian, G. A. (1976). An analogue of Saint-Venant's principle and the uniqueness of solutions of boundary value problems for parabolic equations in unbounded domains. *Uspekhi Mat. Nauk* **31**, 142–166 [*Russian Math. Surveys 31*, 153–178].

Oleinik, O. A., and Yosifian, G. A. (1977a). Boundary value problems for second order elliptic equations in unbounded domains and Saint-Venant's principle. *Ann. Scuola Norm. Sup. Pisa Cl. Sci.*, Serie IV **4**, 269–290.

Oleinik, O. A., and Yosifian, G. A. (1977b). On singularities at the boundary points and uniqueness theorems for solutions of the first boundary value problem of elasticity. *Comm. Partial Differential Equations* **2**, 937–969.

Oleinik, O. A., and Yosifian, G. A. (1977c). Saint-Venant's principle for the mixed boundary value problem of the theory of elasticity and its applications. *Dokl. Akad. Nauk SSSR 233*, 824–827 [*Soviet Phys. Dokl.* **22**, 233–235].

Oleinik, O. A., and Yosifian, G. A. (1978a). On Saint-Venant's principle in plane elasticity theory. *Dokl. Akad. Nauk SSSR 239*, 530–533 [*Soviet Math. Dokl.* **19**, 364–368].

Oleinik, O. A., and Yosifian, G. A. (1978b). The Saint-Venant principle in the two-dimensional theory of elasticity and boundary problems for a biharmonic equation in unbounded domains. *Sibirsk. Mat. Z.* **19**, 1154–1165 [*Siberian Math. J. 19*, 813–822].

Palamà, A. (1976). On Saint-Venant's principle in three-dimensional elasticity. *Meccanica* **11**, 98–101.

Papkovich, P. F. (1940). Über eine Form der Lösung des byharmonischen Problems für des Rechteck. *Dokl. Akad Nauk. SSSR* **27**, 334–338.

Payne, L. E., and Weinberger, H. F. (1960). An optimal Poincaré inequality for convex domains. *Arch. Ration. Mech. Anal.* **5**, 286–292.

Payne, L. E., and Weinberger, H. F. (1961). On Korn's inequality. *Arch. Ration. Mech. Anal.* **8**, 89–98.

Pipkin, A. C. (1979). Stress analysis for fiber-reinforced materials. *In* "Advances in Applied Mechanics" (C.-S. Yih, ed.), vol. 19, pp. 1–51. Academic Press, New York.

Protter, M. H., and Weinberger, H. F. (1967). "Maximum Principles in Differential Equations." Prentice-Hall, Englewood Cliffs, New Jersey.

Rao, N. R., and Valsarajan, K. V. (1980). Saint-Venant's principle in sandwich strip. *Comput. Struct.* **12**, 185–188.

Reissner, E. (1945). The effect of transverse shear deformation on the bending of elastic plates. *J. Appl. Mech (Trans. ASME)* **12**, A69–A77.

Robinson, A. (1966). "Non-Standard Analysis." North-Holland Publ., Amsterdam.

Roseman, J. J. (1966). A pointwise estimate for the stress in a cylinder and its application to Saint-Venant's principle. *Arch. Ration. Mech. Anal.* **21**, 23–48.

Roseman, J. J. (1967). The principle of Saint-Venant in linear and nonlinear plane elasticity *Arch. Ration. Mech. Anal.* **26**, 142–162.

Roseman, J. J. (1973). Phragmén-Lindelöf theorems for some nonlinear elliptic partial differential equations. *J. Math. Anal. Appl.* **43**, 587–602.

Roseman, J. J. (1974). The rate of decay of a minimal surface defined over a semi-infinite strip. *J. Math. Anal. Appl.* **46**, 545–554.

Saint-Venant, A.-J.-C. B. de (1856a). Mémoire sur la torsion des prismes. *Mémoires présentés pars divers Savants a l'Académie des Sciences de l'Institut Impérial de France* **14**, 233–560. (Read to the Academy on June 13, 1853)

Saint-Venant, A.-J.-C. B. de (1856b). Mémoire sur la flexion des prismes. *J. Math. Pures Appl.* **1** (Ser. 2), 89–189.

Shield, R. T. (1965). On the stability of linear continuous systems. *Z. Angew. Math. Phys.* **16**, 649–686.

Shield, R. T., and Anderson, C. A. (1966). Some least work principles for elastic bodies. *Z. Angew. Math. Phys.* **17**, 663–676.

Sigillito, V. G. (1970). On the spatial decay of solutions of parabolic equations. *Z. Angew. Math. Phys.* **21**, 1078–1081.

Sigillito, V. G. (1974). Exponential decay of functionals of solutions of a pseudoparabolic equation. *SIAM J. Math. Anal.* **5**, 581–585.

Simmonds, J. G. (1977) Saint-Venant's principle for semi-infinite shells of revolution. *In* "Recent Advances in Engineering Science," (A. C. Eringen, ed. vol. 8,), Proceedings of the 10th Anniversary Meeting, Society of Engineering Science, Raleigh, N.C., 1973 pp. 367–374. Science Publishers Inc., Boston.

Sokolnikoff, I. S. (1956). "Mathematical Theory of Elasticity," 2nd ed. McGraw-Hill, New York.

Spencer, A. J. M. (1972). "Deformations of Fibre-Reinforced Materials." Oxford Univ. Press. London and New York.

Stanley, R. M., and Bleustein, J. L. (1972). Generalized torsional waves and the non-axisymmetric end problem in a solid circular cylinder. *Int. J. Solids and Structures* **8**, 807–823.

Sternberg, E. (1954). On Saint-Venant's principle. *Q. Appl. Math.* **11**, 393–402.

Sternberg, E. and Al-Khozaie, S. (1964). On Green's functions and Saint-Venant's principle in the linear theory of viscoelasticity. *Arch. Ration. Mech. Anal.* **15**, 112–146.

Sternberg, E., and Knowles, J. K. (1966). Minimum energy characterizations of Saint-Venant's solution to the relaxed Saint-Venant problem. *Arch. Ration. Mech. Anal.* **21**, 89–107.

Synge, J. L. (1945). The problem of Saint-Venant for a cylinder with free sides. *Q. Appl. Math.* **2**, 307–317.

Timoshenko, S. P. (1934). "Theory of Elasticity," 1st ed. McGraw-Hill, New York.

Timoshenko, S. P., and Goodier, J. N. (1970). "Theory of Elasticity," 3rd ed. McGraw-Hill, New York.

Titchmarsh, E. C. (1958). "Eigenfunction Expansions Associated with Second-Order Differential Equations," Part II. Oxford Univ. Press, London and New York.

Toupin, R. A. (1965a). Saint-Venant's principle. *Arch. Ration. Mech. Anal.* **18**, 83–96.

Toupin, R. A. (1965b). Saint-Venant and a matter of principle. *Trans. N.Y. Acad. Sci.* **28**, 221–232.

Truesdell, C. (1959). The rational mechanics of materials—Past, present, future. *Appl. Mech. Rev.* **12**, 75–80.

Truesdell, C. (1966). The rational mechanics of materials—Past, present, future. Reprinted version of Truesdell (1959) with corrections and additions. *In* "Applied Mechanics Surveys" (H. N. Abramson, H. Liebowitz, J. M. Crowley, and S. Juhasz, eds.), pp. 225–236. Spartan Books, Washington, D.C.

Truesdell, C. (1978). Some challenges offered to analysis by rational thermodynamics. *In* "Contemporary Developments in Continuum Mechanics and Partial Differential Equations" (G. M. de la Penha and L. A. J. Medeiros, ed.), pp. 495–603. North-Holland Publ., Amsterdam.

Wan, F. Y. M. (1975). An eigenvalue problem for a semi-infinite pretwisted strip. *Stud. Appl. Math.* **54**, 351–358.

Warren, W. E., Roark, A. L., and Bickford, W. B. (1967). End effect in semi-infinite transversely isotropic cylinders. *AIAA J.* **5**, 1448–1455.

Weck, N. (1976). An explicit St. Venant's principle in three-dimensional elasticity. *In* "Ordinary and Partial Differential Equations" (W. N. Everitt and B. D. Sleeman, eds.), Lecture Notes in Mathematics, Vol. 564, pp. 518–526. Springer-Verlag, Berlin and New York.

Wheeler, L. T., and Horgan, C. O. (1976). A two-dimensional Saint-Venant principle for second-order linear elliptic equations. *Q. Appl. Math.* **34**, 257–270.

Wheeler, L. T., Turteltaub, M. J., and Horgan, C. O. (1975). A Saint-Venant principle for the gradient in the Neumann problem. *Z. Angew. Math. Phys.* **26**, 141–154.

Wu, C. -H. (1970). A Saint-Venant principle for the Neumann problem with a non-thin two-dimensional domain. *Int. J. Eng. Sci.* **8**, 389–402.

Yosifian, G. A. (1978). An analog of Saint-Venant's principle and the uniqueness of the solutions of the first boundary value problem for Stokes' system in domains with noncompact boundaries. *Dokl. Akad. Nauk. SSSR* **242**, 36–39 [*Soviet Math. Dokl.* **19**, 1048–1052].

Yosifian, G. A. (1979). Saint-Venant's principle for the flow of a viscous incompressible liquid. *Uspekhi Mat. Nauk* **34**, 191–192 [*Russ. Math. Surveys* **34**, 166–167].

Zanaboni, O. (1937a). Dimostrazione generale del principio del De Saint-Venant. *Atti Acad. Naz. dei Lincei, Rendiconti* **25**, 117–121.

Zanaboni, O. (1937b). Valutazione dell'errore massimo cui dà luogo l'applicazione del principio del De Saint-Venant in un solido isotropo. *Atti Acad. Naz. dei Lincei, Rendiconti* **25**, 595–601.

Zanaboni, O. (1937c). Sull'approssimazione dovuta al principio del De Saint-Venant nei solidi prismatici isotropi. *Atti Acad. Naz. dei Lincei, Rendiconti* **26**, 340–345.

Nonlinear Elastic Shell Theory

A. LIBAI

Department of Aeronautical Engineering
Technion, Israel Institute of Technology
Haifa, Israel

J. G. SIMMONDS

Department of Applied Mathematics and Computer Science
University of Virginia
Charlottesville, Virginia

I. Introduction 272
II. Cylindrical Motion of Infinite Cylindrical Shells (Beamshells) 274
   A. Geometry of the Undeformed Shell and Planar Motion. 274
   B. Integral Equations of Cylindrical Motion 276
   C. Differential Equations of Cylindrical Motion 278
   D. Virtual Work (Weak Form of the Equations of Motion) 279
   E. The Mechanical Work Identity 280
   F. Strain Measures 280
   G. Alternate Stresses and Strains 282
   H. The Basic Assumption of Beamshell Theory 283
   I. The Mechanical Theory of Elastic Beamshells 283
   J. Constitutive Laws for Elastic Beamshells 284
   K. Elastostatics 287
   L. Elastodynamics 292
   M. Thermodynamics 296
III. The Equations of General Nonlinear Shell Theory 302
   A. Geometry of the Undeformed Shell 303
   B. Integral Equations of Motion 305
   C. Differential Equations of Motion 307
   D. Virtual Work (Weak Form of the Equations of Motion) 307
   E. The Mechanical Work Identity 308
   F. Strain Measures 308
   G. The Finite Rotation Vector 311
   H. Alternate Strain Measures, Rational in the Displacements 312
   I. Constitutive Relations and Boundary Conditions 318
   J. Formulations of the Field Equations 323

IV. Nonlinear Plate Theory.	332
A. The (Simplified) Extended von Karman Equations	332
B. Inextensional Deformation of Plates	335
V. Static One-Dimensional Strain Fields	337
A. Torsionless Axisymmetric Deformation of Shells of Revolution	338
B. Pure Bending of Curved Tubes	340
VI. Approximate Shell Theories	344
A. Introduction	344
B. Quasi-Shallow Shell Theory	347
C. Nonlinear Membrane Theory	349
Appendix: The Equations of Three-Dimensional Continuum Mechanics	362
References.	365

I Introduction

Shells are curved, thin-walled structures. Everyday examples abound: automobile hoods, balloons, beer cans, bells, bladders, bowls, crab carapaces, domes, egg shells, footballs, funnels, inner tubes, light bulb casings, loudspeaker cones, parachutes, peanut hulls, Ping-Pong balls, seashells, skulls, straws, tires, and wine glasses. *Elastic* shells tend to return to their initial shape when a small load is applied and then removed. If sufficiently thin or sufficiently "rubbery," an elastic shell can undergo deformations that are obvious to the unaided eye. Often such deformations are *nonlinear* in the sense that the deflection at a given point on the shell is *not* proportional to the magnitude of the applied load. Such behavior is the subject of this article. Our aim is, within the conceptual framework outlined below, to unify, simplify, and extend a number of results that have appeared in the last 20 years.

The most widespread application of elastic nonlinear shell theory—the prediction of stability and the analysis of postbuckling—is barely mentioned herein. To do it justice would require a book. Fortunately, there are several excellent works in this area: reviews by Stein (1968), Koiter (1968), Hutchinson and Koiter (1970), and Bushnell (1981a), an article by Budiansky (1974), and a book by Brush and Almroth (1975).

Approaches to shell theory range from the purely *direct* to the purely *derived*. A direct theory, as exemplified by parts of Naghi's article in the "Encyclopedia of Physics" (1972), regards a shell, *from the start*, as a two-dimensional continuum whose kinematics are described by a Cosserat surface, that is, a surface endowed at each point with one or more independent vectors called directors. In particular cases the directors may be associated with rotations, transverse shear strains, thickness changes, or other "higher-order" effects. More generally, directors may be identified with the vector coefficients in an expansion of the three-dimensional displacement field in the thickness direction. For an application of the direct approach to sandwich shells, see Malcolm and Glockner (1972a,b).

The derived approach recognizes a shell as a three-dimensional elastic body, but attempts, often by asymptotic analysis, to exploit the special geometry and loading that characterizes a shell to derive in a rational way systems of two-dimensional equations to describe the "interior" behavior of a shell, away from edges and other geometric or load discontinuities. References to and a discussion of the direct approach may be found in Simmonds (1976), Koiter and Simmonds (1973), van der Heijden (1976), and Goldenveiser (1980).

In this monograph we adopt the following *mixed* approach. We start with the *integral* equations of motion and thermodynamics of a three-dimensional body taken over its reference shape. Specializing to shell-like bodies, we immediately obtain definitions of a two-dimensional stress resultant, stress couple, displacement, spin (angular velocity), entropy resultant, temperature mean, temperature drop, and other fields. These definitions provide a concrete, unambiguous tie between two- and three-dimensional continuum mechanics. The two-dimensional integral equations, for sufficiently smooth fields, imply differential equations. Those of conservation of linear and rotational momentum yield in a natural way a mechanical work identity that provides, *automatically*, two-dimensional strains conjugate to the stress resultants and couples. This leads to a *mechanical theory* of shells in which a strain energy density, depending on these strains only, is assumed to exist. Later, a *thermodynamic theory* is formulated from a postulated first law in which the internal energy is a function of the aforementioned two-dimensional strains plus an entropy resultant and couple. Invariance requirements and dimensional analysis are used to simplify the form of the strain energy density (or the free-energy density in the thermodynamic theory).

Any shell theory must deal with the Kirchhoff hypothesis. First, there is ambiguity in the term. The *three-dimensional Kirchhoff hypothesis*, in conventional shell coordinates, states that the three-dimensional transverse shearing and normal strains are zero (Novozhilov, 1953), whereas the two-dimensional or *reduced Kirchhoff hypothesis*, as we use the term, means that the two-dimensional transverse shearing strains and a two-dimensional twist strain are zero. In our approach, the reduced Kirchhoff hypothesis is regarded as a *constitutive assumption*, although we occasionally mention it in the text before we reach constitutive relations, to point out the simplifications it entails in the kinematics.

The prototype of a shell is an arch. So to introduce a number of fundamental notions in as simple a context as possible, we will begin, in Section II, by considering the planar motion of an infinite cylindrical shell, or *beamshell* for short. Because of space limitations, we develop neither nonlinear stress–strain relations (except for membranes in Section VI,C) nor thermodynamics beyond this point. Of course, there are phenomena that never appear in beamshells, such as contracted boundary conditions or special inextensional effects induced by double curvature.

In Section III, we develop a general nonlinear theory of shells that parallels our treatment of beamshells. We offer four different formulations of the field equations: intrinsic, rotational, displacement, and velocity. In the associated kinematics, we retain transverse shearing strains, perhaps to the point of fanaticism in the displacement approach, where we pay a terrible price in complexity for a generality of dubious value in the classical theory. Sections IV–VI deal with special cases of, and approximations to, the general equations.

In addition to neglecting buckling, we confess to further sins of omission: (1) We give no theorems on existence and uniqueness (or nonuniqueness), (2) little mention is of made variational principles, (3) no numerical methods are given (4) there is no discussion of anisotropic or composite materials, and (5) we offer no systematic mathematical development of boundary conditions in general shell theory (although we do give a cursory, qualitive discussion in Section III,I and do discuss them for nonlinear membrane theory). For specific works in the literature, see Antman (1976) for (1), Atluri (1983) for (2), Hughes (1981) for (3), Librescu (1975) for (4), and Pietraszkiewicz (1979, 1980a,b) for (5).

II. Cylindrical Motion of Infinite Cylindrical Shells (Beamshells)

A. Geometry of the Undeformed Shell and Planar Motion

An infinite cylindrical shell is a body that, in its reference shape, occupies a region bounded by two nonintersecting right cylinders and two planes each parallel to the generators of the cylinders. For a quantitative description, consider a fixed right-handed Cartesian reference frame $Oxyz$. Let $\{\mathbf{i}, \mathbf{j}, \mathbf{k}\}$ denote the associated set of unit base vectors with $\mathbf{k}$ parallel to the generators of the cylinders. The reference shape is then the set of all points with a position vector of the form

$$\mathbf{x}(\sigma, \zeta, z) = \mathbf{R}(\sigma, \zeta) + z\mathbf{k}, \qquad 0 \le \sigma \le L, H_-(\sigma) \le \zeta \le H_+(\sigma), -\infty < z < \infty.$$

(2.1)

Here (see Figs. 1 and 2)

$$\mathbf{R}(\sigma, \zeta) = \boldsymbol{\xi}(\sigma) + \zeta \mathbf{v}(\sigma), \tag{2.2}$$

where

$$\boldsymbol{\xi}(\sigma) = \mathbf{i}\xi(\sigma) + \mathbf{j}\eta(\sigma) \tag{2.3}$$

is the parametric equation of a *reference curve* in the xy plane with arc length σ and smoothly turning unit tangent and normal vectors

$$\boldsymbol{\tau} = \boldsymbol{\xi}'(\sigma) = \mathbf{i}\xi'(\sigma) + \mathbf{j}\eta'(\sigma) \equiv \mathbf{i}\cos\alpha(\sigma) + \mathbf{j}\sin\alpha(\sigma), \tag{2.4}$$

$$\mathbf{v} = \mathbf{k} \times \boldsymbol{\tau} = -\mathbf{i}\sin\alpha(\sigma) + \mathbf{j}\cos\alpha(\sigma). \tag{2.5}$$

Here $H_+(\sigma)$ and $H_-(\sigma)$ are assumed to be sufficiently small so that each

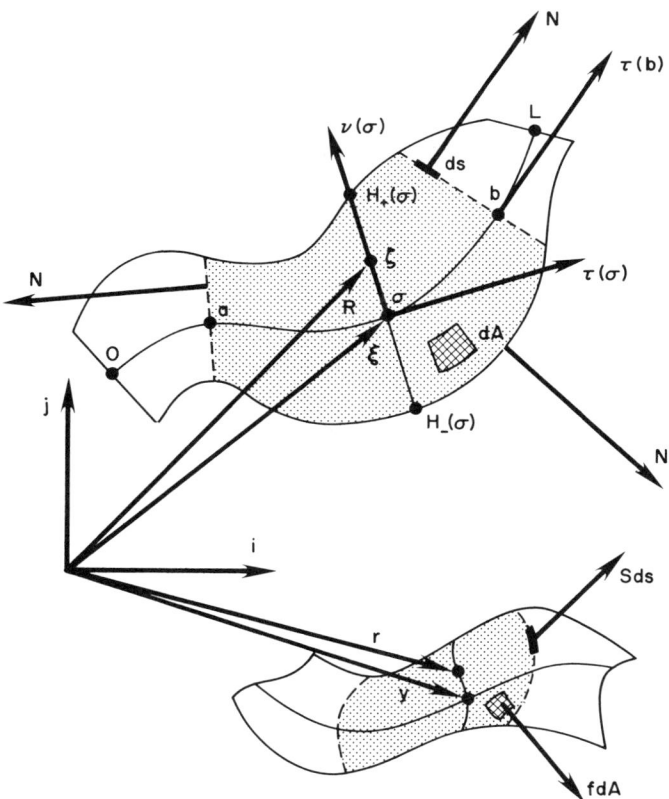

FIG. 1. Cross-sectional geometry of a beamshell.

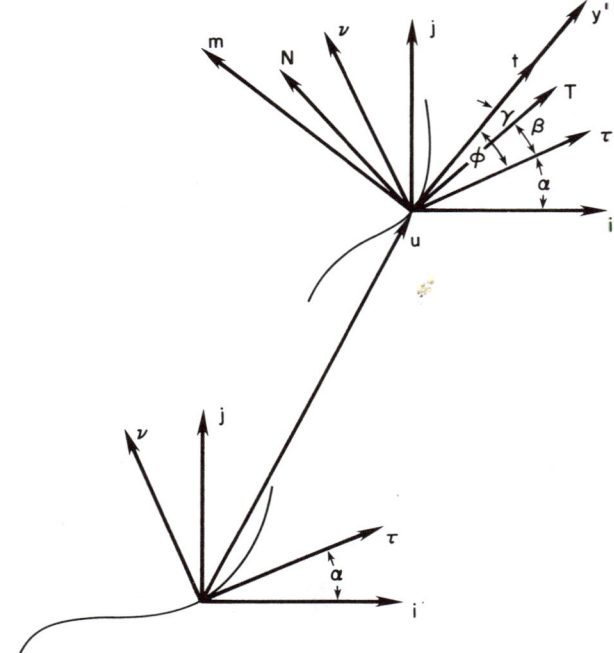

FIG. 2. Vectors and angles associated with the undeformed and deformed beamshell reference curve.

particle in the reference shape is represented by a unique set of coordinates (σ, ζ, z).

We now assume that all material properties, external disturbances, and boundary conditions are independent of z and such that every cross section undergoes the same motion while remaining in its initial plane. We call this *plane* or *cylindrical motion*.

B. INTEGRAL EQUATIONS OF CYLINDRICAL MOTION

Integral equations of cylindrical motion are obtained merely by specializing Eqs. (A4) and (A5), the integral equations of motion for a general body derived in the Appendix. There V denotes an arbitrary volume in the reference shape. The boundary ∂V of V is piecewise smooth with outward unit normal **N**. All vectors in Eqs. (A4) and (A5) lie in a fixed inertial frame. We now assume that the reference shape is at rest in this frame.

To obtain shell equations, we take V to be a truncated panel of the infinite cylindrical shell (2.1). The panel has unit length and cross section

$$A = \{\mathbf{R}(\sigma,\zeta) | 0 \leq a \leq \sigma \leq b \leq L, H_-(\sigma) \leq \zeta \leq H_+(\sigma)\}, \tag{2.6}$$

as indicated in Fig. 1. For cylindrical motion, Eqs. (A3) and (A4) reduce to the form

$$\mathbf{L} \equiv \int_{t_1}^{t_2} \left(\int_{\partial A} \mathbf{S}\, ds + \int_A \mathbf{f}\, dA \right) dt - \int_A \rho \mathbf{v}\, dA \Big|_{t_1}^{t_2} = \mathbf{0}, \tag{2.7}$$

$$R\mathbf{k} \equiv \int_{t_1}^{t_2} \left(\int_{\partial A} \mathbf{r} \times \mathbf{S}\, ds + \int_A \mathbf{r} \times \mathbf{f}\, dA \right) dt - \int_A \mathbf{r} \times \rho \mathbf{v}\, dA \Big|_{t_1}^{t_2} = \mathbf{0}, \tag{2.8}$$

where ∂A, the boundary of A, is assumed to be piecewise smooth with arc length s and outward unit normal $\mathbf{N}$. At time t, $\mathbf{S}\, ds$ is the contact force exerted on the image of ds by the material outside of A, $\mathbf{f}\, dA$ is the body force exerted on the particles initially comprising dA, $\mathbf{r}$ is the position of a particle with initial position $\mathbf{R}$, and $\mathbf{v} = \dot{\mathbf{r}}$.

From Eqs. (2.2)–(2.6),

$$dA = |\mathbf{R}_{,\sigma} \times \mathbf{R}_{,\zeta}|\, d\sigma\, d\zeta = (1 - \zeta\alpha')\, d\sigma\, d\zeta \equiv \mu\, d\sigma\, d\zeta, \tag{2.9}$$

where $\mathbf{R}_{,\sigma} = \partial \mathbf{R}/\partial \sigma$, etc.

Two parts make up ∂A: the *edges* $\sigma = a, b$ on which $\mathbf{N}\, ds = \pm \boldsymbol{\tau}\, d\zeta$ and the *faces* $\zeta = H_\pm(\sigma)$ on which $\mathbf{N}\, ds = \pm(\mu_\pm \mathbf{v} - H'_\pm \boldsymbol{\tau})\, d\sigma$. If we set $\mathbf{S} = (\mathbf{N} \cdot \boldsymbol{\tau})\mathbf{S}_\sigma + (\mathbf{N} \cdot \mathbf{v})\mathbf{S}_\zeta$, then on the edges $\mathbf{S}\, ds = \pm \mathbf{S}_\sigma\, d\zeta$, and on the faces $\mathbf{S}\, ds = \pm(\mu_\pm \mathbf{S}_\zeta - H'_\pm \mathbf{S}_\sigma)\, d\sigma$. Thus Eq. (2.7) takes the form

$$\mathbf{L} \equiv \int_{t_1}^{t_2} \left(\mathbf{F} \Big|_a^b + \int_a^b \mathbf{p}\, d\sigma \right) dt - \int_a^b m\mathbf{w}\, d\sigma \Big|_{t_1}^{t_2} = \mathbf{0}, \tag{2.10}$$

where

$$\mathbf{F} = \int_-^+ \mathbf{S}_\sigma\, d\zeta \tag{2.11}$$

is the *stress resultant* (with units of force), $\int_-^+$ being shorthand for $\int_{H_-(\sigma)}^{H_+(\sigma)}$,

$$\mathbf{p} = \int_-^+ \mathbf{f}\mu\, d\zeta + [\mu \mathbf{S}_\zeta - \zeta' \mathbf{S}_\sigma]_-^+ \tag{2.12}$$

is the *external force*,

$$m = \int_-^+ \rho\mu\, d\zeta \tag{2.13}$$

is the initial mass per unit length, and $\mathbf{w} = \dot{\mathbf{y}}$ is the *weighted velocity*, with

$$\mathbf{y} = \int_-^+ \frac{\rho\mu \mathbf{r}\, d\zeta}{m} \tag{2.14}$$

as the *weighted motion*. We emphasize that $\mathbf{y}$ is an *average* and *not* the first term in an expansion of $\mathbf{r}$ in powers of ζ. In linear beam theory, Cowper (1966) and Jones (1966) were apparently the first to suggest using a weighted vertical deflection.

We now set

$$\mathbf{r} = \mathbf{y} + \boldsymbol{\eta} \tag{2.15}$$

and note from Eqs. (2.13) and (2.14) that

$$\int_-^+ \rho \boldsymbol{\eta} \mu \, d\zeta = \mathbf{0}. \tag{2.16}$$

Thus Eq. (2.8) may be given the form

$$R \equiv \int_{t_1}^{t_2} \left([\mathbf{k} \cdot (\mathbf{y} \times \mathbf{F}) + M]_a^b + \int_a^b [\mathbf{k} \cdot (\mathbf{y} \times \mathbf{p}) + l] \, d\sigma \right) dt$$

$$- \int_a^b [\mathbf{k} \cdot (\mathbf{y} \times m\mathbf{w}) + I\omega] \, d\sigma \Big|_{t_1}^{t_2} = 0, \tag{2.17}$$

where

$$M = \mathbf{k} \cdot \int_-^+ \boldsymbol{\eta} \times \mathbf{S}_\sigma \, d\zeta \tag{2.18}$$

is the *stress couple* (with units of force × length),

$$l = \mathbf{k} \cdot \left(\int_-^+ \boldsymbol{\eta} \times \mathbf{f} \mu \, d\zeta + [\boldsymbol{\eta} \times (\mu \mathbf{S}_\zeta - \zeta' \mathbf{S}_\sigma)]_-^+ \right) \tag{2.19}$$

is the *external torque* per unit initial length,

$$I = \int_-^+ \rho \mu \zeta^2 \, d\zeta \tag{2.20}$$

is the *moment of inertia* per unit initial length, and

$$\omega = \mathbf{k} \cdot \int_-^+ \frac{\rho \mu \boldsymbol{\eta} \times \mathbf{v} \, d\zeta}{I} \tag{2.21}$$

is the *spin*. Note that it is only the *combination* $I\omega$ that appears in Eq. (2.17), so that our definition of I (and hence that of ω) is to some extent arbitrary. However, Eq. (2.20) is certainly reasonable: I can be computed *a priori*, given the reference shape, and by extracting the time-independent factor I from the rotational momentum density, we can write the time rate of change of the rotational kinetic energy density as $\frac{1}{2}(I\omega^2)$.

C. Differential Equations of Cylindrical Motion

In Eqs. (2.10) and (2.17) we have exact integral equations of motion that are formally identical to those for the planar motion of a beam. For conciseness, we shall henceforth refer to a cylindrical shell undergoing cylindrical motion as a *beamshell*. If the various fields in these equations are sufficiently smooth, then differential equations may be obtained as follows. For any function f whose derivative is integrable, set $f|_\alpha^\beta = \int_\alpha^\beta f' \, dx$ and rewrite Eqs. (2.10) and (2.17) in the form $\int_{t_1}^{t_2} \int_a^b g \, d\sigma \, dt = 0$, all $t_1 < t_2$ and all $0 \le a < b \le L$. Then assume that g is continuous in σ and t to conclude that $g = 0$. Thus from Eq. (2.10)

$$\mathbf{F}' + \mathbf{p} - m\dot{\mathbf{w}} = \mathbf{0}, \tag{2.22}$$

and from Eq. (2.17), when Eq. (2.22) is used to simplify the resulting equation,

$$M' + \mathbf{k} \cdot (\mathbf{y}' \times \mathbf{F}) + l - I\dot{\omega} = 0, \tag{2.23}$$

where a prime denotes differentiation with respect to σ.

D. Virtual Work (Weak Form of the Equations of Motion)

Suppose that the differential equations of motion (2.22) and (2.23) are satisfied. Let $\mathbf{W}$ and Ω be arbitrary functions of σ and t, sufficiently smooth for the following operations to make sense. Take the dot product of Eq. (2.22) with $\mathbf{W}$ and multiply Eq. (2.23) by Ω. Add the resulting equations and integrate over any subinterval (a, b) of $(0, L)$ and any interval (t_1, t_2) of time. Finally, integrate by parts to remove spatial derivatives of $\mathbf{F}$ and M and time derivatives of $\mathbf{w}$ and ω to obtain the *virtual work identity*

$$\int_{t_1}^{t_2} \left([\mathbf{F} \cdot \mathbf{W} + M\Omega]_a^b + \int_a^b [\mathbf{p} \cdot \mathbf{W} + l\Omega - \mathbf{F} \cdot (\mathbf{W}' - \Omega \mathbf{k} \times \mathbf{y}') - M\Omega'] \, d\sigma \right) dt$$

$$+ \int_a^b \left(-[m\mathbf{w} \cdot \mathbf{W} + I\omega\Omega]_{t_1}^{t_2} + \int_{t_1}^{t_2} (m\mathbf{w} \cdot \dot{\mathbf{W}} + I\omega\dot{\Omega}) \, dt \right) d\sigma \equiv 0,$$

for all $\mathbf{W}$ and Ω. (2.24)

Because integration is a smoothing process, Eq. (2.24) may hold for a larger class of functions than that which satisfies Eqs. (2.22) and (2.23). Thus Eq. (2.24) also may be called the *weak form of the equations of motion*. Note that if we take $\mathbf{W} = \mathbf{A} + B\mathbf{k} \times \mathbf{y}$ and $\Omega = B$, where $\mathbf{A}$ and B are arbitrary constants, then Eq. (2.24) reduces to $\mathbf{A} \cdot \mathbf{L} + BR \equiv 0$, which implies the integral equations of motion (2.10) and (2.17).

The weak form of the equations of motion is the starting point for the *finite element method*, one of the principal tools of modern structural engineering. For an excellent treatment of static problems, see the books by Strang and Fix (1973) and Becker et al. (1981).

E. The Mechanical Work Identity

If we take $\mathbf{W} = \mathbf{w}$ and $\Omega = \omega$, then the virtual work identity (2.24) can be rewritten as

$$\int_{t_1}^{t_2} \mathscr{W} \, dt \equiv \mathscr{K} \big|_{t_1}^{t_2} + \int_{t_1}^{t_2} \mathscr{D} \, dt, \tag{2.25}$$

where

$$\mathscr{W} = [\mathbf{F} \cdot \mathbf{w} + M\omega]_a^b + \int_a^b (\mathbf{p} \cdot \mathbf{w} + l\omega) \, d\sigma \tag{2.26}$$

is the *apparent external mechanical power*,

$$\mathscr{K} = \tfrac{1}{2} \int_a^b (m\mathbf{w} \cdot \mathbf{w} + I\omega^2) \, d\sigma \tag{2.27}$$

is the *kinetic energy*, and

$$\mathscr{D} = \int_a^b [\mathbf{F} \cdot (\mathbf{w}' - \omega \mathbf{k} \times \mathbf{y}') + M\omega'] \, d\sigma \tag{2.28}$$

is the *deformation power*.

F. Strain Measures

In our approach to shell theory, strains are not defined *a priori* but, rather, fall out naturally from the deformation power. Here we follow Reissner (1962, p. 28).

For simplicity, take the reference curve (2.3) to coincide with the weighted motion (2.14) at $t = 0$[†]:

$$\boldsymbol{\xi}(\sigma) = \mathbf{y}(\sigma, 0) = \int_{-}^{+} \frac{\rho \mathbf{R} \mu \, d\zeta}{m}. \tag{2.29}$$

Furthermore, let

$$\omega = \dot{\beta}, \qquad \beta(\sigma, 0) = 0 \tag{2.30}$$

and set

$$\mathbf{T} = \boldsymbol{\tau} \cos \beta + \mathbf{v} \sin \beta, \qquad \mathbf{N} = -\boldsymbol{\tau} \sin \beta + \mathbf{v} \cos \beta \tag{2.31}$$

(see Fig. 2). (Since we are no longer considering three-dimensional equations of motion, there should be no reason to confuse $\mathbf{N}$ with that of Section II,B and the Appendix.) If $\mathbf{y}' \neq \mathbf{0}$, as we shall henceforth assume, there exist at

[†] Note that Eq. (2.29) implies that $\int_{-}^{+} \zeta \rho \mu \, d\zeta = 0$.

any σ and t unique scalars e and g called the *extensional and shearing strains*, respectively, such that

$$\mathbf{y}' = (1 + e)\mathbf{T} + g\mathbf{N}. \tag{2.32}$$

From Eqs. (2.30) and (2.31),

$$\dot{\mathbf{T}} = \omega \mathbf{N}. \tag{2.33}$$

Hence the time derivative of Eq. (2.32) can be reduced to

$$\begin{aligned}\mathbf{w}' = \dot{\mathbf{y}}' &= \dot{e}\mathbf{T} + (1 + e)\omega\mathbf{N} + \dot{g}\mathbf{N} - \omega g\mathbf{T} \\ &= \dot{e}\mathbf{T} + \dot{g}\mathbf{N} + \omega \mathbf{k} \times \mathbf{y}'.\end{aligned} \tag{2.34}$$

Substituting Eq. (2.34) into Eq. (2.28), we obtain for the deformation power

$$\mathscr{D} = \int_a^b \sigma \cdot \dot{\varepsilon}\, d\sigma, \tag{2.35}$$

where

$$\sigma \cdot \dot{\varepsilon} = N\dot{e} + Q\dot{g} + M\dot{k} \tag{2.36}$$

is the *deformation power density*,

$$k = \beta' \tag{2.37}$$

is the *bending strain*, and

$$N = \mathbf{F} \cdot \mathbf{T}, \qquad Q = \mathbf{F} \cdot \mathbf{N}. \tag{2.38}$$

In a (*reduced*) *Kirchhoff motion* ($g = 0$),[†] N and Q are the components of $\mathbf{F}$ along and normal to the deformed reference curve. Note that in Eq. (2.37) β is differentiated with respect to the arc length of the *undeformed* reference curve. This uncouples, geometrically, bending and stretching. For example, if a circular cylindrical shell undergoes pure radial deflection, $k = 0$.

Reissner (1972a) has developed a beam theory based on Eq. (2.34). The analogous theory for shells, initiated by Simmonds and Danielson (1970, 1972) and elaborated by Pietraszkiewicz (1979), will be discussed in Section III.

If e, g, and β are known, the *displacement*

$$\mathbf{u} = \mathbf{y} - \xi \tag{2.39}$$

follows immediately from Eq. (2.32):

$$\mathbf{u} = \int [(1 + e)\mathbf{T} + g\mathbf{N}]\, d\sigma - \xi. \tag{2.40}$$

[†] We use the adjective *reduced*, because no mention is (or will ever need be) made about the transverse normal strain.

At this point, we have a scalar and a two-dimensional vector equation of motion, three static variables, N, Q, and M, and three kinematic quantities, e, g, and β. Constitutive relations will be discussed in Section II,J. Initial and boundary conditions that we might specify are suggested by the virtual work identity, Eq. (2.24). Conditions on the displacement $\mathbf{u}$ become conditions on e, g, and β via Eq. (2.40). In dynamics, the linear acceleration $\dot{\mathbf{w}} = \ddot{\mathbf{u}}$ that appears in Eq. (2.22) can be expressed in terms of time derivatives of e, g, and β by differentiating Eq. (2.22) with respect to σ, as discussed in Section II,L.

G. Alternate Stresses and Strains

This section is for readers who may prefer to use $\mathbf{u}$ as a basic kinematic variable. To avoid imposing the Kirchhoff hypothesis, we introduce a *new shearing strain* γ as an additional kinematic variable. As the expressions for e and k in terms of $\mathbf{u}$ and γ are awkward, we shall define new stresses $\hat{N}$ and $\hat{M}$ such that the associated extensional and bending strains ε and κ are *rational* (in fact, quadratic) in $\mathbf{u}$ and its derivatives.

Taking note of Eqs. (2.4) and (2.39), we set

$$\mathbf{y}' = \boldsymbol{\tau} + \mathbf{u}' \equiv \lambda(\boldsymbol{\tau}\cos\varphi + \mathbf{v}\sin\varphi) \equiv \lambda\mathbf{t}. \tag{2.41}$$

Then

$$\lambda^2 = \mathbf{y}' \cdot \mathbf{y}' \equiv 1 + 2\varepsilon, \tag{2.42}$$

where

$$\varepsilon = \mathbf{u}' \cdot \boldsymbol{\tau} + \tfrac{1}{2}\mathbf{u}' \cdot \mathbf{u}' \tag{2.43}$$

is the *Lagrangian strain*. Differentiating Eq. (2.41) with respect to t, we have

$$\mathbf{w}' = \dot{\mathbf{y}}' = \dot{\lambda}\mathbf{t} + \dot{\varphi}\mathbf{m}, \quad \mathbf{m} = \mathbf{k} \times \mathbf{y}' = \mathbf{v} + \mathbf{k} \times \mathbf{u}'. \tag{2.44}$$

We call $\mathbf{m}$ the *rational normal* to $\mathbf{y}$; $\mathbf{m} \cdot \mathbf{m} = \lambda^2$. It now follows from Eqs. (2.28), (2.30), and (2.35) that

$$\sigma \cdot \dot{\varepsilon} = \mathbf{F} \cdot (\dot{\lambda}\mathbf{t} + \dot{\gamma}\mathbf{m}) + M\dot{k}, \tag{2.45}$$

where

$$\gamma = \varphi - \beta. \tag{2.46}$$

(The two shearing strains are related by $g = \lambda \sin\gamma$.) If we set

$$\mathbf{F} = \tilde{N}\mathbf{y}' + \tilde{Q}\mathbf{m} \tag{2.47}$$

and note from Eq. (2.42) that $\lambda\dot{\lambda} = \dot{\varepsilon}$, then Eq. (2.45) reduces to

$$\sigma \cdot \dot{\varepsilon} = \tilde{N}\dot{\varepsilon} + \lambda^2\tilde{Q}\dot{\gamma} + M\dot{k}. \tag{2.48}$$

Curvatures arise when we take the second spatial derivative of a position vector. Thus we call $\alpha' = \boldsymbol{\xi}'' \cdot \mathbf{v}$ the curvature of $\boldsymbol{\xi}$. To obtain a curvature measure for the weighted motion **y**, we set

$$\mathbf{y}'' \cdot \mathbf{m} = \lambda^2 \alpha' + \kappa_*, \qquad (2.49)$$

where

$$\kappa_* = \mathbf{u}'' \cdot \mathbf{v} - \alpha' \mathbf{u}' \cdot \boldsymbol{\tau} + \mathbf{k} \cdot (\mathbf{u}' \times \mathbf{u}'') - \alpha' \mathbf{u}' \cdot \mathbf{u}' = \lambda^2 \varphi', \qquad (2.50)$$

which is rational (quadratic) in **u** and its spatial derivatives and is the bending strain of Budiansky (1968).[†] As α is independent of t, Eqs. (2.41) and (2.50) yield $\dot{\kappa}_* = \lambda^2 \dot{\varphi}' + 2\dot{\varepsilon}\varphi'$. But from Eqs. (2.37) and (2.46), $\kappa = \varphi' - \gamma'$. Hence

$$\lambda^2 \dot{k} = \lambda^2 \dot{\varphi}' - (\lambda^2 \gamma')^{\cdot} + 2\dot{\varepsilon}\gamma' = (\kappa_* - \lambda^2 \gamma')^{\cdot} + 2\dot{\varepsilon}(\gamma - \varphi)'. \qquad (2.51)$$

Inserting Eq. (2.51) into Eq. (2.48), we obtain

$$\sigma \cdot \dot{\varepsilon} = \hat{N}\dot{\varepsilon} + \hat{Q}\dot{\gamma} + \hat{M}\dot{\kappa}, \qquad (2.52)$$

where

$$\hat{M} = \lambda^{-2} M, \qquad \hat{Q} = \lambda^2 \tilde{Q}, \qquad \hat{N} = \tilde{N} + 2(\gamma - \varphi)'\hat{M}, \qquad (2.53)$$

and

$$\kappa = \kappa_* - \lambda^2 \gamma'. \qquad (2.54)$$

H. The Basic Assumption of Beamshell Theory

So far all equations have been exact consequences of the equations of motion of three-dimensional continuum mechanics. What, then, is a beamshell? Many definitions have been given. Ours is the following: A beamshell is a one-dimensional continuum in which the apparent external mechanical power (2.26) is taken as the actual external mechanical power. (At this level of generality a beamshell could be a sheet of water!)

To obtain a complete set of field equations, we must introduce constitutive equations. The form of these is delimited by the laws of thermodynamics, which we consider in Section II,M.

I. The Mechanical Theory of Elastic Beamshells

Often, in practice, thermal effects are negligible. An appropriate *mechanical theory* of beamshells may then be derived by assuming the existence of a *strain energy density* $\Phi = \hat{\Phi}(e, g, k; \sigma)$ such that

$$\sigma \cdot \dot{\varepsilon} = N\dot{e} + Q\dot{g} + M\dot{k} \equiv \dot{\Phi}, \qquad \text{for all } \dot{e}, \dot{g}, \dot{k}. \qquad (2.55)$$

Without loss of generality, we may take $\Phi(0, 0, 0) = 0$.

[†] An erroneous factor of 1/2 in the third term of Eq. (85) of Budiansky's paper must be replaced by 1/4.

Suppose that Φ is differentiable. Then, since there always exists a motion such that $\dot{e}$, $\dot{g}$, and $\dot{k}$ may be assigned arbitrarily, it follows that

$$N = \partial\Phi/\partial e, \qquad Q = \partial\Phi/\partial g, \qquad M = \partial\Phi/\partial k. \tag{2.56}$$

In many beamshell problems, the material and loads are apt to produce nearly shearless and inextensional deformations. In such cases, it is best to take N and Q rather than e and g as dependent variables; otherwise the field equations may be ill-conditioned. To effect this change, we assume that the first two equations of Eq. (2.56) can be solved uniquely for e and g. If we then introduce, via a Legendre transformation, a *mixed elastic energy function*

$$\Psi = \Phi - Ne - Qg, \tag{2.57}$$

it follows that

$$e = -\partial\Psi/\partial N, \qquad g = -\partial\Psi/\partial Q, \qquad M = \partial\Psi/\partial k. \tag{2.58}$$

Special beamshell theories emerge when one or more arguments of Φ or Ψ are assumed to be absent. For example, *classical* beamshell theory assumes that $\Psi = \hat{\Psi}(k, N)$, which includes as special cases the *elastica*, $\Psi = \hat{\Psi}(k)$, the *string*, $\Psi = \hat{\Psi}(N)$, and the *idealized chain*, $\Psi = 0$. Section II,J discusses the special forms of Φ and Ψ for rubberlike materials.

J. Constitutive Laws for Elastic Beamshells

In a broad sense, constitutive laws for beamshells are three equations that relate the stress resultants and couples (N, Q, M) to the strains (e, g, k). Once we restrict ourselves to elastic beamshells, the problem reduces to that of constructing an appropriate strain energy density function Φ as defined by Eq. (2.55). Two methods are available.

The *direct method* is a logical consequence of viewing the beamshell as a one-dimensional continuum. Here the form of Φ is postulated to be a function of (e, g, k), the *local* geometry of the beamshell, and the material properties. Various rational arguments and order-of-magnitude considerations, such as those presented by Niordson (1971), can be invoked in order to restrict the possible forms of Φ. These are corroborated by experience and can be supplemented by experiments of the type suggested by Reissner (1972a). The importance of observation and experience cannot be overemphasized; without this correlation with physical reality, the postulated nature of beamshell theory would be vacuous. Once the direct method is supplemented by empirical evidence, it becomes the natural method to use. Indeed, most of the experimental data for constructing constitutive laws for materials comes from tests made on beamshells and similar structures. We proceed with a simple example that displays some of the features of the direct method.

We observe (experimentally) that g has little influence on the behavior of "slender" beamshells ($|h\alpha'|, |h/L| \ll 1$) and delete it from Φ. Here $h = H_+ - H_-$ is the beamshell thickness.

We take ξ to be the unstressed state of the beamshell and consider homogeneous materials only. Furthermore, we assume that Φ depends on the *local geometry* of ξ only. This is reasonable in view of the slenderness of the beamshell, but can be substantiated by experiments or additional arguments if desired (see Niordson, 1971). Continuum mechanics and/or experimental evidence indicate that the basic elastic constant has the dimensions of force × length^{-2}. Coupled with the obvious requirement that $\Phi \to 0$ as $h \to 0$, it follows that Φ must be of the form $\Phi = \bar{E}hf(e, kh, \alpha'h)$, where $\bar{E}$ is an elastic constant.

We require that Φ be invariant under coordinate transformations. Hence the reversal of the signs of k and α' with reversal of coordinate directions implies that f depends on the combinations e, $(kh)^2$, $(\alpha'h)^2$, and $(\alpha'kh^2)$ only.

The experimental observation that Eqs. (2.56) are linear in (e, kh) if these variables are sufficiently small, plus the thermodynamic requirement that $\Phi(e, k) > \Phi(0, 0)$, dictates a Taylor expansion for f of the form $f = a^2e^2 + b^2(kh)^2 + c\alpha'h(ekh) +$ higher-order terms, where a, b, and c are numbers. If $\alpha'h$ is sufficiently small, the c term can be neglected compared to the first two, in which case $f = a^2e^2 + b^2(kh)^2 +$ terms of third and higher order in e, kh, and α'.

This is as far as we can go with the direct method. Experiments should determine values of $\bar{E}$, a, and b [see also Eq. (2.59)].

The major weakness of the direct method, as pointed out by Koiter and Simmonds (1973), is its isolation from three-dimensional theory, of which it should be a special case. This makes it difficult to evaluate its limitations or to estimate its errors, especially for not-so-thin beamshells, rapidly varying loading, boundary effects, or large strains. The main attraction of the alternate *reduction method* is that it starts from the material laws of a three-dimensional continuum. The beamshell constitutive equations are derived by imposing suitable restrictions on the distribution of motion across its thickness. The most common is the *Kirchhoff hypothesis*, which leads to a rather simple form for Φ containing no mixed terms:

$$\Phi = \tfrac{1}{2}\bar{E}h(e^2 + \tfrac{1}{12}h^2k^2), \qquad (2.59)$$

where $\bar{E} = E/(1 - v^2)$ is the *reduced* modulus of elasticity and v is Poisson's ratio.

Implicit in Eq. (2.59) are the requirements that the material obey Hooke's law, that the beamshell be thin ($h/L \ll 1$), that the deformation and loading be slowly varying ($h/\lambda \ll 1$, where λ is the smallest wavelength), and that the strains be small. Validity is restricted to regions far removed from edges and other discontinuities. These requirements were strengthened by qualitative estimates made initially by Koiter (1960), made more precise by John (1965a,b), and summarized by Koiter and Simmonds (1973). Note that for Kirchhoff motion, Φ is independent of g, so that Q must be computed from the equations of motion.

When some of these implict assumptions are not met, one can try to modify Eq. (2.59) by relaxing the Kirchhoff hypothesis. Though this does not necessarily reduce the overall error estimate (Nicholson and Simmonds, 1977), it may give a better indication of the behavior of the beamshell at the boundary of its region of validity. The more plausible relaxation for "short" beamshells is to reintroduce the shear strain g into Φ, whereas the more plausible relaxation for large strains is to retain in the model transverse normal strains and their attendant thickness changes. We conclude this section with a more detailed examination of the latter, taken from Libai and Simmonds (1981a).

The material is taken to be of the incompressible Mooney type. For a beamshell, the strain energy density takes the form

$$\Phi = \frac{1}{2} C \left(\lambda_\sigma - \frac{1}{\lambda_\sigma} \right)^2, \tag{2.60}$$

where λ_σ is the extension ratio of material elements in the beamshell initially in the τ direction and C is a material constant comparable to $\frac{1}{4}\bar{E}$. Incompressibility requires that the transverse extension ratio be $1/\lambda_\sigma$. We relax the Kirchhoff hypothesis by permitting changes in the distances $\bar{\zeta}$ of material points from $\mathbf{y}$ according to the extension ratio in that direction,

$$\bar{\zeta} = \int_0^\zeta \frac{d\hat{\zeta}}{\lambda_\sigma}, \tag{2.61}$$

but we will still require that material normals to $\bar{\zeta}$ be material normals to $\mathbf{y}$. For convenience, we have placed ξ at midthickness, the slight material nonuniformity ($\rho\mu = \rho_0$) being of no consequence. Based on the above, exact integration through the thickness can be performed, yielding

$$\frac{2\Phi}{Ch} = \left(\lambda - \frac{1}{\lambda}\right)^2 + \tfrac{1}{12}(\alpha' h)^2 \left(1 - \frac{\theta}{\lambda^2}\right)(\lambda^2 + 3\lambda^{-2} - 4\theta\lambda^{-4}) + O[(\alpha' h)^4], \tag{2.62}$$

$$\frac{N}{Ch} = (\lambda - \lambda^{-3}) + \tfrac{1}{24}(\alpha' h)^2 \left(1 - \frac{\theta}{\lambda^2}\right)(2\lambda - 6\lambda^{-3} + \theta\lambda^{-1} + 15\theta\lambda^{-5}) + O[(\alpha' h)^4], \tag{2.63}$$

$$\frac{24M}{Ch^2} = \left(1 - \frac{\theta}{\lambda^2}\right)(1 + 7\lambda^{-4})(\alpha' h) + O[(\alpha' h)^3], \tag{2.64}$$

where $\lambda = 1 + e$ and $\theta = \lambda k/\alpha'$ are extensional and bending strain measures. Results are presented here as expansions in $\alpha'h$, but closed-form expressions are available in Libai and Simmonds (1981a).

A study of the results indicates that nonlinear behavior for large strains depends on a combination of three separate effects: nonlinear material laws, changes in thickness, and a shift in the location of the material midsurface with respect to the geometrical one. All these produce comparable effects, including the appearance of strong coupling between extensional strains and moments, and between bending strains and forces.

Differentiation of Φ yields

$$\frac{\partial \Phi}{\partial \lambda} = N + \left[\frac{1}{12}\left(\frac{kh}{\lambda}\right)^2 N + kM\right], \qquad (2.65)$$

$$-\frac{1}{\alpha'}\frac{\partial \Phi}{\partial \theta} = M + \left[\frac{1}{24}\left(\frac{kh}{\lambda}\right)^2 \frac{N}{k}\right]. \qquad (2.66)$$

The appearance of $-\alpha'$ in Eq. (2.66) is due to the choice of coordinate directions and variables. The terms in square brackets indicate that (λ, θ) *are not* exact conjugate strain measures to (N, M).

Conjugate measures $(\bar{\lambda}, \bar{\theta})$ can be obtained by the following transformation:

$$\bar{\theta} = \theta - (\theta - \lambda^2)q(\lambda) - \int_1^\lambda 8u(u^4 + 7)^{-1}\,du, \qquad (2.67)$$

$$\bar{\lambda} = \lambda + \tfrac{1}{24}(\alpha'h)^2(\theta/\lambda^2 - 1)[\tfrac{1}{2}(\theta - \lambda^2)g(\lambda) - \lambda], \qquad (2.68)$$

where

$$q(\lambda) = 1 - d(\lambda)\int_1^\lambda (2u^{10} + 36u^6 - 6u^3)(u^4 + 7)^{-2}\,du, \quad g(\lambda) = \lambda^2 d(\lambda)q(\lambda) - \lambda^{-1},$$

and $d(\lambda) = \lambda^{-3}(\lambda^4 + 7)(\lambda^4 - 1)^{-1}$. These results, though exact within the assumed deformation model, are too cumbersome for practical use. Considering the approximate nature of material models, it appears that simplifications should be acceptable for moderate strains, so long as the deviation from the more exact expressions is not too large. Thus, after some manipulation, we obtain

$$\frac{2\Phi}{Ch} \approx \left(\frac{\lambda^2 - 1}{\lambda}\right)^2 + \frac{1}{3}(\alpha'h)^2\left(\frac{\lambda^2 - \theta}{\lambda}\right)^2, \qquad (2.69)$$

which, despite its approximate nature, would be the model of choice.

K. Elastostatics

The statics of beamshells is as old as that of the mechanics of materials, even in its more rudimentary forms. No attempt will be made here to cover or even discuss the various aspects of beamshell statics. We will merely touch qualitatively on a few points that have some bearing on general shell theory and will conclude with an example related to the *extensible elastica*.

For applications, it is convenient to present the equilibrium equations in component form in the directions of **T** and **N** (although other decompositions are occasionally useful). We represent **p** in the alternate forms

$$\mathbf{p} = p_T \mathbf{T} - p_N \mathbf{N} = p_y \mathbf{y}' - p\mathbf{m} \qquad (2.70)$$

(note that p_y and p are taken to be per unit length of $\mathbf{y}'$). Using simple formulas for the derivatives of **T** and **N**, we obtain from Eq. (2.22) the scalar static equilibrium equations

$$N' - Q(\alpha' + k) + p_T = 0, \qquad (2.71)$$

$$Q' + N(\alpha' + k) - p_N = 0, \qquad (2.72)$$

whereas the moment equation (2.23) reduces to

$$M' + l = gN - (1 + e)Q = -\hat{Q}. \qquad (2.73)$$

1. Pressure-Loaded Beamshells: A Qualitative Discussion

Here we will discuss beamshells subject to a uniform pressure p in a Kirchhoff motion (Fig. 3). First consider an initially straight beamshell held at both ends against translation (hinge supports). If the exact solution is developed as a series in some displacement parameter, we find that the first approximation $\mathbf{y}_0$ (valid for $|\mathbf{u}|/h \ll 1$) can be obtained from elementary beam formulas and that $e_0 = 0$. This is the *inextensional approximation* that is a natural part of the first-approximation theory for straight beamshells under transverse loads. An *inextensional model*, however, which assumes that $e = 0$ and omits it from Φ, has, in this case, a very limited range: As the load increases to where $|\mathbf{u}|/h = O(1)$, the extensions can no longer be ignored. The consideration of this and other geometrical and material nonlinearities forms the basis of the more complex analysis of the nonlinear behavior of beamshells.

At the other extreme of very high loads, the influence of the moments on the solution is small and **y** approaches, asymptotically, a circular arc. The *momentless* or *membrane* beamshell model, which *a priori* assumes that $M = 0$ and omits k from Φ, yields the circular arc configuration directly, with e given by

$$\frac{pL}{2\phi_{,e}} = \sin\left((1 + e)\frac{pL}{2\Phi_{,e}}\right).$$

For *shallowly* deformed beamshells ($kL = pL/\Phi_{,e} \ll 1$), retention of the first two terms in the expansion of the arcsin leads to the simpler expression $\Phi_{,e}(24e)^{1/2} = pL$.

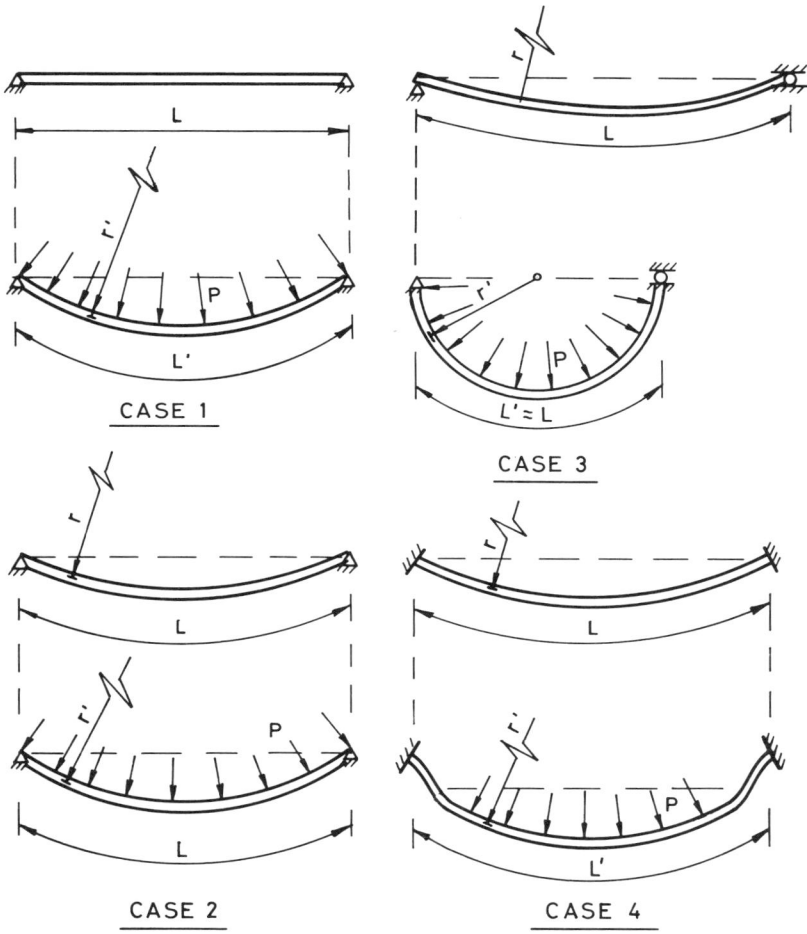

FIG. 3. Large deflections of elastic, pressure loaded beamshells subject to various boundary conditions.

Next consider a beamshell in the shape of a circular arc with an *initial curvature* $\alpha' = 1/r$. The major change in the solution from the flat beamshell is that, owing to *curvature* effects, the membrane model represents a first approximation to the exact solution for all loads, unless the beamshell is shallow ($L/r \ll 1$). In the latter case the solution for light loads is more complex, but the asymptotic behavior at high loads is very similar to the first case.

Consider again a flat beamshell, but with the right support free to move along a line connecting the supports. The situation has changed drastically: It is now the *inextensional* model that provides an excellent first approximation to the exact solution, regardless of the magnitude of the load. This is an elementary illustration of the fact that shells tend to exhibit inextensional motion if the kinematic conditions at the boundaries permit it. For high loads, the effects of the moments decrease too, and y approaches a semicircle. In fact, we could construct a *membrane-inextensional* model that ignores both extensions and moments and where equilibrium and kinematics alone determine the solution. In practice, very high loading would induce some stretching, but the shape of the beamshell would remain semicircular. The normal force would be determined by the equation $N = (pL/\pi)(1 + \Psi_{,N})$. The membrane-inextensional model is not common in beamshell theory, but does find more applications in other forms of shells.

Finally, consider the second (or first) case, but with clamped supports. The situation at *high loads* will be very similar to the corresponding hinged cases, except that now there are narrow *edge zones* or *boundary layers* where the bending stresses may be of the same order or larger than the direct stresses. Within the edge zones there exist solutions that decay rapidly into the interior. These supplement the smoothly varying membrane solutions and permit satisfaction of the clamping condition $\beta = 0$. Edge zone corrections exist in the other three cases, but they are insignificant.

The examples discussed are qualitatively simple, but they do exhibit some of the more important responses of shell models to the effects of boundary conditions, curvature, and loading.

2. Pressure-Loaded Beamshells: A Quantitative Discussion

Most treatments of beamshells found in the literature that deal with geometrical and material nonlinearities are devoted to the formulation and numerical solution of the governing equations. In a few cases, such as the *inextensional elastica*, analytic solutions are available. A detailed collection of many such results may be found in Frisch-Fay's book (1962). A discussion of the *linearly extensible elastica*, in which small extensional strains are admitted, may be found in Pflüger's book (1975). This latter model may be considered as a special case of the *small-strain finite-rotation* model of thin shells.

A more general model of a beamshell involving both geometrical and material nonlinearities, as well as large strains, has been studied extensively by Antman (1968, 1969, 1974), who shows, by obtaining first integrals, how solutions of the governing equations can be reduced to quadratures for the important special cases of straight or circular beamshells under end loads and pressure ($p_y = 0$, $p = constant$). The details are as follows.

Take the dot product of Eq. (2.22) with **F**, multiply Eq. (2.23) by p, and add to obtain

$$\mathbf{F} \cdot \mathbf{F}' + pM' = 0. \tag{2.74}$$

For *constant p*, this equation may be integrated to yield the *equilibrium integral*

$$N^2 + Q^2 + 2pM = A, \tag{2.75}$$

where A is a constant of integration. Note that (2.75) holds regardless of the shape of the beamshell. Combining Eqs. (2.56) and (2.71)–(2.73), and taking Φ to depend on σ only through e, g, and k, we have

$$\left[(1+e)\frac{\partial \Phi}{\partial e} + g\frac{\partial \Phi}{\partial g} + (\alpha' + k)\frac{\partial \Phi}{\partial k} - \Phi \right]' = \alpha'' M. \tag{2.76}$$

If $\alpha'' = 0$ (straight or circular beamshell), then, regardless of the loading, (2.76) implies the *energy integral*

$$(1+e)\frac{\partial \Phi}{\partial e} + g\frac{\partial \Phi}{\partial g} + (\alpha' + k)\frac{\partial \Phi}{\partial k} - \Phi = B. \tag{2.77}$$

where B is a constant of integration.

Equations (2.75), (2.77), and (2.56) [or (2.58)] are five algebraic equations in the six variables e, g, k, M, N, Q. We assume that five of them can be expressed in terms of either M, N or Q, chosen for convenience. Subsequent substitution into that equilibrium equation among (2.71)–(2.73) that contains the derivative of the chosen variable reduces its solution to quadrature. The rectangular coordinates of **y** follow from (2.32) and

$$x' = (1+e)\cos\psi - g\sin\psi, \quad y' = (1+e)\sin\psi + g\cos\psi, \quad \psi = \alpha + \beta. \tag{2.78}$$

The procedure outlined here leads to a complete solution, but it is not devoid of numerical difficulties. Special techniques may have to be devised, because the yet undetermined constants A and B appear inside some of the integrals.

As an application of the equations of this section, consider a ring beamshell subject to two equal and opposite outward radial forces (Fig. 4). The material model is Eq. (2.69). Here the kinematics permit *inextensional motion*, which will dominate the behavior until the ring becomes nearly flat. At this point the *linearly extensional model* becomes a good predictor of the deformation. As the loads increase still further, the full nonlinear model must come into play. At very high loads, the solution approaches asymptotically the *membrane model*, which lets the beamshell straighten without resistance and then stretch according to the material law (2.69), with $\alpha' = 0$. Figure 4 reproduces results from Libai and Simmonds (1981b), giving the exact solution for the

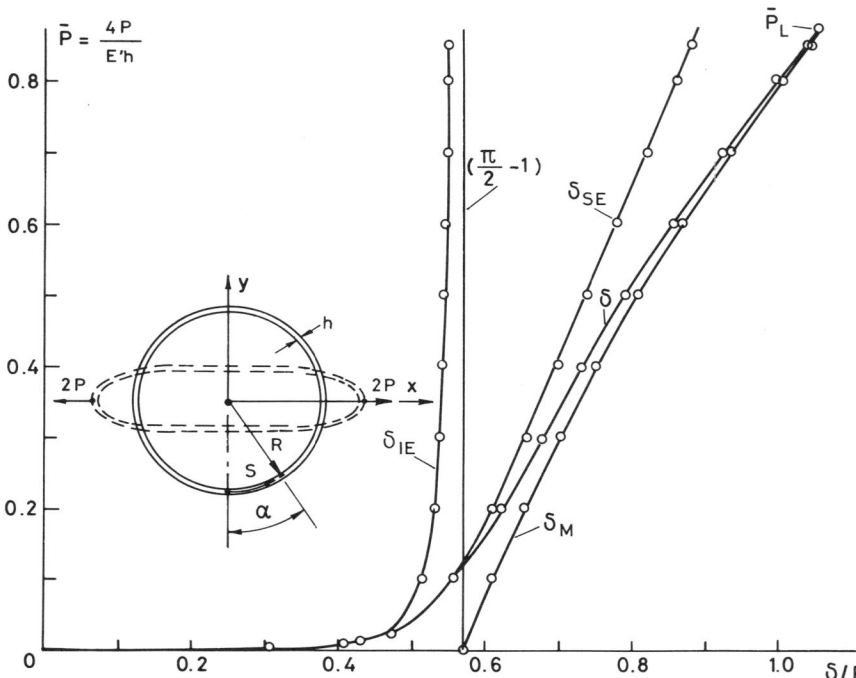

FIG. 4. Load-deflection curves for radially point-loaded beamshells using various constitutive approximations. δ, exact solution; δ_{IE}, classical elastica (inextensional); δ_{SE}, linear-extensional elastica; δ_M, membrane approximation; $h/R = 0.05$.

variation of the outward force P with the displacement δ at its point of application. Also shown are the results predicted by the various simplified models we mentioned, from which their regions of validity can be inferred.

L. ELASTODYNAMICS

In this section, two forms of the equations of motion for a beamshell are discussed. The first and more conventional one takes the displacement components and the shear strain γ as the basic unknowns. The resulting equations have two drawbacks: They are ill-conditioned for motion that is essentially inextensional, and for nonlinear problems they are unduly complicated.

The second form takes β, N, and Q as the basic unknowns. Not only are the full nonlinear equations simpler than those of the first form, but inextensional motion causes no problems. Furthermore, by an extension of an idea of Budiansky et al. for a circular beamshell (see the Appendix of Simmonds, 1979a), we can cast these equations into a form involving a real unknown and a complex-valued unknown of unit modulus that makes the role of nonlinearities transparent.

1. Displacement Shear Strain Form

Set

$$\mathbf{u} = u\boldsymbol{\tau} + w\mathbf{v}. \tag{2.79}$$

Then

$$\begin{aligned}\mathbf{u}' &= (u' - \alpha'w)\boldsymbol{\tau} + (w' + \alpha'u)\mathbf{v} \equiv \bar{e}\boldsymbol{\tau} + \bar{\beta}\mathbf{v}, \\ \mathbf{u}'' &= (\bar{e}' - \alpha'\bar{\beta})\boldsymbol{\tau} + (\bar{\beta}' + \alpha'\bar{e})\mathbf{v},\end{aligned} \tag{2.80}$$

where $\bar{e}$ and $\bar{\beta}$ are a linearized strain and a rotation. From Eqs. (2.41), (2.42), (2.44), (2.49), and (2.54),

$$\begin{aligned} \mathbf{y}' &= (1 + \bar{e})\boldsymbol{\tau} + \bar{\beta}\mathbf{v}, \qquad \mathbf{m} = -\bar{\beta}\boldsymbol{\tau} + (1 + \bar{e})\mathbf{v}, \\ \lambda\cos\varphi &= 1 + \bar{e}, \qquad \lambda\sin\varphi = \bar{\beta}, \qquad \kappa = (1+\bar{e})\bar{\beta}' - \bar{\beta}\bar{e}' - \lambda^2\gamma'. \end{aligned} \tag{2.81}$$

If we set

$$\mathbf{F} = n\boldsymbol{\tau} + q\mathbf{v}, \tag{2.82}$$

then from Eqs. (2.47), (2.53), and (2.81),

$$n = [\hat{N} + 2(\varphi - \gamma)'\hat{M}](1 + \bar{e}) - \lambda^{-2}\bar{\beta}\hat{Q}, \tag{2.83}$$

$$q = \lambda^{-2}(1 + \bar{e})\hat{Q} + \bar{\beta}[\hat{N} + 2(\varphi - \gamma)'\hat{M}]. \tag{2.84}$$

With $\mathbf{p}$ given by Eq. (2.70), the component form of the equation of linear momentum, Eq. (2.22), becomes

$$n' - \alpha'q + (1 + \bar{e})p_y + \bar{\beta}p = \rho\ddot{u}, \tag{2.85}$$

$$q' + \alpha'n - (1 + \bar{e})p + \bar{\beta}p_y = \rho\ddot{w}. \tag{2.86}$$

By Eqs. (2.30), (2.46), (2.47), and (2.53), the equation of rotational momentum, Eq. (2.23), reduces to

$$(\lambda^2\hat{M})' + \hat{Q} + l = I(\varphi - \gamma)^{\cdot\cdot}, \tag{2.87}$$

which may be used to express n and q in the form

$$n = (1+\bar{e})\hat{N} + 2\hat{M}\bar{\beta}' + \bar{\beta}\hat{M}' - 2(1+\bar{e})\hat{M}\gamma' + \bar{\beta}\lambda^{-2}[l - I(\varphi - \gamma)^{\cdot\cdot}], \tag{2.88}$$

$$q = -(1+\bar{e})\hat{M}' - 2\hat{M}\bar{e}' - 2\bar{\beta}\hat{M}\gamma' + \bar{\beta}\hat{N} + (1+\bar{e})\lambda^{-2}[I(\varphi-\gamma)^{\cdot\cdot} - l]. \tag{2.89}$$

To complete the field equations, we must adjoin a set of stress–strain relations of the form

$$\hat{N} = \partial \Phi / \partial \varepsilon, \qquad \hat{Q} = \partial \Phi / \partial \gamma, \qquad \hat{M} = \partial \Phi / \partial \kappa. \tag{2.90}$$

When Eq. (2.90) is substituted into Eqs. (2.87)–(2.89) and the resulting equations substituted into Eqs. (2.85) and (2.86), we obtain three equations for u, w, and γ. They are singularly unenlightening, so we will not write them out.

In the classical theory of beamshells, $\gamma = I = l = 0$. The rotational momentum equation, Eq. (2.87), may then be used to reduce Eqs. (2.88) and (2.89) to

$$n = (1 + \bar{e})\hat{N} + 2\hat{M}\bar{\beta}' + \bar{\beta}\hat{M}', \qquad q = -(1 + \bar{e})\hat{M}' - 2\hat{M}\bar{e}' + \bar{\beta}\hat{N}. \tag{2.91}$$

Furthermore, if ε and $h\kappa$ are small and the beamshell is homogeneous in the thickness direction, we may take $\Phi = \tfrac{1}{2}Eh(\varepsilon^2 + \tfrac{1}{12}h^2\kappa^2)$, in which case Eq. (2.90) yields

$$\hat{N} = Eh[\bar{e} + \tfrac{1}{2}(\bar{e}^2 + \bar{\beta}^2)], \qquad \hat{M} = \tfrac{1}{12}Eh^3[(1 + \bar{e})\bar{\beta}' - \bar{\beta}\bar{e}']. \tag{2.92}$$

Substituting Eq. (2.92) into Eq. (2.91), and Eq. (2.91) into Eqs. (2.85) and (2.86), we obtain two coupled equations for u and w.

Despite the simplifications of the classical theory, these equations are still quite complicated. Their major fault, however, is that they become ill-conditioned when the motion is nearly inextensional, i.e., when there is a near dependence between u and w of the form $\varepsilon = (u' - \alpha'w) + \tfrac{1}{2}[(u' - \alpha'w)^2 + (w' + \alpha'u)^2] \approx 0$. Such a difficulty with a displacement formulation already manifests itself in the linear static theory, as discussed in Section 2 of Simmonds (1966).

2. Stress Resultant Rotation Form

Much simpler and better-conditioned equations of motion are obtained if we take β, N, and Q as the basic unknowns and use the modified elastic energy Ψ defined by Eq. (2.57) to express e and g in terms of β, N, and Q.

To express the inertial term in the linear momentum equation (2.22) in terms of e, g, and β, we differentiate this equation with respect to σ. With the aid of Eqs. (2.34) and (2.38), we obtain the component form

$$N'' - (\alpha + \beta)'^2 N - 2(\alpha + \beta)'Q' - (\alpha + \beta)''Q + p'_T + (\alpha + \beta)'p_N$$
$$= m[\ddot{e} - \dot{\beta}^2(1 + e) - \ddot{\beta}g - 2\dot{\beta}\dot{g}], \tag{2.93}$$

$$Q'' - (\alpha + \beta)'^2 Q + 2(\alpha + \beta)'N' + (\alpha + \beta)''N - p'_N + (\alpha + \beta)'p_T$$
$$= m[\ddot{g} - \dot{\beta}^2 g + \ddot{\beta}(1 + e) + 2\dot{\beta}\dot{e}]. \tag{2.94}$$

The rotational momentum equation (2.23) takes the form

$$M' + (1 + e)Q - gN + l = I\ddot{\beta}. \tag{2.95}$$

When Eq. (2.58) is used to express M and g in terms of $k = \beta'$, N, and Q, we have in Eqs. (2.93)–(2.95) three equations in these three unknowns.

If the center line of the undeformed beamshell coincides with the reference curve and if $e, g, hk \ll 1$, we may take

$$\Psi = \frac{1}{2}\left(E_b h^3 k^2 - \frac{N^2}{E_e h} - \frac{Q^2}{E_s h}\right), \tag{2.96}$$

where E_b, E_e, and E_s are the moduli of bending, extension, and shear. (In classical theory, $E_b = E/12(1 - v^2)$, $E_e = E/(1 - v^2)$, and $E_s/E = \infty$, where E is Young's modulus and v is Poisson's ratio.

3. Classical Flexural Motion

Flexural motion is characterized by the order-of-magnitude relation $\dot\beta = O[R^{-2}(E_b h^3/m)^{1/2}\beta]$, where R is some typical length associated with the reference curve. This suggests the nondimensionalization

$$N, Q = (E_b h^3/R^2)(\bar N, \bar Q), \qquad M = (E_b h^3/R)\bar M,$$
$$p_T, l_N = (E_b h^3/R^3)(\bar p_T, \bar p_N), \qquad \bar l = (E_b h^3/R^2)\bar l, \qquad I = mh^2 S, \tag{2.97}$$
$$\sigma = R\bar\sigma, \qquad t = (m/E_b h^3)^{1/2}R^2 \bar t, \qquad \varepsilon = h/R,$$

where $S = O(1)$ is a shape factor. Calculating e, g, and M from Eq. (2.96), substituting the result into Eqs. (2.93)–(2.95), and introducing the barred variables defined, we obtain three coupled dimensionless equations for β, $\bar N$, and $\bar Q$. If we give E_b, E_e, E_s their classic values, let $\varepsilon \to 0$, and use Eq. (2.95) to eliminate $\bar Q$, we obtain coupled equations for β and $\bar N$. Denoting $\partial/\partial\bar\sigma$ by a prime and dropping bars, we have

$$N'' - (\alpha + \beta)'^2 N + 2(\alpha + \beta)'\beta''' + (\alpha + \beta)''\beta'' + \bar p_T' + (\alpha + \beta)'\bar p_N = -\dot\beta^2, \tag{2.98}$$

$$-\beta'''' + (\alpha + \beta)'^2\beta'' + 2(\alpha + \beta)'N' + (\alpha + \beta)''N - \bar p_N' + (\alpha + \beta)'\bar p_T = \ddot\beta. \tag{2.99}$$

For a circular beamshell of radius R, $\alpha' = 1$, and these equations reduce to Eqs. (35) and (36) of Simmonds (1979a).

We now close this section with a remarkably concise form of Eqs. (2.98) and (2.99) for $\bar p_T = \bar p_N = 0$ that generalizes an equation of Budiansky et al. (see the appendix of Simmonds, 1979a).

Since Eqs. (2.98) and (2.99) represent inextensional motion, we may set $\mathbf{y}' = -\mathbf{i}\sin\psi + \mathbf{j}\cos\psi$. Then with $z = e^{i\psi}$, only two modifications are necessary in the equations of the appendix of Simmonds (1979a). The first is to replace the right-hand side of his Eq. (75) by $-\alpha'' = -z\bar z \alpha'' = -\mathscr{R}(z\bar z \alpha'')$, where $\mathscr{R}$ denotes "the real part of." From this follows the second modification: to replace his Eq. (77) by

$$\ddot z + [z'' + z + (fz) - iz\alpha'']'' = 0, \tag{2.100}$$

where f is an unknown *real* function. *All* nonlinear effects are manifested in the single term in parentheses! The full implications of this deceptively simple equation remain to be explored.

M. Thermodynamics

Thermal effects in shells need to be considered for at least four reasons. First, constrained shells, when heated, may experience large stresses. Second, a proper analysis of crack growth should involve heat flow (Gurtin, 1979). Third, the coupling between deformation and temperature, resulting from the time rate of dilation appearing in the energy equation, produces a slight damping of free vibrations. (Although normally swamped by air and joint damping, thermal damping could be important in large, thin homogeneous structures in outer space). And fourth, as emphasized by Erickson (1966), Koiter (1971, 1969), and Gurtin (1975, 1973a,b), among others, a completely satisfactory definition of stability requires thermodynamics.

The approach in this section was motivated by the work of Coleman and Noll (1963), Truesdell (1969), Müller (1969), and Green and Naghdi (1979). The last authors were concerned with the reconciliation of a direct approach to the thermodynamics of a two-dimensional Cosserat surface (i.e., a shell with directors) with a three-dimensional approach in which *all dependent variables are represented by truncated power series in the thickness coordinate.* In contrast, *no such expansions are used in what follows*; rather, certain two-dimensional thermodynamic variables (an entropy density resultant ι, a heat influx resultant q, an entropy flux resultant p, an "effective" temperature T, and an "effective" temperature drop $h\varphi$) fall out naturally from the integral equations of three-dimensional thermodynamics, while other secondary quantities are a consequence of conservation of energy, the Clausius–Duhem inequality, and the assumed form of the two-dimensional constitutive relations.

Just as the integral equations of motion in the Appendix implied Eqs. (2.7) and (2.8), so Eqs. (A6) and (A8) imply that for a beamshell

$$\mathcal{Q} = \int_{\partial A} v\, ds + \int_A r\, dA, \tag{2.101}$$

$$\int_A \eta\, dA \Big|_{t_1}^{t_2} \geq \int_{t_1}^{t_2} \left(\int_{\partial A} \frac{v}{\theta}\, ds + \int_A \frac{r}{\theta}\, dA \right) dt, \tag{2.102}$$

where, as in Eqs. (2.7) and (2.8), the dimensions of all variables in the Appendix have been multiplied by the z unit of length. Thus, for example, $v\, ds$ is the heat *influx* across the image of ds.

From Eq. (A7) and Fig. 2 it follows that on the edges $-v\, ds = \pm \mathbf{q}\cdot \boldsymbol{\tau}\, d\zeta$, whereas on the faces $v\, ds = [v(\mu^2 + \zeta'^2)^{1/2}]_{\pm}\, d\sigma \equiv (v\tilde{\mu})_{\pm}\, d\sigma$. Thus, with the aid of Eq. (2.9), Eqs. (2.101) and (2.102) reduce to

$$\mathcal{Q} = q|_a^b + \int_a^b s\, d\sigma, \tag{2.103}$$

$$\int_a^b \iota\, d\sigma\Big|_{t_1}^{t_2} \geq \int_{t_1}^{t_2}\left(p|_a^b + \int_a^b v\, d\sigma\right) dt, \tag{2.104}$$

where

$$q = -\boldsymbol{\tau}\cdot\int_-^+ \mathbf{q}\, d\zeta, \qquad s = (v\tilde{\mu})_- + (v\tilde{\mu})_+ + \int_-^+ r\mu\, d\zeta, \tag{2.105}$$

$$\iota = \int_-^+ \eta\mu\, d\zeta, \qquad p = -\boldsymbol{\tau}\cdot\int_-^+ \frac{\mathbf{q}}{\theta}\, d\zeta, \tag{2.106}$$

$$v = \left(\frac{v\tilde{\mu}}{\theta}\right)_- + \left(\frac{v\tilde{\mu}}{\theta}\right)_+ + \int_-^+ \frac{r}{\theta}\mu\, d\zeta. \tag{2.107}$$

We call q and p the *heat and entropy influx resultants*, respectively, and ι the *entropy density resultant*, s and v are, respectively, the *heating* and the *supply of entropy*. If we set

$$\Sigma = (v\tilde{\mu})_- + (v\tilde{\mu})_+, \qquad \chi = (h/2T)[(v\tilde{\mu})_- - (v\tilde{\mu})_+], \tag{2.108}$$

$$\frac{1}{T} = \frac{1}{2}\left(\frac{1}{\theta_-} + \frac{1}{\theta_+}\right), \qquad \varphi = \frac{T^2}{h}\left(\frac{1}{\theta_-} - \frac{1}{\theta_+}\right), \tag{2.109}$$

then the second part of Eq. (2.105) and Eq. (2.107) take the form

$$s = \Sigma + \int_-^+ r\mu\, d\zeta, \qquad v = T^{-1}(\Sigma + \varphi\chi) + \int_-^+ \frac{r}{\theta}\mu\, d\zeta. \tag{2.110}$$

We call T the *effective temperature mean*, and $h\varphi$ the *effective temperature drop*. It is interesting to note that T and φ are *not* weighted averages through the thickness.

Again, we emphasize that the equations of motion (2.10) and (2.17) and the Clausius–Duhem inequality (2.104) are exact; the approximate nature of beamshell theory enters when we take the apparent external mechanical power $\mathcal{W}$ to be the actual external mechanical power.

To obtain a complete thermodynamic theory of beamshells, we *postulate* a law of *conservation of energy* (first law of thermodynamics):

$$\int_{t_1}^{t_2} \mathcal{Q} + \mathcal{W}\, dt = [\mathcal{K} + \mathcal{U}]_{t_1}^{t_2}, \tag{2.111}$$

where $\mathcal{U}$ is the *internal power*. We call a beamshell *thermoelastic* if there exists an *internal energy density* u depending on the generalized state variable

$$\Lambda \equiv (\varepsilon, T, \varphi) = (e, g, k, T, \varphi) \tag{2.112}$$

and its gradient Λ' such that

$$\mathcal{U} = \int_a^b u\, d\sigma. \tag{2.113}$$

It is convenient to write u in the form

$$u = \Phi + \imath T + f\varphi. \tag{2.114}$$

We call Φ the *free-energy density* and f the *entropy density moment*. Both are functions of Λ and Λ'. Inserting Eq. (2.114) into Eq. (2.113) and the resulting expression into Eq. (2.111) and using the mechanical work identity (2.25) with $\mathcal{D}$ given by Eq. (2.35), we have

$$\int_{t_1}^{t_2} \left(q|_a^b + \int_a^b (\sigma \cdot \dot{\varepsilon} + s)\, d\sigma \right) dt = \int_a^b (\Phi + \imath T + f\varphi)\, d\sigma|_{t_1}^{t_2}. \tag{2.115}$$

For sufficiently smooth fields, Eq. (2.115) implies the *reduced differential equation of energy balance*

$$q' + \Xi \cdot \dot{\Lambda} + s = \dot{\Phi} + \imath T + f\dot{\varphi}, \tag{2.116}$$

and Eq. (2.104) implies the *differential Clausius–Duhem inequality*

$$i \geq p' + v, \tag{2.117}$$

where

$$\Xi \equiv (\sigma, -\imath, -f) = (N, Q, M, -\imath, -f). \tag{2.118}$$

If θ were independent of ζ, we would have $T = \theta$, $\varphi = 0$, and $p = q/T$, $v = s/T$. This suggests setting

$$p = T^{-1}q + \varphi w, \qquad v = T^{-1}[s + \varphi(\chi + \lambda)], \tag{2.119}$$

where

$$\lambda = \frac{1}{\varphi} \int_{-}^{+} \left(\frac{T}{\theta} - 1 \right) r\mu\, d\zeta. \tag{2.120}$$

If θ^{-1} were linear in ζ, it would follow that

$$\lambda = \frac{1}{T}\left(\frac{1}{2}(H_+ + H_-) \int_{-}^{+} r\mu\, d\zeta - \int_{-}^{+} \zeta r\mu\, d\zeta \right) \equiv \frac{F[r]}{T}. \tag{2.121}$$

We now assume that Φ, Ξ, q, and w are given by differentiable constitutive functions of the form

$$\Phi = \hat{\Phi}(\Lambda, \Lambda'), \ldots, w = \hat{w}(\Lambda, \Lambda'). \tag{2.122}$$

Thus, in particular,

$$\dot{\Phi} = \Phi_\Lambda \cdot \dot{\Lambda} + \Phi_{\Lambda'} \cdot \dot{\Lambda}', \tag{2.123}$$

where a subscript denotes partial differentiation and Φ_Λ is short for $(\partial\Phi/\partial e, \ldots, \partial\Phi/\partial\varphi)$, etc. (Though it is customary *not* to include strain gradients in the parameter list of the constitutive functions, it actually *simplifies* the form of the resulting equations to do so. As will be seen, the same final form for $\hat{\Phi}$ emerges as if the strain gradients had not been included.) We take Eq. (2.121) as the constitutive equation for λ. Substituting Eq. (2.123) into Eq. (2.116) and noting Eqs. (2.117) and (2.119), we have

$$(\Xi - \Phi_\Lambda) \cdot \dot{\Lambda} - \Phi_{\Lambda'} \cdot \dot{\Lambda}' = iT - q' - s + \dot{f}\varphi$$
$$\geq T^{-1}(-qT' + T^2 w\varphi') + \varphi(\dot{f} + Tw' + \chi + \lambda). \quad (2.124)$$

We now wring a number of consequences from Eq. (2.124). First, since $\dot{f}$ may be expressed in a form similar to Eq. (2.123) and

$$w' = w_\Lambda \cdot \Lambda' + w_{\Lambda'} \cdot \Lambda'', \quad (2.125)$$

Eq. (2.124) can be rewritten as

$$(\tilde{\Xi} - \tilde{\Phi}_\Lambda) \cdot \dot{\Lambda} - \tilde{\Phi}_{\Lambda'} \cdot \dot{\Lambda}' \geq T^{-1}(-qT' + T^2 w\varphi')$$
$$+ \varphi[T(w_\Lambda \cdot \Lambda' + w_{\Lambda'} \cdot \Lambda'') + \chi + \lambda], \quad (2.126)$$

where $\tilde{\Xi} = (\sigma, T, 0)$ and $\tilde{\Phi} = \Phi + f\varphi$ is a modified free-energy density. At any σ and t,

$$\Upsilon \equiv (\Lambda, \dot{\Lambda}, \Lambda', \dot{\Lambda}', \Lambda'') \quad (2.127)$$

can be assigned arbitrarily, provided that **p**, l, and s are chosen so that the equations of momentum and energy balance, Eqs. (2.22), (2.23), and (2.116), are satisfied. By virtue of the *form* of the constitutive equations (2.121) and (2.122), it follows that for a given r a choice of Υ fixes every term in Eq. (2.126) *except* χ. Thus there must be a relation involving χ and at least some of the components of Υ. Otherwise, by the second part of Eq. (2.108), we could give χ any value by prescribing suitable heat fluxes $v_\pm$ and thus violate Eq. (2.126).

The relation involving χ can always be written in the form

$$Tw' + \chi + \lambda = A. \quad (2.128)$$

If we regard A as a new unknown and adopt the *principle of equipresence* (Truesdell & Toupin, 1965, Sect. 293), then this equation must be of the form $A = \hat{A}(\Lambda, \Lambda')$. With this understanding, the right-hand side of Eq. (2.126) reduces to $T^{-1}(-qT' + T^2 w\varphi') + \varphi A$, a function of Λ and Λ' only. Since the coefficients of $\dot{\Lambda}$ and $\dot{\Lambda}'$ on the left-hand side of Eq. (2.126) are independent of $\dot{\Lambda}$ and $\dot{\Lambda}'$, and $\dot{\Lambda}$ and $\dot{\Lambda}'$ may be prescribed arbitrarily at any σ and t, we conclude that, to not violate the inequality, we must have

$$\tilde{\Xi} = (\sigma, T, 0) = \tilde{\Phi}_\Lambda, \quad \tilde{\Phi}_{\Lambda'} = \mathbf{0}. \quad (2.129)$$

Thus $\tilde{\Phi}$ depends on ε and T only. But obviously, if we are to account for so simple a phenomenon as the buildup of stress couples in a plate heated uniformly on one face and constrained to remain flat, we need a constitutive equation for M that depends on φ; that is, $\tilde{\Phi}$ is *not* a suitable free-energy density. We therefore return to Eq. (2.124) and take the relation for χ to be of the form

$$\dot{f} + Tw' + \chi + \lambda = B, \tag{2.130}$$

where

$$B = \hat{B}(\Lambda, \Lambda'). \tag{2.131}$$

By arguments similar to those used before, we conclude that Eq. (2.126) will remain inviolate if and only if $\Phi = \hat{\Phi}(\Lambda)$ and $\Xi = \Phi_\Lambda$, i.e., if and only if

$$\Phi = \hat{\Phi}(e, g, k, T, \varphi) \tag{2.132}$$

and

$$N = \Phi_e, \quad Q = \Phi_g, \quad M = \Phi_k, \quad \iota = -\Phi_T, \quad f = -\Phi_\varphi. \tag{2.133}$$

With these relations, Eqs. (2.116) and (2.124) reduce to

$$q' + s = \dot{\iota}T + \dot{f}\varphi, \tag{2.134}$$

$$-qT' + T^2 w\varphi' + T\varphi B \leq 0. \tag{2.135}$$

It might appear as if the decision to include all of

$$Tw' = T(w_\Lambda \cdot \Lambda' + w_{\Lambda'} \cdot \Lambda'') \tag{2.136}$$

in Eq. (2.130) was arbitrary, but this is not so. Consider the second term on the right-hand side of Eq. (2.136). Had it been left out of Eq. (2.130), then, since Λ'' can be prescribed arbitrarily at any σ and t and since w is independent of Λ'', we would have concluded that to keep Eq. (2.126) inviolate, $w_{\Lambda'} = 0$. But this is a contradiction if $w \neq 0$, since the first part of Eq. (2.119) implies that w must change sign if σ is replaced by $L - \sigma$. The first term on the right-hand side of Eq. (2.136), being a function of Λ and Λ' only, could, if omitted, be reintroduced by redefining B as $B - Tw_\Lambda \cdot \Lambda' = \hat{C}(\Lambda, \Lambda')$, say.

For simplicity, we assume that the constitutive equations for q, w, and B do *not* depend on the strain gradient ε'. Furthermore, we set $T = T_0 + \tau$, where T_0 is a constant, and assume that in the reference state N, Q, M, ι, $f, q, w, B, e, g, k, \tau, \varphi, \tau'$, and φ' are zero.

When σ is replaced by $L - \sigma$, or $\mathbf{v}$ by $-\mathbf{v}$, certain quantities in the constitutive relations change sign as indicated in Table I. The first row indicates

TABLE I
Dimensions and Sign Changes of Constitutive Variables

	(2.5)	(2.32)	(2.37)	(2.105)	(2.109)	(2.114)	(2.119)	(2.130)
$\sigma, \mathbf{v}$	α'	e, g	k	q	$\tau, \tau', \varphi, \varphi'$	Φ	w	B
$L - \sigma, \mathbf{v}$	α'	$e, -g$	k	$-q$	$\tau, -\tau', \varphi, -\varphi'$	Φ	$-w$	B
$\sigma, -\mathbf{v}$	$-\alpha'$	$e, -g$	$-k$	q	$\tau, \tau', -\varphi, -\varphi'$	Φ	$-w$	$-B$
	L^{-1}	$1,1$	L^{-1}	ML^2T^{-3}	$\Theta, L^{-1}\Theta, L\Theta, \Theta$	MLT^{-2}	$ML^3T^{-3}\Theta^{-2}$	$ML^2T^{-3}\Theta^{-1}$

the equation that defines the quantity and the last row, the dimension of the quantity in terms of mass (M), length (L), time (T), and temperature (Θ).

Invariance requirements and dimensional considerations may now be used to limit the possible forms of the constitutive relations for Φ, q, w, and B. In the simplest case, N, Q, M, ι, f, q, w and B are linear in the state variables, which implies that

$$\Phi = \tfrac{1}{2}h(E_{11}e^2 + hE_{13}\alpha'ek - E_{14}e\tau - E_{15}\alpha'e\varphi + E_{22}g^2 + h^2 E_{33}k^2 \\ - hE_{34}k\tau - hE_{35}k\varphi - E_{44}\tau^2 - E_{55}\varphi^2), \tag{2.137}$$

$$q = h(q_0\tau' + h^2 q_1 \alpha'\varphi' + [hq_2\alpha'g]), \tag{2.138}$$

$$w = -h^3(w_0\varphi' + w_1\alpha'\tau') + [-hw_2 g], \tag{2.139}$$

$$B = -h(B_0\varphi + [B_1\alpha'\tau + B_2 k]), \tag{2.140}$$

where $E_{11}, E_{12}, \ldots, B_2$ depend, at most, on σ and α'^2.

Substitution of Eqs. (2.138)–(2.140) into the inequality (2.135) shows that the coefficients in the terms in square brackets must be zero, whereas the remaining coefficients must satisfy

$$q_0, w_0, B_0 \geq 0, \tag{2.141}$$

$$4q_0 w_0 (T_0 + \tau)^2 \geq [q_1 + (T_0 + \tau)^2 w_1]^2 \alpha'^2. \tag{2.142}$$

For a flat beamshell, homogeneous in the thickness, Eqs. (2.137)–(2.142) reduce to

$$\Phi = \frac{1}{2}hE\left(e^2 + \frac{h^2}{12}k^2 - 2\tilde{\alpha}\left(e\tau + \frac{h^2}{12}k\varphi\right)\right) - \frac{1}{2}\left(\frac{cm}{T_0}\right)\left(\tau^2 + \frac{h^2}{12}\varphi^2\right), \tag{2.143}$$

$$q = hq_0\tau', \qquad w = -h^3 w_0 \varphi', \qquad B = -hB_0\varphi, \tag{2.144}$$

where $\tilde{\alpha}$ is the *coefficient of thermal expansion*, c is the *specific heat at constant volume*, and q_0 is the *thermal conductivity*.

If ι, f, q, w, and B are computed from Eqs. (2.133), (2.143), and (2.144), and these expressions substituted into Eqs. (2.134) and (2.130), we obtain equations that, when *linearized*, read

$$(hq_0\tau')' + s = (cm\tau + T_0 hE\tilde{\alpha}e)^{\cdot}, \qquad (2.145)$$

$$(\tfrac{1}{12}T_0)(cmh^2\varphi + T_0 h^3 E\tilde{\alpha}k)^{\cdot} - T_0(hw_0\varphi')' + \chi + \lambda = -hB_0\varphi. \qquad (2.146)$$

The homogeneous parts of these equations agree with Eqs. (10.13) and (10.14) of Green and Naghdi (1979) for a plate, provided that in the former equations we take $w_0 = q_0/12T_0^2$ and $B_0 = q_0$ and in the latter equations we set $(1-v)\gamma_3 = -ve$, $(1-2v)\alpha^* = \tilde{\alpha}$, and $\rho^*h = m$, where γ_3 is the transverse normal strain.

The nonhomogeneous terms in Green and Naghdi's equations come from assuming that the only external heat input is from radiation at the faces (a condition expressed in linearized form). The nonhomogeneous terms in Eqs. (2.145) and (2.146) reduce to these, provided that we set r (and hence λ) $= 0$ and note that "the ambient temperature[s] of the surroundings at the ... surfaces of the plate" must be measured relative to T_0.

Limitations of space prevent us from carrying out an analogous development for general shells. The procedure is straightforward and the reader should have no conceptual difficulties in tracing out these steps.

III. The Equations of General Nonlinear Shell Theory

In this section we derive field equations for a shell, mimicking the steps followed in Section II for a beamshell. Naturally, the jump from the geometry of a curve to that of a surface entails some complexities, but tensor notation minimizes these to some extent.

Until Section III,I on constitutive relations and boundary conditions, *all equations are exact consequences of the equations of motion of a three-dimensional continuum.* Thus we do *not* introduce the Kirchhoff hypothesis, nor do we ignore rotary inertia terms. A minor but unavoidable side effect of using a Lagrangian description while not assuming that normals deform into normals is that the stress couple vector acquires a (small) component about the normal to the deformed reference surface. Aside from this, all goes smoothly until Section III,H, where we define alternate strain measures in terms of the displacement **u** and a *shear-twist* strain γ. Although our definition of γ is geometrical and a natural generalization from the beamshell, it carries a high price of complexity. Therefore, before we even mention the problem of boundary conditions, we offer our presentation as a caveat to those who would include higher-order effects in nonlinear shell theory. If $\gamma = \mathbf{0}$, the results of Section III,H reduce to those of Budiansky (1968).

A. Geometry of the Undeformed Shell

A *shell* is a body whose reference shape is the set of all particles with a position vector of the form

$$\mathbf{X} = \hat{\mathbf{X}}(\boldsymbol{\xi}, \zeta) = \boldsymbol{\xi} + \zeta \hat{\mathbf{n}}(\boldsymbol{\xi}), \qquad \boldsymbol{\xi} \in S, \hat{H}_-(\boldsymbol{\xi}) \leq \zeta \leq \hat{H}_+(\boldsymbol{\xi}), \qquad (3.1)$$

where S is the *reference surface*, assumed to be smooth with an oriented unit normal $\mathbf{n}$ and a piecewise smooth boundary ∂S (see Fig. 5). The notation $\boldsymbol{\xi} \in S$ means that the head of $\boldsymbol{\xi}$ lies in S, where

$$\boldsymbol{\xi} = \xi^1 \mathbf{e}_1 + \xi^2 \mathbf{e}_2 + \xi^3 \mathbf{e}_3 \equiv \xi^i \mathbf{e}_i, \qquad (3.2)$$

and $\{\mathbf{i}, \mathbf{j}, \mathbf{k}\} = \{\mathbf{e}_1, \mathbf{e}_2, \mathbf{e}_3\}$. The bounds $H_\pm$ are assumed to be sufficiently small so that each point in the reference shape is associated with a unique pair $(\boldsymbol{\xi}, \zeta)$.

The *faces* of the shell are the set of points in Eq. (3.1) such that $\zeta = \hat{H}_\pm(\boldsymbol{\xi})$, and the *edges* are those such that $\boldsymbol{\xi} \in \partial S$. (Note that the edges of the shell intersect the reference surface at right angles. A way of handling beveled edges has been suggested by Koiter in some unpublished work.)

To go on, we assume that points on S can be represented in the parametric form

$$\boldsymbol{\xi} = \hat{\boldsymbol{\xi}}(\sigma^\alpha) = \hat{\xi}^i(\sigma^\alpha) \mathbf{e}_i, \qquad \sigma^1, \sigma^2 \in D, \qquad (3.3)$$

where D is some connected set in the $\sigma^1 \sigma^2$ plane and σ^α, $\alpha = 1, 2$, is short for σ^1, σ^2. Then Eq. (3.1) may be taken in the form

$$\mathbf{X} = \hat{\mathbf{X}}(\sigma^\alpha, \zeta) = \hat{\boldsymbol{\xi}}(\sigma^\alpha) + \zeta \hat{\mathbf{n}}(\sigma^\alpha), \qquad \sigma^\alpha \in D, \hat{H}_-(\sigma^\alpha) \leq \zeta \leq \hat{H}_+(\sigma^\alpha). \qquad (3.4)$$

The *covariant base vectors* associated with the Gaussian coordinates σ^α are denoted and defined by

$$\mathbf{a}_\alpha \equiv \boldsymbol{\xi}_{,\alpha} = \xi^i_{,\alpha} \mathbf{e}_i, \qquad (3.5)$$

where $\boldsymbol{\xi}_{,\alpha} = \partial \boldsymbol{\xi}/\partial \sigma^\alpha$. The covariant components of the surface *metric* and *curvature tensor* are denoted and defined by

$$a_{\alpha\beta} \equiv \mathbf{a}_\alpha \cdot \mathbf{a}_\beta, \qquad b_{\alpha\beta} \equiv \boldsymbol{\xi}_{,\alpha\beta} \cdot \mathbf{n}, \qquad (3.6)$$

where

$$\mathbf{n} = \tfrac{1}{2} \varepsilon^{\alpha\beta} \mathbf{a}_\alpha \times \mathbf{a}_\beta \qquad (3.7)$$

and $\varepsilon^{\alpha\beta}$ are the contravariant components of the surface permutation tensor $[a^{1/2} \varepsilon^{12} = 1, a^{1/2} \varepsilon^{21} = -1, \varepsilon^{11} = \varepsilon^{22} = 0, a = \det(a_{\alpha\beta})]$. The derivatives of $\mathbf{a}_\alpha$ and $\mathbf{n}$ are related by the *Gauss–Weingarten equations*

$$\mathbf{a}_{\alpha|\beta} = b_{\alpha\beta} \mathbf{n}, \qquad \mathbf{n}_{,\alpha} = -b_\alpha^\lambda \mathbf{a}_\lambda, \qquad (3.8)$$

where a vertical bar denotes covariant differentiation based on $a_{\alpha\beta}$.

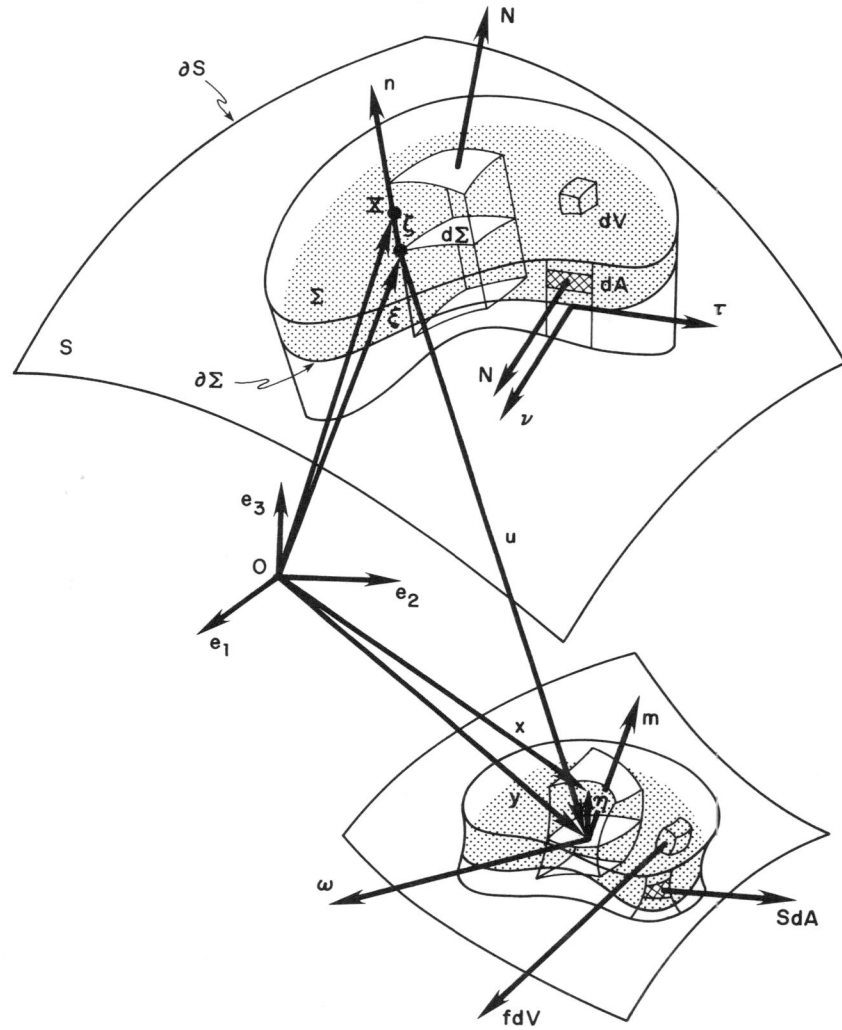

Fig. 5. Geometry of deformation of a shell-like volume.

The covariant base vectors associated with the *shell coordinates* (σ^α, ζ) are

$$\mathbf{G}_\alpha \equiv \mathbf{X}_{,\alpha} = (\delta_\alpha^\lambda - \zeta b_\alpha^\lambda)\mathbf{a}_\lambda, \qquad \mathbf{G}_3 \equiv \mathbf{X}_{,\zeta} = \mathbf{n}. \tag{3.9}$$

From Eq. (3.9),

$$\mathbf{G}_\alpha \times \mathbf{G}_\beta = (1 - 2\zeta M + \zeta^2 K)\varepsilon_{\alpha\beta}\mathbf{n} \equiv \mu\varepsilon_{\alpha\beta}\mathbf{n}, \tag{3.10}$$

which is useful for expressing the volume and boundary elements, dV and

dA, in terms of shell coordinates. Here $M = \frac{1}{2}b_\alpha^\alpha$ and $K = \det(b_\beta^\alpha)$ are, respectively, the *mean* and *Gaussian* curvatures of S.

B. INTEGRAL EQUATIONS OF MOTION

To get integral equations of motion for a shell, we take the arbitrary volume V in Eqs. (A3) and (A4) of the Appendix to consist of all points in Eq. (3.1) such that $\xi \in \Sigma$, where Σ is any connected subregion of S with a piecewise smooth boundary $\partial \Sigma$, having unit tangent τ, outward unit normal $\mathbf{v} = \tau \times \mathbf{n}$, and arc length s.

With $d\Sigma = a^{1/2} d\sigma^1 d\sigma^2$, it follows from Eq. (3.10) that

$$dV = (\mathbf{G}_1 \times \mathbf{G}_2) \cdot \mathbf{G}_3 d\sigma^1 d\sigma^2 d\zeta = \tfrac{1}{2}\varepsilon^{\alpha\beta}(\mathbf{G}_\alpha \times \mathbf{G}_\beta) \cdot \mathbf{n} d\Sigma d\zeta = \mu d\Sigma d\zeta. \quad (3.11)$$

Recall that $\mathbf{N}$ is the outward unit normal to the surface bounding the shell; thus on $\zeta = H_\pm$, $\mathbf{N} dA = \pm \tilde{\mathbf{X}}_{,1} \times \tilde{\mathbf{X}}_{,2} d\sigma^1 d\sigma^2 = \pm \tfrac{1}{2} \varepsilon^{\alpha\beta} \tilde{\mathbf{X}}_{,\alpha} \times \tilde{\mathbf{X}}_{,\beta} d\Sigma$, where $\tilde{\mathbf{X}}_{,\alpha} = \mathbf{G}_\alpha + \zeta_{,\alpha}\mathbf{n}$. Hence, by Eq. (3.10),

$$N_\gamma dA = \mathbf{N} \cdot \mathbf{G}_\gamma dA = \pm \zeta_{,\alpha} \varepsilon^{\alpha\beta}(\mathbf{G}_\beta \times \mathbf{G}_\gamma) \cdot \mathbf{n} d\Sigma = \mp \mu \zeta_{,\gamma} d\Sigma, \quad (3.12)$$

$$N_3 dA = \pm \mu d\Sigma; \quad (3.13)$$

and so the force acting on the surface element dA (see the Appendix) is

$$\mathbf{S} dA = \mathbf{S}^i N_i dA = \pm \mu(\mathbf{S}^3 - \zeta_{,\alpha}\mathbf{S}^\alpha) d\Sigma. \quad (3.14)$$

We assume that $\partial \Sigma$ has the parametric representation $\sigma^\alpha = \hat{\sigma}^\alpha(s)$, so that on the edges, $\mathbf{N} dA = \mathbf{X}_{,s} \times \mathbf{X}_{,\zeta} ds d\zeta = \mathbf{G}_\alpha \times \mathbf{n} \tau^\alpha ds d\zeta$, where $\tau^\alpha = d\sigma^\alpha/ds$. Again by Eq. (3.10),

$$N_\gamma dA = (\mathbf{G}_\gamma \times \mathbf{G}_\alpha) \cdot \mathbf{n} \tau^\alpha ds d\zeta = \mu v_\gamma ds d\zeta, \quad (3.15)$$

$$N_3 dA = 0, \quad (3.16)$$

where $v_\gamma = \varepsilon_{\gamma\alpha} \tau^\alpha$. Since

$$\tau = d\xi/ds = \xi_{,\alpha} d\sigma^\alpha/ds = \mathbf{a}_\alpha \tau^\alpha \quad (3.17)$$

and $\mathbf{v} = \tau \times \mathbf{n}$, it follows that

$$\mathbf{v} \cdot \mathbf{a}_\gamma = (\tau \times \mathbf{n}) \cdot \mathbf{a}_\gamma = (\mathbf{a}_\gamma \times \mathbf{a}_\alpha \tau^\alpha) \cdot \mathbf{n} = \varepsilon_{\gamma\alpha} \tau^\alpha; \quad (3.18)$$

i.e., the v_γ are the surface covariant components of $\mathbf{v}$. Thus on edges

$$\mathbf{S} dA = \mathbf{S}^i N_i dA = \mu \mathbf{S}^\alpha v_\alpha ds d\zeta. \quad (3.19)$$

Inserting Eqs. (3.11), (3.14), and (3.19) into Eq. (A3), we obtain an equation that may be rewritten in the form

$$\mathbf{L} \equiv \int_{t_1}^{t_2} \left(\int_{\partial\Sigma} \mathbf{N}^\alpha v_\alpha ds + \int_\Sigma \mathbf{p} d\Sigma \right) dt - \int_\Sigma m\mathbf{w} d\Sigma \Big|_{t_1}^{t_2} = \mathbf{0}, \quad (3.20)$$

where
$$\mathbf{N}^\alpha = \int_-^+ \mathbf{S}^\alpha \mu \, d\zeta \tag{3.21}$$

is the *stress resultant*,

$$\mathbf{p} = \int_-^+ \mathbf{f} \mu \, d\zeta + [\mu(\mathbf{S}^3 - H_{,\alpha}\mathbf{S}^\alpha)]_-^+ \tag{3.22}$$

is the *external force* per unit initial area,

$$m = \int_-^+ \rho\mu \, d\zeta \tag{3.23}$$

is the initial mass per unit area, and $\mathbf{w} = \dot{\mathbf{y}}$, where

$$\mathbf{y} = \int_-^+ \frac{\rho \mathbf{x} \mu \, d\zeta}{m} \tag{3.24}$$

is the *weighted motion*. For simplicity, we henceforth assume that the reference surface has been chosen so that $\hat{\xi}(\sigma^\alpha) = \hat{\mathbf{y}}(\sigma^\alpha, 0)$.

We now set

$$\mathbf{x} = \mathbf{y} + \boldsymbol{\eta} \tag{3.25}$$

and note from Eq. (3.24) that

$$\int_-^+ \rho \boldsymbol{\eta} \mu \, d\zeta = \mathbf{0}. \tag{3.26}$$

Inserting Eqs. (3.11), (3.14), (3.19), and (3.25) into Eq. (A4), we obtain an equation that may be rewritten in the form

$$\mathbf{R} \equiv \int_{t_1}^{t_2} \left(\int_{\partial\Sigma} (\mathbf{y} \times \mathbf{N}^\alpha + \mathbf{M}^\alpha) v_\alpha \, ds + \int_\Sigma (\mathbf{y} \times \mathbf{p} + \mathbf{l}) \, d\Sigma \right) dt$$
$$- \int_\Sigma (\mathbf{y} \times m\mathbf{w} + I\boldsymbol{\omega}) \, d\Sigma \Big|_{t_1}^{t_2} = \mathbf{0}, \tag{3.27}$$

where

$$\mathbf{M}^\alpha = \int_-^+ \boldsymbol{\eta} \times \mathbf{S}^\alpha \mu \, d\zeta \tag{3.28}$$

is the *stress couple*,

$$\mathbf{l} = \int_-^+ \boldsymbol{\eta} \times \mathbf{f} \mu \, d\zeta + [\mu\boldsymbol{\eta} \times (\mathbf{S}^3 - \zeta_\alpha \mathbf{S}^\alpha)]_-^+ \tag{3.29}$$

is the *external couple* per unit initial area,

$$I = \int_-^+ \rho \zeta^2 \mu \, d\zeta \tag{3.30}$$

is the *moment of inertia* per unit initial area, and

$$\boldsymbol{\omega} = \int_-^+ \frac{\boldsymbol{\eta} \times \rho \dot{\boldsymbol{\eta}} \mu \, d\zeta}{I} \tag{3.31}$$

is the *spin*.

C. Differential Equations of Motion

The integral equations of motion (3.20) and (3.27) for a general shell are exact consequences of three-dimensional continuum mechanics. Their derivation has led in a straightforward way to static and kinematic variables—the stress resultants and couples $\mathbf{N}^\alpha$ and $\mathbf{M}^\alpha$ and the weighted motion and spin $\mathbf{y}$ and $\boldsymbol{\omega}$—defined over the shell reference surface. In Section III,D we shall relate these variables via a principle of virtual work. To do so we need *differential* equations of motion. These follow from the corresponding integral equations by applying arguments similar to those used for beamshells, except that instead of the fundamental theorem of calculus, we use the divergence theorem for a surface, $\int_{\partial \Sigma} v^\alpha v_\alpha ds = \int_\Sigma v^\alpha|_\alpha d\Sigma$ (McConnel, 1957, p. 189), to replace line integrals by surface integrals. Since Σ is essentially arbitrary, it follows by standard continuity arguments that if the various fields are sufficiently smooth in space and time, then the following differential equations must hold at each point on the reference surface at each instant:

$$\mathbf{N}^\alpha|_\alpha + \mathbf{p} - m\dot{\mathbf{w}} = \mathbf{0}, \tag{3.32}$$

$$\mathbf{M}^\alpha|_\alpha + \mathbf{y}_\alpha \times \mathbf{N}^\alpha + \mathbf{l} - I\dot{\boldsymbol{\omega}} = \mathbf{0}, \tag{3.33}$$

where Eq. (3.32) has been used in obtaining Eq. (3.33) from Eq. (3.27) and $\mathbf{y}_\alpha \equiv \mathbf{y}_{,\alpha}$. Further discussion of the differential equations of motion is postponed.

D. Virtual Work (Weak Form of the Equations of Motion)

Let Eqs. (3.32) and (3.33) be satisfied and let $\mathbf{W}$ and $\boldsymbol{\Omega}$ be any arbitrary functions sufficiently smooth for the following operations to make sense. Take the dot product of Eq. (3.32) with $\mathbf{W}$, Eq. (3.33) with $\boldsymbol{\Omega}$, add, and integrate over any subregion Σ of S and over any interval (t_1, t_2) of time. Finally, apply the divergence theorem to eliminate covariant derivatives of $\mathbf{N}^\alpha$ and $\mathbf{M}^\alpha$ and integrate by parts to eliminate time derivatives of $\mathbf{w}$ and $\boldsymbol{\omega}$ to obtain the *virtual work identity*

$$\int_{t_1}^{t_2} \left(\int_{\partial \Sigma} (\mathbf{N}^\alpha \cdot \mathbf{W} + \mathbf{M}^\alpha \cdot \boldsymbol{\Omega}) v_\alpha ds + \int_\Sigma [\mathbf{p} \cdot \mathbf{W} + \mathbf{l} \cdot \boldsymbol{\Omega} - \mathbf{N}^\alpha \cdot (\mathbf{W}_{,\alpha} - \boldsymbol{\Omega} \times \mathbf{y}_\alpha) \right.$$
$$\left. - \mathbf{M}^\alpha \cdot \boldsymbol{\Omega}_{,\alpha}] d\Sigma \right) dt + \int_\Sigma \left(-[m\mathbf{w} \cdot \mathbf{W} + I\boldsymbol{\omega} \cdot \boldsymbol{\Omega}]_{t_1}^{t_2} \right.$$
$$\left. + \int_{t_1}^{t_2} (m\mathbf{w} \cdot \dot{\mathbf{W}} + I\boldsymbol{\omega} \cdot \dot{\boldsymbol{\Omega}}) dt \right) d\Sigma \equiv 0 \quad \text{for all } \mathbf{W} \text{ and } \boldsymbol{\Omega}. \tag{3.34}$$

If **W** and **Ω** are sufficiently smooth, then there may exist fields $\{\mathbf{N}^\alpha, \mathbf{M}^\alpha, \mathbf{w}, \boldsymbol{\omega}, \mathbf{p}, \boldsymbol{l}\}$ that satisfy Eq. (3.34) but are too rough to satisfy the differential equations of motion. Therefore Eq. (3.34) may also be called the *weak form of the equations of motion*. In particular, if we set $\mathbf{W} = \mathbf{A} + \mathbf{B} \times \mathbf{y}$ and $\boldsymbol{\Omega} = \mathbf{B}$, where **A** and **B** are arbitrary constants, then Eq. (3.34) reduces to $\mathbf{A} \cdot \mathbf{L} + \mathbf{B} \cdot \mathbf{R} = 0$, for all **A** and **B**, which implies the integral equations of motion (3.20) and (3.27).

E. The Mechanical Work Identity

With $\mathbf{W} = \mathbf{w}$ and $\boldsymbol{\Omega} = \boldsymbol{\omega}$, Eq. (3.34) can be rewritten as

$$\int_{t_1}^{t_2} \mathscr{W} \, dt = \mathscr{K}\Big|_{t_1}^{t_2} + \int_{t_1}^{t_2} \mathscr{D} \, dt, \tag{3.35}$$

where

$$\mathscr{W} = \int_{\partial \Sigma} (\mathbf{N}^\alpha \cdot \mathbf{w} + \mathbf{M}^\alpha \cdot \boldsymbol{\omega}) v_\alpha \, ds + \int_\Sigma (\mathbf{p} \cdot \mathbf{w} + \boldsymbol{l} \cdot \boldsymbol{\omega}) d\Sigma \tag{3.36}$$

is the *apparent mechanical power*,

$$\mathscr{K} = \frac{1}{2} \int_\Sigma (m \mathbf{w} \cdot \mathbf{w} + I \boldsymbol{\omega} \cdot \boldsymbol{\omega}) d\Sigma \tag{3.37}$$

is the *kinetic energy*, and

$$\mathscr{D} = \int_\Sigma [\mathbf{N}^\alpha \cdot (\mathbf{w}_{,\alpha} - \boldsymbol{\omega} \times \mathbf{y}_\alpha) + \mathbf{M}^\alpha \cdot \boldsymbol{\omega}_{,\alpha}] d\Sigma \equiv \int_\Sigma \boldsymbol{\sigma} \cdot \dot{\boldsymbol{\varepsilon}} \, d\Sigma \tag{3.38}$$

is the *deformation power*.

F. Strain Measures

The first set of strain measures for the beamshell was obtained by expressing the *scalar* ω as the time derivative of an angle β. The analogous step for a general shell is to express $\boldsymbol{\omega}$ in terms of the time derivative of a *finite rotation vector* $\boldsymbol{\psi}$. To do so, however, requires some ground work that will lead us to a set of *intrinsic* strain measures and compatibility conditions first presented by Reissner (1974).

Consider a triad of vectors $\{\mathbf{A}_\alpha, \mathbf{N}\} \equiv \{\mathbf{A}_i\}$ that rotates rigidly with spin $\boldsymbol{\omega}$ and which, at $t = 0$, coincides with $\{\mathbf{a}_\alpha, \mathbf{n}\} \equiv \{\mathbf{a}_i\}$. (The vector **N** is not to be confused with the **N** of the Appendix. In a Kirchhoff motion, the $\mathbf{A}_\alpha$ and **N** are, respectively, tangent and normal to the deformed reference surface.)

Nonlinear Elastic Shell Theory

At each ξ and t, there exists a rotation tensor $\mathbf{Q}$, or *rotator* for short, such that

$$\mathbf{A}_\alpha = \mathbf{Q} \cdot \mathbf{a}_\alpha, \quad \mathbf{N} = \mathbf{Q} \cdot \mathbf{n}, \quad \mathbf{Q} \cdot \mathbf{Q}^T = \mathbf{1}, \tag{3.39}$$

where $\mathbf{1}$ is the identity tensor. Differentiating with respect to t, we have

$$\dot{\mathbf{A}}_i = \dot{\mathbf{Q}} \cdot \mathbf{a}_i = \dot{\mathbf{Q}} \cdot \mathbf{Q}^T \cdot \mathbf{A}_i, \quad \dot{\mathbf{Q}} \cdot \mathbf{Q}^T + \mathbf{Q} \cdot \dot{\mathbf{Q}}^T = \mathbf{0}. \tag{3.40}$$

The second equation shows that $\dot{\mathbf{Q}} \cdot \mathbf{Q}^T$ is skew. Since $\mathbf{Q}$ is three dimensional, this implies that there exists a unique *axis*, which we take to be ω, such that

$$\dot{\mathbf{A}}_I = \omega \times \mathbf{A}_I. \tag{3.41}$$

We now differentiate Eq. (3.39) with respect to σ^β. Taking note of the Gauss–Weingarten equations, we have

$$\mathbf{A}_{\alpha|\beta} = \mathbf{Q}_{,\beta} \cdot \mathbf{a}_\alpha + \mathbf{Q} \cdot \mathbf{a}_{\alpha|\beta} = \mathbf{Q}_{,\beta} \cdot \mathbf{Q}^T \cdot \mathbf{A}_\alpha + b_{\alpha\beta}\mathbf{N}, \tag{3.42}$$

$$\mathbf{N}_{,\beta} = \mathbf{Q}_{,\beta} \cdot \mathbf{n} + \mathbf{Q} \cdot \mathbf{n}_{,\beta} = \mathbf{Q}_{,\beta} \cdot \mathbf{Q}^T \cdot \mathbf{N} - b^\alpha_\beta \mathbf{A}_\alpha, \tag{3.43}$$

$$\mathbf{Q}_{,\beta} \cdot \mathbf{Q}^T + \mathbf{Q} \cdot \mathbf{Q}^T_{,\beta} = \mathbf{0}. \tag{3.44}$$

This last relation shows that $\mathbf{Q}_{,\beta} \cdot \mathbf{Q}^T$ is skew. Hence there exists a unique vector $\mathbf{K}_\beta$ called the *bending* such that

$$\mathbf{A}_{\alpha|\beta} - b_{\alpha\beta}\mathbf{N} = \mathbf{K}_\beta \times \mathbf{A}_\alpha, \quad \mathbf{N}_{,\beta} + b^\alpha_\beta \mathbf{A}_\alpha = \mathbf{K}_\beta \times \mathbf{N}. \tag{3.45}$$

There must be a relation between ω and $\mathbf{K}_\beta$. To find it, let $\mathbf{c}$ be a *constant* and set $\mathbf{v} = \mathbf{Q} \cdot \mathbf{c}$. Then $\dot{\mathbf{v}} = \omega \times \mathbf{v}$ and $\mathbf{v}_{,\beta} = \mathbf{K}_\beta \times \mathbf{v}$. Since $\dot{\mathbf{v}}_{,\beta} = (\mathbf{v}_{,\beta})^\cdot$, it follows that

$$\omega_{,\beta} \times \mathbf{v} + \omega \times (\mathbf{K}_\beta \times \mathbf{v}) = \dot{\mathbf{K}}_\beta + \mathbf{K}_\beta \times (\omega \times \mathbf{v}), \tag{3.46}$$

that is,

$$(\omega_{,\beta} - \dot{\mathbf{K}}_\beta + \omega \times \mathbf{K}_\beta) \times \mathbf{v} = \mathbf{0}, \tag{3.47}$$

which implies that

$$\omega_{,\beta} = \dot{\mathbf{K}}_\beta - \omega \times \mathbf{K}_\beta. \tag{3.48}$$

For if Eq. (3.48) fails to hold at, say, (σ^α_*, t_*), then, to obtain a contradiction, choose $\mathbf{c}_* = \mathbf{Q}^T_* \cdot \mathbf{v}_*$, where $\mathbf{v}_*$ is *any* constant vector such that Eq. (3.47) is violated.

Given the spin ω, $\mathbf{A}_i$ can be determined by integrating the *ordinary differential equations* (3.41) subject to the initial conditions $\mathbf{A}_i = \mathbf{a}_i$ at $t = 0$. But given the bending $\mathbf{K}_\beta$, the same is not true regarding the *partial differential equations* (3.45) unless $\mathbf{K}_\beta$ satisfies *compatibility conditions*. These guarantee that if $\mathbf{A}_i$ is computed at a given point via a line integral over the reference surface S, then the result will be path independent. This, in turn, is guaranteed if and only if around every hole on S

$$\oint \mathbf{A}_{\alpha,\beta}\,d\sigma^\beta = \oint (\mathbf{A}_{\alpha|\beta} + [\alpha\beta,\lambda]\mathbf{A}^\lambda)\,d\sigma^\beta = \mathbf{0}, \qquad \oint \mathbf{N}_{,\beta}\,d\sigma^\beta = \mathbf{0} \quad (3.49)$$

and at every point of S, by the divergence theorem,

$$\varepsilon^{\beta\gamma}\mathbf{A}_{\alpha,\beta\gamma} = \varepsilon^{\beta\gamma}\mathbf{A}_{\alpha|\beta\gamma} + K\varepsilon_{\alpha\lambda}\mathbf{A}^\lambda = \mathbf{0}, \qquad \varepsilon^{\beta\gamma}\mathbf{N}_{,\beta\gamma} = \varepsilon^{\beta\gamma}\mathbf{N}|_{\beta\gamma} = \mathbf{0}. \quad (3.50)$$

In Eq. (3.49), $[\alpha\beta,\lambda] = [\mathbf{a}_{\alpha,\beta} \cdot \mathbf{a}_\lambda]$ is the Christoffel symbol of the first kind. Substituting Eq. (3.45) into Eq. (3.49), we obtain

$$\oint (\mathbf{K}_\beta \times \mathbf{A}_\alpha + b_{\alpha\beta}\mathbf{N} + [\alpha\beta,\lambda]\mathbf{A}^\lambda)\,d\sigma^\beta = \mathbf{0}, \qquad \oint (\mathbf{K}_\beta \times \mathbf{N} - b_\beta^\alpha \mathbf{A}_\alpha)\,d\sigma^\beta = \mathbf{0}. \quad (3.51)$$

Furthermore, substituting Eq. (3.45) into Eq. (3.50) and using the identity

$$\varepsilon^{\beta\gamma}\mathbf{K}_\beta \times (\mathbf{K}_\gamma \times \mathbf{V}) \equiv \tfrac{1}{2}\varepsilon^{\beta\gamma}(\mathbf{K}_\beta \times \mathbf{K}_\gamma) \times \mathbf{V} \qquad \text{for all } \mathbf{V}, \quad (3.52)$$

as well as the Codazzi equation $\varepsilon^{\beta\gamma}b_{\alpha\beta|\gamma} = 0$ and the symmetry of $b_\beta^\alpha b_{\alpha\gamma}$ in β and γ, we obtain

$$\varepsilon^{\beta\gamma}(\mathbf{K}_{\beta|\gamma} + \tfrac{1}{2}\mathbf{K}_\beta \times \mathbf{K}_\gamma) \times \mathbf{A}_\alpha = \mathbf{0}, \qquad \varepsilon^{\beta\gamma}(\mathbf{K}_{\beta|\gamma} + \tfrac{1}{2}\mathbf{K}_\beta \times \mathbf{K}_\gamma) \times \mathbf{N} = \mathbf{0}, \quad (3.53)$$

i.e.,

$$\varepsilon^{\alpha\beta}(\mathbf{K}_{\alpha|\beta} + \tfrac{1}{2}\mathbf{K}_\alpha \times \mathbf{K}_\beta) = \mathbf{0}. \quad (3.54)$$

This remarkably simple formula is the two-dimensional analog of Eq. (18) of Reissner (1975). It was obtained independently by Pietraszkiewicz (1979, p. 37) and Axelrad (1981).

The coefficient of the stress couple that appears in the internal power density defined by Eq. (3.38) is expressed in Eq. (3.48) in terms of a vector $\mathbf{K}_\beta$ that measures the bending of the shell at any point. To obtain a corresponding expression for the coefficient of the stress resultant in Eq. (3.38), we introduce a new vector $\mathbf{E}_\alpha$ called the *stretching* by setting, in analogy with Eq. (2.32) for the beamshell,

$$\mathbf{y}_\alpha = \mathbf{A}_\alpha + \mathbf{E}_\alpha. \quad (3.55)$$

Then, with the aid of Eq. (3.41),

$$\mathbf{w}_{,\alpha} - \boldsymbol{\omega} \times \mathbf{y}_\alpha = \dot{\mathbf{E}}_\alpha - \boldsymbol{\omega} \times \mathbf{E}_\alpha. \quad (3.56)$$

The right-hand side of Eq. (3.55) defines a single-valued function of position on S if and only if around each hole on S

$$\oint \mathbf{y}_\alpha\,d\sigma^\alpha = \oint (\mathbf{A}_\alpha + \mathbf{E}_\alpha)\,d\sigma^\alpha = \mathbf{0}, \quad (3.57)$$

and at each point of S

$$\varepsilon^{\alpha\beta}\mathbf{y}_{\alpha,\beta} = \varepsilon^{\alpha\beta}(\mathbf{A}_{\alpha|\beta} + \mathbf{E}_{\alpha|\beta}) = \varepsilon^{\alpha\beta}(\mathbf{K}_\beta \times \mathbf{A}_\alpha + \mathbf{E}_{\alpha|\beta}) = \mathbf{0}. \quad (3.58)$$

To proceed further, we descend to the component level and set

$$\mathbf{E}_\alpha = E_{\alpha\beta}\mathbf{A}^\beta + E_\alpha\mathbf{N}, \qquad \mathbf{K}_\alpha = K_{\alpha\beta}\mathbf{A}^\beta \times \mathbf{N} + K_\alpha\mathbf{N}. \qquad (3.59)$$

(In a Kirchhoff motion, $E_{\alpha\beta} = E_{\beta\alpha}$ and $E_\alpha = 0$.) From Eq. (3.48) and (3.56) it follows that

$$\boldsymbol{\omega}_{,\alpha} = \dot{K}_{\alpha\beta}\mathbf{A}^\beta \times \mathbf{N} + \dot{K}_\alpha\mathbf{N} \qquad \mathbf{w}_{,\alpha} - \boldsymbol{\omega} \times \mathbf{y}_\alpha = \dot{E}_{\alpha\beta}\mathbf{A}^\beta + \dot{E}_\alpha\mathbf{N} \qquad (3.60)$$

Inserting Eq. (3.60) into Eq. (3.38) and setting

$$\mathbf{N}^\alpha = N^{\alpha\beta}\mathbf{A}_\beta + Q^\alpha\mathbf{N}, \qquad \mathbf{M}^\alpha = M^{\alpha\beta}\mathbf{A}_\beta \times \mathbf{N} + M^\alpha\mathbf{N}, \qquad (3.61)$$

we find that
$$\sigma \cdot \dot{\varepsilon} = N^{\alpha\beta}\dot{E}_{\alpha\beta} + Q^\alpha \dot{E}_\alpha + M^{\alpha\beta}\dot{K}_{\alpha\beta} + M^\alpha \dot{K}_\alpha. \qquad (3.62)$$

The appearance of the normal component of the stress couple, M^α, is not a consequence of admitting stress couples in three dimensions. Rather, as Eq. (3.28) shows, it arises because in general $\boldsymbol{\eta}$ is not parallel to $\mathbf{N}$.

In terms of the components of $\mathbf{E}_\alpha$ and $\mathbf{K}_\alpha$, the two vector compatibility equations (3.54) and (3.58) take the form

$$\hat{K}^{\alpha\beta}{}_{|\beta} + \varepsilon_{\beta\gamma}(b^{\alpha\gamma} + K^{\alpha\gamma})K^\beta = 0, \qquad \varepsilon^{\alpha\beta}K_{\alpha|\beta} - \hat{K}^{\alpha\beta}b_{\alpha\beta} - \tfrac{1}{2}\hat{K}^{\alpha\beta}K_{\alpha\beta} = 0, \qquad (3.63)$$

$$\begin{aligned}\varepsilon^{\beta\gamma}E_{\beta\alpha|\gamma} + \varepsilon^{\beta\gamma}(b_{\alpha\beta} + K_{\alpha\beta})E_\gamma + (1 + E_\beta^\beta)K_\alpha - E_{\alpha\beta}K^\beta = 0, \\ \varepsilon^{\alpha\beta}[E_{\alpha|\beta} + K_{\beta\alpha} + E_{\alpha\cdot}^{\;\gamma}(b_{\beta\gamma} + K_{\beta\gamma})] = 0,\end{aligned} \qquad (3.64)$$

where $\hat{K}^{\alpha\beta} = \varepsilon^{\alpha\lambda}\varepsilon^{\beta\mu}K_{\lambda\mu}$ and $E_{\alpha\cdot}^{\;\gamma} = a^{\beta\gamma}E_{\alpha\beta}$. These equations were first given in extended form by Reissner (1974).

If in (3.62) we set $E_{\alpha\beta} = \gamma_{\alpha\beta} + \varepsilon_{\alpha\beta}\theta_*$ and $K_{\alpha\beta} = R_{\alpha\beta} + \varepsilon_{\alpha\beta}B$, where $\gamma_{\alpha\beta} = \gamma_{\beta\alpha}$ and $R_{\alpha\beta} = R_{\beta\alpha}$, we obtain the alternate form of the internal power density

$$\sigma \cdot \dot{\varepsilon} = \bar{N}^{\alpha\beta}\dot{\gamma}_{\alpha\beta} + \bar{N}\dot{\theta}_* + Q^\alpha \dot{E}_\alpha + \bar{M}^{\alpha\beta}\dot{R}_{\alpha\beta} + M^\alpha \dot{K}_\alpha + \bar{M}\dot{B}. \qquad (3.65)$$

Here
$$N^{\alpha\beta} = \bar{N}^{\alpha\beta} + \varepsilon^{\alpha\beta}\bar{N}, \qquad \bar{N}^{\alpha\beta} = \bar{N}^{\beta\alpha} \qquad (3.66)$$

$$M^{\alpha\beta} = \bar{M}^{\alpha\beta} + \varepsilon^{\alpha\beta}\bar{M}, \qquad \bar{M}^{\alpha\beta} = \bar{M}^{\beta\alpha}. \qquad (3.67)$$

Upon linearization, $R_{\alpha\beta}$ reduces to the bending strain of the "best" linear shell theory of Budiansky and Sanders (1963).

G. The Finite Rotation Vector

The tensor $\mathbf{Q}$ rotates vectors about an axis with unit vector $\mathbf{e}$ through an angle β determined by the right-hand rule. It is a straightforward exercise (e.g., Truesdell, 1977) to show that

$$\mathbf{Q} = P[(1 - \tfrac{1}{4}\boldsymbol{\psi} \cdot \boldsymbol{\psi})\mathbf{1} + \tfrac{1}{2}\boldsymbol{\psi}\boldsymbol{\psi} + \boldsymbol{\psi} \times], \qquad (3.68)$$

where

$$\psi = 2\mathbf{e}\tan(\beta/2), \qquad P = (1 + \tfrac{1}{4}\boldsymbol{\psi}\cdot\boldsymbol{\psi})^{-1} = \cos^2(\beta/2). \qquad (3.69)$$

To relate $\mathbf{K}_\alpha$ to $\boldsymbol{\psi}$, we regard $\mathbf{K}_\alpha \tau^\alpha$ as the "spin" of $\{\mathbf{A}_i\}$ as we move on Σ with unit velocity along an arbitrary smooth curve $\sigma^\alpha = \hat{\sigma}^\alpha(\tau)$ with arc length τ. As the direction of the curve through any point on Σ may be assigned arbitrarily, it follows immediately from Hamel (1967, p. 106) that

$$\mathbf{K}_\alpha = P(\boldsymbol{\psi}_{,\alpha} + \tfrac{1}{2}\boldsymbol{\psi}\times\boldsymbol{\psi}_{,\alpha}) \qquad (3.70)$$

[see also Pietraszkiewicz, 1979, Eq. (4.2.29)].

Since $-\boldsymbol{\psi}$ sends $\mathbf{A}_i$ into $\mathbf{a}_i$,

$$\boldsymbol{\psi} = \psi^\alpha \mathbf{A}_\alpha + \psi\mathbf{N} = \psi^\alpha \mathbf{a}_\alpha + \psi\mathbf{n}. \qquad (3.71)$$

Using Eq. (3.59) to write Eq. (3.45) in the form

$$\mathbf{A}_{\alpha|\beta} = \varepsilon_{\alpha\lambda} K_\beta \mathbf{A}^\lambda + B_{\alpha\beta}\mathbf{N}, \qquad \mathbf{N}_{,\beta} = -B_{\lambda\beta}\mathbf{A}^\lambda, \qquad (3.72)$$

and substituting Eq. (3.71) into Eq. (3.70), we obtain, with the aid of Eqs. (3.59) and (3.72), the strain–rotation relations (Simmonds and Danielson, 1972)

$$K_\beta = P[\psi_{,\beta} + b_\beta^\lambda \psi_\lambda + \tfrac{1}{2}\varepsilon^{\alpha\lambda}(\psi_{\alpha|\beta} - b_{\alpha\beta}\psi)\psi_\lambda] \qquad (3.73)$$

$$K_{\beta\alpha} = P\{\varepsilon_{\lambda\alpha}(\psi^\lambda|_\beta - b_\beta^\lambda \psi) + \tfrac{1}{2}[(\psi_{\alpha|\beta} - b_{\alpha\beta}\psi)\psi - (\psi_{,\beta} + b_\beta^\lambda \psi_\lambda)\psi_\alpha]\}. \qquad (3.74)$$

With $\boldsymbol{\psi}$ instead of $\mathbf{K}_\alpha$ as a basic kinematic variable, the integral and differential vector compatibility equations, Eqs. (3.51) and (3.54), become satisfied identically. In the remaining compatibility equations, Eqs. (3.57) and (3.58), the $\mathbf{A}_i$ are to be expressed in terms of $\boldsymbol{\psi}$ via $\mathbf{A}_i = \mathbf{Q}\cdot\mathbf{a}_i$ and (3.68).

For an alternate development see Reissner (1982).

H. Alternate Strains, Rational in the Displacements

For general shells undergoing large motions, it seems to us that, as with beamshells, there are only minor advantages but many disadvantages in taking the displacement vector

$$\mathbf{u} = \mathbf{y} - \boldsymbol{\xi} \qquad (3.75)$$

as a basic unknown. Yet, because of the limited work on the best way to solve nonlinear shell problems—be this via general numerical procedures or via analytic procedures tailored to special geometries or loads—we do not want to close the door prematurely on a displacement formulation.

Nonlinear Elastic Shell Theory

By defining new stresses, we can make the associated strains that appear in $\sigma \cdot \dot{\varepsilon}$ depend on **u** and a *shear-twist* γ only. The dependence on **u** turns out to be polynomial and γ vanishes in a Kirchhoff motion.

From Eq. (3.75), with $\mathbf{u}_\alpha \equiv \mathbf{u}_{,\alpha}$,

$$\mathbf{y}_\alpha = \mathbf{a}_\alpha + \mathbf{u}_\alpha. \tag{3.76}$$

We call

$$\mathbf{y}_3 \equiv \mathbf{m} = \tfrac{1}{2}\varepsilon^{\alpha\beta}\mathbf{y}_\alpha \times \mathbf{y}_\beta = \mathbf{n} + \tfrac{1}{2}\varepsilon^{\alpha\beta}(\mathbf{a}_\alpha \times \mathbf{u}_\beta + \mathbf{u}_\alpha \times \mathbf{a}_\beta + \mathbf{u}_\alpha \times \mathbf{u}_\beta) \tag{3.77}$$

the *rational normal* to the deformed reference surface because it is *quadratic* in (the derivatives of) **u**. With the exception of $\bar{\mathbf{y}}^\alpha$, defined by

$$\bar{\mathbf{y}}^i \cdot \mathbf{y}_j = \delta^i_j, \tag{3.78}$$

all indices in what follows are raised with respect to $a^{\alpha\beta}$.

We assume that $\{\mathbf{y}_i\}$ is a basis. Then the *polar decomposition theorem* (Truesdell, 1977) implies that there exists a unique rotator **R** and a unique symmetric positive definite *stretch* **W** such that

$$\mathbf{y}_i = \mathbf{R} \cdot \mathbf{W} \cdot \mathbf{a}_i = \mathbf{a}_i \cdot \mathbf{W} \cdot \mathbf{R}^T. \tag{3.79}$$

Since

$$\mathbf{y}_i \cdot \mathbf{y}_j = \mathbf{a}_i \cdot \mathbf{W}^2 \cdot \mathbf{a}_j \tag{3.80}$$

and the $\mathbf{y}_i$ are rational in **u**, it follows that the components of $\mathbf{W}^2$ in the basis $\{\mathbf{a}^i \mathbf{a}^j\}$ are rational in **u**.

To obtain a deformation measure that vanishes in the reference shape and is rational in **u**, we introduce the *Lagrange strain tensor* **L** by setting

$$\mathbf{W}^2 = \mathbf{1} + 2\mathbf{L} \equiv \bar{a}_{\alpha\beta}\mathbf{a}^\alpha\mathbf{a}^\beta + \eta\mathbf{nn}, \tag{3.81}$$

where

$$\bar{a}_{\alpha\beta} = \mathbf{y}_\alpha \cdot \mathbf{y}_\beta \equiv a_{\alpha\beta} + 2L_{\alpha\beta}, \qquad \eta = \mathbf{m} \cdot \mathbf{m}. \tag{3.82}$$

From Eqs. (3.76), (3.77), and (3.80)–(3.82),

$$L_{\alpha\beta} = \tfrac{1}{2}(\mathbf{y}_\alpha \cdot \mathbf{y}_\beta - a_{\alpha\beta}) = \tfrac{1}{2}(\mathbf{u}_\alpha \cdot \mathbf{a}_\beta + \mathbf{u}_\beta \cdot \mathbf{a}_\alpha + \mathbf{u}_\alpha \cdot \mathbf{u}_\beta), \tag{3.83}$$

$$\eta = \bar{a}/a = \det \mathbf{W} = 1 + 2L^\alpha_\alpha + 2\hat{L}^{\alpha\beta}L_{\alpha\beta}, \tag{3.84}$$

where $\bar{a} = \det(\bar{a}_{\alpha\beta})$ and $\hat{L}^{\alpha\beta} = a^{\alpha\beta}L^\gamma_\gamma - L^{\alpha\beta}$.

In what follows we shall need the components of **W** in terms of those of **L**. To this end we set

$$\mathbf{W} = w^\alpha_\beta \mathbf{a}_\alpha \mathbf{a}^\beta + \eta^{1/2}\mathbf{nn} \tag{3.85}$$

and note that $\det \mathbf{W} = \eta$ implies that $\det(w_\beta^\alpha) = \eta^{1/2}$. From the Cayley–Hamilton theorem and Eq. (3.81),

$$w_\gamma^\alpha w_\beta^\gamma = w_\gamma^\gamma w_\beta^\alpha - \eta^{1/2}\delta_\beta^\alpha = \delta_\beta^\alpha + 2L_\beta^\alpha. \tag{3.86}$$

Taking the trace of both sides of Eq. (3.86) yields

$$(w_\gamma^\gamma)^2 = 2(1 + L_\gamma^\gamma + \eta^{1/2}) \equiv D^{-2}. \tag{3.87}$$

Hence

$$w_\beta^\alpha = D[(1 + \eta^{1/2})\delta_\beta^\alpha + 2L_\beta^\alpha] \equiv A\delta_\beta^\alpha + BL_\beta^\alpha. \tag{3.88}$$

To introduce *shear strains* γ_α and a *twist strain* φ (the latter having no counterpart in beamshells), we write Eq. (3.79) in the form

$$\mathbf{y}_i = (\mathbf{R} \cdot \mathbf{W} \cdot \mathbf{R}^T) \cdot \mathbf{R} \cdot \mathbf{a}_i \equiv \mathbf{V} \cdot \mathbf{r}_i \tag{3.89}$$

and call

$$\mathbf{r}_i = \mathbf{R} \cdot \mathbf{a}_i \tag{3.90}$$

the elements of the *rotated basis*; the vectors $\mathbf{A}_i = \mathbf{Q} \cdot \mathbf{a}_i$ of the preceding subsection may be called the *spin* basis. By the polar decomposition theorem and the uniqueness of Eq. (3.89), we conclude that there exists a *shear-twist rotator* $\boldsymbol{\Gamma}$ such that $\mathbf{y}_i = \mathbf{V} \cdot \boldsymbol{\Gamma} \cdot \mathbf{Q} \cdot \mathbf{a}_i$. Thus $\mathbf{R} = \boldsymbol{\Gamma} \cdot \mathbf{Q}$. This means that $\boldsymbol{\Gamma}$ rotates the rigid frame $\{\mathbf{A}_i\}$ into the rigid frame $\{\mathbf{r}_i\}$.

Let γ be the finite rotation vector of $\boldsymbol{\Gamma}$. The spin of $\boldsymbol{\Gamma}$ *relative* to $\{\mathbf{A}_i\}$ is given by

$$\chi \equiv \boldsymbol{\Omega} - \boldsymbol{\omega} = P(\gamma)(\gamma^* - \tfrac{1}{2}\gamma \times \gamma^*), \quad \gamma(\sigma_\alpha, 0) = \mathbf{0}. \tag{3.91}$$

Here

$$\boldsymbol{\Omega} \times = \dot{\mathbf{R}} \cdot \mathbf{R}^T, \tag{3.92}$$

$$\gamma = \gamma_\alpha \mathbf{r}^\alpha \times \mathbf{r}_3 + \varphi \mathbf{r}_3, \quad P = (1 + \tfrac{1}{4}\gamma \cdot \gamma)^{-1}, \tag{3.93}$$

and

$$\gamma^* = \dot{\gamma}_\alpha \mathbf{r}^\alpha \times \mathbf{r}_3 + \dot{\varphi} \mathbf{r}_3 \tag{3.94}$$

is the velocity of γ *relative* to $\{\mathbf{A}_i\}$. [To derive Eq. (3.91), set $\mathbf{R} = \mathbf{Q} \cdot (\mathbf{Q}^T \cdot \boldsymbol{\Gamma} \cdot \mathbf{Q}) \equiv \mathbf{Q} \cdot \mathbf{S}$ and note that if $\boldsymbol{\Gamma} = \Gamma^{ij}\mathbf{A}_i\mathbf{A}_j$, then $\dot{\mathbf{S}} = \dot{\Gamma}^{ij}\mathbf{a}_i\mathbf{a}_j$. Hence $\chi \times = \mathbf{Q} \cdot \dot{\mathbf{S}} \cdot \mathbf{Q}^T \cdot \boldsymbol{\Gamma}^T = \dot{\Gamma}^{ij}\mathbf{A}_i\mathbf{A}_j \cdot \boldsymbol{\Gamma}^T \equiv \boldsymbol{\Gamma}^* \cdot \boldsymbol{\Gamma}^T$.]

We need to express the components of χ in the basis $\{\mathbf{y}_i\}$. From Eqs. (3.79) and (3.90), $\mathbf{r}^i \cdot \mathbf{y}_j = \mathbf{a}^i \cdot \mathbf{R}^T \cdot \mathbf{R} \cdot \mathbf{W} \cdot \mathbf{a}_j = \mathbf{a}^i \cdot \mathbf{W} \cdot \mathbf{a}_j$. Hence, from Eq. (3.85),

$$\mathbf{r}^\alpha = w_\beta^\alpha \bar{\mathbf{y}}^\beta, \quad \mathbf{r}_3 = \mathbf{r}^3 = \eta^{1/2}\bar{\mathbf{m}} = \eta^{-1/2}\mathbf{m}. \tag{3.95}$$

Substituting Eqs. (3.93)–(3.95) into Eq. (3.91), we obtain

$$\eta\chi = P\eta^{1/2}\{w_\lambda^\mu[\delta_\alpha^\lambda \dot{\gamma}_\mu + \tfrac{1}{2}\eta^{-1/2}\varepsilon_{\alpha\beta}T^{\beta\lambda}(\varphi\dot{\gamma}_\mu - \dot{\varphi}\gamma_\mu)]\bar{\mathbf{y}}^\alpha \times \mathbf{m}$$
$$+ (\dot{\varphi} - \tfrac{1}{2}\varepsilon^{\lambda\mu}w_\lambda^\alpha w_\mu^\beta \gamma_\alpha \dot{\gamma}_\beta)\mathbf{m}\}$$
$$\equiv (A_{\alpha\mu}\dot{\gamma}^\mu + B_\alpha \dot{\varphi})\bar{\mathbf{y}}^\alpha \times \mathbf{m} + (C\dot{\varphi} + D^\beta \dot{\gamma}_\beta)\mathbf{m}$$
$$\equiv \chi_\alpha \bar{\mathbf{y}}^\alpha \times \mathbf{m} + \chi\mathbf{m}, \tag{3.96}$$

where $T^{\alpha\beta}$ is given in Eq. (3.114).

Our goal is to express the factors $\dot{\mathbf{y}}_\alpha - \boldsymbol{\omega} \times \mathbf{y}_\alpha$ and $\boldsymbol{\omega}_{,\alpha}$ in $\boldsymbol{\sigma} \cdot \dot{\boldsymbol{\varepsilon}}$ as linear combinations of the time derivatives of $L_{\alpha\beta}$, $k_{\alpha\beta}$, γ_α, and φ. Here $k_{\alpha\beta}$ are the covariant components of Budiansky's bending strain tensor (1968), which is *cubic* in $\mathbf{u}$ and its spatial derivatives. Differentiating Eq. (3.79) with respect to time, we have, using Eq. (3.92),

$$\dot{\mathbf{y}}_i = (\dot{\mathbf{R}} \cdot \mathbf{W} + \mathbf{R} \cdot \dot{\mathbf{W}}) \cdot \mathbf{a}_i = \boldsymbol{\Omega} \times \mathbf{y}_i + \mathbf{R} \cdot \mathbf{W}^{-1} \cdot \mathbf{W} \cdot \dot{\mathbf{W}} \cdot \mathbf{a}_i. \tag{3.97}$$

We next write $\mathbf{W} \cdot \dot{\mathbf{W}}$ as a symmetric tensor plus a skew tensor. From Eq. (3.81),

$$\tfrac{1}{2}(\mathbf{W} \cdot \dot{\mathbf{W}} + \dot{\mathbf{W}} \cdot \mathbf{W}) = \dot{\mathbf{L}}, \tag{3.98}$$

while, from Eqs. (3.85) and (3.88),

$$\tfrac{1}{2}(\mathbf{W} \cdot \dot{\mathbf{W}} - \dot{\mathbf{W}} \cdot \mathbf{W}) = \tfrac{1}{2}B^2 L_\beta^\alpha \dot{L}_{\alpha\lambda}(\mathbf{a}^\beta \mathbf{a}^\gamma - \mathbf{a}^\gamma \mathbf{a}^\beta)$$
$$= (\tfrac{1}{2}B^2 \varepsilon^{\beta\gamma} L_\beta^\lambda \dot{L}_{\gamma\lambda})\mathbf{n}\times$$
$$\equiv \psi\mathbf{n}\times. \tag{3.99}$$

Inserting Eqs. (3.98) and (3.99) into Eq. (3.97), and noting from Eq. (3.79) that $\mathbf{y}_i \cdot \mathbf{R} \cdot \mathbf{W}^{-1} = \mathbf{a}_i$ [which implies that $\mathbf{R} \cdot \mathbf{W} \cdot \mathbf{W}^{-1} \cdot (\mathbf{n} \times \mathbf{a}_i) = \bar{\mathbf{m}} \times \mathbf{y}_i$], we obtain

$$\dot{\mathbf{y}}_i = \boldsymbol{\tau} \times \mathbf{y}_i + \dot{L}_{ij}\bar{\mathbf{y}}^i, \tag{3.100}$$

where

$$\boldsymbol{\tau} \equiv \boldsymbol{\Omega} + \psi\bar{\mathbf{m}}. \tag{3.101}$$

With the aid of Eqs. (3.81), (3.82), and (3.91), we can break Eq. (3.100) into

$$\dot{\mathbf{y}}_\alpha - \boldsymbol{\omega} \times \mathbf{y}_\alpha = (\boldsymbol{\chi} + \psi\mathbf{m}) \times \mathbf{y}_\alpha + \dot{L}_{\alpha\beta}\bar{\mathbf{y}}^\beta, \qquad \dot{\mathbf{m}} = \boldsymbol{\tau} \times \mathbf{m} + \tfrac{1}{2}\dot{\eta}\bar{\mathbf{m}}. \tag{3.102}$$

We can now compute a part of the internal power density. With

$$\mathbf{N}^\alpha = (\tilde{N}^{\alpha\beta} + \varepsilon^{\alpha\beta}\tilde{N})\mathbf{y}_\beta + \tilde{Q}^\alpha \mathbf{m}, \qquad \tilde{N}^{\alpha\beta} = \tilde{N}^{\beta\alpha}, \tag{3.103}$$

Eqs. (3.96), (3.99), and (3.102) yield

$$\mathbf{N}^\alpha \cdot (\dot{\mathbf{y}}_\alpha - \boldsymbol{\omega} \times \mathbf{y}_\alpha) = [\tilde{N}^{\alpha\beta} + \tfrac{1}{2}\tilde{N}B^2(\varepsilon^{\gamma\alpha}L_\gamma^\beta + \varepsilon^{\gamma\beta}L_\gamma^\alpha)]\dot{L}_{\alpha\beta}$$
$$+ (2CN + \eta\tilde{Q}^\alpha B_\alpha)\dot{\varphi} + (\eta\tilde{Q}_\beta A^{\beta\alpha} + 2D^\alpha N)\dot{\gamma}_\alpha$$
$$\equiv \bar{N}^{\alpha\beta}\dot{L}_{\alpha\beta} + \bar{N}\dot{\varphi} + \bar{Q}^\alpha \dot{\gamma}_\alpha. \tag{3.104}$$

To reduce the term $\mathbf{M}^\alpha \cdot \boldsymbol{\omega}_{,\alpha}$ in the internal power density, we set $\boldsymbol{\omega} = \boldsymbol{\tau} - (\boldsymbol{\chi} + \psi\bar{\mathbf{m}})$ and first derive relations for $\boldsymbol{\tau}_{,\alpha}$ rational in $\mathbf{u}$. Differentiating Eq. (3.102), we have

$$\dot{\mathbf{y}}_{\alpha|\beta} = \boldsymbol{\tau}_{,\beta} \times \mathbf{y}_\alpha + \boldsymbol{\tau} \times \mathbf{y}_{\alpha|\beta} + \dot{L}_{\alpha\lambda|\beta}\bar{\mathbf{y}}^\lambda + \dot{L}_{\alpha\lambda}\bar{\mathbf{y}}^\lambda_{|\beta}, \tag{3.105}$$

$$\dot{\mathbf{m}}_\alpha = \boldsymbol{\tau}_{,\alpha} \times \mathbf{m} + \boldsymbol{\tau} \times \mathbf{m}_\alpha + \tfrac{1}{2}(\dot{\eta}/\eta)_{,\alpha}\mathbf{m} + \tfrac{1}{2}(\dot{\eta}/\eta)\mathbf{m}_\alpha, \qquad \mathbf{m}_\alpha \equiv \mathbf{m}_{,\alpha}. \tag{3.106}$$

To proceed, we need the analogs of the Gauss–Codazzi equations for $\mathbf{y}_\alpha$ and $\mathbf{m}$. From Eqs. (3.8) and (3.76),

$$\mathbf{y}_{\alpha|\beta} = b_{\alpha\beta}\mathbf{n} + \mathbf{u}_{\alpha|\beta} = L_{\alpha\beta\gamma}\bar{\mathbf{y}}^\gamma + B_{\alpha\beta}\bar{\mathbf{m}} \tag{3.107}$$

$$\mathbf{m}_\alpha = -B_{\alpha\beta}\bar{\mathbf{y}}^\beta + \tfrac{1}{2}\eta_{,\alpha}\bar{\mathbf{m}}, \tag{3.108}$$

where

$$L_{\alpha\beta\gamma} = \mathbf{y}_{\alpha|\beta} \cdot \mathbf{y}_\gamma = (\mathbf{y}_\alpha \cdot \mathbf{y}_\gamma)_{|\beta} - \mathbf{y}_\alpha \cdot \mathbf{y}_{\gamma|\beta}, \tag{3.109}$$

$$\eta^{1/2}\bar{b}_{\alpha\beta} \equiv B_{\alpha\beta} = \mathbf{m} \cdot \mathbf{y}_{\alpha|\beta} = \mathbf{m} \cdot \mathbf{y}_{,\alpha\beta} = -\mathbf{m}_{,\alpha} \cdot \mathbf{y}_\beta. \tag{3.110}$$

(This $B_{\alpha\beta}$ is not to be confused with the $B_{\alpha\beta}$ of Section III,F.) By permuting indices and noting that $L_{\alpha\beta} = L_{\beta\alpha}$, we find that

$$L_{\alpha\beta\gamma} = L_{\alpha\gamma|\beta} + L_{\beta\gamma|\alpha} - L_{\alpha\beta|\gamma}, \tag{3.111}$$

while from Eqs. (3.76), (3.77), and (3.110),

$$B_{\alpha\beta} = b_{\alpha\beta} + \bar{L}_{\alpha\beta}(\mathbf{u}) + \bar{Q}_{\alpha\beta}(\mathbf{u}) + \bar{C}_{\alpha\beta}(\mathbf{u}) \equiv b_{\alpha\beta} + \tilde{K}_{\alpha\beta}, \tag{3.112}$$

where $\bar{L}_{\alpha\beta}$, $\bar{Q}_{\alpha\beta}$, and $\bar{C}_{\alpha\beta}$, which may be computed immediately from Eqs. (3.76) and (3.77), are, respectively, linear, quadratic, and cubic in (the derivatives of) $\mathbf{u}$. [The notation $\tilde{K}_{\alpha\beta}$ agress with that of Budiansky (1968).] Furthermore, by differentiating $\bar{\mathbf{y}}^\lambda \cdot \mathbf{y}_\gamma = \delta^\lambda_\gamma$ covariantly, we get

$$\bar{\mathbf{y}}^\lambda|_\beta \cdot \mathbf{y}_\gamma = -\eta^{-1}T^{\lambda\mu}L_{\gamma\beta\mu}, \tag{3.113}$$

where

$$T^{\alpha\beta} = a^{\alpha\beta} + 2\hat{L}^{\alpha\beta} = T^{\beta\alpha}. \tag{3.114}$$

Now note that

$$-\dot{B}_{\alpha\beta} = (\mathbf{y}_\beta \cdot \mathbf{m}_\alpha)^\cdot = \dot{\mathbf{y}}_\beta \cdot \mathbf{m}_\alpha + \mathbf{y}_\beta \cdot \dot{\mathbf{m}}_\alpha, \tag{3.115}$$

$$2\varepsilon^{\alpha\gamma}\dot{L}_{\beta\gamma|\alpha} = \varepsilon^{\alpha\gamma}(\mathbf{y}_{\alpha|\beta} \cdot \mathbf{y}_\gamma)^\cdot = \varepsilon^{\alpha\gamma}(\dot{\mathbf{y}}_{\alpha|\beta} \cdot \mathbf{y}_\gamma + \mathbf{y}_{\alpha|\beta} \cdot \dot{\mathbf{y}}_\gamma). \tag{3.116}$$

When Eqs. (3.102), (3.105), (3.106), and (3.113) are substituted into the above equations, all terms containing $\boldsymbol{\tau}$ cancel and the remaining ones can be rearranged to yield

$$\eta(\mathbf{y}_\beta \times \mathbf{m}) \cdot \boldsymbol{\tau}_{,\alpha} = \eta\dot{B}_{\alpha\beta} - \tfrac{1}{2}\dot{\eta}B_{\alpha\beta} - B_{\alpha\gamma}T^{\gamma\lambda}L_{\beta\lambda}, \tag{3.117}$$

$$\eta\mathbf{m} \cdot \boldsymbol{\tau}_{,\alpha} = \varepsilon^{\beta\gamma}(\eta\dot{L}_{\alpha\gamma|\beta} + L_{\alpha\gamma\lambda}T^{\lambda\mu}L_{\beta\mu}). \tag{3.118}$$

From Eq. (3.108) follows

$$\eta(\mathbf{y}_\beta \times \mathbf{m}) \cdot (\psi\mathbf{m})_{,\alpha} = \varepsilon_{\lambda\beta} T^{\lambda\gamma} B_{\alpha\gamma} \psi, \qquad (3.119)$$

$$\eta \mathbf{m} \cdot (\psi\mathbf{m})_{,\alpha} = \eta \psi_{,\alpha} - \tfrac{1}{2} \psi \eta_{,\alpha}. \qquad (3.120)$$

Finally, from (3.96), we derive, with the aid of Eqs. (3.108) and (3.117),

$$\eta(\mathbf{y}_\beta \times \mathbf{m}) \cdot \boldsymbol{\chi}_{,\alpha} = \eta \chi_{\beta|\alpha} + (\chi_\lambda T^{\lambda\gamma} L_{\beta\alpha\gamma} - \tfrac{3}{2} \chi_\beta \eta_{,\alpha} - \varepsilon_{\beta\lambda} T^{\lambda\gamma} B_{\alpha\gamma} \chi), \qquad (3.121)$$

$$\eta \mathbf{m} \cdot \boldsymbol{\chi}_{,\alpha} = \varepsilon^{\beta\gamma} B_{\alpha\beta} \chi_\gamma + \eta \chi_{,\alpha} - \tfrac{1}{2} \chi \eta_{,\alpha}, \qquad (3.122)$$

and note from Eq. (3.84) that

$$\dot{\eta} = 2\dot{L}^\alpha_\alpha + 4\hat{L}^{\alpha\beta} \dot{L}_{\alpha\beta} = 2T^{\alpha\beta} \dot{L}_{\alpha\beta}. \qquad (3.123)$$

If we set

$$\mathbf{M}^\alpha = M^{\alpha\beta} \mathbf{y}_\beta \times \mathbf{m} + \bar{M}^\alpha \mathbf{m} \qquad (3.124)$$

and $\omega_{,\alpha} = [\tau - (\chi + \psi \bar{\mathbf{m}})]_{,\alpha}$, then Eqs. (3.96), (3.99), (3.112), and (3.117)–(3.124) yield

$$\mathbf{M}^\alpha \cdot \omega_{,\alpha} = \bar{M}^{\alpha\beta} \dot{\hat{K}}_{\alpha\beta} + M\dot{\Gamma} + S^{\alpha\beta} \dot{L}_{\alpha\beta} + S^\alpha \dot{\gamma}_\alpha + S\dot{\varphi} + \bar{M}^\alpha \dot{\Gamma}_\alpha, \qquad (3.125)$$

where $M^{\alpha\beta} = \bar{M}^{\alpha\beta} + \varepsilon^{\alpha\beta} M$, $\bar{M}^{\alpha\beta} = \bar{M}^{\beta\alpha}$, and

$$\begin{aligned}\Gamma &= \varepsilon^{\alpha\beta}(A_{\beta\lambda} \gamma^\lambda_{|\alpha} + B_\beta \varphi_{,\alpha}), \\ \Gamma_\alpha &= \varepsilon^{\beta\gamma}(L_{\alpha\gamma|\beta} + \tfrac{1}{2} B^2 L^\lambda_\gamma L_{\lambda\beta|\alpha}) - (C\varphi_{,\alpha} + D^\beta \gamma_{\beta|\alpha}). \end{aligned} \qquad (3.126)$$

The coefficients $S^{\alpha\beta}$, S^α, and S involve $L_{\alpha\beta}, \gamma_\alpha, \varphi$, and their spatial derivatives, as well as $M^{\alpha\beta}$ and $\bar{M}^\alpha$, but are too complicated to write out.

Adding Eqs. (3.104) and (3.125), we get the following expression for the internal power density:

$$\sigma \cdot \dot{\varepsilon} = \tilde{n}^{\alpha\beta} \dot{L}_{\alpha\beta} + \bar{M}^{\alpha\beta} \dot{\hat{K}}_{\alpha\beta} + (M\dot{\Gamma} + \tilde{Q}^\alpha \dot{\gamma}_\alpha + \tilde{N} \dot{\varphi}) + [\bar{M}^\alpha \Gamma_\alpha]. \qquad (3.127)$$

The classical theory of shells assumes the following:

1. $\gamma = \mathbf{0}$ (the reduced Kirchhoff hypothesis).
2. $\mathbf{l} = \mathbf{0}$, $\bar{M}^\alpha = 0$, $\mathbf{m} \cdot I\dot{\omega} = 0$.

Under (1), the term in parentheses in Eq. (3.127) disappears. Under (2), the term in square brackets in Eq. (3.127) disappears and the component along $\mathbf{m}$ of the equation of conservation of rotation momentum, Eq. (3.33), reduces, in view of Eqs. (3.103) and (3.124), to

$$2\eta \tilde{N} + \varepsilon_{\beta\gamma} T^{\lambda\gamma} B_{\lambda\alpha} M^{\alpha\beta} = 0. \qquad (3.128)$$

This makes the term in Eq. (3.125) coming from Eq. (3.119) cancel the dashed underlined term in Eq. (3.104). The term $\tilde{n}^{\alpha\beta}$, by Eqs. (3.117) and (3.123), now reduces to Budiansky's Eq. (73) (1968):

$$\tilde{n}^{\alpha\beta} = \tilde{N}^{\alpha\beta} - \eta^{-1}B_{\gamma\rho}(T^{\rho\beta}M^{\gamma\alpha} + T^{\alpha\beta}M^{\rho\gamma}). \qquad (3.129)$$

The *linear* shell equations of Sanders (1959) and Koiter (1960) are characterized by a bending strain that depends on the linearized rotation vector *only*. To obtain a theory that, when linearized, reduces to Sander's and Koiter's, Budiansky (1968) introduced the new bending strain and stress resultant

$$k_{\alpha\beta} = \tilde{K}_{\alpha\beta} - \tfrac{1}{2}(b_\alpha^\gamma L_{\gamma\beta} + b_\beta^\gamma L_{\gamma\alpha}) - b_{\alpha\beta}L_\gamma^\gamma, \qquad (3.130)$$

$$\bar{n}^{\alpha\beta} = \tilde{n}^{\alpha\beta} + \tfrac{1}{2}(b_\gamma^\alpha \bar{M}^{\beta\gamma} + b_\gamma^\beta \bar{M}^{\alpha\gamma}) + a^{\alpha\beta}b_{\omega\gamma}\bar{M}^{\omega\gamma}. \qquad (3.131)$$

The internal power density, in the classical theory, then reads

$$\sigma \cdot \dot{\varepsilon} = \bar{n}^{\alpha\beta}\dot{L}_{\alpha\beta} + \bar{M}^{\alpha\beta}\dot{k}_{\alpha\beta}. \qquad (3.132)$$

I. Constitutive Relations and Boundary Conditions

The number and form of the boundary conditions in shell theory depends on the assumed form of the deformation power $\mathscr{D}$ defined by Eq. (3.38). Constitutive relations are expressions for stresses as functionals of the associated strains that appear in the internal power density. We note that whereas the equations of motion and compatibility are *exact* (though, from a three-dimensional viewpoint, incomplete), the constitutive relations are, by their very nature, *approximate*. It is a major task to estimate the resulting errors in the field equations (not to mention in the solutions), even for restricted classes of problems and for regions far from boundaries.

For *elastic* shells, the deformation energy is represented by an integral over the reference surface S of a strain energy per unit initial area Φ *plus* an integral over ∂S of a strain energy per unit initial length. The stresses are then partial derivatives of Φ with respect to the associated strains. The integral over ∂S, which obviously contributes directly to the boundary conditions, arises from edge effects analogous to surface tension in a liquid, and was introduced by Koiter and van der Heijden (see Koiter and Simmonds, 1973). More on this integral will be discussed later.

As in the analysis of beamshells, the two major ways for establishing constitutive relations—the direct method and the reduction method—can be used, but they are far more complicated owing to the two-dimensional nature of shell theory.

The direct method, which views the shell *ab initio* as a two-dimensional continuum, uses dimensional analysis and invariance to restrict the possible forms of Φ. Further simplifications result from accounting for small strains, thinness, and the small ratio of the thickness to the wavelengths of the surface loads and the reference surface deformation pattern. Once the form of Φ has been boiled down to its essence, the residual constants may be determined by experiments, as suggested by Reissner (1972a, 1974). As experiments on rods, plates, and shells have helped in the past to establish the constitutive relations of three-dimensional elasticity, a direct approach to finding two-dimensional constitutive relations is natural and fundamental.

If Φ depends on all the strains that appear in the two alternate forms of the internal power density, Eqs. (3.65) or (3.127), one obtains a system of partial differential equations of the twelfth order. A traction boundary value problem then requires specification of the six components of $\mathbf{N}^\alpha v_\alpha$ and $\mathbf{M}^\alpha v_\alpha$ along ∂S. Kinematic boundary conditions have to be expressed in terms of $(\mathbf{K}_\alpha v^\alpha, \mathbf{E}_\alpha v^\alpha)$, $(\boldsymbol{\psi}, \mathbf{E}_\alpha v^\alpha)$, or $(\mathbf{u}, \boldsymbol{\gamma})$, depending on what pairs of vector variables are considered primary. We do not pursue such an elaborate theory for two reasons: First, boundary conditions on thin bodies are rarely known in practice with sufficient precision to warrant a twelfth-order theory, and, second, significant differences between a twelfth-order and an eighth-order classical theory will appear only in regions of high strain gradients, where three-dimensional effects may make *any* two-dimensional theory suspect.

In the classical theory, which is intended to apply to thin shells homogeneous in the thickness direction, the surface energy density is taken as a function of the extensional and bending strains only, and the edge strain energy density is ignored. The principle of stationary potential energy then implies that the *natural* boundary conditions consist of four combinations of the three components of the stress resultant vector and the two tangential components of the stress couple vector. We call these *Kirchhoff* boundary conditions.

Koiter (1960) showed for an elastically isotropic shell that if the strains are small, then the *assumption* of an approximate state of plane stress (and *not* plane strain, as a strict enforcement of the *three-dimensional* Kirchhoff hypothesis would require) implies that Φ may be taken as an *uncoupled* quadratic function of the extensional and bending strains. (Of course, to simplify the final form of the field equations, we may always add small coupling terms to Φ.) John (1965a,b) later showed that the structure of the equations of nonlinear elasticity for a shell-like body under edge loads actually *implies* a state of plane stress to within errors that can be estimated rigorously. John's work was extended by Berger (1973) to include shells of variable thickness under surface as well as edge loads. As Koiter and

Simmonds (1973) observed, the results of John and Berger imply that the stress–strain relations for an elastically isotropic shell may be taken in the form

$$\gamma_{\alpha\beta} = A[(1 + v)\bar{N}_{\alpha\beta} - va_{\alpha\beta}\bar{N}_\gamma^\gamma] + O(\eta\theta^2), \tag{3.133}$$

$$\bar{M}_{\alpha\beta} = D[(1 - v)R_{\alpha\beta} + va_{\alpha\beta}R_\gamma^\gamma] + O(Eh^2\eta\theta^2) \tag{3.134}$$

in terms of the stresses and conjugate strains of Section III,F. Identical relations hold between the stresses and strains of Section III,H. In Eqs. (3.133) and (3.134), $A = 1/Eh$, $D = Eh^3/12(1 - v^2)$, η is the maximum strain in the shell, and

$$\theta^2 = \max(h^2/d^2, h^2/L^2, h^2/L^{*2}, h/R, \eta), \tag{3.135}$$

where d is the distance from the point in question to the edge, L is the minimum wavelength of the deformation pattern, L^* is the wavelength of the curvature pattern of the reference surface $[b_{\alpha\beta|\gamma} = O(R^{-1}, L^{*-2})]$, and R is the minimum local radius of curvature of the reference surface. The introduction of L^* is a refinement of Koiter's (1980).

A direct approach to reducing the possible forms of Φ was made by Niordson (1971), who assumed that Φ depended quadratically on the extensional and bending strains and their derivatives. The main points of his argument were the following:

1. Φ must be invariant to inversion of the direction of the normal, and
2. Φ must be regular at $h = M = K = 0$,

where M and K are, respectively, the mean and Gaussian curvatures of the reference surface. This implies that the only available local length scale of positive power is h! Using a dimensional analysis based on (1) and (2), Niordson was able to show that Φ can be reduced to the form

$$\Phi = Eh[(A_1 a^{\alpha\beta}a^{\gamma\lambda} + A_2 a^{\alpha\lambda}a^{\beta\lambda})\gamma_{\alpha\beta}\gamma_{\gamma\lambda} + h^2(A_3 a^{\alpha\beta}a^{\gamma\lambda} + A_4 a^{\alpha\gamma}a^{\beta\lambda})R_{\alpha\beta}R_{\gamma\lambda} + r\eta^2], \tag{3.136}$$

where r is of the order $h/R + h^2/L^2$. The h/R term results from omission of the various linear invariants of $b_{\mu\nu}\gamma_{\alpha\beta}R_{\gamma\lambda}$. The h^2/L^2 term results from terms depending on the derivatives of the strains. The A_i's are to be found by experiment. The error estimates agree with those implied by Eqs. (3.133) and (3.134).

When the stored energy functional is that of *membrane-inextensional bending theory* (Section VI), then the number of Kirchhoff boundary conditions must be decreased. Already in the linear theory the proper choice of these conditions is subtle (see Novozhilov, 1970, Sections 21 and 32, and Vekua, 1962, Chapter VI, Section 5). The field equations break into two sets of second-order partial differential equations that are elliptic, parabolic, or hyperbolic at a point of S as the Gaussian curvature there is positive, negative, or zero, respectively. If an edge bounds an elliptic region, then one force and one displacement boundary condition are proper; but if it bounds a parabolic or hyperbolic region, the situation is more complicated, depending on, among other things, whether the edge coincides with an asymptotic line of S. As we shall show in Section VI, the nonlinear theory of membranes actually simplifies the selection of proper boundary conditions.

We do not want to leave the false impression that the boundary conditions associated with various restricted theories are themselves necessarily approximate. For example, Koiter (1964) has shown that *the four Kirchhoff boundary conditions of classical shell theory are exact*. The reason is simple. Assuming that the *deformed* edge is a ruled surface normal to the deformed reference surface and integrating the traction boundary conditions of three-dimensional elasticity through the thickness, we obtain five exact conditions on the stress resultants and couples, the stress couple vector now having no component along the normal to the deformed reference surface. The condition on the twisting component of the stress couple involves the skew part $\bar{H}$ of the stress couple tensor. Solving this equation for $\bar{H}$ and inserting the resulting expression into the remaining four conditions, Koiter obtained the four Kirchhoff conditions, without appeal to St. Venant's Principle or integration by parts along the boundary. Koiter's finding suggests that the boundary conditions associated with other special constitutive relations may also be exact. Take those derived by Reissner (1962, 1970) for inextensional deformation. If, following Wan (1968), Novozhilov and Shamina (1975), and Pietraszkiewicz (1979), one rewrites the rate of work of the edge loads in terms of integrals of loads times strain rates, then the proper boundary conditions may be read off from the two terms that are independent of the extensional strains.

The Modified Kirchhoff Boundary Conditions

Though exact, the Kirchhoff boundary conditions do not lead to the "best" possible solution of the shell equations. We infer this from the following arguments of linear theory.

Suppose that the distribution through the thickness of the prescribed edge tractions were known exactly. (In practice, as Koiter noted, this occurs only at a free edge.) If we have a solution of the classical linear shell equations [say, those of Sanders (1959) or Koiter (1960)], then we may ask, How accurate is the three-dimensional stress field that may be inferred from the two-dimensional shell solution? Using the hypercircle method of Prager and Synge (1947), Koiter (1970) and Danielson (1971) estimated the error in an energy (mean-square) sense. This requires the construction, *from the shell theory solution*, of a three-dimensional "statically admissible" stress field, i.e., one that satisfies equilibrium and traction boundary conditions. Satisfaction of the equilibrium equations forces the stress field to have a *specific* thickness variation. Consequently, at a free edge, the stress field does not vanish pointwise, although the associated stress resultants and couples do satisfy the Kirchhoff conditions.

Now consider a solution of the linear three-dimensional elasticity equations with edge tractions equal and opposite to those of the Koiter–Danielson field. An approximate solution of this problem, as first shown by Friedrichs (1950) for a plate, may be obtained by introducing a set of coordinates (s, ξ, ζ) such that ξ measures distance into the shell, orthogonal to an edge. If one introduces dimensionless variables $(s/h, \xi/h, \zeta/h)$ and neglects derivatives with respect to the first variable compared to derivatives with respect to the last two, then the three-dimensional elasticity equations reduce, in a first approximation, to the equations for plane strain and torsion of a semi-infinite strip occupying the region $0 \leq \xi/h < \infty$, $-1 \leq \zeta/h \leq 1$. Golden'veiser (1966) has shown for a plate that when the solutions of these problems are combined and, as $\xi/h \to 0$, matched to the solutions of the classical plate equation $\Delta\Delta w = 0$ as $\xi \to 0$, then the correction to classical plate theory, *away from edges*, is the same as if the solution to the classical plate equation were subject to a set of (slightly) *modified* Kirchhoff boundary conditions. The modifications consist of additional terms of relative order κh, where κ is the local curvature of the edge of the midplane.

In his thesis, van der Heijden (1976) showed that Golden'veiser's method, applied to a shell, yields a supplementary stress field that, when added to the Koiter–Danielson field, produces traction-free edge conditions. Moreover, he showed that the mean-square error in shell theory is then $O(h/R + h^2/L^2)$.

An ingenious shortcut for deriving modified Kirchhoff boundary conditions for a shell was devised by Koiter and van der Heijden (see Koiter and Simmonds, 1973). As noted, the Koiter–Danielson statically admissible stress field constructed from shell theory satisfies, at a traction-free edge, the Kirchhoff conditions only, leaving a residual shear stress. Since Golden'-veiser's analysis applied to shells shows that the three-dimensional stresses predicted by shell theory are seriously in error only in a narrow edge zone of width $O(h)$ (van der Heijden, 1976), we may conclude that shell theory *overestimates* the torsional energy near a free edge. Since the three-dimensional stress field at an edge changes slowly along the edge compared with its change normal to the edge, this torsional energy may be estimated by comparing with the classical plate solution the exact elasticity solution for a flat infinite strip under end torques produced by a shear stress that varies (linearly) across the thickness only. It is found that, to a first approximation, the difference in the total strain energy is proportional to the integral *along the edge* of the square of $w_{,xy}$, the twist of *classical plate theory*. Koiter and van der Heijden then simply applied this result to shell theory; that is, the surface integral for the strain energy of classical linear shell theory is modified by subtracting a line integral along any free edges of a certain constant times the square of the (slightly modified) twist of the deformed reference surface. The principle of stationary potential energy then yields modified Kirchhoff boundary conditions for a shell analogous to those obtained by Golden'veiser for a plate.

In a theory of shells in which the only nonlinearities are geometrical it seems obvious that in a neighborhood of width $O(h)$ of ∂S, the relation between two-dimensional shell theory and three-dimensional elasticity theory must be essentially the same as in linear theory. Thus in the nonlinear theory we must subtract from the stored energy functional a line integral along free edges of the reference surface of the form $\int_{\partial \Sigma} G_* h^4 \omega_*^2 \, ds$, where G_* is an elastic constant of the order of Young's modulus and ω_* is a modified twisting strain (see Section 6 of Koiter and Simmonds, 1973, for details).

Space does not permit us to go into further details of the important problem of relating the constitutive laws of shell theory to those of a three-dimensional continuum, except to note that the basic tools are variational techniques, asymptotic expansions in a small thickness parameter, and error estimates of the field equations of the type done by John (1965a,b) and Berger (1973) (see, for example, Koiter and Simmonds, 1973; Ladeveze, 1976; Koiter, 1980; and Berdichevskii, 1979). A much larger literature, which we make no attempt to cite, exists in the linear theory.

For materials undergoing large strain there has been some work by Biricikoglu and Kalnins (1971) along the lines of our beamshell study in Section II. Current engineering practice is to assume a state of plane stress together with a nonlinear material law, while allowing for transverse normal strains through the introduction of thickness changes.

J. FORMULATIONS OF THE FIELD EQUATIONS

The field equations of shell theory take a variety of forms, depending on the quantities chosen as the primary unknowns and on whether the equations are static or dynamic. Various possibilities are discussed here.

1. *Displacement Form (Classical Theory)*

Here the field equations consist of three *reduced* equations of linear momentum expressed in terms of

$$\mathbf{u} = u^\alpha \mathbf{a}_\alpha + w\mathbf{n}. \tag{3.137}$$

They are the Euler equations associated with the strain energy density (3.132) and the kinetic energy density $\frac{1}{2}m\mathbf{w}\cdot\mathbf{w}$, and are given by Eq. (87) of Budiansky (1968), where, by D'Alembert's principle,

$$\mathbf{f} = f_\gamma \mathbf{a}^\gamma + f\mathbf{n} = (p_\gamma - m\ddot{u}_\gamma)\mathbf{a}^\gamma + (p - m\ddot{w})\mathbf{n}. \tag{3.138}$$

The advantage of the displacement form of the field equations is that they apply as they stand to either static or dynamic problems. But in the words of Budiansky, "unfortunately, even in tensor notation, [they] are very complicated and will not be written out explicitly." We anticipated this in our analysis of beamshells in Section II,L,1.

2. *Rotation Form*

a. Static Equations

Here, following Simmonds and Danielson (1972), we satisfy force equilibrium (3.32) with a vector stress function $\mathbf{F}$:

$$\mathbf{N}^\alpha = \varepsilon^{\alpha\lambda}\mathbf{F}_{,\lambda} - \tfrac{1}{2}a^{-1/2}\int a^{1/2}\mathbf{p}\,d\sigma^\alpha \tag{3.139}$$

$$\equiv \varepsilon^{\alpha\lambda}\mathbf{F}_{,\lambda} + \mathbf{P}^\alpha.$$

(There are other ways to write the particular solution vector.) With

$$\mathbf{F} = F^\beta \mathbf{A}_\beta + F\mathbf{N}, \qquad \mathbf{P}^\alpha = P^{\alpha\beta}\mathbf{A}_\beta + P^\alpha \mathbf{N}, \tag{3.140}$$

it follows from Eqs. (3.61) and (3.66) that the component form of Eq. (3.139) is

$$\bar{N}^{\alpha\beta} + \tfrac{1}{2}\varepsilon^{\alpha\beta}\bar{N} = \varepsilon^{\alpha\lambda}(F^\beta|_\lambda - B^\beta_{.\lambda}F + \varepsilon^{\gamma\beta}K_\lambda F_\gamma) + P^{\alpha\beta}, \tag{3.141}$$

$$Q^\alpha = \varepsilon^{\alpha\lambda}(B_{\beta\lambda}F^\lambda + F_{,\lambda}) + P^\alpha. \tag{3.142}$$

If we introduce the finite rotation vector $\boldsymbol{\psi}$, with components given by Eq. (3.71), then Eqs. (3.73) and (3.74) give K_α and $K_{\alpha\beta}$ in terms of ψ_α and ψ. The remaining field equations are the constitutive relations, the compatibility condition (3.58), and the moment equilibrium equation (3.33). Upon setting $\mathbf{l} = l_\alpha \mathbf{A}^\alpha + l\mathbf{N}$ and using Eqs. (3.67) and (3.72), this latter equation takes the component form

$$\varepsilon_{\alpha\gamma}[\bar{M}^{\gamma\beta}|_\beta + (a^{\beta\gamma} + E^{\beta\gamma})Q_\beta - E_\beta \bar{N}^{\beta\gamma}] + \bar{M}_{,\alpha} - \tfrac{1}{2}E_\alpha \bar{N}$$
$$+ (\bar{M}_{\alpha\beta} + \varepsilon_{\alpha\beta}\bar{M})K^\beta - B_{\alpha\beta}M^\beta + l_\alpha = 0 \tag{3.143}$$
$$\varepsilon_{\alpha\beta}(B^{\alpha\gamma}\bar{M}^\beta_\gamma + E^{\gamma\alpha}\bar{N}^\beta_\gamma) + B^\alpha_\alpha \bar{M} + (1 + \tfrac{1}{2}E^\alpha_\alpha)\bar{N} + \bar{M}^\alpha|_\alpha + l = 0.$$

Because we want field equations that are well conditioned for the extreme states of inextensional bending and membrane stretching, we work with the *mixed* energy density (see Section II,1)

$$\begin{aligned}\Psi &= \hat{\Psi}(\bar{N}^{\alpha\beta}, R_{\alpha\beta}, Q^{\alpha}, K^{\alpha}, \bar{N}, B) \\ &= -\tfrac{1}{2}A[(1+v)\bar{N}^{\alpha\beta}\bar{N}_{\alpha\beta} - v(\bar{N}_\gamma^\gamma)^2] \\ &\quad + \tfrac{1}{2}D[(1-v)R^{\alpha\beta}R_{\alpha\beta} \\ &\quad + v(R_\gamma^\gamma)^2] + O(Eh\eta^2\theta^2),\end{aligned} \quad (3.144)$$

which implies Eqs. (3.133) and (3.134). Substituting Eq. (3.133) and (3.134) into the compatibility equations (3.64) and the moment equilibrium equations (3.143) and noting that Eq. (3.144) implies that $\theta_* = -\partial\Psi/\partial\bar{N} = O(\eta\theta^2)$, $E_\alpha = -\partial\Psi/\partial Q^\alpha = O(\eta\theta)$, $M^\alpha = \partial\Psi/\partial K^\alpha = O(Eh^2\eta\theta)$, and $\bar{M} = \partial\Psi/\partial B = O(Eh^2\eta\theta)$, we obtain

$$\varepsilon_{\alpha\beta}K^\beta - A\bar{N}^\beta_{\beta',\alpha} = O(\eta\theta^2/\lambda), \quad (3.145)$$

$$\varepsilon^{\alpha\beta}[\delta^\gamma_\alpha + A(1+v)\bar{N}^\gamma_\alpha]B_{\gamma\beta} = O(\eta\theta^4/h), \quad (3.146)$$

$$DR^\beta_{\beta,\alpha} + Q_\alpha + l_\alpha = O(Eh^2\eta\theta^2/\lambda), \quad (3.147)$$

$$\bar{N} + \varepsilon_{\alpha\beta}\{D(1-v)B^{\alpha\gamma}R^\beta_\gamma + A[(1+v)\bar{N}^\alpha_\gamma\bar{N}^{\gamma\beta} - v\bar{N}^{\alpha\beta}\bar{N}^\gamma_\gamma]\} + l = O(Eh\eta\theta^4), \quad (3.148)$$

where $R_{\alpha\beta} = \text{sym } K_{\alpha\beta}$, $B_{\beta\alpha} = b_{\alpha\beta} + K_{\alpha\beta}$, and K_α are to be expressed in terms of ψ_α and ψ via (3.73) and (3.74) and $\bar{N}^{\alpha\beta}$, $\bar{N}$, and Q^α are to be expressed in terms of F^α and F via (3.141) and (3.142). In the error terms,

$$\lambda = \min(d, L, L^*, h/\sqrt{\eta}) \quad (3.149)$$

measures the size of derivatives along the reference surface $[df/ds = O(f/\lambda)]$. Its value is based on the work of John (1965a,b) and Berger (1973), as modified by Koiter (1980). Note that at a distance $d = O(h)$ from an edge, $\theta = O(1)$ and each of the error terms in (3.145)–(3.148) becomes of the same order as the terms on the left. If the error terms are set to zero, we obtain an eight-order system.

Appropriate boundary conditions to accompany (3.145)–(3.148) are discussed by Simmonds and Danielson (1972), Pietraszkiewicz (1979, 1980), and Pietrazkiewicz and Szwabowicz (1981).

b. Dynamic Equations

To obtain dynamic field equations that do not involve the displacement, we differentiate the equation of linear momentum with respect to σ^β, so obtaining

$$\mathbf{N}^\alpha|_{\alpha\beta} + \mathbf{p}_{,\beta} = m\ddot{\mathbf{y}}_\beta. \quad (3.150)$$

Introducing the * derivative,

$$\mathbf{v}^* \equiv \dot{\mathbf{v}} - \boldsymbol{\omega} \times \mathbf{v}, \quad (3.151)$$

which represents the time rate of change of **v** relative to the frame $\{\mathbf{A}_i\}$, and noting that $\boldsymbol{\omega}^* = \dot{\boldsymbol{\omega}}$, we have, by Eqs. (3.45), (3.55), and (3.57),

$$\ddot{\mathbf{y}}_\beta = \boldsymbol{\omega}^* \times (\mathbf{A}_\beta + \mathbf{E}_\beta) + 2\boldsymbol{\omega} \times \mathbf{E}_\beta^* + \mathbf{E}_\beta^{**}. \tag{3.152}$$

The spin $\boldsymbol{\omega}$ is related to the finite rotation vector introduced in Section III.G by

$$\begin{aligned}\boldsymbol{\omega} &= P(\boldsymbol{\psi})(\dot{\boldsymbol{\psi}} + \tfrac{1}{2}\boldsymbol{\psi} \times \dot{\boldsymbol{\psi}}) \\ &= P(\boldsymbol{\psi})(\boldsymbol{\psi}^* - \tfrac{1}{2}\boldsymbol{\psi} \times \boldsymbol{\psi}^*),\end{aligned} \tag{3.153}$$

where $P(\boldsymbol{\psi}) = (1 + \tfrac{1}{4}\boldsymbol{\psi} \cdot \boldsymbol{\psi})^{-1}$ (Hamel, 1967). Thus

$$\boldsymbol{\omega}^* = -\tfrac{1}{2}\boldsymbol{\psi} \cdot \boldsymbol{\psi}^* P^2(\boldsymbol{\psi}^* - \tfrac{1}{2}\boldsymbol{\psi} \times \boldsymbol{\psi}^*) + P(\boldsymbol{\psi}^{**} - \tfrac{1}{2}\boldsymbol{\psi} \times \boldsymbol{\psi}^{**}). \tag{3.154}$$

When Eq. (3.154) is substituted into Eq. (3.152), and the resulting equation is substituted into Eq. (3.150), we obtain, upon introducing the component form of $\mathbf{E}_\beta, \boldsymbol{\psi}$, and $\mathbf{N}^\alpha$, three equations in the unknowns $\bar{N}^{\alpha\beta}, \bar{N}, Q^\alpha, \gamma_{\alpha\beta}, \theta_*, E_\alpha, \psi_\alpha$, and ψ.

In classical theory, we take $\mathbf{l} = \mathbf{0}$ and $I = E_\alpha = \theta_* = 0$ and use Eqs. (3.147) and (3.148) to eliminate Q^α and $\bar{N}$. Furthermore, if we use the strain–stress relation (3.133) to express $\gamma_{\alpha\beta}$ in terms of $\bar{N}^{\alpha\beta}$, then Eq. (3.150) combined with the compatibility conditions (3.145) and (3.146) leaves us with six equations in the six unknowns $\bar{N}^{\alpha\beta}, \psi^\alpha$, and ψ. The equations are too complicated to write out. To recover the beamshell equations of motion, Eqs. (2.93) and (2.94), we take $\sigma^1 = z$, $\sigma^2 = \sigma$, $\mathbf{N}^1 = \mathbf{0}$, $\mathbf{N}^2 = N\mathbf{A}_2$, and $\boldsymbol{\psi} = 2\mathbf{k}\tan(\beta/2)$. In this case the compatibility equations are satisfied identically.

3. Intrinsic Form

Here we take the field equations in a form that makes no reference to displacements or rotations.

a. Static Equations

Based on the work of John (1965a,b) and Danielson (1970), Koiter and Simmonds (1973) developed a set of static canonical intrinsic equations for edge-loaded shells. Later, Koiter (1980) introduced the surface load

$$\mathbf{p} = p^\alpha \mathbf{y}_\alpha + p\bar{\mathbf{n}} \tag{3.155}$$

and, extending the error estimates of Berger (1973) for shells of variable thickness with surface loads, gave these equations the following form:

$$An^\alpha_\alpha|^\beta_\beta - b^\alpha_\beta \rho^\beta_\alpha + b^\alpha_\alpha \rho^\beta_\beta - \tfrac{1}{2}\rho^\alpha_\beta \rho^\beta_\alpha + \tfrac{1}{2}\rho^\alpha_\alpha \rho^\beta_\beta + (1+v)Ap^\alpha|_\alpha = O(\eta\theta^2/\lambda^2), \tag{3.156}$$

$$\rho^\beta_\alpha|_\beta - \rho^\beta_\beta|_\alpha - \tfrac{1}{2}(1+v)A(b^\lambda_\alpha n^\beta_\lambda - b^\lambda_\beta n^\beta_\alpha)|_\beta + Ab^\beta_\alpha n^\lambda_\lambda|_\beta + (1+v)A(\rho^\beta_\alpha n^\lambda_\lambda|_\beta + \rho^\lambda_\beta n^\beta_{\beta|\alpha})$$
$$- vA\rho^\lambda_\alpha n^\beta_\beta|_\alpha + 2(1+v)A(b^\beta_\alpha + \rho^\alpha_\beta)p_\beta = O(\eta\theta^4/k\lambda), \tag{3.157}$$

$$D\rho^\alpha_\alpha|^\beta_\beta - (b^\alpha_\beta + \rho^\alpha_\beta)n^\beta_\alpha - p = O(Eh^2\eta\theta^2/\lambda^2), \tag{3.158}$$

$$n^\beta_\alpha|_\beta + \tfrac{1}{2}(1-v)D(b^\lambda_\alpha \rho^\beta_\lambda - b^\beta_\alpha \rho^\lambda_\alpha)|_\beta + Db^\beta_\alpha \rho^\lambda_\lambda|_\beta + D(\rho^\beta_\alpha \rho^\lambda_\lambda - \tfrac{1}{2}\delta^\beta_\alpha \rho^\kappa_\kappa \rho^\lambda_\lambda)|_\beta$$
$$+ 2A(n^\lambda_\alpha n^\beta_\lambda)|_\beta - \tfrac{1}{2}A[(1-v)n^\kappa_\lambda n^\lambda_\kappa + vn^\kappa_\kappa n^\lambda_\lambda]|_\alpha + (1-2vAn^\lambda_\lambda)p_\alpha$$
$$+ 2(1+v)An^\lambda_\alpha p_\lambda = O(Eh\eta\theta^4/\lambda). \tag{3.159}$$

Here, first in Koiter's notation and then in ours [see Eqs. (3.103), (3.110), (3.124), and (3.130)],

$$n^{\alpha\beta} = (\bar{a}/a)^{1/2}[\bar{n}^{\alpha\beta} + \text{sym}(\bar{\bar{b}}^\alpha_\kappa \bar{m}^{\kappa\beta})]$$
$$= \tilde{N}^{\alpha\beta} + \eta^{-1}\text{sym}(T^\alpha_\kappa \bar{M}^{\kappa\beta}), \qquad \bar{M}^{\alpha\beta} = \text{sym}(M^{\alpha\beta}), \tag{3.160}$$

$$\rho_{\alpha\beta} = \bar{b}_{\alpha\beta} - b_{\alpha\beta} - \text{sym}(b^\kappa_\alpha L_{\kappa\beta})$$
$$= \eta^{-1/2}k_{\alpha\beta} + (\eta^{-1/2} - 1)\text{sym}(b^\kappa_\alpha L_{\kappa\beta})$$
$$+ [\eta^{-1/2}(1+L^\gamma_\gamma) - 1]b_{\alpha\beta} \tag{3.161}$$

and all indices are raised or lowered with respect to the metric coefficients $a_{\alpha\beta}$ of the undeformed reference surface. When linearized, $\rho_{\alpha\beta}$, like Budiansky's $k_{\alpha\beta}$, reduces to the bending strain of the Sanders–Koiter linear shell theory. The advantages of Eqs. (3.156)–(3.159), as well as a detailed discussion of their derivation, may be found in Koiter and Simmonds (1973) and Koiter (1980). Associated boundary conditions are derived and discussed in Danielson (1970) and Pietrazkiewicz (1980a,b).

b. Dynamic Equations

The dynamic intrinsic approach considers the *internal* geometry of the evolving surface y, as represented by its metric and curvature coefficients $\bar{a}_{\alpha\beta}$ and $\bar{b}_{\alpha\beta}$, as the object of shell analysis. It relies on the fact that, at least in a Kirchhoff motion, the internal geometry of y completely determines the state of stress and deformation, modulo a rigid-body motion.

In intrinsic *statics* we added three compatibility equations to the equilibrium equations to obtain a complete set of field equations. In intrinsic *dynamics* we must eliminate the inerita term $m\dot{w}$ from the equation conservation of linear momentum in favor of time derivatives of $\bar{a}_{\alpha\beta}$ and $\bar{b}_{\alpha\beta}$. This procedure will be summarized in this section. For more details, see Libai (1981). Henceforth overbars indicate quantities associated with the deformed reference surface.

In the sequel, we assume Kirchhoff motion and consider only shells whose evolving reference surfaces y have nonzero Gaussian curvature $\bar{K}$. For simplicity, the rotary inertia $I\dot{\omega}$ and the external couple l will be dropped. This is consistent with the errors implied by the Kirchhoff hypothesis.

Since we focus on the geometry of y, all geometry and tensor operations in the field equations are now to be referred to y and the material coordinate system it carries with it. The initial reference surface $\hat{\xi}(\sigma^\alpha) = \hat{y}(\sigma^\alpha, 0)$ will enter the formulation indirectly through the constitutive laws (which essentially involve increments of the metric and curvature coefficients from the reference state).

i. Intrinsic Longitudinal Rod Dynamics. To clarify some of the notions in intrinsic dynamics, we consider, briefly, a simple example.

Let $y = \hat{y}(\xi, t)$ be the time-dependent position of a material point of a rod, where ξ is its reference (unstressed) position. Let $u = y - \xi$ be the displacement, ε the strain, N the longitudinal force, A the cross-sectional area, ρ the mass per unit initial volume, and p the longitudinal load per unit initial volume. The the rod equations are

$$\text{motion: } N_{,\xi} = \rho A y_{,tt} - p \tag{3.162}$$

$$\text{kinematics: } \varepsilon = y_{,\xi} - 1 = u_{,\xi} \tag{3.163}$$

$$\text{constitutive: } \varepsilon = N/EA \tag{3.164}$$

$$\text{initial conditions: } \hat{y}(\xi,0), \hat{y}_{,t}(\xi,0) \text{ given} \tag{3.165}$$

$$\text{boundary conditions: } N \text{ or } y \text{ given at } \xi = 0, L \quad \text{ for all } t \geq 0, \tag{3.166}$$

where L is the initial length of the rod.

In the "displacement form" we use u as unknown and eliminate N and ε to obtain the equation of motion

$$u_{,\xi\xi} = (\rho/E)u_{,tt} - p/EA, \tag{3.167}$$

with u and $u_{,t}$ given at $t = 0$ and u or $u_{,\xi}$ given at $\xi = 0, L$ for all $t \geq 0$.

In the "intrinsic form" we use the metric coefficient as represented by ε as unknown. To do so, the equation of motion must be first differentiated with respect to ξ, yielding

$$\varepsilon_{,\xi\xi} = (\rho/E)\varepsilon_{,tt} - (1/EA)p_{,\xi}. \tag{3.168}$$

The initial conditions, ε, $\varepsilon_{,t}$ given at $t = 0$, are also obtained by differentiation. The kinematic boundary conditions need elaboration.

If y is given at $\xi = 0, L$, so is $y_{,tt}$. This implies that $\varepsilon_{,\xi} = (\rho/E)y_{,tt} - p/EA$ is known at the ends. Hence the boundary conditions become ε or $\varepsilon_{,\xi}$ given at $\xi = 0, L$.

We make note of the following: (1) We need a spatial differentiation for the derivation of the intrinsic equations; (2) the kinematic conditions transform into a condition on the *derivative* of the metric coefficient; and (3) spatially constant loads drop out of the field equations, but reappear in the boundary conditions. We now return to our main problem.

ii. Preliminary Kinematics. We start with the *velocity gradient tensor* whose covariant components $d_{\alpha\beta}$ have half of the "metric" time rate for their symmetric part:

$$d_{\alpha\beta} = \mathbf{y}_\alpha \cdot \mathbf{w}_{,\beta} = \tfrac{1}{2}\bar{a}_{\alpha\beta} + \bar{\varepsilon}_{\alpha\beta}\Omega. \tag{3.169}$$

The $d_{\alpha\beta}$ (or $\dot{\bar{a}}_{\alpha\beta}$ and Ω) are the basic kinematic unknowns. For $\bar{K} \neq 0$, Eq. (3.169) has an inverse, obtained by forming from it $d_{\alpha\beta;\gamma} = \bar{b}_{\alpha\gamma}\bar{\mathbf{n}} \cdot \mathbf{w}_{,\beta} + \mathbf{y}_\alpha \cdot \mathbf{w}_{;\beta\gamma}$. Multiplying by $\bar{\varepsilon}^{\beta\gamma}$ and extricating $\bar{\mathbf{n}} \cdot \mathbf{w}_{,\beta}$, we obtain

$$\mathbf{w}_{,\beta} = d_{\alpha\beta}\mathbf{y}^\alpha - \bar{K}^{-1}\bar{b}_{\alpha\beta}\hat{d}^{\alpha\gamma}_{;\gamma}\bar{\mathbf{n}}. \tag{3.170}$$

In the above, the hat ($\hat{}$) notation now denotes the raising of indices with respect to the contravariant components of the permutation tensor of $\mathbf{y}$, the semicolon ($;$) denotes covariant differentiation based on $\bar{a}_{\alpha\beta}$, and $\bar{\mathbf{n}}$ is the unit normal to $\mathbf{y}$. The time rates of the deformed metric and curvature coefficients can be expressed in terms of $d_{\alpha\beta}$ via Eqs. (3.169) and (3.170) as

$$\dot{\bar{a}}_{\alpha\beta} = d_{\alpha\beta} + d_{\beta\alpha},$$
$$\dot{\bar{b}}_{\alpha\beta} = \bar{\mathbf{n}} \cdot \dot{\mathbf{y}}_{\alpha;\beta} = \bar{\mathbf{n}} \cdot \mathbf{w}_{;\alpha\beta} = -(\bar{K}^{-1}\bar{b}_{\alpha\lambda}\hat{d}^{\alpha\lambda}_{;\gamma})_{;\beta} + b^\gamma_\beta d_{\gamma\alpha}. \tag{3.171}$$

Introduction of Eq. (3.169) into Eq. (3.171) gives the curvature rates in terms of the metric rates and Ω, which justifies our calling Ω the *curvature rate function*. Note from Eq. (3.100) that $-\Omega$ is just the component of $\boldsymbol{\tau}$ along $\bar{\mathbf{n}}$.

To assure the existence of $\mathbf{w}$ in Eq. (3.170), we must satisfy the compatibility condition $\varepsilon^{\alpha\beta}\mathbf{w}_{;\alpha\beta} = \mathbf{0}$. The $\mathbf{y}_\alpha$ components of this equation were used in deriving Eq. (3.170); the normal component is

$$\bar{\varepsilon}^{\alpha\beta}\dot{\bar{b}}_{\alpha\beta} = \bar{\varepsilon}^{\alpha\beta}[-\bar{K}^{-1}\bar{b}_{\alpha\gamma}\hat{d}^{\lambda\gamma}_{;\gamma})_{;\beta} + b^\gamma_\beta d_{\gamma\alpha}] = 0. \tag{3.172}$$

The *acceleration gradient tensor* with covariant components $f_{\alpha\beta}$ is defined analogously to the velocity gradient tensor. Simple vector operations relate $f_{\alpha\beta}$ to $(\bar{a}_{\alpha\beta}, \Omega)$ and their time derivatives:

$$f_{\alpha\beta} = \mathbf{y}_\alpha \cdot \dot{\mathbf{w}}_{,\beta} = \dot{d}_{\alpha\beta} - C_{\alpha\beta} = \tfrac{1}{2}\ddot{\bar{a}}_{\alpha\beta} + (\bar{\varepsilon}_{\alpha\beta}\Omega)\dot{} - C_{\alpha\beta}, \tag{3.173}$$

where

$$C_{\alpha\beta} = \mathbf{w}_{,\alpha} \cdot \mathbf{w}_{,\beta} = d^\gamma_\alpha d_{\gamma\beta} + \bar{K}^{-2}\bar{b}_{\gamma\alpha}\bar{b}_{\lambda\beta}\hat{d}^{\gamma\kappa}_{;\kappa}d^{\lambda\mu}_{;\mu}. \tag{3.174}$$

iii. Equations of Motion. For the remainder of this section we consider Eq. (3.32) to be referred to $\mathbf{y}$, and use this equation to express $f_{\alpha\beta}$ in terms of the stress resultant $\mathbf{N}^\alpha$, the loading $\bar{\mathbf{p}}$, and the (now variable) mass $\bar{m}$ as follows:

$$f_{\alpha\beta} = \mathbf{y}_\alpha \cdot \{\bar{m}^{-1}[(\eta^{-1/2}\mathbf{N}^\gamma)_{;\gamma} + \bar{\mathbf{p}}]\}_{;\beta}. \tag{3.175}$$

In terms of the symmetrical stress resultant and couple components $\bar{n}^{\alpha\beta}$ and $\bar{m}^{\alpha\beta}$ of Koiter (1966, Section 6),

$$\eta^{-1/2}\mathbf{N}^\alpha = (\bar{n}^{\alpha\beta} + \bar{b}^\beta_\lambda \bar{m}^{\alpha\lambda})\mathbf{y}_\beta - \bar{m}^{\beta\alpha}_{;\beta}\bar{\mathbf{n}}, \tag{3.176}$$

$$f_{\alpha\beta} = [\bar{m}^{-1}(\bar{n}^\gamma_{\alpha;\gamma} + \bar{b}_{\alpha\lambda;\gamma}\bar{m}^{\gamma\lambda} + 2\bar{b}_{\alpha\lambda}\bar{m}^{\gamma\lambda}_{;\gamma})]_{;\beta}$$
$$- \bar{m}^{-1}\bar{b}_{\alpha\beta}[\bar{n}^{\gamma\lambda} + \bar{b}^\lambda_\kappa \bar{m}^{\gamma\kappa})\bar{b}_{\gamma\lambda} - \bar{m}^{\gamma\lambda}_{;\gamma\lambda}] + p_{\alpha\beta}, \tag{3.177}$$

with

$$p_{\alpha\beta} = \mathbf{y}_\alpha \cdot (\bar{m}^{-1}\mathbf{p})_{,\beta} = (\bar{m}^{-1}p_\alpha)_{;\beta} - \bar{m}^{-1}\bar{b}_{\alpha\beta}p, \tag{3.178}$$

the components of **p** being given by Eq. (3.155).

The final equations of motion, obtained by combining Eqs. (3.173) and (3.175) [or (3.177)] and separating symmetric and skew parts, are as follows:

$$\ddot{\bar{a}}_{\alpha\beta} = f_{\alpha\beta} + f_{\beta\alpha} + 2C_{\alpha\beta}, \tag{3.179}$$

$$\frac{1}{\sqrt{\bar{a}}}(\sqrt{\bar{a}}\Omega)^\cdot = \bar{\varepsilon}^{\alpha\beta}f_{\alpha\beta}. \tag{3.180}$$

To complete the system, we append constitutive relations of the form

$$\bar{n}^{\alpha\beta} = \bar{n}^{\alpha\beta}(\bar{a}_{\gamma\lambda},\bar{b}_{\gamma\lambda}), \qquad \bar{m}^{\alpha\beta} = \bar{m}^{\alpha\beta}(\bar{a}_{\gamma\lambda},\bar{b}_{\gamma\lambda}) \tag{3.181}$$

and the mass conservation law

$$(\sqrt{\bar{a}}\bar{m})^\cdot = 0. \tag{3.182}$$

Equations (3.171), (3.179), and (3.180), together with the auxiliary set (3.174), (3.177), (3.181), and (3.182), constitute a full set of field equations for $\bar{a}_{\alpha\beta}$ and $\bar{b}_{\alpha\beta}$. The equivalence of the set to the original equations is discussed in Libai (1981). It is shown there that Eq. (3.180) can be replaced with the compatibility condition (3.172). This may be useful for some applications. Notable is inextensional motion for which $\dot{\bar{a}}_{\alpha\beta} = \ddot{\bar{a}}_{\alpha\beta} = 0$. Then Eq. (3.172) takes on the simple form

$$\hat{\bar{b}}^{\alpha\beta}(\bar{K}^{-1}\Omega_{,\alpha})_{;\beta} + \bar{b}^\alpha_\alpha \Omega = 0. \tag{3.183}$$

iv. *Initial and Boundary Conditions.* Intrinsic dynamics is a convenient recasting of *existing* shell theories. Hence its boundary and initial conditions are those of the underlying theory. The main problem is the representation of these conditions in terms of the intrinsic unknowns.

The initial conditions are to specify the initial position $\hat{\mathbf{y}}(\sigma^\alpha, 0) \equiv \mathbf{y}_0$ and initial velocity $\hat{\mathbf{w}}(\sigma^\alpha, 0)$. Using the six components of these vectors, we can calculate directly the 10 initial values of $\bar{a}_{\alpha\beta}$, $\bar{b}_{\alpha\beta}$, $\dot{\bar{a}}_{\alpha\beta}$, and Ω via Eq. (3.169), if we recall the definitions of $\bar{a}_{\alpha\beta}$ and $\bar{b}_{\alpha\beta}$ in terms of $\mathbf{y}_0$ and its derivatives. Thus all the required initial values are available, and compatibility is satisfied initially.

Stress boundary conditions present no problems, since they are exactly those of the underlying theory. The only difficulty is encountered in the recasting of the kinematic boundary conditions (as we saw in the rod example).

Linear kinematic constraints specify **y** on the boundary; however, **ÿ** is also known there (from differentiating **y** with respect to t), so that the conservation of linear momentum (3.32) *applied at the boundary* yields three boundary conditions on the covariant derivatives of the components of $\mathbf{N}^\alpha$, such as those in Eq. (3.176).

Rotational constraints can be transformed into prescribed values of the component $\tau_{\bar{s}}$ of $\boldsymbol{\tau}$ along the unit tangent $\lambda s\, d\sigma^\alpha/ds\, \mathbf{y}_\alpha$ to $\partial \bar{S}$, where λs is the extension of the edge of the deformed reference surface. We have, with the aid of Eq. (3.102),

$$\dot{\bar{\mathbf{n}}} = -(\bar{\mathbf{n}} \cdot \mathbf{w}_{,\alpha})\mathbf{y}^\alpha = \bar{K}^{-1}\bar{b}_{\gamma\alpha}\hat{d}^{\gamma\lambda}_{;\lambda}\bar{\mathbf{y}}^\alpha = \boldsymbol{\tau} \times \bar{\mathbf{n}}. \tag{3.184}$$

Hence

$$\tau_{\bar{s}} = (\bar{K}\lambda s)^{-1}\bar{\varepsilon}_{\beta\alpha}\bar{b}^\beta_\gamma\hat{d}^{\gamma\lambda}_{;\lambda}\, d\sigma^\alpha/ds. \tag{3.185}$$

We have thus cast all of the kinematic constraints in terms of our intrinsic unknowns.

4. Velocity Form

The velocity form occupies a position between the displacement and the dynamic intrinsic formulations. As in the latter, the deforming surface $\mathbf{y}$ is taken as the reference surface to which all quantities and tensor operations are referred. This is an example of an "updated Lagrangian" approach to shell dynamics.

For simplicity we consider Kirchhoff motion only. The components of the velocity vector $\mathbf{w} = \dot{\mathbf{y}}$ are

$$w_\alpha = \dot{\mathbf{y}} \cdot \mathbf{y}_\alpha, \qquad w^\alpha = \dot{\mathbf{y}} \cdot \mathbf{y}^\alpha, \qquad w_n = \dot{\mathbf{y}} \cdot \bar{\mathbf{n}}, \tag{3.186}$$

where, as before, $\bar{\mathbf{n}}$ is the unit normal to $\mathbf{y}$.

Expressions for the components of the velocity gradients $\dot{\mathbf{y}}_\alpha$ follow from differentiation of Eq. (3.186):

$$d_{\alpha\beta} = \mathbf{y}_\alpha \cdot \dot{\mathbf{y}}_\beta = w_{\alpha;\beta} - \bar{b}_{\alpha\beta}w_n, \tag{3.187}$$

$$\tilde{\omega}_\alpha = \bar{\mathbf{n}} \cdot \dot{\mathbf{y}}_\beta = w_{n,\alpha} + \bar{b}_{\alpha\beta}w^\beta. \tag{3.188}$$

The time rates of the components of the metric and curvature tensors follow immediately from the above and Eq. (3.171) as

$$\dot{\bar{a}}_{\alpha\beta} = d_{\alpha\beta} + d_{\beta\alpha}, \qquad \dot{\bar{b}}_{\alpha\beta} = \tilde{\omega}_{\alpha;\beta} + d^\gamma_\alpha \bar{b}_{\beta\gamma}. \tag{3.189}$$

To complete the kinematics, we get the components of the acceleration vector by time differentiation of Eq. (3.186) and use of Eqs. (3.187) and (3.188):

$$\ddot{\mathbf{y}} \cdot \mathbf{y}^\alpha = \dot{w}^\alpha + d^\alpha_\beta w^\beta - \tilde{\omega}^\alpha w_n, \qquad \ddot{\mathbf{y}} \cdot \bar{\mathbf{n}} = \dot{w}_n + \tilde{\omega}_\alpha w^\alpha.$$

The accelerations form the right sides of the equations of motion. The left sides are the static terms that can, for example, be expressed in terms of symmetrical stress resultants (Koiter, 1966). Combined, the result is

$$\dot{w}^\alpha = \bar{m}^{-1}[\bar{n}^{\beta\alpha}_{;\beta} + \bar{b}^\alpha_{\lambda;\beta}\bar{m}^{\lambda\beta} + 2\bar{b}^\alpha_\lambda \bar{m}^{\beta\lambda}_{;\beta} + \bar{p}^\alpha] - d^\alpha_\beta w^\beta + \tilde{\omega}^\alpha w_n \quad (3.190)$$

$$\dot{w}_n = \bar{m}^{-1}[\bar{n}^{\alpha\beta}\bar{b}_{\alpha\beta} + \bar{b}^\lambda_\beta \bar{b}_{\alpha\lambda}\bar{m}^{\alpha\beta} - \bar{m}^{\alpha\beta}_{;\alpha\beta} + \bar{p}] - \tilde{\omega}^\alpha w_\alpha. \quad (3.191)$$

Here $\bar{p}^\alpha$, $\bar{p}$, $\bar{m}$ are loading components and mass, respectively, per unit area of **y**.

Equations (3.187)–(3.191) together with the constitutive relations and mass conservation of Eqs. (3.181) and (3.182) are the field equations of the velocity form.

Boundary Conditions. These are acceptable if they can be related to the basic triad $\{\bar{\mathbf{n}}, \bar{\mathbf{v}}, \bar{\boldsymbol{\tau}} = \bar{\mathbf{n}} \times \bar{\mathbf{v}}\}$ of the evolving boundary curve $\partial \bar{S}$, as is the case in many physical applications. Kinematic data, which consist of the components $w^\alpha \bar{v}_\alpha$, $w^\alpha \bar{\tau}_\alpha$, w_n of the velocity vector and the rotational velocity $\tilde{\omega}^\alpha \bar{v}_\alpha$ around the unit tangents to $\partial \bar{S}$, are easy to specify. Stress boundary data are again those of the underlying theory. Their duality with the corresponding kinematic quantities is assured by the mechanical power (or virtual work) identities referred to **y**.

IV. Nonlinear Plate Theory

The characteristic feature of a shell is its curvature. Thus linear theory predicts, typically, that a plate carries a load (say, a uniform external pressure) quite differently than a shell does. In nonlinear theory, the sharp distinction between plate and shell behavior blurs, because a deformed plate can bend considerably. Our aim in this section is to show, in outline, what further simplifications of the static intrinsic equations (3.156)–(3.159) are possible when $b_{\alpha\beta}$ is zero. In Section IV,A we discuss the *extended von Karman equations* derived by Simmonds (1979b) and show that they admit an additional simplification. These equations are valid for all types of deformations, except those in which the angles of rotation are $O(1)$ *and* the spatial gradients are very large, as when the edge of a plate is curled over. The extended von Karman equations hold in many cases where the conventional von Karman equations fail. They are exact for inextensional deformation, although in this case, as discussed in Section IV,B there is a better, geometrical approach.

A. The (Simplified) Extended von Karman Equations

The von Karman plate equations are based on the assumption that fibers rotate through angles whose squares can be neglected compared to unity. Simmonds (1979b) attempted to remove this deficiency. The result was a set of coupled, von Karman-like equations for a stress function F, a curvature function W, and an auxiliary, symmetric tensor $\mathbf{P}$. The divergence of $\mathbf{P}$ is equal to a certain vector-valued function of the derivatives of F and W. In the von Karman approximation, $\mathbf{P}$ may be set to zero and W identified with the midplane normal displacement.

Simmonds showed that the extended von Karman equations accurately described three special highly nonlinear problems: the shape of the *elastica*, the deformation of an infinite strip under an axial tension and twist, and the buckling under axial compression of an infinite circular cylindrical shell, made by bending and butt welding the two sides of an infinite strip. These are benchmark problems in the sense that their governing equations may be formulated or verified directly, *without* starting from the nonlinear shell equations (3.156)–(3.159) on which the von Karman equations are based.

1. Simplifications

First, it will be shown that the compatibility equation (3.10) of Simmonds (1979b),

$$A[\Delta\Delta F - (1+v)\{W|^\alpha_\beta \hat{W}|^\beta_\lambda \hat{F}|^\lambda_\alpha\} + W|^\alpha_\beta P^\beta_\alpha] + \tfrac{1}{2}W|^\alpha_\beta \hat{W}|^\beta_\alpha = O(\eta\theta^2/\lambda^2), \quad (4.1)$$

can be simplified by ignoring the term in braces. Here

$$\hat{W}|^\alpha_\beta = \delta^\alpha_\beta \Delta W - W|^\alpha_\beta, \quad (4.2)$$

the P^α_β are the mixed components of the auxiliary tensor $\mathbf{P}$, and Δ is the Laplacian. The error term on the right-hand side of Eq. (4.1) comes from Eq. (3.156).

Using Eq. (4.2), we set

$$W|^\alpha_\beta \hat{W}|^\beta_\lambda \hat{F}|^\lambda_\alpha = \tfrac{1}{2}W|^\alpha_\beta \hat{W}|^\beta_\lambda \hat{F}|^\lambda_\alpha + \tfrac{1}{2}W|^\alpha_\beta \hat{W}|^\beta_\lambda (\delta^\lambda_\alpha \Delta F - F|^\lambda_\alpha)$$
$$= \tfrac{1}{2}W|^\alpha_\beta \hat{W}^\beta_\lambda (\hat{F}|^\lambda_\alpha - F|^\lambda_\alpha) + \tfrac{1}{2}W|^\alpha_\beta \hat{W}|^\beta_\alpha \Delta F. \quad (4.3)$$

But by Eq. (4.2) and the Cayley–Hamilton theorem,

$$W|^\alpha_\beta \hat{W}|^\beta_\lambda = W|^\alpha_\lambda \Delta W - W|^\alpha_\beta W|^\beta_\lambda = \delta^\alpha_\lambda \det(W|^\beta_\gamma). \quad (4.4)$$

Thus, the first term in the second line of Eq. (4.3) reduces to

$$\tfrac{1}{2}\det(W|^\beta_\gamma)(\hat{F}|^\alpha_\alpha - F|^\alpha_\alpha) = 0, \quad (4.5)$$

and Eq. (4.1) may be written as

$$A(\Delta\Delta F + W|^\beta_\alpha P^\beta_\beta) + \tfrac{1}{2}W|^\alpha_\beta \hat{W}^\beta_\alpha \{1 - [(1+v)A\,\Delta F]\} = O(\eta\theta^2/\lambda^2). \quad (4.6)$$

But from Eq. (4.1), $\tfrac{1}{2}\hat{W}|^\alpha_\beta W|^\beta_\alpha = O(A\,\Delta\Delta F) = O(\eta/\lambda^2)$. Hence the term in square brackets in Eq. (4.6) is $O(\eta^2/\lambda^2) = O(\eta\theta^2/\lambda^2)$ and is therefore negligible, being of the same order as the error term.

Dropping this error term and introducing the notation

$$\langle W, F \rangle = W|_\beta^\alpha \hat{F}|_\alpha^\beta, \tag{4.7}$$

we have, from Eq. (4.6) and Eqs. (3.8) and (3.11) of Simmonds (1979b), the following (simplified) *extended von Karman equations* for edge-loaded plates:

$$A(\Delta\Delta F + W|_\beta^\alpha P_\alpha^\beta) + \tfrac{1}{2}\langle W, W \rangle = 0, \tag{4.8}$$

$$D\{\Delta\Delta W + [\tfrac{1}{2}(\Delta W)^2 - \langle W, W \rangle]\Delta W\} - \langle W, F \rangle = 0, \tag{4.9}$$

$$[P^{\alpha\beta} + \tfrac{1}{2}(1+\nu)\varepsilon^{\alpha\beta}\varepsilon_{\gamma\mu}F|_\lambda^\gamma W|^{\lambda\mu}]|_\beta = \Delta W \Delta F|^\alpha. \tag{4.10}$$

Equations for stress resultants in terms of F and W and bending strains in terms of W, F, and **P** are given by Eqs. (3.6) and (3.9) of Simmonds (1979b). To within the errors inherent in linear stress–strain relations, the stress couples are given by

$$M_\beta^\alpha = D[(1-\nu)W|_\beta^\alpha + \nu\delta_\beta^\alpha \Delta W]. \tag{4.11}$$

For ease (and in hope) of future applications, we will next list the extended von Karman equations in Cartesian and polar coordinates.

2. Cartesian Coordinates (x, y)

In this system, covariant derivatives reduce to partial derivatives, indicated by a comma. Thus

$$\Delta W = W_{,xx} + W_{,yy}, \quad \langle W, F \rangle = W_{,xx}F_{,yy} + W_{,yy}F_{,xx} - 2W_{,xy}F_{,xy}, \tag{4.12}$$

and, with $P_1^1 = P_{xx}$, $P_2^1 = P_1^2 = P_{xy}$, $P_2^2 = P_{yy}$, etc.,

$$W|_\beta^\alpha P_\alpha^\beta = W_{,xx}P_{xx} + 2W_{,xy}P_{,xy} + W_{,yy}P_{yy}. \tag{4.13}$$

The auxiliary equation (4.10) and the expressions for the stress resultants, bending strains, and stress couples take the expanded forms

$$P_{xx,x} + \{P_{xy} + \tfrac{1}{2}(1+\nu)[W_{,xy}(F_{,xx} - F_{,yy}) - F_{,xy}(W_{,xx} - W_{,yy})]\}_{,y} = \Delta W \Delta F_{,x}, \tag{4.14}$$

$$\{P_{xy} - \tfrac{1}{2}(1+\nu)[W_{,xy}(F_{,xx} - F_{,yy}) - F_{,xy}(W_{,xx} - W_{,yy})]\}_{,y} + P_{yy,y} = \Delta W \Delta F_{,y}, \tag{4.15}$$

$$\begin{aligned} N_{xx} = F_{,yy} &+ \tfrac{1}{2}D(W_{,yy} - W_{,xx})\Delta W \\ &+ A\{\tfrac{1}{2}[(\Delta F)^2 + (1+\nu)\langle F, F \rangle] - 2F_{,yy}\Delta F\}, \end{aligned} \tag{4.16}$$

$$\begin{aligned} N_{yy} = F_{,xx} &+ \tfrac{1}{2}D(W_{,xx} - W_{,yy})\Delta W \\ &+ A\{\tfrac{1}{2}[(\Delta F)^2 + (1+\nu)\langle F, F \rangle] - 2F_{,xx}\Delta F\}, \end{aligned} \tag{4.17}$$

$$N_{xy} = -F_{,xy} - DW_{,xy}\Delta W + 2AF_{,xy}\Delta F, \tag{4.18}$$

$$K_{xx} = W_{,xx} + A[-(1+v)(W_{,xx}F_{,xx} + W_{,xy}F_{,xy}) + P_{yy}], \tag{4.19}$$

$$K_{yy} = W_{,yy} + A[-(1+v)(W_{,yy}F_{,yy} + W_{,xy}F_{,xy}) + P_{xx}], \tag{4.20}$$

$$K_{xy} = W_{,xy} - A[\tfrac{1}{2}(1+v)(\Delta W F_{,xy} + \Delta F W_{,xy}) + P_{xy}], \tag{4.21}$$

$$M_{xx} = D(W_{,xx} + vW_{,yy}), \qquad M_{yy} = D(W_{,yy} + vW_{,xx}), \tag{4.22}$$

$$M_{xy} = D(1-v)W_{,xy}. \tag{4.23}$$

3. Polar Coordinates (r, θ)

In this system the only nonzero Christoffel symbols are $\Gamma^2_{12} = r^{-1}$ and $\Gamma^1_{22} = -r$. Hence

$$W|^1_1 = W_{,rr}, \qquad W|^2_2 = r^{-2}W_{,\theta\theta} + r^{-1}W_{,r}, \tag{4.24}$$

$$W|^1_2 = W_{,r\theta} - r^{-1}W_{,\theta}, \qquad W|^2_1 = r^{-2}W_{,r\theta} - r^{-3}W_{,\theta}, \tag{4.25}$$

which implies that

$$\Delta W = W_{,rr} + r^{-1}W_{,r} + r^{-2}W_{,\theta\theta},$$
$$\langle W, F \rangle = W_{,rr}(r^{-2}F_{,\theta\theta} + r^{-1}F_{,r}) + F_{,rr}(r^{-2}W_{,\theta\theta} + r^{-1}W_{,r}) \tag{4.26}$$
$$- 2(r^{-1}W)_{,r\theta}(r^{-1}F)_{,r\theta}.$$

The physical and mixed components of **P** are related as follows: $P_{rr} = P^1_1$, $P_{r\theta} = r^{-1}P^1_2 = rP^2_1$, $P_{\theta\theta} = P^2_2$. Thus

$$W|^\alpha_\beta P^\beta_\alpha = W_{,rr}P_{rr} + 2(r^{-1}W)_{,r\theta}P_{r\theta} + (r^{-2}W_{,\theta\theta} + r^{-1}W_{,r})P_{\theta\theta}, \tag{4.27}$$

while Eq. (4.10) takes the expanded form

$$(rP_{rr})_{,r} + [P_{r\theta} + \tfrac{1}{2}(1+v)T]_{,\theta} - P_{\theta\theta} = r\Delta W(\Delta F)_{,r},$$
$$P_{\theta\theta,\theta} + r^{-1}\{r^2[P_{r\theta} - \tfrac{1}{2}(1+v)T]\}_{,r} = \Delta W(\Delta F)_{,\theta}, \tag{4.28}$$

where

$$T = (r^{-1}W)_{,r\theta}(F_{,rr} - r^{-1}F_{,r} - r^{-2}F_{,\theta\theta}) - (r^{-1}F)_{,r\theta}(W_{,rr} - r^{-1}W_{,r} - r^{-2}W_{,\theta\theta}).$$

B. Inextensional Deformation of Plates

A check on the approximations introduced in deriving the extended von Karman equations shows that they are exact for inextensional deformation. Formally, we obtain the governing field equations by setting $A = 0$ in Eq. (4.8), which leaves $\tfrac{1}{2}\langle W, W \rangle = 0$. This is the Monge–Ampere equation that has, in Cartesian coordinates, the well-known general solution $W_{,x} = f(W_{,y})$, where f is an arbitrary differentiable function. Rather than follow this course to derive the field equations for inextensional deformation, it is better, as was first shown by Mansfield (1955), to start from the fact that a plane, if deformed smoothly without stretching, must go into a *developable* surface. Aside from the degenerate cases of cones and cylinders, any developable surface can be represented in the form

$$\mathbf{y} = \hat{\mathbf{x}}(\xi) + \eta \hat{\mathbf{u}}(\xi). \tag{4.29}$$

where $\mathbf{x}$ is an involute of the *edge of regression* of the surface (Struik, 1961) and η and $\mathbf{u}$ are, respectively, distance and a unit vector along a generator. Building on earlier work of Mansfield (1955), whose analysis incorporated the same assumptions as the conventional von Karman equations, Simmonds and Libai (1979a) exploited Eq. (4.29) to reduce the *exact* equations of inextensional deformation of a plate to a system of 13 first-order *ordinary* differential equations. The plate is assumed to have two opposite curved edges, each stress free, and two straight edges, one built in and the other subject to a net force and torque of magnitude P and T, respectively. The system of equations includes the following classical Frenet–Serret equations (Struik, 1961) with initial conditions:

$$\mathbf{x}' = \mathbf{t}, \qquad \mathbf{x}(0) = \mathbf{0}, \tag{4.30}$$

$$\mathbf{t}' = \kappa \mathbf{n}, \qquad \mathbf{t}(0) = \mathbf{i}, \tag{4.31}$$

$$\mathbf{n}' = -\kappa \mathbf{t} + \tau \mathbf{b}, \qquad \mathbf{n}(0) = -\mathbf{j} \sin \psi_0 + \mathbf{k} \cos \psi_0, \tag{4.32}$$

$$\mathbf{b}' = -\tau \mathbf{n}, \qquad \mathbf{b}(0) = -(\mathbf{j} \cos \psi_0 + \mathbf{k} \sin \psi_0). \tag{4.33}$$

Here $\{\mathbf{t}, \mathbf{n}, \mathbf{b}\}$ is a triad of orthonormal vectors (the tangent, normal, and binormal vectors) and $\hat{\kappa}(\xi)$ and $\hat{\tau}(\xi)$ are the curvature and torsion of $\hat{\mathbf{x}}(\xi)$. In the initial conditions, ψ_0 is an unknown angle that must be determined as part of the solution. The vector $\mathbf{u}$ in Eq. (4.29) is related to $\mathbf{n}$ and $\mathbf{b}$ by

$$\mathbf{u} = -\mathbf{n} \sin \psi - \mathbf{b} \cos \psi, \tag{4.34}$$

where

$$\psi = \int_0^\xi \tau(x)\, dx + \psi_0. \tag{4.35}$$

Simmonds and Libai (1979a) erroneously took $\psi_0 = 0$. However, aside from their equations (3), (4), (6), and (7), whose correct form is given by Eqs. (4.32)–(4.35), respectively, none of the rest of their other analysis is affected. We also note that $\mathbf{n}$ in their Eq. (6) should be $-\mathbf{n}$, as in Eq. (4.34).

Simmonds and Libai (1979b) have shown, by a change of variables, that a group of 7 independent equations can be split off from their original 13. When nondimensionalized, these seven equations are shown to contain a single parameter $\varepsilon = \max(PL/D, T/D)$, where L is a typical dimension of the undeformed midplane and D is the bending stiffness. As $\varepsilon \to 0$, Mansfield's equations emerge. Remarkably, these seven equations reduce to fewer for special geometries and loads. For a plate of quadrilateral planform under a pure torque acting on and with axis along the (necessarily straight) end, there are only four equations. These were solved numerically in Simmonds and Libai (1979b).

In dynamic problems, Simmonds (1981) has shown, for plates that twist as they bend, that one can obtain a set of 14 equations that are first order in a spatial variable. The inclusion of inertial terms is nontrivial, because the natural set of surface coordinates to use with a dynamically bending inextensible surface is neither Lagrangian nor Eulerian.

Numerical solutions of the exact static or dynamic equations of inextensional deformation for nontrivial problems remain to be constructed. Fortunately, in static problems, we have as a guide solutions of Mansfield's equations for the bending of strips with oblique ends (Mansfield, 1955) and triangular plates under uniform and tip loads (Mansfield and Kleeman, 1955).

V. Static One-Dimensional Strain Fields

In this section we will consider some of the consequences of the answer to the following question: When do the partial differential equations of nonlinear shell theory reduce to *ordinary* differential equations? A necessary geometrical condition is that the reference surface admit a one-dimensional strain field; that is, the deformation must be such that the deformed metric and curvature coefficients $\bar{a}_{\alpha\beta}$ and $\bar{b}_{\alpha\beta}$ are functions of one independent variable only. Simmonds (1979c) has shown that the only such surfaces are general helicoids. These include, as limiting cases, planes, general cylinders, and surfaces of revolution. Whether a one-dimensional strain field is actually obtained depends on the form of the constitutive relations and the external loads. A shell with a one-dimensional strain field is shown in Fig. 6, where an elastically isotropic helical tube of constant thickness but arbitrary cross section is subject to an axial force and torque and a constant internal pressure.

The simplified form that the general equations must assume when a general helicoid suffers a one-dimensional strain are yet to be stated. However, there is extensive literature on three special cases of great practical interest: the bending and extension of general cylinders, the torsionless axisymmetric deformation of shells of revolution, and the pure bending of curved tubes. The first case was studied in Section II; here, in Sections V,A and V,B, we discuss the second two.

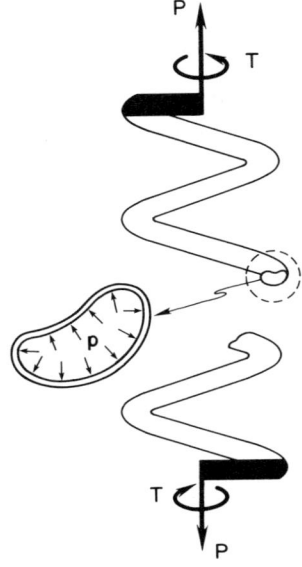

FIG. 6. A helicoidal shell with a one-dimensional strain field.

A. Torsionless Axisymmetric Deformation of Shells of Revolution

The linear theory of shells of revolution has a long history that we make no attempt to trace. The nonlinear theory was pioneered by Reissner (1950), who has continued to develop and refine his equations (1963a, 1969, 1972b). Reissner's equations consist of two coupled second-order ordinary differential equations for the meridional angle of deformation β and a stress function f. We shall now show that a simplified form of Reissner's equations, derived by Koiter (1980) directly from the intrinsic equations (3.156)–(3.159), is a special case of Eqs. (3.145)–(3.148), the "rotation" form of the static field equations discussed in Section III,J,2,a. Koiter showed that his simplified equations, which he dubbed the RMR equations, in honor of H. Reissner, E. Meissner, and E. Reissner, were fully equivalent, in the absence of surface loads, to Reissner's equations of 1963, in the sense that the differences were of the same order as the inherent errors in the intrinsic equations.

Let (r, θ, z) be a set of circular cylindrical coordinates with associated unit base vectors $(\mathbf{e}_r, \mathbf{e}_\theta, \mathbf{k})$, and let

$$\boldsymbol{\xi} = \hat{r}(s)\mathbf{e}_r + \hat{z}(s)\mathbf{k}, \qquad 0 \leq s \leq L, \quad 0 \leq \theta \leq 2\pi, \tag{5.1}$$

describe the reference surface, where s is *meridional* arc length. The reference surface coordinates are $\sigma^1 = s$ and $\sigma^2 = \theta$, and the associated covariant base vectors are

$$\mathbf{a}_1 = \boldsymbol{\xi}_{,s} = \cos\varphi \mathbf{e}_r + \sin\varphi \mathbf{k}, \tag{5.2}$$

$$\mathbf{a}_2 = \boldsymbol{\xi}_{,\theta} = r\mathbf{e}_\theta, \tag{5.3}$$

where

$$\cos\varphi = r' = d\hat{r}(s)/ds, \qquad \sin\varphi = z' = d\hat{z}(s)/ds. \tag{5.4}$$

For a shell of revolution undergoing torsionless axisymmetric deformation, Eq. (3.39) reduces to

$$\mathbf{A}_s \equiv \mathbf{A}_1 = \cos\bar{\varphi}\mathbf{e}_r + \sin\bar{\varphi}\mathbf{k}, \qquad \mathbf{A}_2 = r\mathbf{e}_\theta,$$
$$\mathbf{N} = -\sin\bar{\varphi}\mathbf{e}_r + \cos\bar{\varphi}\mathbf{k}, \qquad \bar{\varphi} = \varphi + \beta, \tag{5.5}$$

and it follows from Eqs. (3.69) and (3.71) that the finite rotation vector $\boldsymbol{\psi}$ takes the one component form

$$\boldsymbol{\psi} = \psi^2 \mathbf{A}_2 = -2\tan(\beta/2)\mathbf{e}_\theta. \tag{5.6}$$

From Eqs. (3.59) and (3.70), we have

$$\mathbf{K}_1 = K_{11}\mathbf{A}^1 \times \mathbf{N} \equiv -K_s \mathbf{e}_\theta, \tag{5.7}$$

$$\mathbf{K}_2 = K_{22}\mathbf{A}^2 \times \mathbf{N} + K_2 \mathbf{N} \equiv r(K_\theta \mathbf{A}_s + K\mathbf{N}), \tag{5.8}$$

where K_s, K_θ, and K are *physical* components given by

$$K_s = \beta', \qquad rK_\theta = \sin\bar{\varphi} - \sin\varphi, \tag{5.9}$$

$$rK = \cos\bar{\varphi} - \cos\varphi. \tag{5.10}$$

One of Reissner's insights was that the equilibrium equations simplify if *force* vectors lying in the rz plane are resolved into horizontal ($\mathbf{e}_r$) and vertical ($\mathbf{k}$) components. Thus we set

$$\mathbf{p} = p_H \mathbf{e}_r + p_V \mathbf{k} \tag{5.11}$$

and introduce the following alternate component forms of the stress resultants:

$$\mathbf{N}^1 = \bar{N}^{11}\mathbf{A}_1 + Q^1\mathbf{N} \equiv N_r \mathbf{A}_s + Q\mathbf{N} \equiv H\mathbf{e}_r + V\mathbf{k}, \tag{5.12}$$

$$\mathbf{N}^2 = \bar{N}^{22}\mathbf{A}_2 \equiv r^{-1} N_\theta \mathbf{e}_\theta. \tag{5.13}$$

Here N_r, N_θ, and Q are physical components. As we remarked after Eq. (3.139), there are other forms of this equation that satisfy the force equilibrium equations identically. For the case at hand, we follow Reissner and take

$$\mathbf{F} = -f\mathbf{e}_\theta, \qquad r\mathbf{P}^1 = -\left(\int rp_V \, ds\right)\mathbf{k}, \qquad \mathbf{P}^2 = p_H \mathbf{e}_\theta. \qquad (5.14)$$

It then follows from Eqs. (3.139) and (5.12)–(5.14) that

$$rH = f, \qquad rV = -\int rp_V \, ds, \qquad N_\theta = f' + rp_H, \qquad (5.15)$$

$$N_r = H \cos \bar{\varphi} - V \sin \bar{\varphi}, \qquad Q = -H \sin \bar{\varphi} + V \cos \bar{\varphi}. \qquad (5.16)$$

When these expressions are substituted into Eqs. (3.145)–(3.148), we see that Eq. (3.145) for $\alpha = 2$, Eqs. (3.146) and (3.147) for $\alpha = 2$, and Eq. (3.148) are satisfied trivially, whereas the remaining equations reduce to

$$rA(f' + r^{-1}f \cos \bar{\varphi} - V \sin \bar{\varphi} + rp_H)' = \cos \bar{\varphi} - \cos \varphi, \qquad (5.17)$$

$$rD[\beta' + r^{-1}(\sin \bar{\varphi} - \sin \varphi)]' + V \cos \bar{\varphi} = f \sin \bar{\varphi}. \qquad (5.18)$$

These equations, less the load terms p_H and V, are precisely Eqs. (5.6) of Koiter (1980). Remarkably, in contrast to Reissner's equations (1963), *they are completely free of Poisson's ratio*. In the linear theory, a justification of these simplified equations was given by Simmonds (1975), who based his arguments on the structure of the Reissner–Meissner equations themselves.

B. Pure Bending of Curved Tubes

We consider a toroid formed by rotating a closed meridional curve, $r = \rho(s), z = \zeta(s), 0 \le s \le L$, around the z axis of a fixed cylindrical coordinate system (r, θ, z). Here s is meridional arc length and $0 \le \theta \le \Theta \le 2\pi$. A curved tube is a shell having such a toroid as its reference surface. A straight tube is a limiting case of a curved tube and will be included in the analysis.

The edges of the shell, $\theta = 0$ and Θ, are loaded in such a way that the following conditions are met:

1. The stresses and strains in the shell (but not necessarily displacements or rotations) are independent of θ.
2. The only nonzero resultant of the total stress distribution over each of the edges is a couple $M\mathbf{k}$.

This is a statement of the problem of the in-plane pure bending of a tube by end couples. Uniqueness is an open question except in the linear case. There any two solutions would differ by a self-equilibrating state of stress on each of the edges that *must* decay away from them and therefore *cannot* be axisymmetric (here we have invoked St. Venant's principle; see also Simmonds, 1977). Hence there exists at most one axisymmetric solution. Intuitively, we expect that this is true in the weakly nonlinear case also.

In practice, the most common tube is the circular pipe,

$$r = a + b\sin(s/b), \quad z = -b\cos(s/b), \quad 0 \le s \le 2\pi b, \tag{5.19}$$

where b is the radius of the pipe and a is the radius of curvature of the pipe center line.

The nonlinear bending of straight pipes and the linear and nonlinear bending of curved pipes have important engineering applications. The major phenomenon is the loss of circularity of the pipe's cross sections. This "ovality" is accompanied by increased flexibility, stress concentrations, appearance of shell-like meridional bending moments, and, finally, collapse of the pipe. A common engineering approach has been to apply "correction factors" to the beam-type formulas for pipe bending. These were even incorporated into some of the Engineering Codes of Practice (see ASME, Boiler and Pressure Vessel Code, Part III). During the last decade, the need has arisen for more accurate information on pipe bending for nuclear power plants, oil pipes, and aerospace uses. This has led to many tests and numerical analyses. "Pipe elbow" routines have been added to several special and general-purpose computer program (see Bushnell, 1981b, and Sobel and Newman, 1980, where additional references can be found). Some of the present-day computer codes can also handle inelastic behavior and circumferentially varying states of stress, including edge effect corrections.

The *nonlinear* loss of rigidity and collapse of *straight* tubes was first analyzed by Brazier (1927). Additional studies were made by Wood (1958), who included internal pressure effects, Axelrad (1965), who considered bifurcation-type instability together with ovality, and Thurston (1977), who made a more precise calculation of the collapse moment using the equations developed by Reissner and Weinitchke (1963).

The first *linear* study of the bending of curved tubes was made by von Karman (1911). A review and extension of the linear analysis was made by Clark and Reissner (1951) and many papers have appeared since. The literature on the corresponding nonlinear theory is, however, far less extensive and more approximate in nature. Reissner (1959a, 1961) introduced nonlinear effects into an approximate analysis of the bending of tubes, and presented a unified approach to curved and straight tubes. Axelrad (1962) introduced "beam bending" terms into the nonlinear axisymmetric analysis of shells of revolution. The most recent studies, which start with the finite rotation theory of Simmonds and Danielson (1970, 1972) and Reissner (1974) (discussed in Chapter III), are those of Boyle (1981) and Reissner (1981). Though differing in details, both papers include meridional transverse shearing strain that is later dropped. The simplified derivation presented here, which we develop from first principles, is closer to Reissner's approach and uses strain compatibility and moment equilibrium as basic equations.

1. Kinematics

An important property of a curved tube undergoing bending deformation is that it behaves as a beam with no out-of-plane distortions of its cross sections. That is, material meridional planes of the undeformed shell remain meridional planes of the deformed shell. If (r, θ, z) are the cylindrical coordinates of a material point on ξ, then $\bar{r} = \bar{\rho}(s)$, $\bar{z} = \bar{\zeta}(s)$, and $\bar{\theta} = (1 + k)\theta$ are its cylindrical coordinates on $\mathbf{y}$, where k is a constant that measures, in effect, the change of curvature of the tube beam. From a geometrical point of view, we can regard the deformation as composed of two parts: an in-plane deformation of meridional curves plus the rigid rotation of the meridians around the z axis by the variable amount $k\theta$. The two are obviously commutative.

We next note that surface coordinates (s, θ) are *lines of principal curvature* of both ξ and $\mathbf{y}$. It follows immediately that all the shearing $(s\theta)$ strains vanish. Let φ and $\bar{\varphi}$ be the angles between the meridional tangents and the xy plane before and after deformation. Then expressions for a *deformed* meridional element of length $d\bar{s}$ and circumferential elements of length $d\bar{s}_\theta$ are

$$d\bar{s} = \sec \bar{\varphi}\, d\bar{r}, \qquad d\bar{s}_\theta = \bar{r}\, d\bar{\theta}, \tag{5.20}$$

from which the corresponding expressions for the extensional strains ε_s and ε_θ are

$$\varepsilon_s = \sec \bar{\varphi}\, \bar{r}' - 1, \qquad \varepsilon_\theta = (1 + k)(\bar{r}/r) - 1, \qquad r' = dr/ds = \cos \varphi. \tag{5.21}$$

Elimination of $\bar{r}$ results in the strain compatibility equation

$$(r\varepsilon_\theta)' = (1 + \varepsilon_s)(1 + k)\cos \bar{\varphi} - \cos \varphi. \tag{5.22}$$

The meridional rotation of the material normal differs from that of the tangent by the small shearing angle γ. We suppress γ *at this stage*. Hence the expressions for the bending strains are the common ones for shells of revolution and reduce to

$$\kappa_s = (\bar{\varphi} - \varphi)', \qquad r\kappa_\theta = (1 + k)\sin \bar{\varphi} - \sin \varphi. \tag{5.23}$$

2. Equilibrium

The nonvanishing physical resultants for axisymmetric deformation are the membrane stress resultants (N_s, N_θ), the meridional transverse shear Q, and the stress couple M_s, M_θ, all measured per unit undeformed length. Lack of axial external forces permits an integral statement of the equation of axial equilibrium:

$$r(N_s \sin \bar{\varphi} - Q \cos \bar{\varphi}) = C. \tag{5.24}$$

The appearance of a (possibly) nonzero constant C reflects the fact that the meridians are closed. The equation of radial equilibrium can be expressed in terms of the horizontal *radial* force per unit length $H = N_s \cos \bar{\varphi} + Q \sin \bar{\varphi}$ as follows:

$$N_\theta = (1 + k)^{-1}(rH)'. \tag{5.25}$$

The other force resultants can also be expressed in terms of H and $\bar{\varphi}$ as

$$N_s = H \cos \bar{\varphi} + (C/r) \sin \bar{\varphi}, \qquad Q = H \sin \bar{\varphi} - (C/r) \cos \bar{\varphi} \tag{5.26}$$

[compare with Eq. (5.16)]. Finally, we have the equation of moment equilibrium

$$(rM_s)' - (1 + k) \cos \bar{\varphi} M_\theta = rQ(1 + \varepsilon_s). \tag{5.27}$$

3. Constitutive Relations

We consider elastic isotropic materials undergoing *small* strains and finite rotations and take Eqs. (3.133) and (3.134) in the form

$$E_s = A(N_s - vN_\theta), \qquad E_\theta = A(N_\theta - vN_s), \tag{5.28}$$

$$M_s = D(\kappa_s + v\kappa_\theta), \qquad M_\theta = D(\kappa_\theta + v\kappa_s). \tag{5.29}$$

4. Field Equations

Direct substitution into the equations of *strain compatibility* and *moment equilibrium* reduces them to two second-order differential equations in $\bar{\varphi}$ and H, exact within the scope of small-strain finite-rotation theory. Additional simplifying assumptions (which are occasionally made) are very helpful, yield good results, but are not absolutely necessary. The most prominent are $M_\theta = \varepsilon_s = 0$. With these, the two equations become

$$\left[\frac{Eh^3}{12} r(\bar{\varphi} - \varphi)' \right]' = rH \sin \bar{\varphi} - C \cos \bar{\varphi}, \tag{5.30}$$

$$\left[\frac{1 - v^2}{(1 + k)Eh} r(rH)' \right]' = (1 + k) \cos \bar{\varphi} - \cos \varphi. \tag{5.31}$$

5. Boundary Conditions

Kinematic and static continuity requirements around the closed meridians dicate that both $\bar{\varphi}$ and H should be single valued together with their first derivatives. These supply four conditions. A fifth is needed for the determination of C and is given by the requirement that there be no radial component of the total couple resultant at the edge:

$$\oint (N_\theta \bar{z} - M_\theta \cos \bar{\varphi}) \, ds = 0. \tag{5.32}$$

Integration by parts using previous equations yields

$$\oint [(1 + k)M_\theta \cos \bar{\varphi} + rH(1 + \varepsilon_s) \sin \bar{\varphi}] \, ds = 0. \tag{5.33}$$

A sixth condition, needed in order to relate the moment resultant at the edge to the applied external couple, is

$$\oint (N_\theta \bar{r} + M_\theta \sin \bar{\varphi}) \, ds = M, \tag{5.34}$$

which yields, upon integration by parts,

$$\oint [(1 + k)M_\theta \sin \bar{\varphi} - rH(1 + \varepsilon_s) \cos \bar{\varphi}] \, ds = (1 + k)M. \tag{5.35}$$

This completes the formulation.

More detailed calculations for the special case of circular pipes exist in the literature. In Reissner's article (1981) the equations were presented and the straight pipe was retrieved as a special case. Boyle (1981) provided parametric calculations for different tube geometries. Typical M–k plots were shown and a definite maximum (collapse) moment was seen to exist for most geometries (see Fig. 4 in the aforementioned paper).

VI. Approximate Shell Theories

A. Introduction

The formidable complexities of the general nonlinear shell equations have led researchers and engineers to try to simplify these equations as far as possible without changing their essential features or impairing their accuracy.

Special approximate shell theories are usually good for restricted geometries, boundary conditions, loadings, or restricted magnitudes of the various components of the deformation. These theories have been used indiscriminately again and again, beyond their range of validity. Nevertheless, they often provide not unreasonable and practical solutions for otherwise almost intractable problems.

The chief tool for developing approximate theories is order-of-magnitude comparisons among the various terms in the equations of equilibrium and compatibility (or, equivalently, in the strain–rotation–dislacement relations). The quantities that appear in the comparisons are the following:

1. *Geometric:* The shell thickness h, a typical radius of curvature R of the reference surface, the Gaussian curvature K, and the loading (or deformation) wavelength L;
2. *Kinematic:* Extensional strains of magnitude ε, curvature changes of magnitude κ, rotations of magnitude β, and displacements $\mathbf{u}$;
3. *Dynamics:* Stress resultants and couples of respective magnitude n and m.

Koiter (1966, Part III) has presented a classification of *small-strain* shell theories based on the ratio of the relative magnitude of the bending strain to extensional strain, $h\kappa/\varepsilon$, compared with h/R and $(h\kappa)^{-1}$. The main result of his analysis is that when

$$h/R \ll h\kappa/\varepsilon \ll \min[R/h, (h\kappa)^{-1}], \tag{6.1}$$

some of the terms in the static intrinsic field equations (3.156)–(3.159) can be suppressed, resulting in the following system of shell equations for "comparable bending and extensional strains":

$$An_\alpha^\beta|_\beta + (b_\beta^\alpha + \tfrac{1}{2}\rho_\beta^\alpha)\hat\rho_\alpha^\beta = 0, \tag{6.2}$$

$$\hat\rho_\alpha^\beta|_\beta = 0, \tag{6.3}$$

$$D\rho_\alpha^\beta|_\beta - (b_\beta^\alpha + \rho_\beta^\alpha)n_\alpha^\beta - p = 0, \tag{6.4}$$

$$n_\alpha^\beta|_\beta + p_\alpha = 0. \tag{6.5}$$

We note that the suppressed terms are essentially of three types:

1. Extensional strain terms in the Gauss–Codazzi equations (6.2) and (6.3),
2. Transverse shear terms in the equation of tangential equilibrium (6.5), and
3. Nonlinear terms in the equation of tangential equilibrium (6.5) resulting from referring the equations to the undeformed configuration and coming from the incremental Christoffel symbols (3.111)

Some caution should be exercised regarding the latter omissions. In certain cases of small deformations superimposed on large ones (such as may arise in stability and vibration analyses of prestressed shells) or in cases where *length* effects produce cumulative deformations (such as may arise in long cylindrical shells and slender curved tubes), some of the omitted terms may become significant. It would be wise in these cases to refer equilibrium to the deformed configuration and supress unimportant terms by direct comparisons.

To see the relative magnitudes of the various terms and their effects on compatibility, consider Eq. (6.2), which can be written in the order-of-magnitude form

$$O(\varepsilon/L^2) + O(\kappa/R) + O(\kappa^2) = 0. \tag{6.2a}$$

Here the first term is the extensional deformation, the second is the linear inextensional bending deformation, and the third term is the nonlinear inextensional response.

A similar order-of-magnitude equation can be written in place of the normal equilibrium equation (6.4). Dividing by Eh, we have

$$O(h^2\kappa/L^2) + O(\varepsilon/R) + O(\varepsilon\kappa) + O(p/Eh) = 0. \tag{6.4a}$$

Here the first term is the bending response, the second is the linear membrane response, the third is the nonlinear membrane response, and the fourth term is the reduced load.

Order-of-magnitude estimates can yield important information on the range of validity of approximate shell theories. For example, if instead of the left part of inequality (6.1), we have $\kappa = O(\varepsilon/R)$, then *linear membrane theory* holds, whereas if $\kappa = O(R^{-1})$, then *nonlinear membrane theory* (for small strains) holds, provided that, in addition, $\varepsilon \gg h^2/L^2$. If an edge zone of width $O(L)$ develops in nonlinear membrane theory, Eq. (6.2a) shows that $\varepsilon/L^2 = O(\kappa/R)$, whence Eq. (6.4a) yields $L = O(R\sqrt{\varepsilon})$.

For *both* bending and nonlinear effects to be comparable, we must have $\varepsilon = O(h^2/L^2)$. In an edge zone of width $O(\sqrt{hR})$, this requires that $\varepsilon = O(h/R)$. Lack of space prevents us from discussing this particular point in more detail. For examples on the interplay of bending versus nonlinear membrane effects, see Reissner (1959b,c, 1963b), Rossettos and Sanders (1965), Jordan and Shelley (1968), Libai (1972a), and Wittrick (1964a,b).

When, instead of the right-hand side of Eq. (6.1), $h\kappa/\varepsilon = O(\min[R/h, (h\kappa)^{-1}])$, we have *inextensional theory*. (Nearly) inextensional deformations tend to occur when the shell is inadequately constrained at its edge, that is, when the kinematic boundary conditions are such that Eqs. (6.2) and (6.3) admit nontrivial solutions with $A = 0$. A qualitative explanation of this phenomena can be given with the aid of the theorem of minimum total potential. We regard the shell problem as that of minimizing $\int_\Sigma (\Phi - \mathbf{p}\cdot\mathbf{u})\,d\Sigma$, subject to the compatibility conditions and to the kinematic boundary data. The order-of-magnitude relation for Φ is

$$\Phi = \Phi_\varepsilon + \Phi_\kappa \approx Eh[\varepsilon^2 + (h/R)^2(\kappa R)^2]. \tag{6.6}$$

For $h/R \ll 1$ and in the absence of (sufficient) kinematic constraints, the minimization process would tend to distribute the energy in such a way that $(\Phi, \Phi_\varepsilon) = O(\Phi_\kappa)$, so that $h\kappa/\varepsilon \leq O(1)$.

In a typical inextensional analysis, in the absence of constraints, we find that $h\kappa/\varepsilon = O(L/h)$. We call this *strong inextensionality*. The first term in Eq. (6.2) may be deleted over the entire reference surface, but the bending terms that were suppressed in the full, intrinsic, tangential equilibrium equation, Eq. (3.159), have to be put back. The large bending of cantilevered plates is like this.

The range $h\kappa/\varepsilon = O(1)$ defines *weak inextensionality*. It occurs often in nonlinear deformations. Since it satisfies Eq. (6.1), Eqs. (6.2)–(6.5) hold. The deformation is such that over the major parts of the shell, where $L/R = O(1)$, the first term in Eq. (6.2) can be suppressed and the kinematics are inextensional. However, regions near the edge or near high strain gradients, $L = O(\sqrt{hR})$ and the full equation (6.2) must be used. The postbuckling of very thin, axially loaded cylindrical shells is like this.

Nonlinear membrane theory and inextensional theory are not mutually exclusive. For example, consider the combined order-of-magnitude inequalities:

$$(h/L)^2 \ll \varepsilon = O(h\kappa) \qquad \text{with } L/R = O(1). \tag{6.7}$$

Here the requirements for nonlinear membrane and inextensional theories are met simultaneously (except in edge zones). The conditions (6.7) lead to a *nonlinear membrane-inextensional* theory for very thin shells, where the geometry is inextensional but the statics are momentless.

For the main class of comparable extensional and bending strains, additional simplifications of Eqs. (6.2)–(6.5) can be made for special geometries. The important class of *quasi-shallow shells* is a good example and will be discussed next; nonlinear membrane theory will then follow.

B. QUASI-SHALLOW SHELL THEORY

A surface is shallow in the strict sense if it differs but little from a plane. A more precise definition would require (1) that a plane π exists such that the square of the angle between the normals to ξ and π are small compared to 1, and (2) that the largest linear dimension of the projection of ξ on π be small compared with the smallest radius of curvature of ξ.

Marguerre (1938) was the first to develop a general nonlinear theory for strictly shallow shells with moderate rotations $[\beta^2 = O(\varepsilon)]$. This theory has become the cornerstone of the analysis of nearly flat shells. An extensive literature exists that we shall not try to record. For additional discussion, see, for example, Vlasov (1958), Mushtari and Galimov (1961). Sanders (1963), and Koiter (1966). Reissner (1969) extended the equations to include shear deformations and independent translation and rotation components, and Naghdi (1960) discussed the corresponding elastodynamic equations.

An important extension of the theory can be made if we relax the definition of "shallowness" to imply "weak curvatures" in the sense that $L/R \ll 1$. The restriction of moderate rotations is retained. The nonlinear theory based on this extension is the Donnell–Mushtari–Vlasov (DMV) theory. It was apparently first used by Donnell (1934) in his studies on the buckling of cylindrical shells. The assumptions underlying the theory were further developed by Vlasov and Mushtari (1961). Discussions of the assumptions of the theory within the framework of general nonlinear shell theory may be found in Sanders (1963) and Koiter (1966). The DMV theory is successful in shell buckling because in most cases the characteristic wavelength during buckling is small, so that $L/R \ll 1$. Still the theory is rather restricted by its assumptions that $L/R \ll 1$ and that the tangential components of the displacement can be ignored in the expressions for the bending strains and linear rotations.

The next important extension, which further relaxed some of the restrictions, was made in the 1960s, when the theory of quasi-shallow shells was developed.

A quasi-shallow shell is characterized by its *weak Gaussian curvature*: $|K|L^2 \ll 1$. This restriction is far less demanding than $L/R \ll 1$ of shallow shell theory. For example, developable or "almost developable" shells are quasi-shallow regardless of the size of L or R. Many other shells with small wavelengths of deformation are also quasi-shallow. Libai (1962) was the first to construct a theory for such shells. His development had some restrictive assumptions, but included elastodynamics. The theory was brought to its present state by Koiter (1966), who removed the earlier restrictive assumptions and established its position within the family of approximate shell theories. A variational principle for quasi-shallow shells is discussed by Huang (1973).

The starting point of the derivation is the field equations (6.2)–(6.5), valid when the extensional and bending strains are comparable. Consider the formula for interchanging the order of covariant differentiation on a surface:

$$\varepsilon^{\lambda\mu} V_{\beta|\lambda\mu} - K\varepsilon_{\alpha\beta} V^{\alpha} = 0. \tag{6.8}$$

For a quasi-shallow shell, the second term is much smaller than the first, so that $\varepsilon^{\lambda\mu} V_{\beta|\lambda\mu} \approx 0$, indicating that the order of covariant differentiation is immaterial. This permits us to use a curvature function

$$\rho_{\beta\lambda} = W|_{\beta\lambda} \tag{6.9}$$

to satisfy Eq. (6.3). Likewise, assuming that $p_{\alpha} = 0$, we can satisfy Eq. (6.5) by introducing a stress function:

$$n^{\alpha\beta} = \hat{F}|^{\alpha\beta}. \tag{6.10}$$

As before, the hat notation denotes the raising of indices by $\varepsilon^{\alpha\beta}$. Substitution of Eqs. (6.9) and (6.10) into Eqs. (6.2) and (6.4) yields the following two coupled nonlinear equations:

$$A \Delta\Delta F + \hat{W}|^{\alpha\beta}(b_{\alpha\beta} + \tfrac{1}{2}W|_{\alpha\beta}) = 0, \tag{6.11}$$

$$-D \Delta\Delta W + \hat{F}|^{\alpha\beta}(b_{\alpha\beta} + W|_{\alpha\beta}) + p = 0. \tag{6.12}$$

It should be emphasized that the development does not pose any restrictions on the displacement field beyond those inherent in Eqs. (6.2)–(6.5) and those implied by $|K|L^2 \ll 1$. In this respect, quasi-shallow theory has a wider scope than DMV theory. Its main disadvantage is that the "stress–curvature" form developed here is intrinsic, and therefore more difficult to use with the conventional form of the kinematic boundary conditions.

Classical DMV or shallow shell theory is more approximate but can be related directly to the displacement field. The major step is to identify W with the displacement w normal to ξ. Thus

$$\rho_{\alpha\beta} \approx w|_{\alpha\beta}. \tag{6.13}$$

This substitution reduces Eqs. (6.11) and (6.12) to standard form. In addition, the covariant components of the Lagrangian strain given by Eq. (3.83) and the rotations can be expressed in terms of the displacements u^α and w as follows:

$$L_{\alpha\beta} = \tfrac{1}{2}(u_{\alpha|\beta} + u_{\beta|\alpha} + w_{,\alpha}w_{,\beta}) - b_{\alpha\beta}w, \qquad \varphi_\alpha = -w_{,\alpha}. \tag{6.14}$$

These expressions follow from the shallow shell assumptions stated earlier.

The Marguerre shallow shell equations follow upon representing the reference surface in the form $z = \hat{z}(\xi^1, \xi^2)$, where ξ^α are coordinates in the plane π. The shallowness approximation implies that $b_{\alpha\beta} \approx z|_{\alpha\beta}$ and, in addition, that the metric coefficients $a_{\alpha\beta}$ are those of the ξ^α coordinate system in the plane. When $z \equiv 0$, we get the von Karman equations. In practice, we often work with Cartesian or polar coordinates.

C. Nonlinear Membrane Theory

The membrane theory of shells is obtained formally by suppressing the stress couples and consequently the transverse stress resultants in the equations of motion. It is one of the simpler and more important shell theories, both as a shell model and as an interior approximation to the full shell equations, valid in many practical situations. We will confine our presentation to statics. Lack of space and literature prevents inclusion of the important topic of membrane dynamics.

1. Equilibrium

The equilibrium equations of membranes are, superficially, simplest when referred to the *deformed* configuration, and take the form

$$\bar{n}^{\alpha\beta}_{;\alpha} + \bar{p}^\beta = 0, \tag{6.15}$$

$$\bar{n}^{\alpha\beta}\bar{b}_{\alpha\beta} + \bar{p} = 0, \tag{6.16}$$

$$\varepsilon_{\alpha\beta}\bar{n}^{\alpha\beta} = 0. \tag{6.17}$$

The last equation shows that in membrane theory, the contravariant components of the stress resultant tensor are symmetric. Here we have used Koiter's (1980) notation, and $\bar{\mathbf{p}} = \bar{p}^\alpha \mathbf{y}_\alpha + \bar{p}\bar{\mathbf{n}}$ is the external surface load per unit *deformed* area.

Notwithstanding their formal simplicity, these equations are not easy to deal with, since the covariant derivative in Eq. (6.15) is based on the unknown metric coefficients $\bar{a}_{\alpha\beta}$ of $\mathbf{y}$. An alternative is to shift it to the undeformed metric coefficients $a_{\alpha\beta}$ and at the same time to introduce the surface force per unit undeformed area and the modified stress resultants given by Eqs. (3.155) and (3.160) respectively. This gives

$$n^{\beta\alpha}|_\beta + \bar{\bar{a}}^{\alpha\kappa}(2L_{\kappa\lambda|\beta} - L_{\lambda\beta|\kappa})n^{\lambda\beta} + p^\alpha = 0, \tag{6.18}$$

$$n^{\alpha\beta}\bar{b}_{\alpha\beta} + p = 0, \tag{6.19}$$

where $\bar{\bar{a}}^{\alpha\kappa} = \eta^{-1} T^{\alpha\kappa}$, with η and $T^{\alpha\kappa}$ given by Eqs. (3.84) and (3.114), respectively. For small extensional strains, $\bar{\bar{a}}^{\alpha\kappa}$ can be replaced by $a^{\alpha\kappa}$.

Another form is obtained if, following Budiansky (1968), we take components of the equilibrium equations along the undeformed triad $\{\mathbf{a}_\alpha, \mathbf{n}\}$. This theory has the advantage of being rational in the displacements and is convenient to use when the loading and boundary conditions are specified with respect to $\mathbf{a}_\alpha$ and $\mathbf{n}$. The resulting equations are

$$[(a_{\alpha\beta} + s_{\alpha\beta})n^{\alpha\beta}]|_\alpha + b_{\gamma\alpha}\varphi_\beta n^{\alpha\beta} + \hat{p}_\gamma = 0, \tag{6.20}$$

$$b^\gamma_\alpha(a_{\gamma\beta} + s_{\gamma\beta})n^{\alpha\beta} - (n^{\alpha\beta}\varphi_\beta)|_\alpha + \hat{p} = 0, \tag{6.21}$$

where

$$\mathbf{p} = \hat{p}_\alpha \mathbf{a}^\alpha + \hat{p}\mathbf{n}. \tag{6.22}$$

The deformation measures that appear in the equations are expressed in terms of the components of the displacement,

$$\mathbf{u} = u_\alpha \mathbf{a}^\alpha + w\mathbf{n}, \tag{6.23}$$

in the following way:

$$s_{\alpha\beta} = u_{\alpha|\beta} - b_{\alpha\beta}w, \tag{6.24}$$

$$\varphi_\alpha = -w_{,\alpha} - b_{\alpha\beta}u^\beta. \tag{6.25}$$

Equations (6.18) and (6.19) or (6.20) and (6.21) in conjunction with Eqs. (6.23)–(6.25) are a part of the "displacement formulation" of nonlinear membrane theory. The advantages of this formulation for extrinsic data are obvious. However, it becomes more cumbersome when the data is intrinsic. In addition, it is heavily dependent on the initial configuration, whereas in many nonlinear membrane problems the curvature of the reference surface is immaterial (as we shall see later).

2. Compatibility

This brings us to the "intrinsic formulation" that uses, in place of displacements, the Gauss–Codazzi equations for the deformed surface. We can take these either as Eqs. (3.63) and (3.64) with $E_\alpha = 0$ and $K_{\alpha\beta}$ replaced by $R_{\alpha\beta} - B$, with B coming from the second part of Eq. (3.64), or else use the following semi-incremental form [cf. Eqs. (5.6) and (5.7) of Koiter (1966)]:

$$\varepsilon^{\lambda\mu}(\overline{b}_{\beta\lambda|\mu} - \overline{a}^{\kappa\nu}\overline{b}_{\kappa\lambda}L_{\nu\beta\mu}) = 0, \tag{6.26}$$

$$K(1 + L_\gamma^\gamma) + \varepsilon^{\alpha\beta}\varepsilon^{\lambda\mu}(L_{\alpha\mu|\beta\lambda} + \tfrac{1}{2}\overline{b}_{\alpha\mu}\overline{b}_{\beta\lambda} + \tfrac{1}{2}\overline{a}^{\kappa\nu}L_{\kappa\alpha\mu}L_{\nu\beta\lambda}) = 0, \tag{6.27}$$

where $L_{\alpha\beta\gamma}$ is given by Eq. (3.111).

If the displacements from the reference shape are small, we can omit the nonlinear terms in Eq. (3.83) for $L_{\alpha\beta}$ and $s_{\alpha\beta}$ and φ_β in Eqs. (6.20) and (6.21), to obtain *linear* membrane theory. In doing so, we reduce the order of the differential system from 6 to 4, with the attendant difficulties in satisfying the boundary conditions discussed in Section III,J. Before considering constitutive laws and boundary conditions in nonlinear membrane theory, a short digression is in order.

There are two distinct justifications for ignoring moments: Either the shell cannot support them or the momentless equations provide good approximations to the full shell equations.

In the first category we include the asymptotic behavior of shells as $h/R \to 0$. Shells made from very thin sheets, such as biological membranes, can be approximated by this model. As $h/R \to 0$, the shell cannot support rotational constraints. Its response, should they be imposed, would be to form angular discontinuities. Since the depth of the transition zone, where bending and three-dimensional effects come into play, approaches zero with h, we may regard these discontinuities as representing, mathematically, the "inner (edge zone) limit of the outer (membrane) expansion." Another feature of the asymptotic behavior is the inability to support compressive forces that lead to wrinkling. Further discussion of this point may be found at the end of this section.

In the second category we find approximate interior solutions for shells of finite thickness. It has long been taken as an experimental fact that away from boundaries and discontinuities, the neglect of moments in the equilibrium equations yields a good approximation to the response of the shell to external loads. (Note that we do not say that the moments are zero, but only that their effects on equilibrium are not important.) The following qualitative support can be given to this statement.

(a) The wavelength L of the deformation pattern should be $O(R)$. This excludes effects of rapidly varying loads, instabilities, short-wavelength transverse vibrations, and regions near discontinuities where bending boundary layers of depth $O(hR)^{1/2}$ might exist. All of this is valid for small strains. For large strains, $\varepsilon = O[(h/R)^{1/2}, (h/L)^{1/2}]$, the *nonlinear membrane response* may even dominate the edge zone behavior. The examples cited in the section on the behavior of beamshells (Section II,K) demonstrate this type of behavior.

(b) The shape of the shell (it must not be too long) and its kinematic boundary conditions must be such that substantial inextensional deformations cannot occur. We note in our beamshell examples how the relaxation of a kinematic boundary condition transforms the large deformation response of a beamshell from an essentially membranous one to the other extreme of an inextensional one.

3. Constitutive Laws

We will now continue with our main development and discuss constitutive laws. For small strains, the uncoupled form of the strain energy density can be used, with a constitutive law such as Eq. (3.133) for isotropic materials. This has been the cornerstone for most "geometrically nonlinear" membrane analyses. However, the considerable simplicity that membrane theory entails has made possible the introduction of extensive material nonlinearities that are of importance in rubberlike materials, biological membranes (or other very thin structures such as arterial and uterine walls), and inflatable structures. Confining our discussion to isotropic materials only and to membranes in the sense of (1) ($h/R \to 0$), we find that Φ depends on the two extensional strain invariants $J_1 = 2(1 + L_\alpha^\alpha)$ and $J_2 = \eta = \bar{a}/a$ [see Eq. (3.84)].

The form of $\Phi(J_1, J_2)$ has been studied by several authors. The book by Green and Adkins (1970) summarizes earlier work. Wu (1979) expressed the constitutive relations for the general elastically isotropic material in the form

$$\frac{\bar{n}^{\alpha\beta}}{h} = \frac{1}{IJ} \frac{\partial \Phi}{\partial I} a^{\alpha\beta} + \left(\frac{1}{J} \frac{\partial \Phi}{\partial I} + \frac{\partial \Phi}{\partial J} \right) \bar{a}^{\alpha\beta}, \qquad (6.28)$$

where

$$J = J_2^{1/2}, \quad I = (J_1 + 2J)^{1/2}, \quad I_1 = J_1 + J_2^{-1}, \quad I_2 = J_2 + J_1 J_2^{-1}. \tag{6.29}$$

Forms of Φ for many materials capable of large elastic strains are given in the article. Two important examples are the neo-Hookean type:

$$\Phi = C_1(I_1 - 3) \tag{6.30}$$

and the Mooney type:

$$\Phi = C_1(I_1 - 3) + C_2(I_2 - 3). \tag{6.31}$$

Also given in Wu's article are some asymptotic solutions for plane and axisymmetric membranes.

We continue our discussion of constitutive laws by noting that although results for category 1 membranes may serve as a first approximation for moderate strains and curvatures to the behavior of category 2 membranes, finite thickness effects may cause $\bar{n}^{\alpha\beta}$ to depend on $b_{\alpha\beta}$ and $\bar{b}_{\alpha\beta}$. In fact, the results of our beamshell studies in Section II,K show a significant dependence of the membrane force on the curvatures. The possibility of regions in a shell with moderate strains and very large curvatures must also be considered. The development of large-strain constitutive laws for shells of finite thickness remains open.

4. Boundary Conditions

The boundary conditions and field equations of nonlinear membrane theory can be obtained by suppressing the stress couples in the virtual work principle of general shell theory. It is important to note that *there is no a priori reason to suppress the normal component of the applied external boundary force* **T**. The fact that it is subsequently to be set to zero is a *result* and not an assumption. The virtual rotations, however, vanish with the stress couples.

We can formulate the principle either in terms of the deformed shape and get conditions to be used with Eqs. (6.15) and (6.16) or (6.18) and (6.19). Alternatively, we can relate the principle to the reference shape and get conditions to be used with Eqs. (6.20) and (6.21). Using the first approach and integrating by parts, we have

$$\int_{\bar{\Sigma}} [(\bar{n}^{\beta\alpha}_{;\beta} + \bar{p}^{\alpha})\delta\bar{u}_{\alpha} + (\bar{n}^{\alpha\beta}\bar{b}_{\alpha\beta} + \bar{p})\delta\bar{w}] d\Sigma = \int_{\partial\bar{\Sigma}} [(\bar{n}^{\beta\alpha}\bar{v}_{\beta} - \bar{T}^{\alpha})\delta u_{\alpha} - \bar{T}\delta\bar{w}] d\bar{s}, \tag{6.32}$$

where

$$\bar{\mathbf{T}} = \bar{T}^{\alpha}\mathbf{y}_{\alpha} + \bar{T}\bar{\mathbf{n}}, \quad \delta\mathbf{u} = \delta\bar{u}^{\alpha}\mathbf{y}_{\alpha} + \delta\bar{w}\bar{\mathbf{n}} \tag{6.33}$$

are the boundary force per unit deformed length and the virtual displacement, respectively. All terms relate to the geometry of **y**. Standard arguments yield differential equations and boundary conditions. The kinematic conditions are the specification of the three components of **u**. The corresponding force conditions are

$$\bar{n}^{\beta\alpha}\bar{v}_\beta = \bar{T}^\alpha, \qquad \bar{T} = 0. \tag{6.34}$$

The last condition on the right-hand side has turned up somewhat unexpectedly. It is actually a geometrical condition stating that for a given force $\bar{T}$, the membrane has to deform and rotate in such a way that $\bar{T}$ will lie in the tangent plane of **y** on the boundary. We have in this way reintroduced rotational boundary conditions "through the back door." We note that this condition does not exist in linear membrane theory, since the direction of the surface normal is known there. The vanishing of $\bar{T}$ is an *a priori requirement of linear theory* and not a boundary condition. Budiansky (1968) used a similar approach to get the following conditions to be used with Eqs. (6.20) and (6.21):

$$(a_{\gamma\beta} + s_{\gamma\beta})n^{\alpha\beta}v_\alpha = T_\gamma, \qquad -\varphi_\beta n^{\alpha\beta}v_\alpha = T, \tag{6.35}$$

where all quantities now refer to the geometry of the reference surface and

$$\mathbf{T} = T^\alpha \mathbf{a}_\alpha + T\mathbf{n}. \tag{6.36}$$

The corresponding kinematic conditions are also in the direction of the triad $\{\mathbf{a}_\alpha, \mathbf{n}\}$. We note the interesting fact that at a free edge, $\mathbf{T} = \mathbf{0}$ and the second parts of Eqs. (6.34) and (6.35) are satisfied automatically. Hence the number of boundary conditions at a free edge in both approaches *degenerates to two*:

$$\bar{n}^{\beta\alpha}\bar{v}_\beta = 0. \tag{6.37}$$

This fact is further supported by mathematical arguments, as will be discussed presently.

5. The Nonlinear Membrane Problem

We are now in a position to state *the nonlinear membrane problem* (using the intrinsic set as an example). Find $\bar{n}^{\alpha\beta}$ and $\bar{b}_{\alpha\beta}$ that satisfy the following equations: equilibrium, Eqs. (6.18) and (6.19); compatibility, Eqs. (6.26) and (6.27); constitutive, Eq. (3.133) or (6.28), as the case may be; and boundary, Eq. (6.34) or appropriate kinematic conditions. We note that $b_{\alpha\beta}$ is not present in any of these equations. Hence the formulation of the nonlinear membrane problem does not depend on the curvature of the reference surface (unless used, in some fashion, in the loads or boundary conditions). A highly elastic inflatable bag is an example where the final shape does not depend on the deflated shape.

We conclude with some remarks on the mathematical nature of the equations. Nonlinear partial differential equations can be studied by linearizing them in the neighborhood of a point in the solution domain (Petrovsky, 1954). Libai (1972a) examined this problem for geometrically nonlinear membranes and showed that the membrane equations near a solution point $_0\bar{n}^{\alpha\beta}$, $_0\bar{b}_{\alpha\beta}$ can be reduced, qualitatively, to an equation of the form

$$(L_b^2 - AL_n \Delta\Delta)F = G(\mathbf{p}, \bar{\Delta}\mathbf{p}) + r, \tag{6.38}$$

where $L_b(\ ) = _0\hat{b}^{\alpha\beta}(\ )_{;\alpha\beta}$ is a curvature operator, $L_n(\ ) = _0\bar{n}^{\alpha\beta}(\ )_{;\alpha\beta}$ is a stress operator, F is a stress function such that $\bar{\Delta}\bar{n}^{\alpha\beta} = \hat{F}^{;\alpha\beta} + r$, r denotes terms with lower-order derivatives, $G(\mathbf{p}, \bar{\Delta}\mathbf{p})$ is a function of the loading and its increments, and $\bar{\Delta}$ indicates an increment from the solution point.

The following conclusions can be drawn from Eq. (6.38):

(a) The membrane system is of sixth order, except for the unstressed state when the operator on the left-hand side degenerates to L_b^2 of linear membrane theory.

(b) The coefficients of the higher-order operator are $O(\varepsilon)$, where ε is a typical strain at the solution point. Hence if boundary zones develop, $O(R\sqrt{\varepsilon})$ will be their depth.

(c) The system is always twice elliptic, but the nature of the third operator L_n depends on the state of stress. For example, at a free edge where $_0\bar{n}^{\alpha\beta}\bar{v}_\beta = 0$, the stress operator is parabolic. Hence the third boundary condition $(6.34)_2$ cannot be applied (compare with the results from the virtual work principle).

Further work on general nonlinear membranes remains to be done.

a. Axisymmetric Membranes

We consider membranes of revolution subject to axisymmetric normal loading $\bar{p} = -p_n$ and to axisymmetric boundary forces acting on meridional planes. The deformed membranes, if stable, is also a membrane of revolution. This situation is important in practice and as a relatively simple illustration of nonlinear membrane theory.

If the strains are small and the rotations moderate, linear membrane theory is usually adequate, except near boundaries or discontinuities, where nonlinear boundary layer effects compete with bending effects in thick membranes (i.e., in shells). Small strains mean that the stress-strain relations can usually be taken as linear.

If the strains or rotations are large and the stress-strain relations nonlinear, numerics are needed, although analytic treatments do exist in special cases (see Green and Adkins, 1970; Wu, 1970a,b, 1979; Yang and Feng, 1970; Schmidt and DaDeppo, 1975; Rigbi and Hiram, 1981; and many others).

A summary of the formulation for the strongly nonlinear case follows. Let $\tilde{r}, \theta, \tilde{z}$ be a fixed cylindrical coordinate system. Let $P_0(r_0, \theta, z_0)$ be a material point on ξ, and let β_0 be the angle between the $\tilde{z}$ axis and the tangents to the meridians. As the shell deforms, P_0 moves to P_1 on y with coordinates z, θ, r and angle β. We have

$$\cos \beta_0 = dr_0/ds_0, \quad \cos \beta = dr/ds, \quad \tan \beta = dz/dr, \quad (6.39)$$

where s_0 and s are the corresponding meridional length measures.

Let $\bar{n}_r$ and $\bar{n}_\theta$ be the physical force resultants (per unit deformed length) in the meridional and circumferential directions, respectively. The corresponding extensions are

$$\lambda_r = ds/ds_0, \quad \lambda_\theta = r/r_0. \quad (6.40)$$

The equations of equilibrium follow from Eqs. (6.15) and (6.16):

$$\frac{d}{dr}(r\bar{n}_r) = \bar{n}_\theta, \quad \bar{n}_r \frac{d}{dr} \sin \beta + \frac{1}{r} \sin \beta \bar{n}_\theta = p_n. \quad (6.41)$$

An integral of the equations is obtained by considering axial equilibrium of a portion of the shell between its boundary $r = a$ and any circumferential cross section

$$r\bar{n}_r \sin \beta = \int_a^r p_n r \, dr + \frac{1}{2\pi} T_a, \quad (6.42)$$

where T_a is the total axial force on the boundary. Finally, the force resultants are expressed in terms of the strain energy density function Φ (Wu, 1970a):

$$\bar{n}_r = \frac{1}{\lambda_\theta} \frac{\partial \Phi}{\partial \lambda_r}, \quad \bar{n}_\theta = \frac{1}{\lambda_r} \frac{\partial \Phi}{\partial \lambda_\theta}. \quad (6.43)$$

To reduce the system to two first-order equations in $\beta(r_0)$ and $r(r_0)$, we use Eq. (6.40) and rewrite Eq. (6.43) (if possible) in the form

$$\lambda_r = \hat{\lambda}_r(\bar{n}_r, r/r_0), \quad \bar{n}_\theta = \hat{\bar{n}}_\theta(\bar{n}_r, r/r_0), \quad (6.44)$$

which, in view of Eq. (6.42), are known functions of $r, r_0, \sin \beta, T_a$, and p_n. Introduction of $\bar{n}_r$ and $\bar{n}_\theta$ into the second part of Eq. (6.41) reduces it to a first-order equation in $\sin \beta$. The second equation of the reduced system is simply

$$\frac{dr}{dr_0} = \frac{\cos \beta}{\cos \beta_0} \hat{\lambda}_r. \quad (6.45)$$

After their solution, $z(r_0)$ is obtained by integrating Eq. (6.39).

Four boundary conditions are needed for the system: One is for axial location of the shell in Eq. (6.39); one is T_a (or the total axial extension), and two are needed for the second part of Eq. (6.41) and Eq. (6.45). These can be either r or β at each boundary (see also the discussion of the boundary conditions for the nonlinear membrane).

In special cases, the problem can be reduced to quadratures. An example is the cylindrical membrane. Here, following Pipkin (1968), we form the expression

$$\frac{d}{dr}\left(\Phi - \lambda_r \frac{\partial \Phi}{\partial \lambda_r}\right) = \lambda_r \left(\bar{n}_\theta \frac{d\lambda_\theta}{dr} - \frac{d}{dr}(\lambda_\theta \bar{n}_r)\right). \tag{6.46}$$

In view of the first part of Eq. (6.41), the right-hand side vanishes for constant r_0 so that

$$\Phi - \lambda_r \frac{\partial \Phi}{\partial \lambda_r} = C. \tag{6.47}$$

We further assume that Eq. (6.47) can be written in the form $\lambda_r = \hat{\lambda}_r(r, C)$. Then, from Eq. (6.43), $\bar{n}_r = \hat{\bar{n}}_r(r, C)$. Introduction into Eq. (6.42) yields $\beta = \hat{\beta}(r, C, T_a)$. The equation

$$dr/dz_0 = \hat{\lambda}_r \cos \beta \tag{6.48}$$

can now be integrated to find $z_0(r)$, and $z(r)$ is found from Eq. (6.39). Several cases were studies by Wu (1970b, 1979). Note the strong analogy with the equilibrium and energy integrals of beamshell analysis! As in beamshells, the determination of the constants of integration is not simple and special techniques may have to be used.

b. Annular Membranes

The annular membrane is a special membrane of revolution. Its geometrical simplicity, its importance for the understanding of the transition from plate to shell, and its engineering uses have led to a wealth of literature. To conform with our main topic, we shall restrict our discussion to annular membranes that undergo axisymmetric nonplanar deformations. We consider small strains and finite rotations only. The case of large strains is similar to that of the general axisymmetric membrane and the reader is referred to the Section VI,C,5.

We consider an annular membrane whose reference surface is defined by $z = 0$, $b \leq r \leq a$. The membrane is subjected to radial edge loads (or displacements) at $r = a, b$ and to an axial force T_b at the inner edge $r = b$. The axial displacement at $r = a$ is taken as zero. In addition, normal pressure p (per unit undeformed area) is applied to the membrane.

This problem was treated by Schwerin (1929), who assumed moderate rotations ($\beta^2 \ll 1$). This makes it a special case of the Föppl–Hencky theory of small finite deflections of flat membranes. When the inner edge carries no axial load, the governing equations reduce to

$$\frac{d^2 U}{dy^2} + \left(\frac{y-\varepsilon^2}{U}\right)^2 = 0, \tag{6.49}$$

where

$$\varepsilon = b/a, \qquad y = (r/a)^2, \qquad U = 4(2Ehp^2a^2)^{-1/3}\bar{n}_r y. \tag{6.50}$$

The case of a completely free inner edge was solved by a power-series expansion.

For edge-loaded shells, the problem simplifies further. Let T be the total axial force: then the equation becomes (Nachbar, 1969)

$$\frac{d^2 F}{dy^2} + \frac{1}{2F^2} = 0, \tag{6.51}$$

where

$$F^2 = \left(\frac{2T}{vEha}\right)^{2/3} (2\sec\beta - 2)^{-1} y. \tag{6.52}$$

The last equation can be brought to quadratures. In the general case, the constants of integration are determined by a transcendental equation, but an explicit solution is available for $v = 1/3$.

Later studies, using the equations developed by Reissner (1950, 1963a) for the axisymmetric deformations of shells of revolution, have put the Föppl problem for the annulus on a firmer mathematical basis and extended the results to large rotations, with the moment and transverse shear terms suppressed.

Clark and Narayanaswamy (1967) studied an annular membrane with an axial load at its inner edge. This problem *must* involve large rotations, even for slight loads, since the membrane has to adjust to the direction of the load at the inner edge. The boundary conditions are $\bar{n}_r = T/2\pi b$ and $\beta = \pi/2$ at $r = b$.

Nachbar (1969) analyzed general edge loadings. Differential equations for large loadings were derived and the existence and behavior of the corresponding Schwerin problem were considered in detail.

Weinitschke (1980) investigated the annular membrane subjected to transverse normal loadings. The corresponding Schwerin problem was analyzed in detail. Proof of convergence for the power series and existence of solutions, extension to large rotations, as well as numerical results were included in the study.

Finally, we consider the case where the *deformed* shell is an annular membrane. Curiously, this is a (theoretically) common situation that occurs when purely radial loads are applied to a membrane of revolution. Consider a radial edge load applied to the boundary of an otherwise unloaded membrane. At the point of application, $\beta = 0$ is one of the boundary conditions. In addition, since there is no total axial load, either $\beta = 0$ or $\bar{n}_r$ (and hence $\bar{n}_\theta$) = 0, signifying an unstressed shell. Starting from the point of load application, we have $\beta = 0$ as the solution. This region of the shell has deformed into an annular membrane that extends until a point is reached where $\bar{n}_r = \bar{n}_\theta = 0$ and where an *angular discontinuity* in β may occur. From this point on, the membrane is unstressed and retains its original shape. It is interesting to note how a discontinuous solution may arise in a seemingly innocuous problem. For additional material, including the case of radial distributed loads, see Simmonds (1961), Reissner and Wan (1965), and Clark and Narayanaswamy (1967).

c. General Cylindrical Membranes

General cylindrical membranes include some of the simpler and more useful membranes. Nonlinear membrane theory finds its use here not only in shells that cannot support stress couples—such as inflated cylindrical roofs or blood vessels—but also in thin cylindrical panels and shells of aerospace structures where it complements bending effects wherever linear membrane theory fails. Nonlinear membrane (or bending) effects arise mostly at the shell boundaries or near discontinuities. There exist, however, several cases where nonlinear corrections have to be applied "in the large": One is a long oval membrane where linear theory exaggerates the stresses and deformations; a second is a cylindrical panel with flexible supports along its generators. As is well known, linear membrane theory cannot accept prescribed data along boundaries that coincide with generators.

We present here equations for pressure-loaded cylinders undergoing small strains. The coordinates on the reference surface are x, measured along the generators, and arc length s, measured along orthogonal trajectories. It follows that the metric coefficients of the reference surface are Cartesian and use of Eqs. (6.18)–(6.21), which refer to the undeformed shape, is advantageous. The small-strain approximation is used for deleting terms, but only through direct comparisons. We have *equilibrium*,

$$[N_x(1 - \nu N_s) - \tfrac{1}{2}N_s^2]_{,x} + [N_t(1 + 2N_x - 2\nu N_s)]_{,s} = 0, \quad (6.53)$$

$$[N_s(1 - \nu N_x) - \tfrac{1}{2}N_x^2]_{,s} + [N_t(1 + 2N_s - 2\nu N_x)]_{,s} = 0, \quad (6.54)$$

$$N_x b_x + 2N_t b_t + N_s b_s = Ap, \quad (6.55)$$

and *compatibility*,

$$[b_x(1 - N_x + vN_s)]_{,s} - \{b_t[1 + (1 + v)(N_s - N_x)]\}_{,x}$$
$$+ b_s[2(1 + v)N_{t,x} - (N_x - vN_s)_{,s}] = 0, \qquad (6.56)$$

$$[b_s(1 - N_s + vN_x)]_{,x} - \{b_t[1 + (1 + v)(N_x - N_s)]\}_{,s}$$
$$+ b_x[2(1 + v)N_{t,s} - (N_s - vN_x)_{,x}] = 0, \qquad (6.57)$$

$$b_x b_s - b_t^2 + (N_x - vN_s)_{,ss} + (N_s - vN_x)_{,xx} - 2(1 + v)N_{t,xs} = 0, \quad (6.58)$$

where

$$(N_x, N_t, N_s) = A(n^{xx}, n^{xs}, n^{ss}), \qquad (b_x, b_t, b_s) = (b_{xx}, b_{xs}, b_{ss}). \qquad (6.59)$$

The small-strain approximation can be used for further simplification, but care must be taken because of the possibility of localized strain gradient effects. We delete, in the equations, strains compared with unity and arrive at the simplified form:

$$(N_x - \tfrac{1}{2}N_s^2)_{,x} + N_{t,s} = 0, \qquad (6.60)$$

$$(N_s - \tfrac{1}{2}N_x^2)_{,s} + N_{t,x} = 0, \qquad (6.61)$$

$$N_x b_x + 2N_t b_t + N_s b_s = Ap, \qquad (6.62)$$

$$b_{x,s} - b_{t,x} - b_s[(2 + v)N_s + N_x]_{,s} = 0, \qquad (6.63)$$

$$b_{s,x} - b_{t,s} - b_x[(2 + v)N_x + N_s]_{,x} = 0, \qquad (6.64)$$

$$b_x b_s - b_t^2 + \Delta(N_x + N_s) = 0. \qquad (6.65)$$

The tangential equilibrium equations are satisfied by a *membrane stress function* ψ such that

$$N_x = \psi_{,ss} + \tfrac{1}{2}N_s^2 \approx \psi_{,ss} + \tfrac{1}{2}\psi_{,xx}^2, \qquad (6.66)$$

$$N_s = \psi_{,xx} + \tfrac{1}{2}N_x^2 \approx \psi_{,xx} + \tfrac{1}{2}\psi_{,ss}^2, \qquad N_t = -\psi_{,xs}. \qquad (6.67)$$

A corresponding deformation function *does not* follow from the Codazzi equations without further approximations that do not appear to be justified in the general case.

As noted before, the undeformed curvature $b_s^{(0)}$ is not represented in the field equations. For rigid edges, it becomes a part of the *boundary conditions* along $x = x_0, x_0 + L$. It has been observed, however, that sufficiently far from the boundaries all closed cylinders (and open cylinders with straight rigid generators) approach a circular shape with increasing pressure. The relevant parameter that determines the shell behavior is $\gamma = pL^4/EhR_0^3$, where R_0 is a typical dimension in the s direction (say, the radius of the isoparametric circle of the noncircular cylinder).

The fact that a circular shape is approached with increasing pressure makes it a convenient point for starting a perturbation series solution to the membrane equations. Problems related to noncircular shapes or flexible generators can then be treated by *boundary perturbation* schemes. Libai (1972a, 1972b) used this approach for the analysis of oval membranes and of cylindrical panels with flexible supports. His first perturbation solution for oval membranes has the following properties:

1. The linear membrane stress and deformation patterns are approached as $\gamma \to 0$.
2. The circular shape is approached as γ becomes large.
3. The solution smooths irregularities and discontinuities in the undeformed shape.

It is in (1) and (3) that linear membrane theory has its greatest difficulties. This demonstrates the capability of nonlinear membrane theory to treat problems where the corresponding linear theory "fails in the large."

Large-strain problems for the nonaxisymmetric deformations of cylindrical membranes have also been considered. A useful method for this class of problems is the superposition of small nonsymmetric deformation on the large-strain axisymmetric response. Koga (1972) used this method to obtain the bending rigidity of an inflated rubbery circular cylindrical membrane.

d. Wrinkled Membranes

An important property of very thin membranes is their inability to support compressive stresses, which leads to instability and wrinkling. The appearance of wrinkles is accompanied by increased deformations, although subsequent stabilization may occur with the formation of tension fields.

To avoid this undesirable (but not necessarily catastrophic) phenomenon, thin membranes are often stabilized by the superposition of tension. In plane membranes, tensile forces can be imposed through the boundaries, and in curved membranes, by internal pressure. Pressure-stabilized inflated membranes are quickly becoming useful engineering structures, providing an important application of nonlinear membrane theory. Even here the appearance of wrinkles is not always avoidable (Wu, 1974), unless the shape of the membrane is specially designed (Mansfield, 1981).

The study of plane wrinkled membranes was initiated by Wagner (1929), who applied the results to the postbuckling analysis of plates in shear. Reissner (1938) studied the wrinkling of an annular membrane in shear. Wrinkles in inflated cylindrical beams was studied by Stein and Hedgepeth (1961), Comer and Levy (1963), and Koga (1972). A nonlinear analysis of the wrinkling of membranes of revolution was made by Wu (1974, 1978).

The basic concept is that of a "wrinkly region." The wrinkles are not studied in detail but, rather, the region they occupy is replaced by an equivalent region with different mechanical properties. In this "smoothed out" region not only must the membrane equations continue to hold, but the minimum principal stress must vanish as well. In addition, the load deformation response in the wrinkly region has to be modified. Reissner (1938) introduced the notion of a curvilinearly orthotropic membrane with a vanishing modulus normal to the wrinkle directions. Stein and Hedgepeth (1961) used a variable Poisson's ratio. Mansfield (1969, 1970) applied energy considerations and also showed the duality of the plane wrinkle problem with inextensional theory.

Wu (1978) analyzed meridianly wrinkled axisymmetric nonlinear membranes by introducing an unwrinkled pseudosurface that carries the loads. A "wrinkle strain" quantity was also introduced; it is defined as the difference in circumferential strains between the true surface and the pseudosurface. Additional assumptions were needed to relate the two surfaces and complete the formulation.

The details of the behavior of wrinkled nonlinear membranes are not completely resolved and further study may be needed.

Appendix. The Equations of Three-Dimensional Continuum Mechanics

A. The Integral Equations of Motion

A *body* is represented by a collection of points, called *particles*, that move through three-dimensional Euclidean space. The set of all such points at time t, $\mathscr{S}(t)$, is called the *shape* of the body; $\mathscr{S}(0)$ is called the *reference shape* and is assumed to be a connected region.

In classical mechanics a set of n particles may be denoted by $\{p_1, \ldots, p_n\}$, where an integer index identifies a particle. In continuum mechanics a particle is indexed (i.e., identified) by its *position* $\mathbf{X}$ in $\mathscr{S}(0)$, where $\mathbf{X}$ is taken with respect to a given frame with origin 0. Let $\mathbf{x}$ denote the position of this same particle at time t. Then its *motion* is a relation of the form

$$\mathbf{x} = \hat{\mathbf{x}}(\mathbf{X}, t), \qquad -\infty < t < \infty, \tag{A1}$$

where $\hat{\mathbf{x}}(\mathbf{X}, 0) = \mathbf{X}$. We assume that $\hat{\mathbf{x}}$ is one to one and sufficiently smooth and note that along boundaries or wave fronts, certain derivatives of $\hat{\mathbf{x}}$ may be discontinuous or fail to exist. The velocity and acceleration of a particle

are denoted and defined by
$$\mathbf{v} = \dot{\mathbf{x}} = \partial \hat{\mathbf{x}}(\mathbf{X}, t)/\partial t, \qquad \mathbf{a} = \ddot{\mathbf{x}} = \partial^2 \hat{\mathbf{x}}(\mathbf{X}, t)/\partial t^2. \tag{A2}$$

The equations of motion state that in an *inertial frame*, the change in the linear (rotational) momentum of a *fixed* set of particles over any time interval (t_1, t_2) is equal to the net linear (rotational) impulse acting on these particles. To be precise, let $\rho(\mathbf{X})$ denote the mass density of the body in $\mathscr{S}(0)$ and, for simplicity, assume that the impulses arise from body and contact forces. Furthermore, let V denote any subregion of $\mathscr{S}(0)$ having a piecewise smooth boundary ∂V with outward unit normal $\mathbf{N}$, and let dV and dA denote differential elements of volume and area within V and ∂V. If at time t, $\mathbf{f}\,dV$ denotes the force on the particles initially in dV and $\mathbf{S}\,dA$ denotes the force on the particles initially in dA exerted by the particles initially on the outward side of dA, then the equations of motion take the form

$$\int_{t_1}^{t_2} \left(\int_{\partial V} \mathbf{S}\, dA + \int_V \mathbf{f}\, dV \right) = \int_V \rho \mathbf{v}\, dV \bigg|_{t_1}^{t_2}, \tag{A3}$$

$$\int_{t_1}^{t_2} \left(\int_{\partial V} \mathbf{x} \times \mathbf{S}\, dA + \int_V \mathbf{x} \times \mathbf{f}\, dV \right) = \int_V \mathbf{x} \times \rho \mathbf{v}\, dV \bigg|_{t_1}^{t_2}. \tag{A4}$$

If $\mathbf{f}$ and $\rho \mathbf{v}$ are bounded in a neighborhood of $(\mathbf{X}, t)$ and if $\mathbf{S}$ is a continuous function of $(\mathbf{X}, t, \mathbf{N})$ at $(\mathbf{X}, t)$, then, by applying Eq. (A3) to an arbitrarily small tetrahedron containing $\mathbf{X}$, over an arbitrarily small interval containing t, it follows by a standard argument (Truesdell and Toupin, 1965, Section 203) that

$$\mathbf{S} = \mathbf{N} \cdot \mathbf{S} = \mathbf{S}^T \cdot \mathbf{N}, \tag{A5}$$

where $\mathbf{S} = \hat{\mathbf{S}}(\mathbf{X}, t)$ is the *nominal* stress tensor and $\mathbf{S}^T$ is the *first Piola–Kirchhoff* stress tensor.

B. Heat

The net heat that flows into a body over the interval (t_1, t_2) is given by

$$\int_{t_1}^{t_2} \mathcal{Q}\, dt \equiv \int_{t_1}^{t_2} \left(\int_{\partial V} v\, dA + \int_V r\, dV \right) dt, \tag{A6}$$

where $v\, dA$ is the heat *influx* across the image of dA and $r\, dV$ is the rate of heat production by the particles that initially comprised dV. By arguments similar to those that establish Eq. (A5), it may be shown that

$$v = -\mathbf{q} \cdot \mathbf{N}, \tag{A7}$$

where $\mathbf{q}$ is the *heat efflux* vector.

C. The Clausius–Duhem Inequality

The generally accepted form of the second law of thermodynamics is[†]

$$\int_V \eta \, dV \bigg|_{t_1}^{t_2} \geq \int_{t_1}^{t_2} \left(\int_{\partial V} \frac{v}{\theta} \, dA + \int_V \frac{r}{\theta} \, dV \right) dt, \tag{A8}$$

where $\eta \, dV$ is the *entropy* of the particles that initially comprised dV and $\theta > 0$ is the *absolute temperature*.

D. Components of the Nominal Stress Tensor

If the position $\mathbf{X}$ is given as a function of a set of curvilinear coordinates σ^i, $i = 1, 2, 3$, then the associated *covariant base vectors* are denoted and defined by

$$\mathbf{G}_i \equiv \partial \mathbf{X}/\partial \sigma^i \equiv \mathbf{X}_{,i}. \tag{A9}$$

Denoting the direct product of two vectors $\mathbf{u}$ and $\mathbf{v}$ by $\mathbf{uv}$, we may represent $\mathbf{S}$ in the form

$$\mathbf{S} = \mathbf{G}_i S^i, \tag{A10}$$

where we sum over a repeated index. We call the S^i the *contravariant vector components* of $\mathbf{S}$. Inserting Eq. (A10) into Eq. (A5), we have

$$\mathbf{S} = S^i N_i, \tag{A11}$$

where $N_i = \mathbf{G}_i \cdot \mathbf{N}$ are the covariant components of $\mathbf{N}$.

Since $\mathbf{S} \, dA$ is the force acting on the *image* of dA in the *deformed* body, it is natural to resolve S^i into components along the *deformed covariant base vectors*

$$\mathbf{g}_i \equiv \mathbf{x}_{,i}. \tag{A12}$$

Thus with

$$S^i = S^{ij} \mathbf{g}_j \tag{A13}$$

Eqs. (A5) and (A11) take the form

$$\mathbf{S} = S^{ij} \mathbf{G}_i \mathbf{g}_j, \tag{A14}$$

$$\mathbf{S} = S^{ij} N_i \mathbf{g}_j. \tag{A15}$$

For sufficiently smooth fields, it may be shown that Eqs. (A3) and (A4)

[†] There is still a quest for the final word.

imply that

$$\mathbf{g}_i \times \mathbf{S}^i = \mathbf{0}. \quad (A16)$$

Thus, by Eq. (A13), $S^{ij} = S^{ji}$, although, in general, $\mathbf{S} \neq \mathbf{S}^T$. The symmetry of the components S^{ij} is a consequence of representing $\mathbf{S}$ in the mixed dyadic basis $\{\mathbf{G}_i \mathbf{g}_j\}$.

REFERENCES

Antman, S. S. (1968). General solutions for plane extensible elasticae having nonlinear stress–strain laws. *Q. Appl. Math.* **26**, 35–47.
Antman, S. S. (1969). Equilibrium states of nonlinearly elastic rods. In "Bifurcation Theory and Nonlinear Eigenvalue Problems" (J. B. Keller and S. Antman, eds.), pp. 331–358. Benjamin, New York.
Antman, S. S. (1974). Kirchhoff's problem for nonlinearly elastic rods. *Q. Appl. Math.* **32**, 221–240.
Antman, S. S. (1976). Ordinary differential equations of non-linear elasticity I: Foundations of the theories of non-linearly elastic rods and shells. *Arch. Rat. Mechs. Anal.* **61**, 307–351.
Atluri, S. N. (1983). Alternate stress and conjugate strain measures, and mixed variational formulations involving rigid rotations, for computational analyses of finitely deformed solids, with application to plates and shells—Part I: Theory, *Comput. Struct.* **16** (to appear).
Axelrad, E. L. (1962). Equations of deformation for shells of revolution and for the bending of thin-walled tubes subjected to large elastic displacements. *Am. Rocket Soc. J.* **32**, 1147–1151.
Axelrad, E. L. (1965). Refinement of the upper critical loading of pipe bending taking account of the geometric nonlinearity. *Izv. AN SSSR, OTN Mek.* **4**, 133–139.
Axelrad, E. L. (1981). On vector description of arbitrary deformation of shells. *Int. J. Solids Struct.* **17**, 301–304.
Becker, E. B., Carey, G. F., and Oden, T. J. (1981). "Finite Elements." vol 1. Prentice-Hall, Englewood Cliffs, New Jersey.
Berdichevskii, V. L. (1979). Variational-asymptotic method of constructing a theory of shells. *Appl. Math. Mech.* **43**, 711–736.
Berger, N. (1973). Estimates for stress derivatives and error in interior equations for shells of variable thickness with applied forces. *SIAM J. Appl. Math.* **24**, 97–120.
Biricikoglu, V., and Kainins, A. (1971). Large elastic deformations of shells with the inclusion of transverse normal strain. *Int. J. Solids Struct.* **7**, 431–444.
Boyle, J. T. (1981). The finite bending of curved pipes. *Int. J. Solids Struct.* **17**, 515–529.
Brazier, L. G. (1927). On the flexure of thin cylindrical shells and other 'thin' sections. *Proc. R. Soc. London, Ser. A* **116**, 104–114.
Brush, D. O. and Almroth, B. O. (1975). "Buckling of Bars, Plates and Shells." McGraw-Hill, New York.
Budiansky, B. (1968). Notes on nonlinear shell theory. *J. Appl. Mech.* **35**, 393–401.
Budiansky, B. (1974). Theory of buckling and post-buckling of elastic structures. *Adv. Appl. Mech.* **14**, 1–64.
Budiansky, B., and Sanders, J. L., Jr. (1963). On the "best" first-order linear shell theory. In "Progress in Applied Mechanics" (Prager Anniversary Volume), pp. 129–140. Macmillian, New York.
Bushnell, D. (1981a). Buckling of shells—Pitfalls for designers. *AIAA J.* **19**, 1183–1226.
Bushnell, D. (1981b). Elastic–plastic bending and buckling of pipes and elbows. *Comput. Struct.* **13**, 241–248.

Clark, A. E., and Narayanaswamy, O. S. (1967). Nonlinear membrane problems for elastic shells of revolution. *In* Proceedings of the Symposium on the Theory of Shells to Honor L. H. Donnell (D. Muster, ed.). Univ. of Houston Press, Houston.

Clark. R. A., and Reissner, E. (1951). Bending of curved tubes. *Adv. Appl. Mech.* **2**, 93–122.

Coleman, B. D., and Noll, W. (1963). The thermodynamics of elastic materials with heat conduction and viscosity. *Arch. Rat. Mech. Anal.* **51**, 1–53.

Comer, R. L., and Levy, S. (1963). Deflection of an inflated circular cylindrical cantilever beam. *AIAA J.* **1**, 1652–1655.

Cowper, G. R. (1966). The shear coefficient in Timoshenko's beam theory. *J. Appl. Mech.* **33**, 335–340.

Danielson, D. A. (1970). Simplified intrinsic equations for arbitrary elastic shells. *Int. J. Eng. Sci.*, **8**. 251–259.

Danielson, D. A. (1971). Improved error estimates in the linear theory of thin elastic shells. *Proc. Ned. Ak. Wet.* **B74**, 294–300.

Donnell, L. H. (1934). A new theory for the buckling of thin cylinders under axial compression and bending. *Trans. ASME* **56**, 795–806.

Ericksen, J. L. (1966). A thermo-kinetic view of elastic stability theory. *Int. J. Solids Struct.* **2**, 573–580.

Ferrarese, G. (1979). Dinamica dei continui sottili bidimensionali: formulazione intrinseca del problema di Cauchy. *Riv. Math. Univ. Parma* **5**, 321–335.

Friedrichs, K. O. (1950). Kirchhoff's boundary conditions and the edge effects for elastic plates. "Proceedings of the Third Symposium on Applied Mathematics." pp. 117–124. McGraw-Hill, New York.

Frisch-Fay, R. (1962). "Flexible Bars." Butterworth, London.

Goldenveiser, A. L. (1966). The principles of reducing three-dimensional problems of elasticity to two-dimensional problems of the theory of plates and shells. *In* "Proceedings of the Eleventh International Congress, Munich 1964," (H. Gortler, ed.), pp. 306–311. Springer-Verlag, Berlin and New York.

Goldenveiser, A. L. (1980). Asymptotic method in the theory of shells, *Theoretical and Applied Mechanics, Proc. 15th Int. Cong. Theor. & Appl. Mechs.*, Toronto 1980 (F. P. J. Rimrott and B. Tabarrok, eds.) pp. 91–104. North-Holland Publ., Amsterdam.

Green, A. E. and Adkins, J. E. (1970). "Large Elastic Deformations," 2nd ed., Oxford. Univ. Press, London and New York.

Green, A. E., and Naghdi, P. M. (1979). On thermal effects in the theory of shells. *Proc. R. Soc. London, Ser. A* **365**, 161–190.

Gurtin, M. E. (1973a). Thermodynamics and the potential energy of an elastic body. *J. Elasticity* **3**, 23–26.

Gurtin, M. E. (1973b). Thermodynamics and the energy criteria for stability. *Arch. Rat. Mech. Anal.* **52**, 93–103.

Gurtin, M. E. (1975). Thermodynamics and stability. *Arch. Rat. Mech. Anal.* **59**, 63–96.

Gurtin, M. E. (1979). Thermodynamics and the Griffith criterion for brittle fracture. *Int. J. Solids Struct.* **15**, 553–560.

Hamel, G. (1967). "Theoretische Mechanik." Springer-Verlag, Berlin and New York.

Huang, N. C. (1973). A variational principle for finite deformations of quasi-shallow shells. *Int. J. Mech. Sci.* **15**, 457–464.

Hughes, T. J. R., ed. (1981). "Nonlinear Finite Element Analysis of Plates and Shells," AMD-Vol. 48. Am. Soc. Mech. Eng., New York.

Hutchinson, J. W., and Koiter, W. T. (1970). Postbuckling theory. *Appl. Mech. Rev.* **23**, 1353–1366.

John, F. (1965a). A priori estimates applied to nonlinear shell theory. "Proceedings of the Symposium in Applied Mathematics," Vol. 17, pp. 102–110. Am. Math. Soc.

John, F. (1965b). Estimates for the derivatives of the stresses in a thin shell and interior shell equations. *Comm. Pure Appl. Math.* **18**, 235–267.
Jones, J. P. (1966). Thermoelastic vibrations of beams. *J. Acoustical Soc. Am.* **39**, 542–548.
Jordan, P. F., and Shelley, P. E. (1968). Effect of small bending stiffness on deformation of inflated structures. *In* "Proceedings of the I.A.F., 18th International Astronautics Congress, Belgrade 1967 (M. Lune, ed.), Vol. 2, pp. 209–217, Pergamon, Oxford.
Koga, T. (1972). Bending rigidity of an inflated circular cylindrical membrane of rubbery material. *AIAA J.* **10**, 1485–1489.
Koiter, W. T. (1960). A consistent first approximation in the general theory of thin elastic shells. *In* "The Theory of Thin Elastic Shells," Proceedings of I.U.T.A.M. Symposium, Delft, 1959 (W. T. Koiter, ed.), pp. 12–33. North-Holland Publ., Amsterdam.
Koiter, W. T. (1964). On the dynamic boundary conditions in the theory of thin shells. *Proc. Kon. Ned. Ak. Wet.* **B67**, 117–126.
Koiter, W. T. (1966). On the nonlinear theory of thin elastic shells, *Proc. Kon. Ned. Ak. Wet.* **B69**, 1–54.
Koiter, W. T. (1968). Purpose and achievements of research in elastic stability. *In* "Recent Advances in Engineering Science," Proc. 4th Tech. Meeting Soc. Engr. Sci., Raleigh, N. C., 1966, pp. 197–218.
Koiter, W. T. (1969). On the thermodynamic background of elastic stability theory. Problems of Hydrodynamics and Continuum Mechanics (Sedov Anniversary Volume). *SIAM*. Philadelphia, 423–433.
Koiter, W. T. (1970). On the mathematical foundation of shell theory. *Actes. Congr. Int. Math.* **3**, 123–130.
Koiter, W. T. (1971). Thermodynamics of elastic stability. "Proceedings of the 3rd Canadian Congress in Applied Mechanics., Calgary," pp. 29–37.
Koiter, W. T. (1980). The intrinsic equations of shell theory with some applications. *In* "Mechanics Today" (E. Reissner Anniversary Volume) (S. Nemat-Nasser, ed), Vol. 5, pp. 139–154. Pergamon, Elmsford, N.Y.
Koiter, W. T., and Simmonds, J. G. (1973). Foundations of shell theory. *In* "Proceedings of the Thirteenth International Congress on Theoretical and Applied Mechanics, Moscow, 1972. (E. Becker and G. K. Mikhailov, eds.), pp. 150–176, Springer-Verlag, Berlin and New York.
Ladeveze, P. (1976). Validity criteria of the nonlinear theory of elastic shells. *In* Lecture Notes in Mathematics, Vol. 503, pp. 384–394. Springer-Verlag, Berlin and New York.
Libai, A. (1962). On the nonlinear elastokinetics of shells and beams. *J. Aero. Sci.* **29**, 1190–1195.
Libai, A. (1972a). The nonlinear membrane shell with application to noncircular cylinders. *Int. J. Solids Struct.* **8**, 923–943.
Libai, A. (1972b). Pressurized cylindrical membranes with flexible supports. *In* I.A.S.S. Int. Symp. Pneumatic Structures, Delft, 1972.
Libai, A., (1981). On the nonlinear intrinsic dynamics of doubly curved shells. *J. Appl. Mech.* **48**, 909–914.
Libai, A., and Simmonds, J. G. (1981a). Large-strain constitutive laws for the cylindrical deformation of shells. *Int. J. Nonlinear Mech.* **16**, 91–103.
Libai, A., and Simmonds, J. G. (1981b). Highly nonlinear cylindrical deformations of rings and shells. Technion, Israel Institute of Technology. TAE Rep. No. 467.
Librescu, L. (1975). "Elastostatics and Kinetics of Anisotropic and Heterogeneous Shell-Type Structures." Noordhoff, Leyden.
Malcolm, D. J., and Glockner, P. G. (1972a). Cosserat surface and sandwich shell equilibrium. *Proc. A.S.C.E. J. Eng. Div.* **98** (EM5), 1075–1086.
Malcolm, D. J. and Glockner, P. G. (1972b). Nonlinear sandwich shell and Cosserat surface theory. *Proc. A.S.C.E. J. Eng. Mech. Div.* **98** (EM5), 1183–1203.

Mansfield, E. H. (1955). The inextensional theory for thin flat plates. *Q. J. Mech. Appl. Math.* **8**, 338–352.

Mansfield, E. H. (1969). Tension field theory, Proc. 12th Int. Congr. Appl. Mech., (M. Hetényi and W. G. Vincenti, eds.), 305–320. Springer-verlag, Berlin and New York.

Mansfield, E. H. (1970). Load transfer via a wrinkled membrane, *Proc. R. Soc., London, Ser. A*, **316**, 269–289.

Mansfield, E. H. (1981). An optimum surface of revolution for pressurized shells. *Int. J. Mech. Sc.* **23**, 57–62.

Mansfield, E. H., and Kleeman, P. W. (1955). Stress analysis of triangular cantilever plates. *Aircraft Eng.* **27**, 287–291.

Marguerre, K. (1938). Zur Theorie der gekrümmten Platte grosser Formänderung. In "Proceedings of the Fifth International Congress on Applied Mechanics," pp. 93–101. Wiley, New York.

McConnel, A. J. (1957). "Application of Tensor Analysis." Dover, New York.

Müller, I. (1969). On the entropy inequality. *Arch. Rat. Mech. Anal.* **26**, 118–141.

Mushtari, K. M. and Galimov K. Z. (1961). "Non-linear theory of thin elastic shells." Israel Program for Scientific Translations, Jerusalem (NASA–TT–F62).

Nachbar, W. (1969). Finite deformation of membranes and shells under localized loading. *In* "Theory of Thin Shells" (IUTAM Symp., Copenhagen, 1967) (F. I. Niordson, ed.), pp. 176–211. Springer-Verlag, Berlin and New York.

Naghdi, P. M. (1960). On the general problem of elastokinetics in the theory of shallow shells. *In* "The Theory of Thin Elastic Shells," Proceedings of I.U.T.A.M. Symposium Delft, 1959 (W. T. Koiter, ed.), pp. 301–330. North-Holland Publ. Amsterdam.

Naghdi, P. M. (1972). The theory of plates and shells. *In* "Encyclopedia of Physics," 2nd ed., Vol. Vla/2 (S. Flügge, ed.), pp. 425–640. Springer-Verlag, Berlin and New York.

Nicholson, J. W., and Simmonds, J. G. (1977). Timoshenko beam theory is not always more accurate than elementary beam theory. *J. Appl. Mech.* **44**, 337–338.

Niordson, F. I. (1971). A note on the strain energy of elastic shells. *Int. J. Solids Struct.* **7**, 1573–1579.

Novozhilov, V. V. (1953). "Foundations of the Nonlinear Theory of Elasticity." Graylock Press, Rochester, New York.

Novozhilov, V. V. (1970). "Thin Shell Theory," 2nd ed. Wolters-Nordhoff, the Netherlands.

Novozhilov, V. V., and Shamina, V. A. (1975). On kinematic boundary conditions in nonlinear elasticity problems. *Mech. Solids* **5**, 63–74.

Petrovsky, I. G. (1954). "Lectures on Partial Differential Equations." Wiley, New York.

Pflüger, A. (1975). "Stabilitätsprobleme der Elastostatik," 3rd ed. Springer-Verlag, Berlin, and New York.

Pietraszkiewicz, W. (1979). "Finite Rotations and Lagrangian Description in the Non-Linear Theory of Shells." Polish Scientific Pub., Warsaw.

Pietraszkiewicz, W. (1980a). Finite rotations in the non-linear theory of thin shells. *In* "Thin Shell Theory, New Trends and Applications" (W. Olszak, ed.), pp. 153–208. Springer Verlag, Vienna and New York.

Pietraszkiewicz, W. (1980b/a). Finite rotations in shells, *In* "Theory of Shells" (W. T. Koiter and G. K. Mikhailov, eds.), pp. 445–471. North-Holland Publ. Amsterdam.

Pietraszkiewicz., W. and Szwabowicz, M. L. (1981). Entirely Lagrangian non-linear theory of thin shells. *Arch. Mech.* **33**, 273–288.

Pipkin, A. D. (1968). Integration of an equation in membrane theory. *Z.A.M.P.* **19**, 818–819.

Prager, W., and Synge, J. L. (1947). Approximations in elasticity based on the concept of function space. *Q. Appl. Math.* **5**, 241–269.

Reissner, E. (1938). On tension field theory, Proc. 5th Int. Cong. Appl. Mech., pp. 359–361, Wiley, New York.

Reissner, E. (1950). On axisymmetrical deformations of thin shells of revolution. *Proc. Sym. Appl. Math.* **3**, 27–52.
Reissner, E. (1959a). On finite bending of pressurized tubes. *J. Appl. Mech.* **26**, 386–392.
Reissner, E. (1959b). On influence coefficients and nonlinearity for thin shells of revolution. *J. Appl. Mech.* **26**, 69–72.
Reissner, E. (1959c). The edge effect in symmetric bending of shallow shells of revolution. *Comm. Pure Appl. Math.* **12**, 385–398.
Reissner, E. (1961). On finite pure bending of cylindrical tubes. *Österreichisches Ingenieur* **15**, 165–172.
Reissner, E. (1962). Variational considerations for elastic beams and shells. *Proc. ASCE, Eng. Mech. Div.* **8**, 23–57.
Reissner, E. (1963a). On the equations for finite symmetrical deflections of thin shells of revolution. *In* "Progress in Applied Mechanics" (The Prager Anniversary Volume) (D. C. Drucker, ed.), pp. 171–178, Macmillian, New York.
Reissner, E. (1963b). On stresses and deformations in toroidal shells of circular cross sections which are acted upon by uniform normal pressure. *Q. Appl. Math.* **21**, 177–188.
Reissner, E. (1969). On finite symmetrical deflections of thin shells of revolution. *J. Appl. Mech.* **36**, 267–270.
Reissner, E. (1970). Variational methods and boundary conditions in shell theory. *In* "Studies in Optimization," pp. 78–94. SIAM, Philadelphia.
Reissner, E. (1972a). On one-dimensional finite-strain beam theory. *Z.A.M.P.* **23**, 795–804.
Reissner, E. (1972b). On finite symmetrical strain in thin shells of revolution. *J. Appl. Mech.* **39**, 1137.
Reissner, E. (1974). Linear and nonlinear theory of shells. *In* "Thin-Shell Structures" (Sechler Anniversary Volume) (Y. C. Fung and E. E. Sechler, eds.), pp. 29–44, Prentice-Hall, Englewood Cliffs. New Jersey.
Reissner, E. (1975). Note on the equations of finite-strain force and moment stress elasticity. *Stud. Appl. Math.* **54**, 1–8.
Reissner, E. (1981). On finite pure bending of curved tubes. *Int. J. Solids Struct.* **17**, 839–844.
Reissner, E. (1982). A note on two-dimensional, finite-deformation theories of shells. *Int. J., Non-linear Mech.* **17**, 217–221.
Reissner, E., and Wan, F. Y. M. (1965). Rotating shallow elastic shells of revolution. *J. Soc. Indust. Appl. Math* **13**, 333–352.
Reissner, E., and Weinitschke, H. J. (1963). Finite pure bending of circular cylindrical tubes. *Q. Appl. Math.* **20**, 305–319.
Rigbi, Z., and Hiram, Y. (1981). An approximate method for the study of large deformations of membranes. *Int. J. Mech. Sci.* **23**, 1–10.
Rossettos, J. N., and Sanders, J. L., Jr. (1965). Toroidal shells under internal pressure in the transition range. *AIAA J.* **3**. 1901–1909.
Sanders, J. L., Jr. (1959). An improved first-approximation theory for thin shells. NASA Rep. No. 24.
Sanders, J. L. Jr. (1963). Nonlinear theories for thin shells. *Q. Appl. Math.* **21**, 21–36.
Schmidt, R. and DaDeppo, D. (1975). Theory of axisymmetrically loaded membranes of revolution undergoing arbitrarily large deflections. *Indus. Math.* **25**, 9–16.
Schwerin, E. (1929). Über Spannungen und Formänderungen Kreisringförmiger Membranen. *Z. Tech. Phys.* **12**, 651–659.
Simmonds, J. G. (1961). The general equations of equilibrium of rotationally symmetric membranes and some static solutions for uniform centrifugal loading. NASA Rep. No. D-816.
Simmonds, J. G. (1966). A set of simple, accurate equations for circular cylindrical elastic Shells. *Int. J. Solids Struct.* **2**, 525–541.

Simmonds, J. G. (1975). Rigorous expunction of Poisson's ratio from the Reissner–Meissner equations. *Int. J. Solids Struct.* **11**, 1051–1056.
Simmonds, J. G. (1976). Recent advances in shell theory. *In* "Advances in Engineering Science." NASA CP-2001, pp. 617–626.
Simmonds, J. G. (1977). Saint-Venant's principle for semi-infinite shells of revolution. *Recent Advances in Engineering Science* **8**, 367–374, Scientific Publishers, Boston.
Simmonds, J. G. (1979a). Accurate nonlinear equations and a perturbation solution for the free vibrations of a circular elastic ring. *J. Appl. Mech.* **46**, 156–160.
Simmonds, J. G. (1979b). Special cases of the nonlinear shell equations. *In* "Trends in Solid Mechanics" (J. F. Besseling and A. M. A. van der Heijden, eds.), pp. 211–223. Delft Univ. Press, Sijthoff & Noordhoff.
Simmonds, J. G. (1979c). Surfaces with metric and curvature tensors that depend on one coordinate only are general helicoids. *Q. Appl. Math.* **37**, 82–85.
Simmonds. J. G. (1981). Exact equations for the large inextensional motion of elastic plates. *J. Appl. Mechs.* **48**, 109–112.
Simmonds, J. G. and Danielson, D. A. (1970). Nonlinear shell theory with a finite rotation vector, *Kon. Ned. Ak. Wet.* **B 73**, 460–478.
Simmonds, J. G. and Danielson, D. A. (1972). Nonlinear shell theory with finite rotation and stress-function vectors. *J. Appl. Mech.* **39**, 1085–1090.
Simmonds, J. G. and Libai, A. (1979a). Exact equations for the inextensional deformation of cantilevered plates. *J. Appl. Mech.* **46**, 631–636.
Simmonds, J. G. and Libai, A. (1979b). Alternate exact equations for the inextensional deformation of arbitrary, quadrilateral, and triangular plates. *J. Appl. Mech.* **46**, 895–900.
Sobel, L. H. and Newman, S. Z. (1980). Comparison of experimental and simplified analytical results for the in-plane plastic bending and buckling of an elbow. *J. Press. Vess. Tech.* **102**, 400–409.
Stein, M. (1968). Some recent advances in the investigation of shell buckling. *AIAA Journal* **6**, 2339–2345.
Stein, M. and Hedgepeth, J. M. (1961). Analysis of partly wrinkled membranes. *NASA* TN D-813.
Strang, G., and Fix, G. J. (1973). "An Analysis of the Finite Element Method." Prentice-Hall, Englewood Cliffs, N.J.
Struik, D. J. (1961). "Differential Geometry," 2nd Ed. Addison-Wesley, Reading, Mass.
Thurston, G. A. (1977). Critical bending moment of circular cylindrical tubes. *J. Appl. Mech.* **44**, 173–175.
Truesdell, C. A. (1969). "Rational Thermodynamics." McGraw-Hill, New York.
Truesdell, C. A. (1977). "A First Course in Rational Continuum Mechanics," Vol 1. Academic Press, New York.
Truesdell, C. A., and Toupin, R. A. (1965). The classical field theories. *In* "Encyclopedia of Physics," (S. Flugge, ed.) vol. III/1. Springer-Verlag, Berlin and New York.
Van der Heijden, A. M. A. (1976). On modified boundary conditions for the free edge of a shell. Thesis, University of Delft.
Vekua, I. N. (1962). "Generalized Analytic Functions," Pergamon, Oxford.
Vlasov, V. Z. (1958). Allgemeine Schalentheorie und Ihre Anwendung ln der Technik, *Akademie Verlag. Berlin. NASA* TT F-99, April 1964.
Von Karman, T. (1911). Über die formänderung dünnwandiger Rohre. *Z. Vereines deutscher Ingenieure* **55**, 1889–1895.
Wagner, H. (1929). Flat sheet metal girders with a very thin metal web. *Z. Flugtechn. Motorluftschiffahrt* **20**. 200–314 [NASA TM, 604–606].
Wan, F. Y. M. (1968). Two variational theorems for thin shells. *J. Math. & Phys.* **47**, 429–431.

Weinitschke, H. J. (1980). On axisymmetric deformations of nonlinear elastic membranes. *In* "Mechanics Today" (Reissner Anniversary Volume). (S. Nemat-Nasser, ed.), Vol. 5, pp. 523–542. Pergamon, Oxford.

Wempner, G. (1969). Finite elements, finite rotations and small strain of flexible shells. *Int. J. Solids Struct.* **5**, 117–153.

Wittrick, W. H. (1964a). A non-linear theory for the edge stresses in thin shells. *Int. J. Eng. Sci.* **21**, 155–177.

Wittrick, W. H. (1964b). Non-linear discontinuity stresses in shells of revolution under internal pressure. *Int. J. Eng. Sci*, **21**, 179–188.

Wood, J. D. (1958). The flexure of a uniformly pressurized, circular cylindrical shell. *J. Appl. Mech.* **25**, 453–458.

Wu, C. H. (1970a). On the solutions of a nonlinear membrane problem. *SIAM J. Appl. Math.* **18**, 738–747.

Wu, C. H. (1970b). On certain integrable nonlinear membrane solutions. *Q. Appl. Math.* **28**, 81–90.

Wu, C. H. (1974). The wrinkled axisymmetric bags made of inextensible membrane. *J. Appl. Mech.* **41**, 963–968.

Wu, C. H. (1978). Nonlinear wrinkling of nonlinear membranes of revolution. *J. Appl. Mech.* **45**, 533–538.

Wu, C. H. (1979). Large finite strain membrane problems. *Q. Appl. Math.* **36**, 347–359.

Yang, W. H., and Feng, W. W. (1970). On axisymmetrical deformations of nonlinear membranes. *J. Appl. Mech.* **37**, 1002–1011.

Note Added in Proof

A semiintrinsic form of the dynamic equations of general shell theory has been developed by Ferrarese (1979) in which angular velocities are the basic field variables. The approach of Libai (1981), presented in Section III,J,3,b was developed independently and may be considered as an extension of Ferrarese's work. The boundary conditions in the two approaches are similar, but the main concern of Ferrarese is the proper formulation of an initial value problem in the Cauchy sense.

Elastic Surface Waves with Transverse Horizontal Polarization

GÉRARD A. MAUGIN

Laboratoire de Mécanique Théorique Associé au CNRS
Université Pierre-et-Marie-Curie
Paris, France

I. Introduction	373
II. Some General Features of Surface Waves	380
III. The Notion of Resonance Coupling between Modes	385
IV. A Typical SH Surface Mode: The Bleustein–Gulyaev Wave	391
V. SH Surface Waves in Nonhomogeneous Elastic Media	395
A. The Generalized Sturm–Liouville Problem	398
B. The Simple Sturm–Liouville Problem	400
C. Other Problems	402
VI. Electroacoustic SH Surface Waves in Ferroelectrics	407
A. General Features	407
B. Coupled Bulk Modes and Splitting of the Surface Wave Problem	409
C. Piezoelectric Rayleigh Modes in Elastic Ferroelectrics	411
D. Bleustein–Gulyaev Modes in Elastic Ferroelectrics	415
VII. Magnetoacoustic SH Surface Waves in Magnetoelasticity	421
A. General Features	421
B. The Nonreciprocity of Surface Modes	424
C. SH Surface Waves of the Bleustein–Gulyaev Type	426
References	431

I. Introduction

Generally speaking, surface waves are time-varying, spatially nonuniform perturbations that exhibit spatial variation in the field amplitude markedly confined to the vicinity of the limiting surface of a body and practically nil outside this relatively narrow zone. The treatment of surface waves is the simplest dynamic problem concerning a body of finite extent in one of its dimensions. It requires considering boundary conditions so that, in principle at least, any field theory and resulting continuous field equations that are accompanied by well-set boundary conditions can be made to exhibit a surface wave phenomenon to a greater or lesser degree. The stability in time of such a phenomenon is not guaranteed in all cases.

The paragon of surface waves remains the *Rayleigh wave mode* (Rayleigh, 1885), which represents a nondispersive, mixed, longitudinal–transverse, time-harmonic plane elastic mode of propagation and propagates at a speed less than that of volume shear elastic waves at the *free* surface of a semi-infinite *isotropic, homogeneous* linear elastic space. Its amplitude, polarized in the *sagittal plane* (the plane spanned by the direction of the real wave vector and the unit normal to the limiting plane), decreases exponentially away from the surface in the elastic material, with a penetration depth of the order of a few wavelengths (Fig. 1). It was later remarked that for *anisotropic* homogeneous linear elastic bodies (e.g., crystals under small and moderate loadings), the exponential decay is accompanied by a sinusoidal oscillation in the amplitude with depth (e.g., Buckens, 1958; Stoneley, 1955; Synge, 1956–1957; Currie, 1974; Ingebrigten and Tonning, 1977) and the varying elastic displacement generally remains *fully three dimensional* and not confined to the sagittal plane. This is well understood if one recalls that for volume elastic waves in most crystals, the notions of pure longitudinality and transversality are lost for an arbitrary direction of propagation (see Fig. 2). Numerous technical books contain chapters devoted to Rayleigh

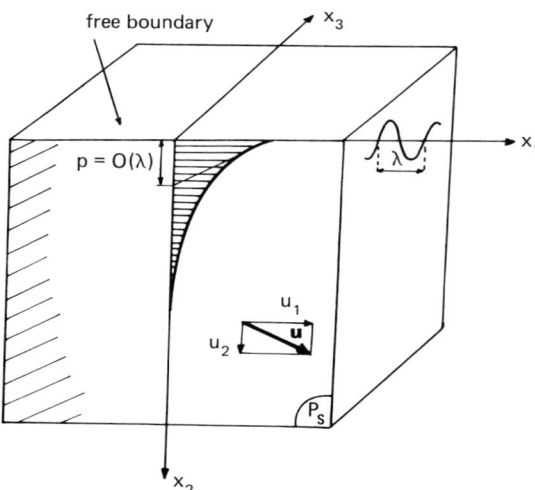

FIG. 1. Rayleigh waves in isotropic linear elastic homogeneous, solids.

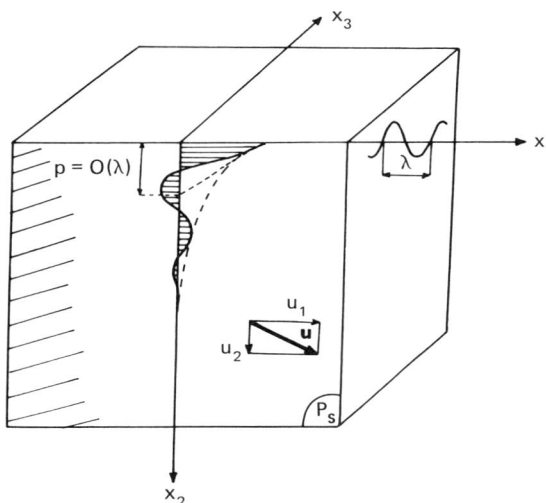

FIG. 2. Generalized Rayleigh waves (anisotropic linear elastic homogeneous solids or prestressed isotropic elastic homogeneous solids).

waves in isotropic media and crystals (e.g., Achenbach, 1973; Dieulesaint and Royer, 1974; Eringen and Suhubi, 1975; Graff, 1975; Musgrave, 1970; Viktorov, 1967). Several reviews offer extensive discussions on Rayleigh waves and their technological applications (e.g., Chadwick and Smith, 1979; De Klerk, 1972; Farnell, 1970, 1978; Farnell and Adler, 1972; Pajewski, 1978; Smith, 1970; Viktorov, 1979; White, 1970). Pure Rayleigh waves (elastic displacement confined to the sagittal plane) and generalized Rayleigh waves (elastic displacement with three rectangular components) receive many applications, among which we may cite seismology and geological prospecting, at low frequencies (see, e.g., Ewing *et al*, 1957; Colloque International du C.N.R.S., 1961); nondestructive inspection of sample surfaces and flaw detection, at high frequencies (e.g., 10^6 Hz); and *acoustoelectronics*, at higher frequencies, about 10^6–10^{10} Hz, (see, for instance, Dieulesaint and Royer, 1974; White, 1970).

Pure Rayleigh waves may also travel along curved surfaces, two striking examples of this phenomenon being those of Rayleigh waves on cylinders (see Viktorov, 1979) and Rayleigh waves along a periodically curly-shaped interface (Fokkema, 1979). Generalized Rayleigh waves may also propagate in an isotropic but *prestressed* elastic half space (see, e.g., Buckens, 1958; Nalamwar and Epstein, 1976; Wilson, 1973, 1974). The obvious reason for this is the elastic *anisotropy inducement* caused by initial stresses (a phenomenon akin to optical anisotropy inducement in photoelasticity).

So-called Stoneley waves—after their discovery by Stoneley in 1924 (see Graff, 1975)—provide another example of elastic surface wave where the elastic displacement is polarized in the sagittal plane (for instance, at the common interface of an isotropic homogeneous linear elastic body and a liquid, as is the case at the bottom of oceans for seismic propagation) (Fig. 3). These can also be used in ultrasonic flaw detection and acoustoelectronics, but they exist only for a definite range of ratios between densities and elastic parameters of the adjoining media.

In the case of *linear elastic piezoelectric* bodies, where an electric field coexists with strain and stress fields, a study of surface waves is also possible. In a general manner, the influence of piezoelectricity on the propagation of elastic surface waves materializes in the possibility of *piezoelectric Rayleigh waves* (see Fig. 4). In the classical linear theory of piezoelectricity these waves correspond to propagation of the displacement component in the sagittal plane and of the electric potential (quasi-electrostatic approximation; see Balakirev and Gorchakov, 1977; Coquin and Tiersten, 1967; Dieulesaint and Royer, 1974; Taylor and Crampin, 1978; Tiersten, 1969). Obviously, the crystal symmetry must allow for the existence of piezoelectricity, and generalized Rayleigh waves are more often encountered than pure Rayleigh waves. Piezoelectric Rayleigh waves (or *piezoelectrically stiffened* Rayleigh waves) present no dispersion, have a penetration depth of

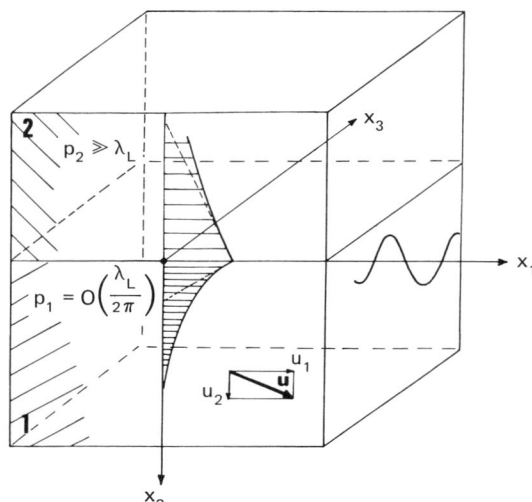

FIG. 3. Stoneley waves (rigid bond at the interface, e.g., medium 1: isotropic linear elastic solid, medium 2: liquid; wave energy mainly concentrated in the liquid, $p_2 \gg p_1$).

Elastic Surface Waves with Transverse Horizontal Polarization

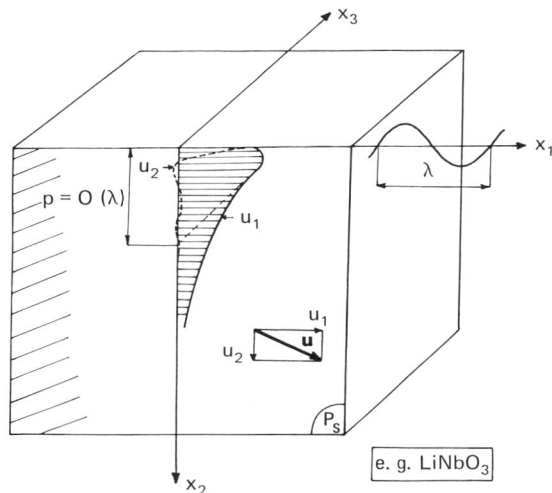

FIG. 4. Piezoelectrically stiffened Rayleigh wave (piezoelectric linear elastic crystals).

a few wavelengths, and their speed is somewhat greater than the classical Rayleigh speed, owing to a slight positive correction resulting from electromechanical couplings. Remarkably enough, the electric boundary condition that complements the bulk electrostatic equations (grounded boundary, matching with the external potential) has little influence on the displacement behavior. Acoustoelectronics is the privileged domain of application of these waves. Magnetoelastic bodies with a paramagnetic or soft ferromagnetic type of magnetic behavior allow for the existence of "*magnetically stiffened*" Rayleigh waves in the same manner, piezoelectricity being replaced by *magnetostriction* (an effect quadratic in the magnetic field, and thus permitted in a wider class of crystals than piezoelectricity) (see Parekh and Bertoni, 1974a,b; Tsutsumi *et al*, 1975; Viktorov, 1975).

In the previously mentioned piezoelectric and magnetostrictive cases the types of surface waves considered are modifications of the surface waves known to exist in purely elastic homogeneous materials (pure and generalized Rayleigh waves for a semi-infinite space). Given the wealth of the technical literature on these waves and their piezoelectric and magnetostrictive generalizations, we have chosen here to focus on the case of elastic surface waves, and their generalizations to electromagnetic continua, which do *not* exist (for they are unstable) in classical linear homogeneous elasticity, namely, waves or combined waves associated with an elastic displacement parallel to the limiting surface but *orthogonal* to the sagittal plane. These are so-called *SH surface waves* or *shear* elastic surface waves with *transverse horizontal* polarization. The SH denomination is in accordance with seismological usage (see Bullen, 1965). Mechanical and civil engineering practices refer, rather, to an *out-plane wave motion* (see Achenbach, 1973, p. 58).

It was A. E. H. Love in 1911 who put forward the first theory of SH surface waves. A plane shear wave grazing along the boundary of a semi-infinite isotropic homogeneous linear elastic space strictly satisfies the volume field equations and stress-free boundary conditions of linear elasticity. This grazing transverse volume wave is *unstable* in the sense that a slight variation of the boundary conditions or properties of the medium converts it into a surface wave. In the case of Love waves (see Fig. 5), this variation is built up in the structure, which is composed of an elastic half space covered by a layer of elastic material wherein the velocity of bulk shear waves is lower than that of those waves in the substrate (the so-called *slow-wave layer*). The bond at the interface is rigid. As for elastic plates, a series of modes of propagation is possible. Only the amplitude behavior of the first mode is sketched out in Fig. 5. We may say that Love waves are *guided* by the superimposed slow-wave layer, but that they simultaneously, penetrate the substrate. They can be generalized to both linear piezoelectric (see, e.g., Sinha and Tiersten, 1973; Tiersten, 1963) and magnetostrictive (see, e.g., Van de Vaart, 1971) cases. They can also travel along curved layers, as is the case in seismology (see Gérard, 1979, 1980).

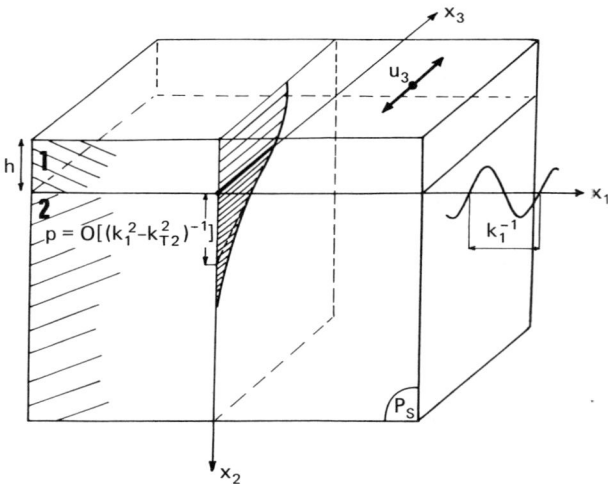

FIG. 5. First mode of Love waves (rigid bond at the interface between two isotropic linear elastic solids; $c_{T1} < c_{T2}$; slow-wave layer).

A second possibility giving rise to the existence of SH surface waves is a symmetry condition relative to the anisotropy of a piezoelectric material, (Fig. 6). These are the so-called *Bleustein–Gulyaev waves*, after their simultaneous discovery by Bleustein (1968) and Gulyaev (1969). Such nondispersive SH surface waves exist when the sagittal plane is orthogonal to an axis d of symmetry of order 6 (for instance, d may be the direction of the sixfold axis for a crystal in class $C_{6v} = 6mm$ or the poling direction of a poled ferroelectric ceramic). A further property of these waves is of great interest: Their penetration depth is large compared to the wavelength λ (between 10λ and 1000λ) and their decay with depth essentially depends on piezoelectricity, whereas usual Rayleigh waves have a markedly smaller penetration depth. Such waves prove to be useful in high-frequency surface wave devices (see Dieulesaint and Royer, 1974; Pajewski, 1977).

Love waves and Bleustein–Gulyaev waves provided the impetus for the recent advances reported in this article. A third factor is that acoustic surface waves on solids may couple with other elementary excitations accessible to the surface acoustic wave frequency range. Two such elementary excitations are *spin waves* or *magnons* (collective oscillations in the ordered magnetic spin system typical of ferromagnets such as yttrium–iron garnets and other magnetically ordered crystals) and *soft ferroelectric modes* or the quasi-electrostatic version of *polaritons* (collective oscillations related to the internal motion in molecular ferroelectric crystals of the displacive type

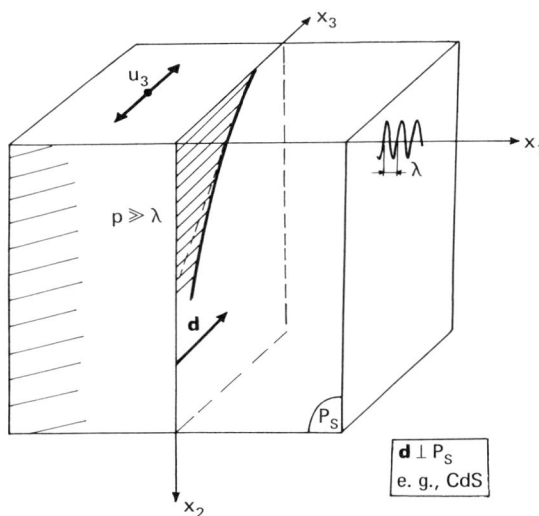

FIG. 6. Bleustein–Gulyaev surface waves in piezoelectric crystals (SH bulk waves perturbed to meet the electrical contribution to the boundary condition).

such as barium titanate). Ferromagnetic and ferroelectric states imply the existence of relatively strong static fields, which in turn modify the essential properties of wave modes. Surface spin waves are already known to exist independently of magnetoacoustic couplings (see Damon and Eshbach, 1961; Filipov, 1967). The same holds true of surface soft ferroelectric modes (see Fischer and Lagois, 1979). Coupled magnetoacoustic surface waves then provide means for exciting a spin system in ferromagnets and a unique possibility for studying the bulk spin wave structure and the surface spin waves with a probe of short wavelength. Similarly, coupled acoustoelectric surface waves, with elastic polarization of the Rayleigh or Love type, provide a means for analyzing soft ferroelectric modes in elastic ferroelectrics.

By way of prerequisites, we recall a few mathematical features of surface wave propagation in Section II and the notion of resonance coupling between modes in Section III. The exemplary case of Bleustein–Gulyaev waves in *linear* piezoelectrics is then briefly examined in Section IV. SH surface waves in *inhomogeneous* isotropic linear elasticity are the subject of Section V. Surface modes with SH elastic polarization in elastic ferroelectrics and *elastic* ferromagnets are examined in Sections VI and VII, respectively. Most of the models and methods reported rely heavily on the advances made in rational continuum mechanics during the last 15 years, *especially* insofar as nonlinear theories with finite strains provide the starting point in several cases and systematic use of the method of superimposition of small dynamical signals on finite bias static fields and the anisotropy inducement by these bias fields are considered.

II. Some General Features of Surface Waves

Consider the half space $\mathscr{D} = \{x_2 > 0\}$ of Fig. 1 and assume that an indexed array of scalar fields u_j, $j = 1, 2, \ldots, n$, satisfies in $\mathscr{D}$ the set of linear, or linearized, homogeneous equations

$$D_t u_k - P_{kj} u_j = 0, \qquad k = 1, 2, \ldots, n \tag{2.1}$$

where D_t is a time differential operator (e.g., second-order time derivative) and the P_{kj}'s are polynomials of differentiation with respect to the space coordinates x_i, $i = 1, 2, 3$. For time-harmonic plane solutions of the type

$$g = \tilde{g} \exp[i(\omega t - \mathbf{k} \cdot \mathbf{x})] \tag{2.2}$$

for all u_k's, where ω is the angular frequency and $\mathbf{k}$ is the wave vector, system (2.1) is *nondispersive* if and only if the joint space–time polynomial of differentiation in (2.1) is homogeneous. This is the case of the classical "wave" equation

$$\left(\frac{\partial^2}{\partial t^2} - c^2 \Delta\right) u = 0. \tag{2.3}$$

Otherwise, system (2.1) is *dispersive*. The problem (2.1), (2.2) in the absence of boundary conditions constitutes the search for free *bulk* or *volume* wave modes.

In a well-constructed field theory the bulk equations (2.1) are complemented by boundary conditions of the type

$$B_{kj} u_j = 0, \quad k = 1, 2, \ldots, n, \tag{2.4}$$

on $\partial \mathscr{D} = \{x_2 = 0\}$. On selecting a wave propagating along the x_1 direction in Fig. 1, a typical *time-harmonic plane surface wave* is a solution of Eqs. (2.1) and (2.4) that persists or decays in time and has the form

$$u_j = \hat{u}_j \exp[i(\omega t - k_1 x_1 - k_2 x_2)], \tag{2.5}$$

where k_1 is the *real* wave number and k_2 is the complex attenuation coefficient with depth such that

$$\mathrm{Im}(k_2) \leq 0 \quad \text{for } x_2 > 0 \tag{2.6}$$

if the nonzero solution is to be localized in the vicinity of $\partial \mathscr{D}$, hence decays exponentially with depth away from $\partial \mathscr{D}$.

The study of surface wave solutions, their dispersion relation, and their behavior with depth in the substrate $\mathscr{D}$ follows a now classic routine. On substituting from Eq. (2.5) into Eq. (2.1), one is led to a linear homogeneous system for the u_j's such that

$$L_{ij}(\Omega, q_1, q_2) \hat{u}_j = 0, \quad i, j = 1, 2, \ldots, n, \tag{2.7}$$

where Ω, q_1, and q_2 are nondimensional quantities defined by

$$\Omega = \frac{\omega}{\omega_0}, \quad q_1 = \frac{k_1}{k_0} = \lambda_0 k_1, \quad q_2 = \frac{k_2}{k_0} = \lambda_0 k_2, \tag{2.8}$$

if ω_0, k_0, and λ_0 are characteristic frequencies, wavenumbers, and wavelengths, respectively (in general associated with bulk wave modes). System (2.7) admits nontrivial solutions if and only if

$$\mathscr{P}(\Omega, q_1, q_2) = \det|L_{ij}(\Omega, q_1, q_2)| = 0. \tag{2.9}$$

The polynomial $\mathscr{P}$ has real coefficients in the absence of damping. Equation (2.9) can be solved for q_2. The n corresponding roots are either real or complex conjugates; thus

$$q_2^\alpha = G_\alpha(\Omega, q_1), \quad \alpha = 1, 2, \ldots, n. \tag{2.10}$$

Condition (2.6) therefore imposes that

$$\text{Im}[G_\alpha(\Omega, q_1)] \leq 0, \qquad \alpha = 1, 2, \ldots, n. \tag{2.11}$$

Each of these inequalities delineates an admissible domain $\mathscr{D}_\alpha$ in the real plane $\mathbb{R}^2$ such that Eq. (2.11) be satisfied for a given α if the pair (Ω, q_1) represents a point inside $\mathscr{D}_\alpha$. For the entire set of n conditions (2.11), Ω and q_1 must be related in such a way that

$$(\Omega, q_1) \subseteq \mathscr{D}_A = \bigcap_{\alpha=1}^{n} \mathscr{D}_\alpha. \tag{2.12}$$

The *dispersion relation* is the relation between Ω and q_1. It is established as follows. For each value q_2^α satisfying Eq. (2.10), system (2.7) provides an elementary solution indexed $\hat{u}_j^\alpha$, hence a corresponding u_j^α. In view of the linearity of the problem, the global solution u_j is a linear combination of these elementary solutions:

$$u_j = \left(\sum_{\alpha=1}^{n} C_\alpha \hat{u}_j^\alpha \exp(-ik_2^\alpha x_2) \right) \mathscr{F}, \qquad k_2^\alpha = k_0 q_2^\alpha, \tag{2.13}$$

where $\mathscr{F}$ is the propagation factor $\exp[i(\omega t - k_1 x_1)]$ and $\hat{u}_j^\alpha$ is an n-dimensional eigenvector of the system (2.7) corresponding to an admissible q_2^α for a given Ω. Accounting now for the set (2.4) of boundary conditions at $x_2 = 0$ and substituting for Eq. (2.13), we obtain a linear homogeneous system for the C_α's in the form

$$\sum_{\alpha=1}^{n} D_{\beta\alpha}(q_1, q_2^\gamma) C_\alpha = 0, \qquad \alpha, \beta, \gamma = 1, 2, \ldots, n. \tag{2.14}$$

Again this system possesses nontrivial solutions if and only if

$$\det |D_{\beta\alpha}(q_1, q_2^\gamma)| = 0. \tag{2.15}$$

This is an implicit equation in Ω when we account for Eq. (2.10). Therefore Eq. (2.15) can also be written as

$$\mathscr{D}(\Omega, q_1) = 0, \tag{2.16}$$

which is the *dispersion relation*. The curve (or curves) represented by Eq. (2.16) must be entirely situated in the allowed domain (2.12) if we want the solutions to die out away from the boundary. On carrying now the coefficients C_α, solved from Eq. (2.14) up to a multiplicative constant, into Eq. (2.13) and accounting for Eq. (2.16), we obtain the behavior of surface wave modes with depth as [compare Eq. (2.2)]

$$\tilde{g}(x_2) = \sum_{\alpha=1}^{n} \Gamma_\alpha(q_1) \exp[-\zeta_\alpha(q_1) X], \tag{2.17}$$

where we have set
$$X = k_1 x_2, \qquad \zeta_\alpha(q_1) = i\frac{k_2^\alpha}{k_1} = i\frac{q_2^\alpha}{q_1}. \tag{2.18}$$

Solution (2.17) has the general behavior
$$\tilde{g}(x_2) \propto \exp(-x_2/p) \tag{2.19}$$

for large positive x_2's. The parameter p is the *penetration depth* of the surface wave in the substrate.

Crucial in the preceding reasoning is the selection of the set of admissible q_2^α's such that Eq. (2.6) or (2.12) be satisfied. For example, in the classical Rayleigh wave problem the elastic displacement $\mathbf{u} = \{u_1, u_2, 0\} \subset P_S$, where P_S is the sagittal plane. Equations (2.1) and (2.4) are given by

$$\begin{aligned}\ddot{\mathbf{u}} - (c_L^2 - c_T^2)\nabla(\nabla \cdot \mathbf{u}) - c_T^2 \nabla^2 \mathbf{u} = \mathbf{0} & \quad \text{for} \quad x_2 > 0, \\ \left.\begin{array}{r}u_{1,2} + u_{2,1} = 0 \\ (c_L^2 - 2c_T^2) u_{1,1} + c_L^2 u_{2,2} = 0\end{array}\right\} & \quad \text{for} \quad x_2 = 0 \end{aligned} \tag{2.20}$$

where
$$\nabla = \left(\frac{\partial}{\partial x_1}, \frac{\partial}{\partial x_2}\right), \qquad \nabla^2 = \nabla \cdot \nabla,$$

and c_L and c_T are the velocities of bulk longitudinal and transverse waves, respectively, in isotropic homogeneous linear elasticity. Condition (2.9) reads
$$\begin{aligned}\mathscr{P}_R(\Gamma, K) = (\Gamma - c_T^2 K)(\Gamma - c_L^2 K) = 0, \\ \Gamma = \omega^2, \qquad K = k_1^2 + k_2^2.\end{aligned} \tag{2.21}$$

The *two* roots corresponding to Eq. (2.10) are given by
$$K_I = \Gamma/c_T^2, \qquad K_{II} = \Gamma/c_L^2, \qquad K_J = k_1^2 + (k_2^J)^2, \quad J = 1, 2. \tag{2.22}$$

Condition (2.6) is equivalent to $(k_2^J)^2 \leq 0$ for $J = 1, 2$, hence
$$(k_2^I)^2 = (\Gamma/c_T^2) - k_1^2 \leq 0, \qquad (k_2^{II})^2 = (\Gamma/c_L^2) - k_1^2 \leq 0. \tag{2.23}$$

Thus with $c_L > c_T$, in the positive quadrant of $\mathbb{R}^2$, $\mathscr{D}_A$ is the domain situated below the straight line of slope c_T. Hence the propagation speed of Rayleigh waves is less than that of bulk shear waves in the same material. On looking for nondispersive solutions such that ω and k_1 be related by $\omega^2 = c_R^2 k_1^2$, on account of Eq. (2.20) one finds that the dispersion relation $\mathscr{D}_R(\omega, k_1) = 0$ [compare Eq. (2.16)] is equivalent to finding the roots of the celebrated Rayleigh equation
$$4(1 - \xi)^{1/2}\left(1 - \frac{\xi}{\gamma_0}\right)^{1/2} = (2 - \xi)^2, \tag{2.24}$$

where

$$\gamma_0 = (c_L/c_T)^2, \quad \xi = (c_R/c_T)^2. \tag{2.25}$$

The case where $\xi = 0$ is an obvious solution of Eq. (2.24), which therefore reduces to a cubic in ξ. Only the nonzero positive root of this cubic which is less than unity is an admissible solution, so that

$$c_R = \sqrt{\xi} c_T, \quad 0 < \xi < 1. \tag{2.26}$$

The exact value of ξ depends on γ_0, and therefore on Poisson's ratio. The penetration depth p is of the order of $(|k_2^I|)^{-1}$ or $(|k_2^{II}|)^{-1}$, and therefore, according to Eq. (2.23), of the order of the wavelength λ of bulk, shear, or dilatational elastic waves in the solid.

In many cases the generalized field u_j in Eq. (2.1) has many more components than in the pure Rayleigh case (for instance, in the case of magnetoelasticity in ferromagnets); other less tedious methods, such as the "*matrix method*" developed by Fahmy and Adler (1973), are more efficient then. This latter method can be briefly sketched out as follows. In this method system (2.1) is written as a first-order matrix differential equation in x_2 for a one-column state vector **s** once the propagation factor $\mathcal{F}$ has been accounted for:

$$\partial \mathbf{s}/\partial X = \mathbf{A}\mathbf{s}, \quad X = x_2/\lambda, \tag{2.27}$$

where λ is a characteristic length and $\mathbf{A}$ is an $m \times m$ matrix involving k_1. A similarity operation involving a matrix $\mathbf{M}$ such that

$$\mathbf{s}' = \mathbf{M}\mathbf{s}, \quad \mathbf{A}' = \mathbf{M}\mathbf{A}\mathbf{M}^{-1},$$
$$\partial \mathbf{s}'/\partial X = \mathbf{A}'\mathbf{s}' \tag{2.28}$$

can help simplify the boundary conditions if necessary (e.g., produce transformed boundary conditions of the very simple type $s_\alpha = 0$ for $\alpha = 1, 2, \ldots, n \leq m$). Then a formal solution of Eq. (2.27) is

$$\mathbf{s}'(X) = \mathbf{\Phi}(X)\mathbf{s}'(0), \tag{2.29}$$

where $\mathbf{\Phi}(X)$ is the matrix exponential

$$\mathbf{\Phi}(X) = \exp(\mathbf{A}'X). \tag{2.30}$$

The state vector $\mathbf{s}'$ can be expressed in terms of the uncoupled normal-state vector $\mathbf{Y}$ in the substrate (bulkwave mode problem). The latter is related to $\mathbf{s}'$ by the transformation

$$\mathbf{Y} = \mathbf{P}^{-1}\mathbf{s}', \quad \mathbf{P}^{-1}\mathbf{A}'\mathbf{P} = \bar{\mathbf{A}}, \tag{2.31}$$

where $\bar{\mathbf{A}}$ is a diagonal matrix with the eigenvalues of $\mathbf{A}'$ as its diagonal elements. Hence the columns of $\mathbf{P}$ are the eigenvectors of the matrix $\mathbf{A}'$. From Eqs. (2.29)–(2.31), we obtain

$$\mathbf{Y}(X) = \exp(\bar{\mathbf{A}}X)\mathbf{Y}(0). \tag{2.32}$$

Only the eigenvectors, say k, of $\mathbf{A}'$ that have negative real parts will correspond to allowed surface modes decaying with depth in the substrate. Let $\mathbf{Y}_p(X)$ denote that part of the normal-state vector representing the depth-decaying modes and $\mathbf{P}_p$ the matrix ($m \times k$) of eigenvectors belonging to the said k eigenvalues. Then, from Eq. (2.31), we have for the allowed modes

$$\mathbf{s}'(X) = \mathbf{P}_p \mathbf{Y}_p(X). \tag{2.33}$$

Substituting now Eq. (2.32) into Eq. (2.33) and applying the boundary conditions at $X = 0$, we obtain the following dispersion relation.

$$\det|\mathbf{P}_p|_u = 0, \tag{2.34}$$

where the left-hand side stands for the determinant of the upper part of the $m \times k$ matrix $\mathbf{P}_p$. In practice, for given values of ω, eigenvalues and eigenvectors of the matrix $\mathbf{A}'$ are determined from an initial guess of the phase velocity $v_\varphi = \omega/k_1$ and the determinant of the matrix $\mathbf{P}_p$ is calculated. If the determinant is nonzero, a new value of v_φ is obtained and the whole process repeated until the determinant becomes vanishingly small. An application of this method is given in Ganguly *et al.* (1978).

III. The Notion of Resonance Coupling between Modes

A second prerequisite for the study of surface waves in the present context is the notion of *resonance coupling* between modes or, in other words, excitation by resonance of one wave mode by another (see, for instance, Maugin, 1980, 1981). This can be examined first in the case of bulk waves on the basis of Eqs. (2.1) and (2.2). Let, for instance, three scalar fields u_α, $\alpha = 1, 2, 3$, be of the time-harmonic plane type (2.2). In carrying Eq. (2.2) in a given undamped system (2.1) of field equations, we assume that for nontrivial solutions $\hat{u}_\alpha$, we are led to the following type of *dispersion relation*:

$$\mathscr{D}(\omega, k) = [\omega - \omega_3(k)]\{[\omega - \omega_1(k)][\omega - \omega_2(k)] - \varepsilon\bar{\omega}^2\} = 0, \tag{3.1}$$

where the ω_α's are known functions of the wavenumber k and material parameters and bias fields; eventually $\bar{\omega}$ is a characteristic (eventually wavenumber-dependent) angular frequency and ε is an infinitesimally small parameter. The relations

$$\mathcal{D}_\alpha(\omega, k) \equiv \omega - \omega_\alpha(k) = 0, \qquad \alpha = 1, 2, 3, \tag{3.2}$$

are the dispersion relations of uncoupled modes. Equation (3.1) tells us that the amplitude $\hat{u}_3$ is not coupled with the other two and that its associated dispersion relation—Eq. (3.2) for $\alpha = 3$—is influenced only by the material parameters and, possibly, initial static fields that may appear in $\omega_3(k)$. The remaining two solutions of Eq. (3.1) are coupled via ε. The moduli of the ratios $|\hat{u}_1/\hat{u}_2|$ and $|\hat{u}_2/\hat{u}_1|$ can be cast in the form

$$\begin{aligned}|\hat{u}_1/\hat{u}_2| &\propto \sqrt{\varepsilon\bar{\omega}} |\omega(k) - \omega_2(k)|^{-1}, \\ |\hat{u}_2/\hat{u}_1| &\propto \sqrt{\varepsilon\bar{\omega}} |\omega(k) - \omega_1(k)|^{-1}.\end{aligned} \tag{3.3}$$

Let (ω_0, k_0) denote the intersection point of the two curves $\mathcal{D}_\alpha(\omega, k) = 0$, $\alpha = 1, 2$, in the positive quarter of the (ω, k) plane. In the neighborhood of this critical point C, which is said to define a *crossover region* for the coupled modes, we have

$$\begin{aligned}\omega_1(k) &\simeq \omega_0 + v_1(k - k_0), \\ \omega_2(k) &\simeq \omega_0 + v_2(k - k_0),\end{aligned} \tag{3.4}$$

where

$$\omega_0 = \omega_1(k_0) = \omega_2(k_0)$$

and v_1 and v_2 are the group velocities ($v_g = \partial\omega/\partial k$) of the uncoupled modes at C. In the neighborhood of C the remaining factor of (3.1) yields two approximate coupled solutions:

$$\omega_\pm(k) - \omega_0(k) = \tfrac{1}{2}\{(v_1 + v_2)(k - k_0) \pm [(k - k_0)^2(v_1 - v_2)^2 + 4\varepsilon\bar{\omega}^2]^{1/2}\} \tag{3.5}$$

or

$$k_\pm(\omega) - k_0 = \frac{1}{2v_1 v_2}\{(v_1 + v_2)(\omega - \omega_0) \pm [(\omega - \omega_0)^2(v_1 - v_2)^2 + 4\varepsilon\bar{\omega}^2 v_1 v_2]^{1/2}\}. \tag{3.6}$$

To sketch out the picture of the coupled dispersion diagram in the neighborhood of C, four cases must be distinguished. These are

1. $\varepsilon > 0$, $v_1 v_2 > 0$,
2. $\varepsilon > 0$, $v_1 v_2 < 0$,
3. $\varepsilon < 0$, $v_1 v_2 > 0$,
4. $\varepsilon < 0$, $v_1 v_2 < 0$.

They are pictured in Fig. 7. In the cases of interest in the present work v_1, v_2, and ε are all positive, so that we always are in case (1). The other cases can

Elastic Surface Waves with Transverse Horizontal Polarization

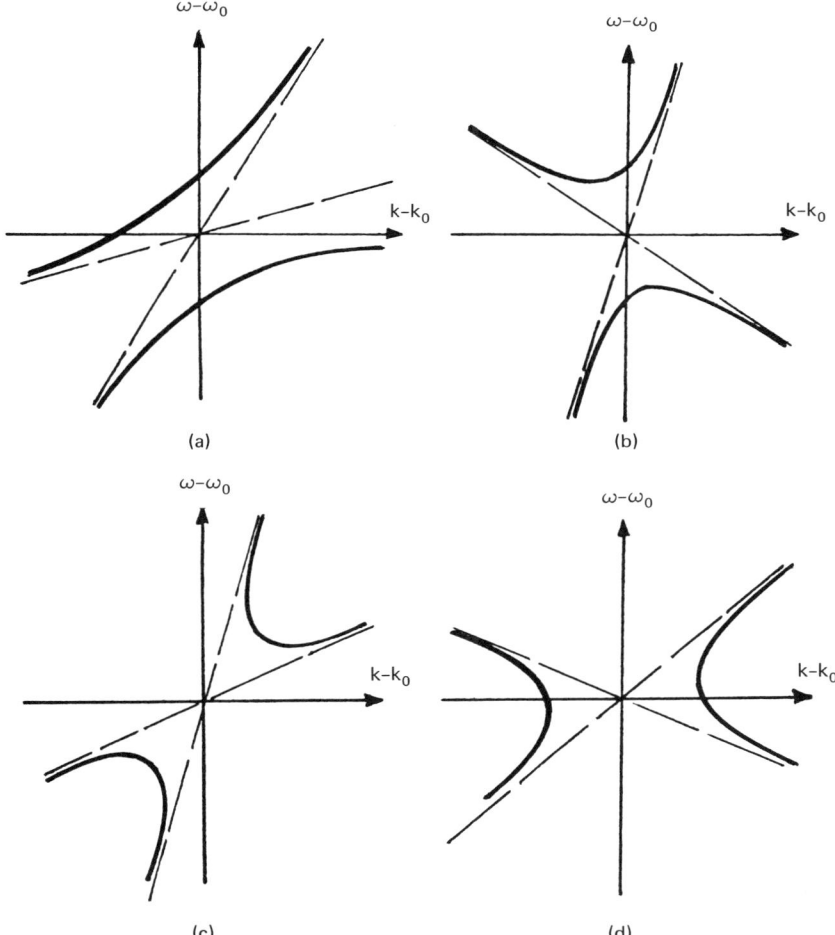

FIG. 7. Various types of crossover regions.

be met in other physical theories such as plasma physics and, more generally, in physical kinetics (see Lifshitz and Pitaevskii, 1979). The overall behavior of the remaining two coupled solutions in Eq. (3.1) is therefore as follows. For $k \in [0, \infty) = \mathbb{R}^+$, we have

$$\omega_I(k) = f_I[\omega_1(k), \omega_2(k); \varepsilon],$$
$$\omega_{II}(k) = f_{II}[\omega_1(k), \omega_2(k); \varepsilon],$$
(3.7)

with, e.g., $\omega_I \simeq \omega_1$ and $\omega_{II} \simeq \omega_2$ for $k \ll k_0$ and the reverse situation for

$k \gg k_0$, whereas in the neighborhood of C,

$$\omega_{\mathrm{I}} \simeq \omega_0 + \omega_+, \qquad \omega_{\mathrm{II}} = \omega_0 + \omega_-. \tag{3.8}$$

We see that the critical point C in fact no longer belongs to the coupled dispersion diagram. It has acted like a *repulsion* point (or a saddle point) for the dispersion curves. This repulsion or separation has the essential property to be such that

$$\left|\frac{\Delta\omega}{\bar{\omega}}\right| = \left|\frac{\omega_{\mathrm{I}}^2 - \omega_{\mathrm{II}}^2}{\bar{\omega}(\omega_{\mathrm{I}} + \omega_{\mathrm{II}})}\right|_{k=k_0} \simeq \left|\frac{\omega_{\mathrm{I}}^2 - \omega_{\mathrm{II}}^2}{2\omega_0 \bar{\omega}}\right|_{k=k_0} = O(\sqrt{\varepsilon}). \tag{3.9}$$

Simultaneously, a *resonance effect* takes place in the crossover region, as is clear from Eq. (3.3). Indeed, for example, for $\omega = \omega_1(k)$ approaching $\omega_2(k)$, the ratio $|\hat{u}_1/\hat{u}_2|$ diverges. The same holds true for $|\hat{u}_2/\hat{u}_1|$ when $\omega = \omega_2(k)$ approaches $\omega_1(k)$, hence at the coupling point C. Obviously, the resonance is smoothed out if dissipative processes associated with u_1 and u_2 are taken into account. Moreover, in following continuously one of the coupled dispersion curves ω_{I} or ω_{II} with increasing k, we have an *energy conversion* from one type of oscillation to the other one (frequencies and energies are related proportionally by the reduced Planck constant of quantum mechanics). Two examples of such resonance couplings involving an elastic mode and one of the elementary excitations cited in the introduction are the *magnetoacoustic resonance* coupling in elastic ferromagnets and the *electroacoustic resonance* coupling in elastic ferroelectrics. In the first case $\omega_1(k)$ is a nondispersive elastic mode and $\omega_2(k)$ is a dispersive spin wave mode and ε represents either magnetostriction couplings or, on the basis of a nonlinear theory (Maugin, 1979a), all magnetoelastic couplings of any order starting with piezomagnetism induced in a symmetry breaking by an intense bias magnetic field [see Maugin (1979b) for the analytical and qualitative study and Maugin and Pouget (1981) for the numerical study]. In the second case $\omega_{(1)}$ is again a nondispersive shear elastic mode, but $\omega_2(k)$ is a slightly dispersive, almost nonpropagating, soft ferroelectric mode (see Maugin and Pouget, 1980) and ε, on the basis of a nonlinear theory, represents all electroacoustic couplings starting with piezoelectricity induced by a symmetry breaking by an intense bias electric polarization [see Pouget and Maugin (1980) for the analytical, qualitative, and quantitative study in the case of ferroelectric barium titanate]. Typical dimensionless real dispersion relations exhibiting resonance couplings are reproduced in Fig. 8 for ferromagnetic yttrium–iron garnet and ferroelectric barium titanate. A magnification of the crossover regions is necessary in reason of the very small separation of coupled branches at C. This coupling results from the common Bose–Einstein nature of acoustic and spin vibrations in quantum statistical mechanics for the ferromagnetic case. For ferroelectrics, a lattice dynamics approach justifies the continuum approach (see Pouget, Askar and Maugin, 1983). Obviously, the resonance is somewhat smoothed out when combined dissipative effects are taken into account (viscosity and spin relaxation in the

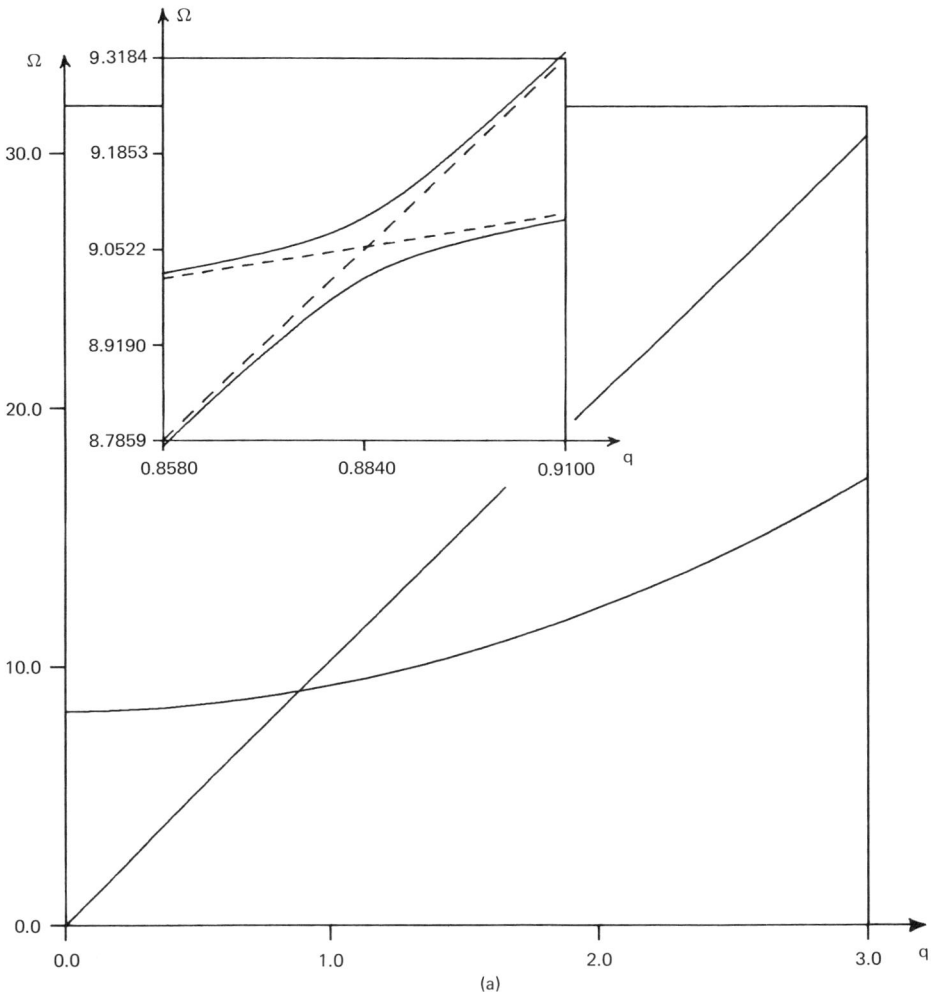

FIG. 8. Elastic-electromagnetic resonance couplings. (a) Dimensionless dispersion diagram for coupled magnons and transverse phonons in YIG (no dissipation, longitudinal setting of the bias magnetic field) (after Maugin and Pouget, 1981). (b) Numerical dispersion curve for $BaTiO_3$ (mixed transverse acoustic and and polariton modes (no dissipation, longitudinal setting of the initial polarization) (after Pouget and Maugin, 1980).

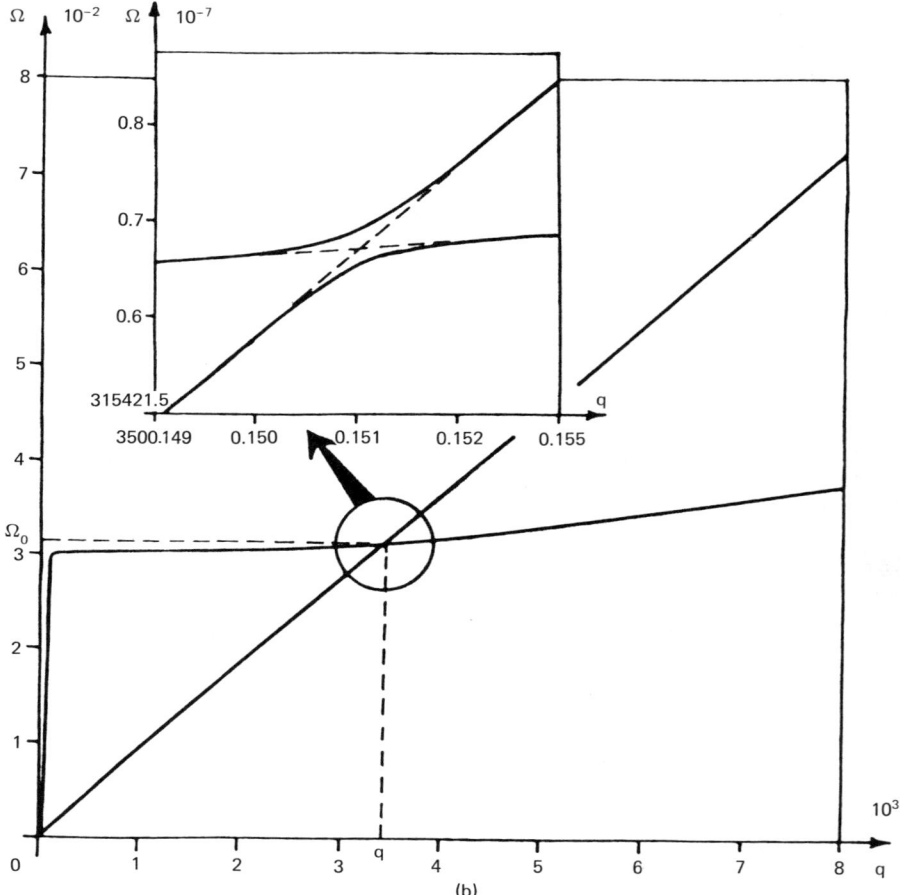

FIG. 8. *(continued)*

first case, and viscosity and electric relaxation in the second case.) Typically ω_0 and k_0 are of the order of 10^{10} Hz and 10^5 cm^{-1}, respectively, in the first case and 10^{10} Hz and 10^6 cm^{-1}, respectively, in the second case. These resonance couplings therefore concern the *hypersound* region—hence acoustoelectronics. Applications include high-frequency magnetostrictive transducers, pumping of magnetoelastic waves, design of wave filters and delay lines with electronically variable delay, means of analyzing the internal magnetic field in the first case, devices using acousto-optical phenomena in the second case.

In the case of surface waves the characteristic resonance frequency ω_0 and wavenumber k_0 obviously provide the adequate means for nondimensionalizing various quantities [compare Eq. (2.8)]. Whereas both modes ω_I and ω_{II} in Eq. (3.7) are admissible for coupled bulk modes, it is clear that a constraint such as Eq. (2.12) will in fact restrict the possibilities of coupling, eliminate some of these modes, and finally cause a *unilateral* repulsion of the dispersion branches at C. This is examined in Sections VI and VII.

IV. A Typical SH Surface Mode: The Bleustein–Gulyaev Wave

In the usual Cartesian tensor notation in rectangular coordinate frames, the equations of linear piezoelectricity are the

field equations

$$t_{ij,j} = \rho \ddot{u}_i, \quad t_{[ij]} = 0$$
$$\nabla \cdot \mathbf{D} = 0, \quad \nabla \times \mathbf{E} = 0 \quad \mathbf{E} = -\nabla \phi \quad \text{in } \mathcal{D}, \qquad (4.1)$$

with the *boundary conditions*

$$t_{ij} n_j = 0$$
$$\mathbf{n} \cdot (\mathbf{D}^{\text{out}} - \mathbf{D}^{\text{in}}) = 0, \quad \phi^{\text{out}} - \phi^{\text{in}} = 0 \quad \text{on } \partial \mathcal{D}, \qquad (4.2)$$

with $\phi \to 0$ at infinity;

and the *constitutive equations*

$$t_{ij} = c^E_{ijkl} u_{(k,l)} - e_{kij} E_k,$$
$$D_i = \varepsilon_{ij} E_j + e_{ijk} u_{(j,k)}. \qquad (4.3)$$

In these equations t_{ij} is the symmetric Cauchy stress tensor, $\mathbf{u}$ is the three-dimensional elastic displacement, $\mathbf{D}$ is the electric displacement, $\mathbf{E}$ is the electric field, ϕ is the electrostatic potential, c^E_{ijkl} is the tensor of elastic moduli at constant electric field, ε_{ij} is the symmetric tensor of dielectric constants at zero strains, and e_{ijk} is the tensor of piezoelectric moduli. The unit normal to the body is denoted $\mathbf{n}$. The superscripts *out* and *in* indicate the uniform limit in approaching the boundary from the outside and inside faces, respectively. The present theory being linear from the start, there are no ponderomotive contributions such as body forces and couples due to $\mathbf{E}$ in Eq. (4.1), since these quantities would be of second order in the electromagnetic fields. The quasi-electrostatic approximation is sufficient, since we are not interested in optical phenomena. Equation (4.1) yields

$$c^E_{ijkl}u_{k,lj} + e_{kij}\phi_{,kj} = \rho\ddot{u}_i,$$
$$e_{ijk}u_{j,ki} - \varepsilon_{ij}\phi_{,ji} = 0 \tag{4.4}$$

on account of Eq. (4.3) and the last equation of (4.1). The important parameter in piezoelectricity is the *electromechanical coupling factor*. This notion can be introduced as follows. For a one-dimensional dynamic problem in a unit direction d_k we can set $(\partial/\partial x_k) = d_k D$, where $D = \mathbf{d} \cdot \nabla$ is the gradient in the direction of $\mathbf{d}$. Then Eq. (4.4) takes on the form

$$c^E_{ijkl}d_j d_l D^2 u_k + e_{kij}d_k d_j D^2\phi = \rho\ddot{u}_i,$$
$$e_{ijk}d_k d_i D^2 u_j - \varepsilon_{ij}d_i d_j D^2\phi = 0. \tag{4.5}$$

On eliminating $D^2\phi$ from the second of these equations, the first yields

$$\rho\ddot{u}_i = \bar{c}_{ijkl}d_j d_l D^2 u_k, \tag{4.6}$$

where

$$\bar{c}_{ijkl} = c^E_{ijkl} + \frac{d_a e_{aij} d_b e_{bkl}}{d_p \varepsilon_{pq} d_q} \tag{4.7}$$

is the tensor of *piezoelectrically stiffened* elastic moduli. The electromechanical coupling factor K^2 is a relative measure of the two contributions on the right-hand side of Eq. (4.7); that is, if c^E, e, and ε are typical components of $\mathbf{c}^E$, $\mathbf{e}$, and $\boldsymbol{\varepsilon}$, we have

$$K^2 = \frac{\text{piezoelectric contribution}}{\text{pure elastic contribution}} = \frac{e^2}{\varepsilon c^E}. \tag{4.8}$$

Bleustein (1968) was the first to recognize that for a crystal in the class C_{6v} with the sixfold axis directed along the x_3 direction in Fig. 6 system (4.4) for the three components $u_j, j = 1, 2, 3$, of the displacement and ϕ uncouples in two systems, one of which governing u_3 and ϕ. In the latter system the only surviving material coefficients (in Voigt's short-hand notation) are c^E_{44}, e_{15}, and ε_{11}. The system (4.4) for u_3 and ϕ is thus reduced to

$$c^E_{44}\nabla^2 u_3 + e_{15}\nabla^2\phi = \rho\ddot{u}_3,$$
$$e_{15}\nabla^2 u_3 - \varepsilon_{11}\nabla^2\phi = 0, \tag{4.9}$$

where ∇^2 is the two-dimensional Laplacian defined after Eq. (2.20). The surviving components of the constitutive equations (4.3) are

$$t_{23} = c^E_{44}u_{3,2} + e_{15}\phi_{,2}, \quad t_{31} = c^E_{44}u_{3,1} + e_{15}\phi_{,1},$$
$$D_1 = e_{15}u_{3,1} - \varepsilon_{11}\phi_{,1}, \quad D_2 = e_{15}u_{3,2} - \varepsilon_{11}\phi_{,2}. \tag{4.10}$$

Elastic Surface Waves with Transverse Horizontal Polarization 393

For a stress-free, electrically grounded boundary $\partial \mathscr{D} = \{x_2 = 0\}$, the boundary conditions (4.2) take on the form

$$t_{22} \equiv 0, \qquad t_{23} = t_{21} = 0, \qquad \phi = 0 \quad \text{on } x_2 = 0. \tag{4.11}$$

On defining a new electrostatic potential ψ by

$$\psi = \phi - (e_{15}/\varepsilon_{11})u_3 \tag{4.12}$$

and accounting for Eq. (4.10), we find that Eqs. (4.9) and (4.11) become

$$\begin{aligned}\bar{c}_{44}\nabla^2 u_3 = \rho \ddot{u}_3, \qquad \nabla^2 \psi = 0 \qquad &\text{for } x_2 > 0, \\ \bar{c}_{44} u_{3,2} + e_{15}\psi_{,2} = 0, \qquad (e_{15}/\varepsilon_{11})u_3 + \psi = 0 \qquad &\text{at } x_2 = 0,\end{aligned} \tag{4.13}$$

where [compare Eqs. (4.7) and (4.8)]

$$\bar{c}_{44} = c_{44}^E + (e_{15}^2/\varepsilon_{11}) = c_{44}^E(1 + K^2), \qquad K^2 = e_{15}^2/\varepsilon_{11}c_{44}^E. \tag{4.14}$$

The first equation of (4.13) says that piezoelectrically stiffened bulk shear waves travel with a speed $\bar{c}_T$ that is slightly more than the usual shear wave speed in a nonpiezoelectric medium, since

$$\bar{c}_T^2 = c_T^2(1 + K^2), \qquad c_T^2 \equiv c_{44}^E/\rho. \tag{4.15}$$

Substituting a solution of the type (2.5) in Eq. (4.13), for $\hat{u}_3$ and $\hat{\psi} \neq 0$ we obtain

$$(k_1^2 + k_2^2) = \omega^2/\bar{c}_T^2, \qquad k_1^2 + k_2^2 = 0, \tag{4.16}$$

from which, in agreement with Eq. (2.6), there follow the two solutions (2.10), with negative imaginary parts, given by

$$k_2 = -ik_1\zeta, \quad k_2 = -ik_1 \qquad (k_1 > 0), \tag{4.17}$$

where

$$\zeta = (1 - \xi)^{1/2}, \qquad \xi = (c_{BG}/\bar{c}_T)^2, \qquad c_{BG}^2 = (\omega^2/k_1^2) < \bar{c}_T^2, \tag{4.18}$$

of which the last equation corresponds to the constraint (2.12). Thus *the speed c_{BG} of Bleustein–Gulyaev modes can only be less than the speed of piezoelectrically stiffened bulk shear waves*. With $X = k_1 x_2$, the general solution (2.13) reads

$$\begin{aligned}u_3 &= A\exp(-\zeta X)\mathscr{F}, \\ \phi &= [B\exp(-X) + (e_{15}/\varepsilon_{11})A\exp(-\zeta X)]\mathscr{F},\end{aligned} \tag{4.19}$$

which, carried in the boundary conditions (4.13), gives

$$\bar{c}_{44}\zeta A + e_{15}B = 0, \qquad (e_{15}/\varepsilon_{11})A + B = 0. \tag{4.20}$$

A nontrivial solution is obtained for A and B whenever

$$\zeta = \bar{K}^2 = \frac{K^2}{1+K^2} = \frac{e_{15}^2}{\varepsilon_{11}\bar{c}_{44}}. \tag{4.21}$$

Inserting this result in Eq. (4.18), we find that

$$c_{BG}^2 = \bar{c}_T^2(1 - \bar{K}^4). \tag{4.22}$$

The speed c_{BG} is very close to $\bar{c}_T$, and the smaller $\bar{K}$ is, the closer it is to $\bar{c}_T$. For a material such as CdS, $\bar{K} \simeq 0.19$, $\bar{c}_{BG} = 0.9994$, and $\bar{c}_T = 1788$ m/sec. The penetration depth of the elastic component u_3 is given by $p = (\zeta k_1)^{-1} = \lambda/\zeta$ if $\lambda = k_1^{-1}$. For $\bar{K} = 0.19$, $\zeta = 0.036$ and $p \simeq 27.7\lambda$. The amplitude variation with depth for CdS is reproduced in Fig. 9. From the elasticity point of view, it is clear that the Bleustein–Gulyaev wave is an SH bulk wave *perturbed* to meet the electrical contribution to the boundary condition (see the introduction), since for K and $\bar{K}$ going to zero, ζ and $|k_2|$ go to zero and the Bleustein–Gulyaev wave degenerates into an SH bulk wave, the displacement amplitude no longer decaying as we move away from the surface. We leave to the reader by way of exercise the case where the boundary $\partial\mathcal{D}$

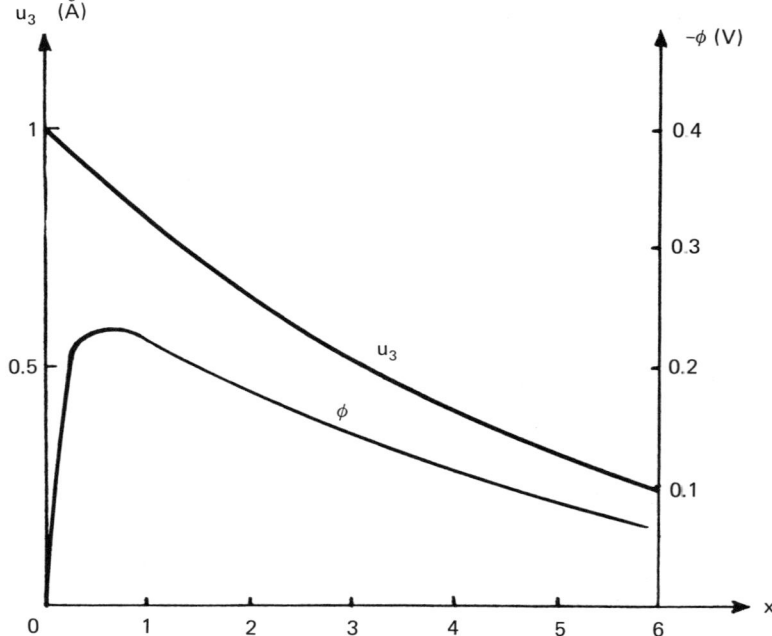

FIG. 9. Behavior with depth of u_3 and ϕ for Bleustein–Gulyaev waves in CdS (grounded boundary) (after Dieulesaint and Royer, 1974, p. 238).

is not grounded but the potential ϕ in the substrate matches an external potential for $x_2 < 0$. In this case, Eq. (4.22) is replaced by

$$c_{BG}^2 = \bar{c}_T^2 \left(1 - \frac{\bar{K}^4}{(1 + \varepsilon_{11})^2}\right). \tag{4.23}$$

Then the Bleustein–Gulyaev wave penetrates more deeply in the substrate than for a grounded boundary, since

$$\zeta = \bar{K}^2/(1 + \varepsilon_{11}). \tag{4.24}$$

For CdS, $\varepsilon_{11} = 9$ (the vacuum dielectric constant ε_0 is taken equal to one), $\zeta = 0.0036$, and $p \simeq 278\lambda$. The fact that Bleustein–Gulyaev waves penetrate much more deeply in the substrate than classical Rayleigh waves explains why their speed does not differ much from (but is lower than) the speed of bulk shear waves.

V. SH Surface Waves in Nonhomogeneous Elastic Media

As was shown in Section IV, one way of perturbing an SH bulk wave so that it yields an SH surface wave is to impose a slight variation in the boundary conditions in the form of an electrical contribution. Another way (see Section I) is to modify the boundary condition at the limiting boundary by superimposing a thin "slow-wave" layer of elastic material. The latter situation can be equivalently represented by a *jump profile* (at $x_2 = 0$ in Fig. 5) in the material properties of a unique medium. Less drastic would be a smoother change in these properties, which should nonetheless yield the same class of surface phenomena. Let s be a typical compliance, the inverse of an elasticity coefficient. Then the situations I–III and IV–VII pictured in Fig. 10, where a continuous change in s is observed from the limiting plane toward the interior of the substrate with $s(x_2 = 0) > s(x_2 > 0)$, so that the Love condition of a slow-wave layer is respected by the inhomogeneity present near the surface, may represent the material properties more realistically. Power profiles ($n = 1$, linear profile; $n = 2$, parabolic profile), case (V): exponential profile, case (VI): Gaussian profile, case (VII): complementary error function profile, may be envisaged. This kind of material surface inhomogeneity is fairly typical and occurs, for instance, in the ion implantation ordering of the surfaces of glasses, in the irradiation of a semiconducting crystal by strongly absorbed light, and in the metallization of the surface of piezoelectric crystals (hence in manufacturing processes typical of microelectronics and metallurgy). A similar situation prevails in soil mechanics and seismology, owing to the marked stratification of certain

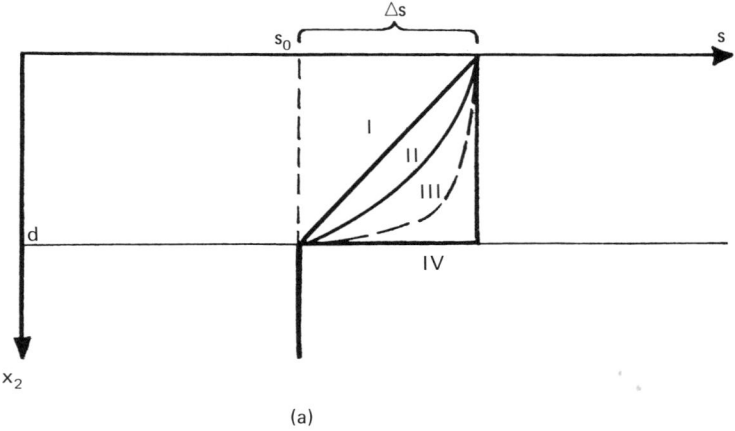

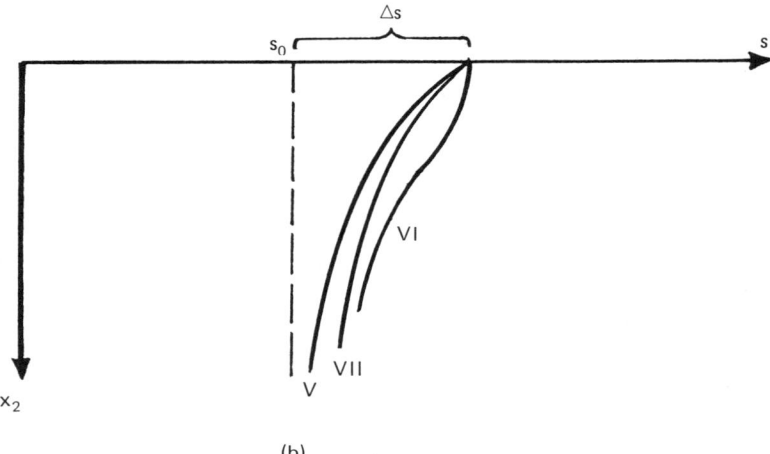

FIG. 10. Dependence of compliance s upon depth x_2: (a) I, linear profile ($n = 1$); II, parabolic profile ($n = 2$); III, power profile ($n = 10$); and IV, jump profile typical of Love waves ($n = \infty$); (b) V, exponential profile; VI, Gaussian profile; and VII, complementary error function profile.

soils near the surface. This is illustrated in Fig. 11, where the measurements made some years ago of the elastic modulus with depth at Rasattepe's site (nearby Ankara, Turkey, a reputed active seismic region) in view of the building of a mausoleum are reported.

By analogy with the setting in favor of the existence of Love waves, it is clear that an inhomogeneity in the material parameters localized near

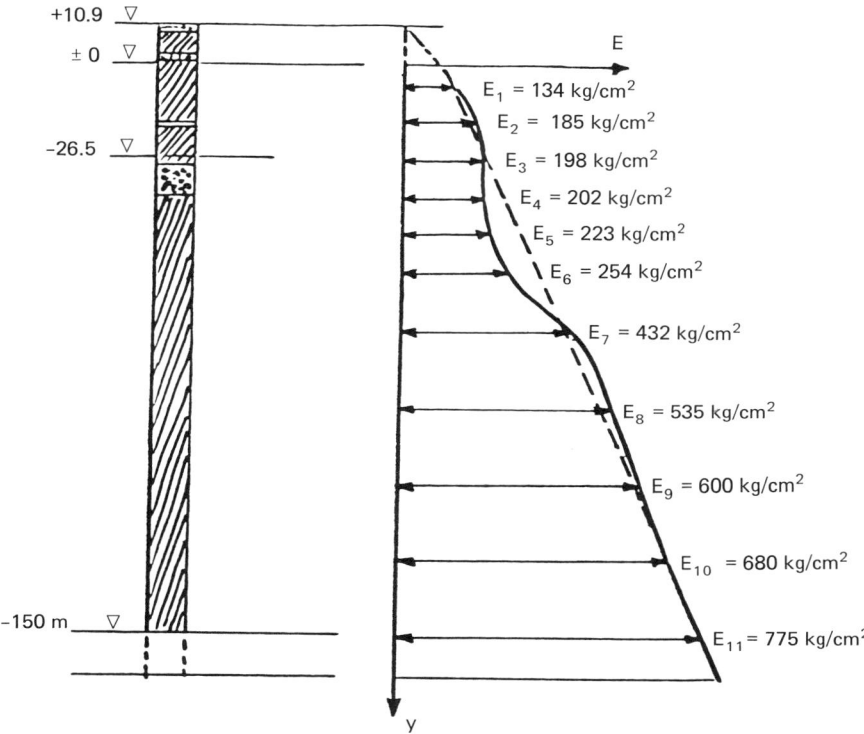

FIG. 11. Variation with depth of the elasticity modulus on Rasattepe's site (Atatürk's mausoleum, Ankara, Turkey) (after Peynircioğlu, 1981, p. 92).

the limiting surface will favor the existence of pure SH surface waves localized in a surface layer, waves that will therefore be "guided" by the material inhomogeneity. The existence of such a wave mode of propagation can be used to analyze the subsurface layer of metallurgically treated materials, to characterize subsurface anomalies by elastic surface wave propagation (see Tittman, 1974), and therefore provides a nondestructive means for subsurface gradient determination (see Szabo, 1974). Similar phenomena occur in optics in materials with a continuously variable optical index (see Meunier, 1980). By appropriately selecting the inhomogeneity, a delay line of desired spectral characteristics can be designed and, conversely, as already mentioned, from the knowledge of the velocity dispersion of these waves, one can conclude on the shape of the inhomogeneity (this constitutes the inverse Sturm–Liouville problem). In the solid mechanics case, an incoming SH bulk wave will materialize in an SH surface wave of which the bulk of the energy will be concentrated near the surface, thus causing damaging effects on structures.

A. The Generalized Sturm–Liouville Problem

Consider, for instance, the equations of *isotropic nonhomogeneous* linear elasticity for an *out-plane* motion

$$\mathbf{u} = [(0, 0, u_3(x_\alpha, t)], \qquad \alpha = 1, 2. \tag{5.1}$$

For $x_2 > 0$, Eq. (4.1) then yields

$$\mu \nabla^2 u_3 + (\nabla \mu) \cdot \nabla u_3 = \rho \ddot{u}_3, \tag{5.2}$$

where $\mu(x_2)$ is the space-dependent shear modulus and ∇ is the gradient operator in the sagittal plane. On considering for Eq. (5.2) solutions of the type

$$u_3 = \phi(x_2) \exp[i(\omega t - k_1 x_1)], \tag{5.3}$$

we obtain

$$\frac{d}{dx_2}\left(\mu \frac{d\phi}{dx_2}\right) + (\rho\omega^2 - \mu k_1^2)\phi = 0. \tag{5.4}$$

For a surface wave solution in an elastic half space $\mathscr{D} = \{x_2 > 0\}$ with a stress-free boundary $\partial \mathscr{D} = \{x_2 = 0\}$, Eq. (5.4) is supplemented by the boundary condition

$$t_{32}(x_2 = 0) = 0 \quad \text{or} \quad \frac{d\phi}{dx_2}(0) = 0, \tag{5.5}$$

while the condition of vanishingly small displacement far away from $\partial \mathscr{D}$ requires that

$$\phi(x_2 \to \infty) = 0. \tag{5.6}$$

The problem represented by Eqs. (5.4)–(5.6) is a *generalized Sturm–Liouville problem* (see Coddington and Lewison, 1955). Such a problem consists in finding values of the squared wavenumber k_1^2, as well as the corresponding function $\phi(x_2)$ for which Eq. (5.4) has a nontrivial solution. Equation (5.4) can also be cast in the form

$$\left[\frac{d}{dx_2}\left(\mu \frac{d}{dx_2}\right) + \rho\omega^2\right]\phi = (\mu k_1^2)\phi. \tag{5.7}$$

The eigenvalues of this problem then define the propagation parameter $\kappa = k_1^2$, whereas the eigenvectors describe the wave amplitude distribution with depth in the substrate. It can be noted that an entirely analogous problem arises in the study of even TH modes in planar optical wave guides. For the inhomogeneity profiles represented in Fig. 10, the generalized Sturm–Liouville problem is *well posed*. In fact, following Viktorov (1978) and considering inhomogeneities in rigidity and density of the exponential type

$$\mu = \mu_0 \left[1 - \frac{\Delta\mu}{\mu_0} \exp\left(\frac{-x_2}{p_\mathrm{I}}\right) \right],$$
$$\rho = \rho_0 \left[1 + \frac{\Delta\rho}{\rho_0} \exp\left(\frac{-x_2}{p_\mathrm{I}}\right) \right],$$
(5.8)

where p_I is a characteristic depth and $\mu_0 \simeq \mu(x_2 = \infty)$ and $\rho_0 \simeq \rho(x_2 = \infty)$ with

$$0 < \frac{\Delta\mu}{\mu_0} \ll 1, \qquad 0 < \frac{\Delta\rho}{\rho_0} \ll 1, \tag{5.9}$$

an *exact* solution to the problem (5.4)–(5.6) is found in the following analytical form:

$$u_3 = A\mathscr{F}(\alpha, \beta, \gamma, \xi) \exp(-\zeta X) \exp[i(\omega t - k_1 x_1)], \tag{5.10}$$

where A is an arbitrary constant and $\mathscr{F}$ is the hypergeometric function (see Abramowitz and Stegun, 1968), with arguments and parameters defined by

$$\alpha = X_\mathrm{I}(\zeta - 1), \qquad \beta = X_\mathrm{I}(\zeta + 1),$$
$$\gamma = 1 + 2\zeta X_\mathrm{I}, \qquad \xi = \left(\frac{\Delta\mu}{\mu_0} + \frac{\Delta\rho}{\rho_0}\right) \exp\left(\frac{-X}{X_\mathrm{I}}\right),$$
(5.11)

and

$$k_1 = k_{\mathrm{T}0} \left[1 + \frac{X_{\mathrm{I}0}^2}{2} \left(\frac{\Delta\mu}{\mu_0} + \frac{\Delta\rho}{\rho_0} \right) \right],$$
$$k_{\mathrm{T}0}(\omega) = \omega/c_{\mathrm{T}0}, \quad c_{\mathrm{T}0}^2 = \mu_0/\rho_0, \quad \zeta = [1 - (k_{\mathrm{T}0}/k_1)^2]^{1/2}, \quad (5.12)$$
$$X = k_1 x_2, \quad X_\mathrm{I} = k_1 p_\mathrm{I}, \quad X_{\mathrm{I}0} = k_{\mathrm{T}0} p_\mathrm{I}.$$

The surface wave solution (5.10) is localized in a surface layer of thickness $p = \lambda/\zeta$ with $\lambda = k_1^{-1}$ and, according to Eqs. (5.12) and (5.9), its phase velocity $v_\varphi = \omega/k_1$; therefore

$$v_\varphi = \frac{c_{\mathrm{T}0}}{1 + (X_{\mathrm{I}0}^2/2)(\Delta\mu/\mu_0 + \Delta\rho/\rho_0)} \tag{5.13}$$

is somewhat smaller than the bulk shear wave velocity $c_{\mathrm{T}0}$.

B. The Simple Sturm–Liouville Problem

For the other profiles given in Fig. 10, under certain assumptions the generalized Sturm–Liouville problem (5.4)–(5.6) can be transformed into a *simple* Sturm–Liouville problem. For instance, for an elastic body having a 6mm hexagonal symmetry $\mu(x_2)$ is replaced by the elasticity coefficient $c_{44}(x_2)$. On substituting $\psi = \phi\sqrt{c_{44}}$, dropping terms involving derivatives of c_{44}, and introducing the compliance $s_{44} = 1/c_{44}$, we transform the problem (5.4)–(5.6) into the *simple* Sturm–Liouville problem

$$\frac{d^2\psi}{dx_2^2} + (K_T - \kappa)\psi = 0,$$

$$\frac{d\psi}{dx_2}(0) = 0, \qquad \psi(x_2 \to \infty) = 0,$$

(5.14)

with

$$\kappa = k_1^2, \qquad K_T = (\omega/c_T)^2, \qquad c_T^2(x_2) = [\rho s_{44}(x_2)]^{-1}. \qquad (5.15)$$

The first equation of (5.14) is analogous to the time-independent one-dimensional Schrödinger equation and to the propagation equation of so-called TE modes in planar optical wave guides. Problem (5.14) can be handled by various approximate methods such as the Galerkin method, the finite element method (FEM), the finite difference method (FDM), the Green function method, etc. For the sake of illustration, we report results obtained by Kielczynski (1981) by the FEM. This author considers an elastic medium with parameters typical of PZT-4 ceramics: $\rho_0 = 7.5$ kg/m^3, $s_0 = s_{44}(x_2 = \infty) = 39 \times 10^{-9}$ m^2/N, and $\Delta s/s_0 = 0.1$. Amplitudes and phase velocities are calculated for propagating SH surface waves for the inhomogeneity profiles pictured in Fig. 10. Insofar as the wave amplitude is concerned, a modal structure is found for the acoustic field u_3 in the wave guide. The dependence of the amplitudes of those modes upon depth is of an oscillatory nature, but tends to zero with increasing depth. The amplitude of the Nth kind of vibration performs $N - 1$ oscillations. Figure 12 shows the dependence of the amplitude on the normalized depth for a power distribution profile ($n = 10$) and $d/\lambda_0 = 4$ and $d/\lambda_0 = 2$ ($\lambda_0 = k_{T0}^{-1}$). Diagrams showing the variation of the phase velocity with normalized depth in the substrate are given in Fig. 13 for power distribution profiles up to the depth d. Since the sequence of inhomogeneity structures in Fig. 10a tends, in the limit, to a jump structure typical of Love waves, it can be expected that the properties of SH surface waves should also approximate the properties of Love waves for high powers of n. As a matter of fact, for $n = 10$, the amplitude

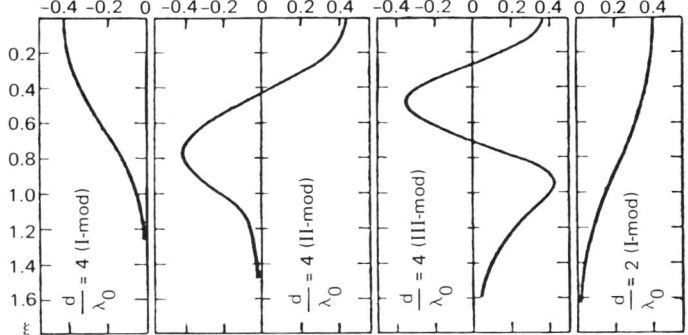

FIG. 12. Dependence of the amplitude of SH surface waves upon normalized depth $\xi = x_2/d$ for a power distribution of the surface inhomogeneity (after Kielczynski, 1981, p. 76).

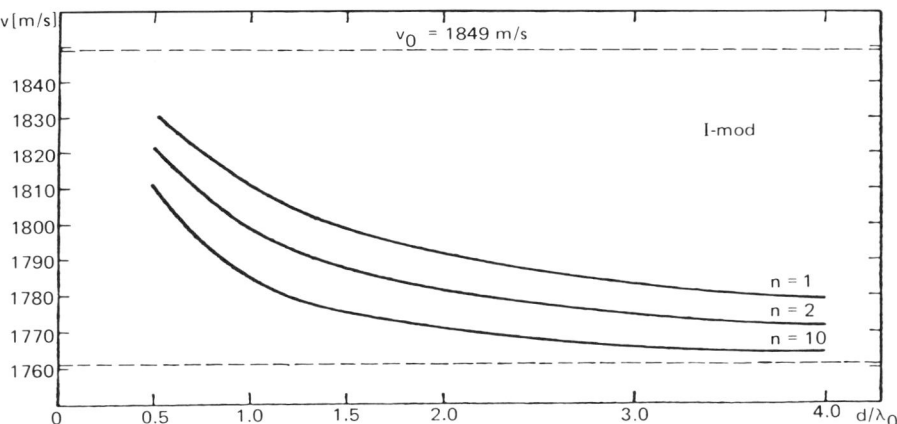

FIG. 13. Dispersion curves for a power profile of the surface inhomogeneity ($n = 1$, $n = 2$, $n = 10$) (after Kielcynski, 1980, p. 76).

of the SH surface wave differs from that of the Love case by less than 2%, whereas the phase velocities coincide with an accuracy of ± 1 m/sec. In the case of a Gaussian profile for the inhomogeneity (case VI in Fig. 10b), the variation in phase velocity as a function of normalized depth is given in Fig. 14. As d/λ_0 increases, frequency increases and there appear successively higher modes. Each of these modes, except for the first one, has a cutoff frequency below which it does not propagate. Similar variations can be exhibited for exponential and complementary error function profiles.

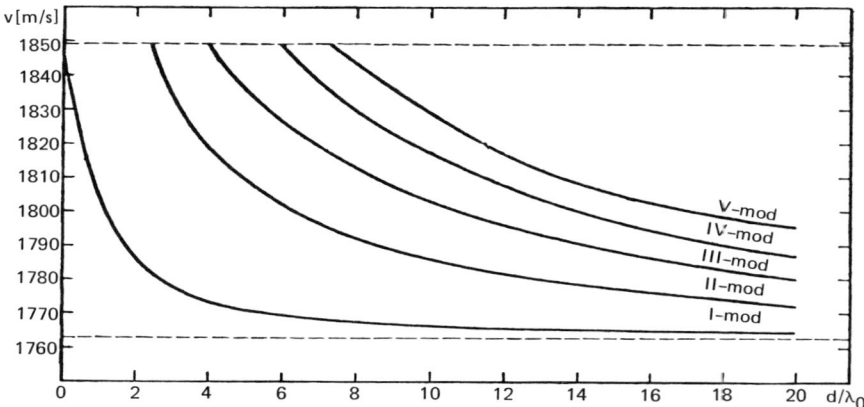

FIG. 14. Dispersion curves for a Gaussian profile of the surface inhomogeneity (after Kielczynski, 1981, p. 76).

C. OTHER PROBLEMS

Bakirtas and Maugin (1982) have analytically solved Eq. (5.4) for the problem of an inhomogeneous layer superimposed on a homogeneous substrate (a generalized Love problem) and the problem of a free inhomogeneous layer (a generalized Lamb problem) for linear and exponential inhomogeneity profiles. For an exponential profile in rigidity and density distribution

$$\mu(x_2) = \mu_0 \exp(hx_2), \qquad \rho(x_2) = \rho_0 \exp(rx_2), \tag{5.16}$$

Eq. (5.4) transforms to

$$\frac{d^2\phi}{dx_2^2} + h\frac{d\phi}{dx_2} + [K_c^2 \exp(-mx_2) - k_1^2]\phi = 0, \tag{5.17}$$

where

$$m = h + r, \qquad K_c = k_1(c/c_{T0}), \qquad c_{T0}^2 = \mu_0/\rho_0, \qquad c = \omega/k_1, \tag{5.18}$$

whereas for linear distributions

$$\mu(x_2) = \mu_0(1 + hx_2), \qquad \rho(x_2) = \rho_0(1 + rx_2), \tag{5.19}$$

we have

$$(1 + hx_2)\frac{d^2\phi}{dx_2^2} + h\frac{d\phi}{dx_2} + k_1^2\left\{\left[\left(\frac{c}{c_{T0}}\right)^2 - 1\right] + \left[r\left(\frac{c}{c_{T0}}\right)^2 - h\right]x_2\right\}\phi = 0. \tag{5.20}$$

Elastic Surface Waves with Transverse Horizontal Polarization

The formal solutions of Eqs. (5.17) and (5.20) are (Bakirtaş and Maugin, 1982)

$$\phi(x_2) = \left[2\left(\frac{c}{c_{T0}}\right)\frac{k_1}{m}\right]^p \exp\left(\frac{-hx_2}{2}\right)[AJ_p(\xi) + BY_p(\xi)],$$

$$\xi = 2\left(\frac{c}{c_{T0}}\right)\left(\frac{k_1}{m}\right)\bigg/[\exp(mx_2)]^{1/2}, \quad p = \frac{1}{m}(4k_1^2 + h^2)^{1/2}, \quad (5.21)$$

and

$$\phi(x_2) = \exp(-\delta k_1 x_2)[AM(a, 1; \xi) + BU(a, 1; \xi)],$$

$$\delta = \left[1 - \left(\frac{r}{h}\right)\left(\frac{c}{c_{T0}}\right)^2\right]^{1/2}, \quad a = \left[k_1(h - r)\left(\frac{c}{c_{T0}}\right)^2\right]\bigg/2\delta h^2, \quad (5.22)$$

$$\xi = 2\delta k_1\left(x_2 + \frac{1}{h}\right),$$

where A and B are integration constants, $J_p(\xi)$ and $Y_p(\xi)$ are Bessel functions of the first and second kinds, respectively, of index p, and $M(a, 1; \xi)$ and $U(a, 1; \xi)$ are confluent hypergeometric functions of parameters a and $b = 1$ and variable ξ (see Abramowitz and Stegun, 1968). For the generalized Love problem (Fig. 15), with a nonhomogeneous layer of thickness H we have

$$\phi_I = A_I \phi_{11}(x_2) + B_I \phi_{21}(x_2), \quad 0 < x_2 < H,$$
$$\phi_{II} = A_{II} \exp(-\gamma x_2), \quad x_2 > H, \quad (5.23)$$

where ϕ_1 and ϕ_2 are two independent solutions and

$$\gamma = k_1 \left[1 - \left(\frac{c}{c_T^{II}}\right)^2\right]^{1/2}, \quad (c_T^{II})^2 = \frac{\mu_{II}}{\rho_{II}} = \text{const.}, \quad (5.24)$$

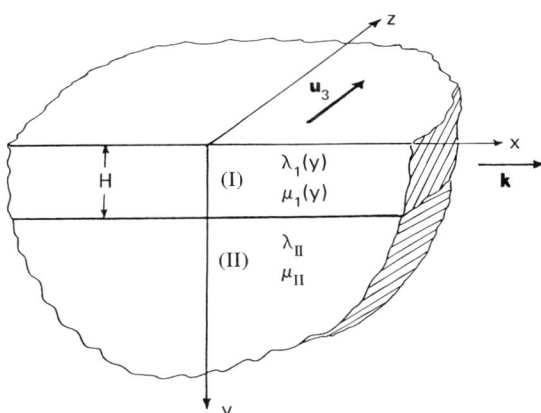

FIG. 15. Love wave problem with an inhomogeneous layer.

along with the boundary and matching conditions

$$t_{32(I)} = 0 \quad \text{at } x_2 = 0,$$
$$\phi_I = \phi_{II}, \quad t_{32(I)} = t_{32(II)} \quad \text{at } x_2 = H. \tag{5.25}$$

Nontrivial solutions for the triplet (A_I, B_I, A_{II}) are obtained through Eq. (5.25) whenever we have the dispersion relation of the general form

$$\Psi_1(\xi_1)\Omega_2(\xi_2) - \Psi_2(\xi_1)\Omega_1(\xi_2) = 0,$$
$$\xi_1 = \xi(x_2 = 0), \quad \xi_2 = \xi(x_2 = H), \tag{5.26}$$

where Ψ_α and Ω_α are expressed in terms of Bessel functions $J_p(\xi)$ and $J_{p+1}(\xi)$ in the case of an exponential inhomogeneity, and in terms of hypergeometric functions $M(a, 1; \xi)$, $M(a + 1, 2; \xi)$, $U(a, 1; \xi)$, and $U(a + 1, 2; \xi)$ in the case of a linear inhomogeneity [see the exact expressions of Eq. (5.26) in Bakirtaş and Maugin (1982)]. Whenever the superimposed layer becomes homogeneous with shear elastic modulus μ_I and shear wave speed c_T^I, Eq. (5.26) reduces to the classic equation for Love waves (see Eringen and Suhubi, 1975, p. 545):

$$\tan\left\{\left[\left(\frac{c}{c_T^I}\right)^2 - 1\right]^{1/2}(k_1 H)\right\} - \left(\frac{\mu_{II}}{\mu_I}\right)\left(\frac{1 - (c/c_T^{II})^2}{(c/c_T^I)^2 - 1}\right)^{1/2} = 0, \quad c_T^I < c < c_T^{II}, \tag{5.27}$$

which provides the sequence of wavenumbers

$$(k_1)_n = \frac{\pi(n-1)}{4H[(c_T^{II}/c_T^I)^2 - 1]^{1/2}}, \quad n = 1, 2, \ldots. \tag{5.28}$$

In the case of Eq. (5.26), a numerical evaluation (see Bakirtaş and Maugin, 1982) allows one to show that the first-mode wavenumber $(k_1)_1$ is practically unaffected for a wide range of exponential parameters. The second wavenumber, $(k_1)_2$, however, increases considerably with the inhomogeneity exponential parameter. In the case of a linear homogeneity for $0 < x_2 < H$, $(k_1)_1$ is practically unaltered, but the following modes, e.g., $(k_1)_2$ and $(k_1)_3$, are the more affected the "bigger" the inhomogeneity. The dispersion of the fundamental mode as a function of the parameter $(\lambda/H)^{-1}$ is given in Fig. 16 for various values of $\beta = hH$. This dispersion is relatively important for large wavelengths (i.e., small $2k_1 H/\pi$), and the more important, the bigger the inhomogeneity parameter $\beta = hH$.

For a free inhomogeneous layer (Fig. 17), one takes

$$\phi = A\phi_1(x_2) + B\phi_2(x_2), \quad -H < x_2 < +H, \tag{5.29}$$

where ϕ_1 and ϕ_2 are two independent solutions of the type (5.21) or (5.22), depending on whether the inhomogeneity is exponential or linear along the

Elastic Surface Waves with Transverse Horizontal Polarization 405

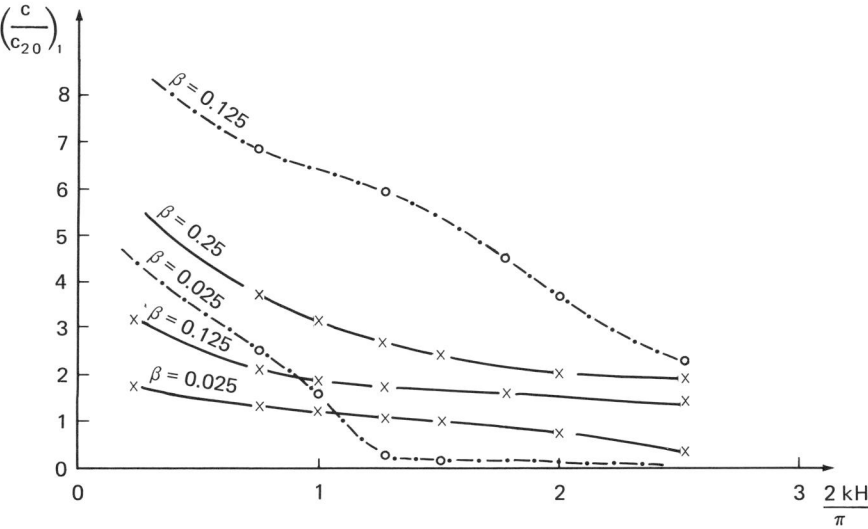

FIG. 16. Dispersion of the fundamental mode of Love waves for an inhomogeneous layer, $\beta = hH$, after Bakirtaş and Maugin (1982).——, exponential case; —·—·—, linear case.

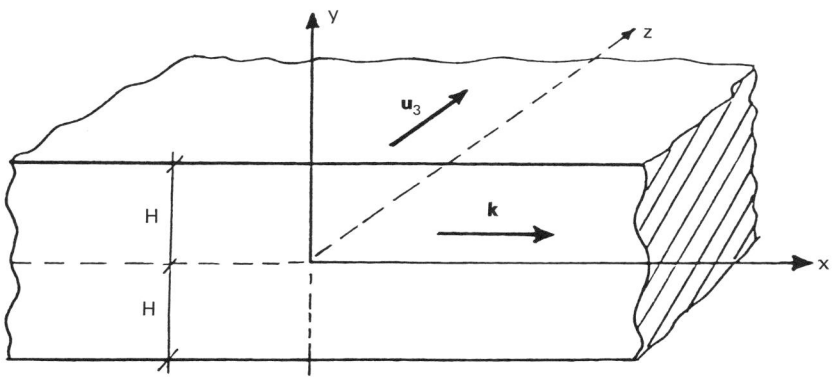

FIG. 17. The inhomogeneous layer.

normal to the layer. The boundary conditions to be met at the free surfaces of the layer are

$$t_{32} = 0 \quad \text{at} \quad x_2 = \pm H \qquad (5.30)$$

or

$$A\frac{d\phi_1}{dx_2} + B\frac{d\phi_2}{dx_2} = 0 \quad \text{at} \quad x_2 = \pm H \qquad (5.31)$$

The compatibility condition for solving this system for A and B yields an expression of the type (5.26), with $\xi_1 = \xi(x_2 = +H)$ and $\xi_2 = \xi(x_2 = -H)$. Such an expression reduces to the well-known sequence of proper phase velocities (see Achenbach, 1973, p. 203)

$$\left(\frac{c}{c_{T0}}\right)_n = \left[1 + \left(\frac{\pi}{2k_1 H}\right)^2 n^2\right]^{1/2}, \quad n = 0, 1, 2, \ldots, \quad (5.32)$$

for a homogeneous layer. We have reproduced in Fig. 18 numerical results concerning the phase velocity of the first few modes as a function of hH for various wavelengths. For an exponential inhomogeneity and small wavelengths $\lambda = k_1^{-1}$ (e.g., $\lambda = H/5$), the modes do not differ much from those provided by (5.32) for a wide range of the inhomogeneity parameter. However, the speed is smaller for β's of the order of 0.1–0.2 and larger for β's in the range 0.7–0.8. Variations with wavelength (e.g., $\lambda = H/5$) are not marked. In the case of a linear inhomogeneity, the fundamental mode ($n = 0$) is strongly affected for $\beta \gtrsim 0.1$ and a wavelength of the order of the half thickness of the layer with a phase velocity markedly increased (e.g., multiplied by a factor 4–7) as compared to the homogeneous case for inhomogeneity factors of the order of 0.3 to 0.8.

Before leaving purely mechanical descriptions, we note that the study of SH surface waves at the *periodically* shaped interface of two elastic media follows exactly the same line as that of transverse electromagnetic waves along a reflection grating in optics (see Van den Berg, 1971).

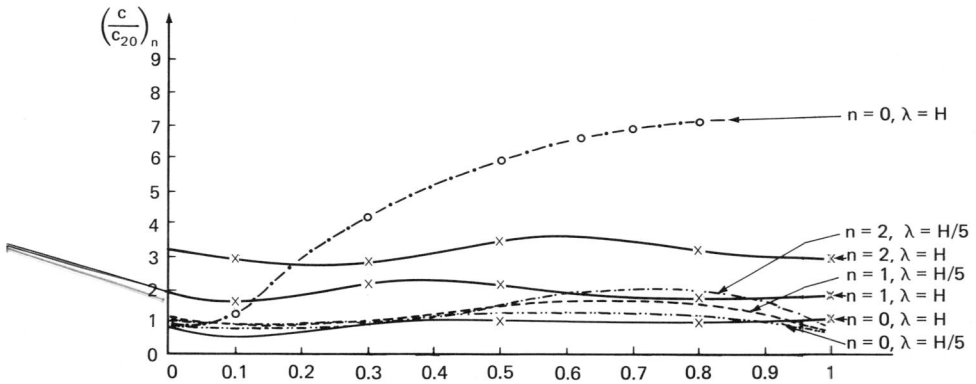

FIG. 18. Phase velocity as a function of the parameter in rigidity inhomogeneity for the layer problem: ———, exponential case, $\lambda = k^{-1} = H$; -----, exponential case, $\lambda = k^{-1} = H/5$; -·-·-, linear case, $\lambda = H$ (after Bakirtaş and Maugin, 1982).

VI. Electroacoustic SH Surface Waves in Ferroelectrics

A. General Features

Elastic ferroelectrics such as barium titanate provide a substratum for essentially two types of vibrations; acoustic or elastic vibrations and, outside the electromagnetic optical range, so-called soft ferroelectric vibrations. The couplings that may occur between these two are of the resonance type, as presented in a general framework in Section III. We shall not give all the details of the theory of elastic ferroelectrics here; rather, we content ourselves with the essential features that bear interest to the present problem and refer the reader to the original works for more on the subject (see Maugin and Pouget, 1980). Ferroelectrics are electric materials that belong in the class of *dielectrics* (i.e., electrically insulating materials) which exhibit a relatively strong bias electric polarization field below a certain phase transition temperature θ_C. Under these conditions it is understood that a correct theory of elastic ferroelectrics that takes these intense bias fields into account can only be *nonlinear* at the start. Then the method for treating small dynamic fields can be sketched out as follows (see Maugin, 1980, 1981). In the *exact nonlinear* theory, nonlinear coupled field equations for a set of primary variables (finite motion, electric polarization, and electromagnetic fields in the case of deformable ferroelectrics) are complemented by *nonlinear* coupled constitutive equations that are constructed for a specified material symmetry (e.g., isotropy, transverse isotropy) with respect to a stress-free, field-free reference configuration K_R, with the help of the theory of invariants, by a group of space transformations in K_R, a now classic method in finitely deformable solids. Then a one-domain (i.e., spatially uniform) fundamental ferroelectric phase at temperature $\theta \ll \theta_C$ is obtained for the whole body by application of a strong static electric field $\mathbf{E}_0$. In the process all domains (regions of small extent in the specimen, which already presented a spatially uniform spontaneous electric polarization) that could exist in a naturally ordered phase of the body coalesce, since those domains in which the spontaneous field was aligned with the applied field grow at the expense of others. This process yields the *initial*, spatially uniform configuration K_0, which, in general, is now endowed not only with initial fields, but also with initial *internal* (mechanical and electromagnetic) forces. Such an initial state, in general, can be maintained in equilibrium only by means of mechanical and electromagnetic actions on the boundary, except for such bodies as ellipsoids and their degenerate forms (e.g., infinitely long cylinder, half space, etc.). The present configuration K_t that corresponds to the actual dynamic process (the wavelike disturbances), is by definition, in an electromagnetomechanical "neighborhood" of K_0 (in some topological sense; stability requires that any of these disturbances remain in this neighborhood) (Fig. 19). The perturbation—e.g., **u** for the elastic displacement about K_0, **p** and **e** for the perturbations in polarization and electric fields—is then shown to take place about K_0 in a material body whose *linearized* constitutive equations have now acquired a symmetry induced by the bias field; that is, the initially applied field has provoked a *symmetry breaking* or *inducement* in the same way as a static stress field causes an optical anisotropy inducement and birefringence in a transparent material in photoelasticity. The restricted material symmetry (as compared to the one in the reference configuration) reached is richer in coupled phenomena than the initial one. Moreover, a control in the applied fields provides a control in the coupled dynamic phenomena, something that escaped us, for instance, in the case examined in Section IV. The procedure of linearization is generally tedious and implies so-called *Lagrangian* (at fixed material coordinates) and/or *Eulerian* (at fixed spatial coordinates) variations of the nonlinear field and constitutive equations about K_0 [see Maugin (1981) for an introduction to this variational procedure]. During this procedure three types of characteristically small parameters emerge: first, those noted ε_1, which measure, for instance, the strength of the initial state of electric polarization P_0. Classically $\varepsilon_1 = P_0^2/\rho c_T^2$, where c_T is a typical bulk acoustic speed and ρ is the initial matter density. The parameters ε_1 thus defined are of the order of 10^{-5}. These small parameters generally cause a slight alteration in the speed of otherwise uncoupled acoustic modes. The second category of small parameters, globally denoted ε_2, is related to the fact that dynamic disturbances in electromagnetic fields correspond to small spatial deviations from the ordering prevailing in the configuration K_0; that is, in the ferroelectric case, ε_2 is related to polariza-

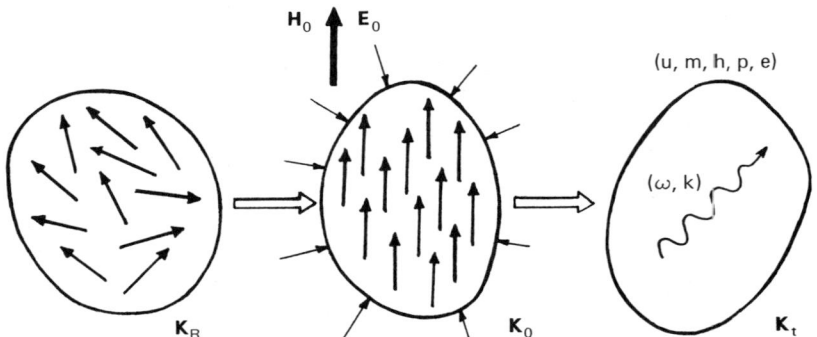

FIG. 19. Superimposition of small dynamical fields on bias fields.

tion gradients. In addition to causing a *dispersion* of the otherwise uncoupled ferroelectric modes, the parameter ε_2 may cause *boundary layer* effects where steep field gradients can be expected, e.g., nearby boundaries. Finally, the symmetry breaking, which sets evidence for privileged directions, favors the existence of *electromechanical couplings* starting with piezoelectricity, which are globally represented by a small parameter ε_3. The latter plays the role of the parameter ε in Eq. (3.1). The important role played by bias fields is recognized experimentally. For instance, it is well known, since the discovery by the Soviet scientist Rzhanov in the 1940s, that a material such as barium titanate, $BaTiO_3$, although ferroelectric in all respects, becomes strongly piezoelectric only after a strong external electric field $\mathbf{E}_0$ has been applied to it.

B. Coupled Bulk Modes and Splitting of the Surface Wave Problem

On using the linearization method outlined in the previous paragraph, one can show that the linearized bulk equations that govern small disturbances within a fully dynamic electromagnetic framework and when both viscosity and dielectric relaxation are accounted for, have the following general symbolic form once the dynamic magnetic induction **b** has been eliminated (see Maugin and Pouget, 1980):

$$\ddot{\mathbf{u}} = L_u [\nabla\nabla\mathbf{u}, \nabla\mathbf{p}, \nabla\nabla\mathbf{p}, \dot{\mathbf{p}}, \nabla\mathbf{e}; \nabla\nabla\dot{\mathbf{u}}, \nabla\dot{\mathbf{p}} | \mathbf{P}_0],$$
$$\ddot{\mathbf{p}} = L_p [\nabla\mathbf{u}, \nabla\nabla\mathbf{u}, \dot{\mathbf{u}}, \mathbf{p}, \nabla\nabla\mathbf{p}, \mathbf{e}; \nabla\dot{\mathbf{u}}, \dot{\mathbf{p}} | \mathbf{P}_0], \quad (6.1)$$
$$\ddot{\mathbf{e}} = L_e [\nabla\nabla\nabla\mathbf{u}, \nabla\ddot{\mathbf{u}}, \nabla\nabla\mathbf{p}, \ddot{\mathbf{p}}, \nabla\nabla\mathbf{e} | \mathbf{P}_0].$$

Here **u**, **p**, and **e** are the perturbations in the elastic displacement and polarization (per unit mass) and electric fields. The following notational device has been introduced: $L[A, B, \ldots, C | D]$ denotes a vector-valued operator that acts linearly on the variables $A, B, \ldots, C$, with parameters D. The set (6.1) of equations exhibits beforehand the different bulk wave modes than can propagate according to this model: we have coupled *acoustic modes* (vibrations in **u**), *soft ferroelectric modes* (vibrations in **p**), and electromagnetic modes (vibrations in **e** and **b**; the latter are always transverse). The nonhomogeneity in the polynomials of differentiation in Eq. (6.1) indicates that these modes will be *dispersive* (small parameter ε_2). The presence of the variables to the right of the semicolons further indicates *damping*. The contributions linear in $\dot{\mathbf{p}}$ and $\dot{\mathbf{u}}$ in the first two equations of (6.1) indicate the presence of a so-called *magnetoacoustic dragging* effect, which will not manifest itself in the absence of bias magnetic field. Finally, the presence of the parameter $\mathbf{P}_0$ to the right of the vertical bars indicates that all mode characteristics will depend on it (small parameters ε_1). The small parameter ε_3 (induced piezoelectricity) is hidden in the formalism. We refer the reader to Pouget and Maugin (1980) for an exhaustive study of the consequences of system (6.1), the essential result for the present contribution being already summarized in Fig. 8b, i.e., the resonance coupling between acoustic and soft ferroelectric modes. All we need for the following development is the *quasi-electrostatic version of Eq. (6.1) in the absence of damping*, that is,

$$\ddot{\mathbf{u}} = L_u [\nabla\nabla\mathbf{u}, \nabla\mathbf{p}, \nabla\nabla\mathbf{p}, \nabla e | \mathbf{P}_0],$$
$$\ddot{\mathbf{p}} = L_p [\nabla\mathbf{u}, \nabla\nabla\mathbf{u}, \mathbf{p}, \nabla\nabla\mathbf{p}, e | \mathbf{P}_0], \qquad (6.2)$$
$$0 = L_e [\nabla\nabla\nabla\mathbf{u}, \nabla\nabla\mathbf{p}, \nabla\nabla e | \mathbf{P}_0].$$

Accompanying these bulk equations valid for $x_2 > 0$ (Fig. 20) are boundary

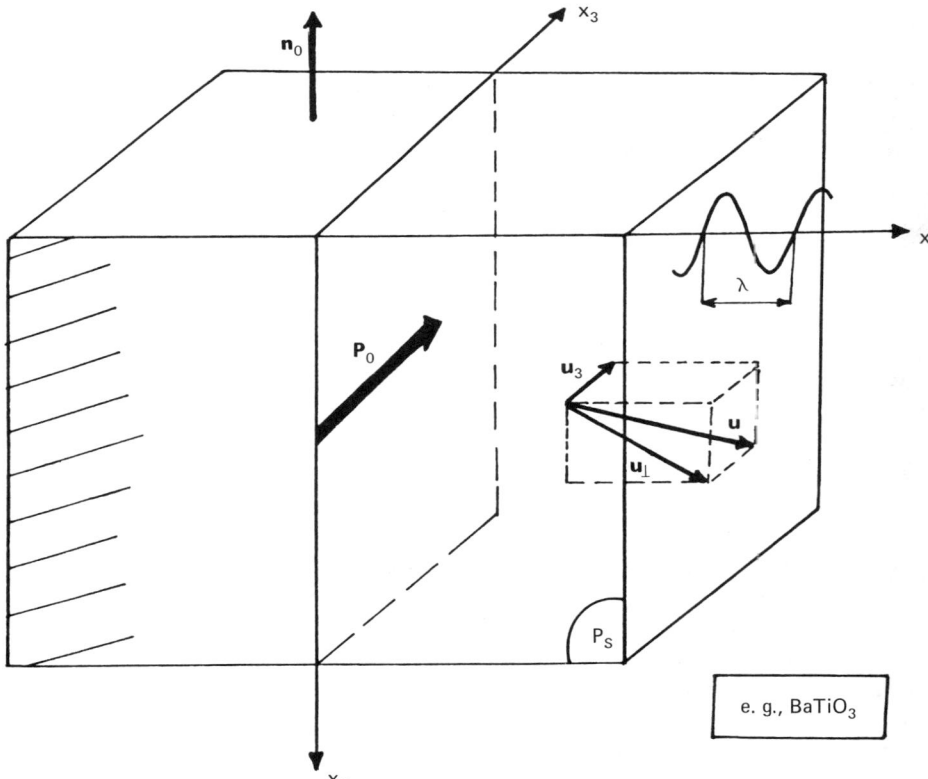

FIG. 20. Ferroelectric elastic crystal with broken symmetry ($\mathbf{P}_0 \perp P_S$).

conditions at $x_2 = 0$ that are obtained by variation of the exact nonlinear boundary conditions at the boundary of a body in the configuration K_0. These can be expressed symbolically as

$$B_u[\nabla \mathbf{u}, \mathbf{p}, \mathbf{e} | \mathbf{P}_0] = 0,$$
$$B_p[\nabla \mathbf{p}, \nabla \mathbf{e} | \mathbf{P}_0] = 0, \qquad (6.3)$$
$$B_e[\mathbf{e}, \mathbf{p} | \mathbf{P}_0] = 0,$$

where the B's are linear operators on the variables to the left of the vertical bars, $\mathbf{P}_0$ being again a parameter.

It is a remarkable feature of exact nonlinear theory that the surface wave problem associated with Eqs. (6.2) and (6.3) for a body that behaves isotropically with respect to K_R and a setting of $\mathbf{E}_0$ and $\mathbf{P}_0$ orthogonal to the sagittal plane (see Fig. 20) uncouples into two surface wave problems, one of the *Rayleigh type*, with solution

$$S_{PR} = \{\mathbf{u} = (u_1, u_2, 0) \subset P_S, \mathbf{p} = (0, 0, p_3) \perp P_S\}, \qquad (6.4)$$

and one of the *Bleustein–Gulyaev type*, with solution

$$S_{BG} = \{\mathbf{u} = (0, 0, u_3) \perp P_S, \mathbf{p} = (p_1, p_2, 0) \subset P_S, \varphi\}, \qquad (6.5)$$

where φ is the electrostatic potential such that $\mathbf{e} = -\nabla\varphi$. Clearly, the second solution (6.5) will critically depend on the electric boundary conditions, for which, as in Section IV, two cases must be considered: the electrically free boundary and the grounded (or electroded) boundary. Only solution (6.5) resorts to the present contribution. Nonetheless, we shall first briefly comment on solution (6.4) in reason of its illustrative interest for the contents of Section III.

C. Piezoelectric Rayleigh Modes in Elastic Ferroelectrics

After introduction of the Helmholtz decomposition of the displacement field

$$\mathbf{u} = (u_1, u_2, 0) = \nabla\Phi + \nabla \times \mathbf{U} = (\Phi_{,1} + U_{,2}, \Phi_{,2} - U_{,1}, 0), \qquad (6.6)$$

system (6.2) for solution (6.4) takes on the form (see Pouget and Maugin, 1981a)

$$\rho\ddot{\Phi} = (C_1 + C_2)\nabla^2\Phi + \rho C_6 P_0 p_3,$$
$$\rho\ddot{U} = C_2 \nabla^2 U,$$
$$d\ddot{p}_3 = -C_6 P_0 \nabla^2 \Phi - \rho(2C_4 + C_7 P_0^2)p_3 + (\bar{C}_{14} + P_0^2 \bar{C}_{17})\nabla^2 p_3 \qquad (6.7)$$

for $x_2 > 0$, whereas the boundary conditions (6.3) at $x_2 = 0$ read

$$(C_1 - C_2)\Phi_{,11} + (C_1 + C_2)\Phi_{,22} - 2C_2 U_{,12} + \rho C_6 P_0 p_3 = 0,$$
$$2\Phi_{,12} + U_{,22} - U_{,11} = 0, \qquad (6.8)$$
$$p_{3,2} = 0.$$

In these equations C_1 and C_2 are elasticity coefficients, C_6 is a piezoelectric coefficient, C_4 and C_7 are related to electric susceptibility, and $\bar{C}_{14}$ and $\bar{C}_{17}$ refer to polarization gradients (they are connected with the small parameter ε_2 mentioned in the foregoing paragraphs). All these coefficients are defined thermodynamically at K_0 by second-order partial derivatives of the free-energy density. From Eq. (6.7) for solutions of the type (2.5), it is found that Eq. (2.9) reads

$$(\varepsilon_1' Q - \Gamma)[(\varepsilon_1 Q - \Gamma)(\Gamma_1 + \hat{\varepsilon}_T^p Q - \Gamma) - \varepsilon_3 \zeta Q] = 0, \qquad (6.9)$$

wherein

$$Q = q_1^2 + q_2^2, \qquad \Gamma = \Omega^2, \qquad q_1 = ck_1/\omega_0, \qquad q_2 = ck_2/\omega_0,$$
$$\omega_0^2 = \rho(1 + 2C_4 + P_0^2 C_7)/d, \qquad \Omega = \omega/\omega_0, \qquad (6.10)$$
$$\Gamma_1 = \rho(2C_4 + P_0^2 C_7)/d\omega_0^2 = \Omega_1^2,$$

and

$$\varepsilon_1 = (C_1 + C_2)/\rho c^2, \qquad \varepsilon_1' = C_2/\rho c^2,$$
$$\hat{\varepsilon}_T^p = (\bar{C}_{14} + P_0^2 \bar{C}_{17})/dc^2, \qquad \varepsilon_3 = P_0^2/\rho c^2, \qquad (6.11)$$
$$\zeta = C_6^2(1 + 2C_4 + P_0^2 C_7)^{-1},$$

where c is the velocity of light in a vacuum. The admissible domain of dispersion determined in agreement with the reasoning sketched out in Eqs. (2.10)–(2.12) is that situated below the hatched region in Fig. 21a, where (a) represents an uncoupled transverse elastic mode (**u**) and the parabola (c) represents an uncoupled soft ferroelectric mode. The intersection point of abscissa q_1^* represents the bulk electroacoustic resonance between an elastic longitudinal mode and the ferroelectric mode. When one pursues the general procedure given in Section II by considering general relations of the type (2.13) and accounting for the boundary conditions (6.8), one is finally led to the *dispersion relation*, which we will reproduce here [see Eq. (46) in Pouget and Maugin (1981b)]. In any case, the admissible dispersion curve, sketched out as (d) in Fig. 21b, is very close to a straight line of slope c_R (the Rayleigh wave velocity in classical piezoelectricity), situated below the straight line of slope c_T (bulk shear wave velocity) for $q_1 < q_1^0$, while it almost coincides with the pure soft ferroelectric mode for large q_1's. The piezoelectric Rayleigh mode thus placed in evidence starts to present a strong *dispersion* when q_1 approaches q_1^0 from below. The curvature of the

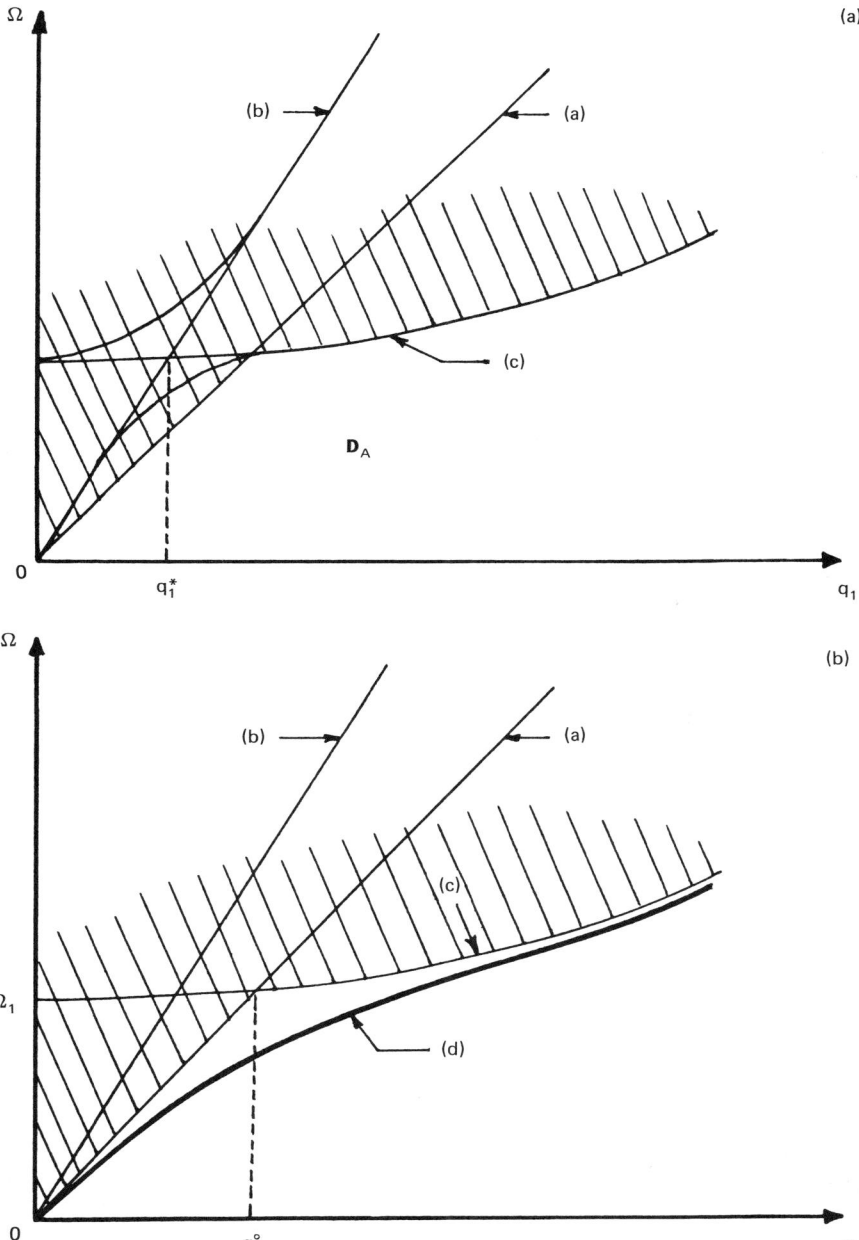

Fig. 21. Piezoelectric Rayleigh waves in ferroelectrics: (a) the allowed domain of dispersion; (b) qualitative sketch of the dispersion relation.

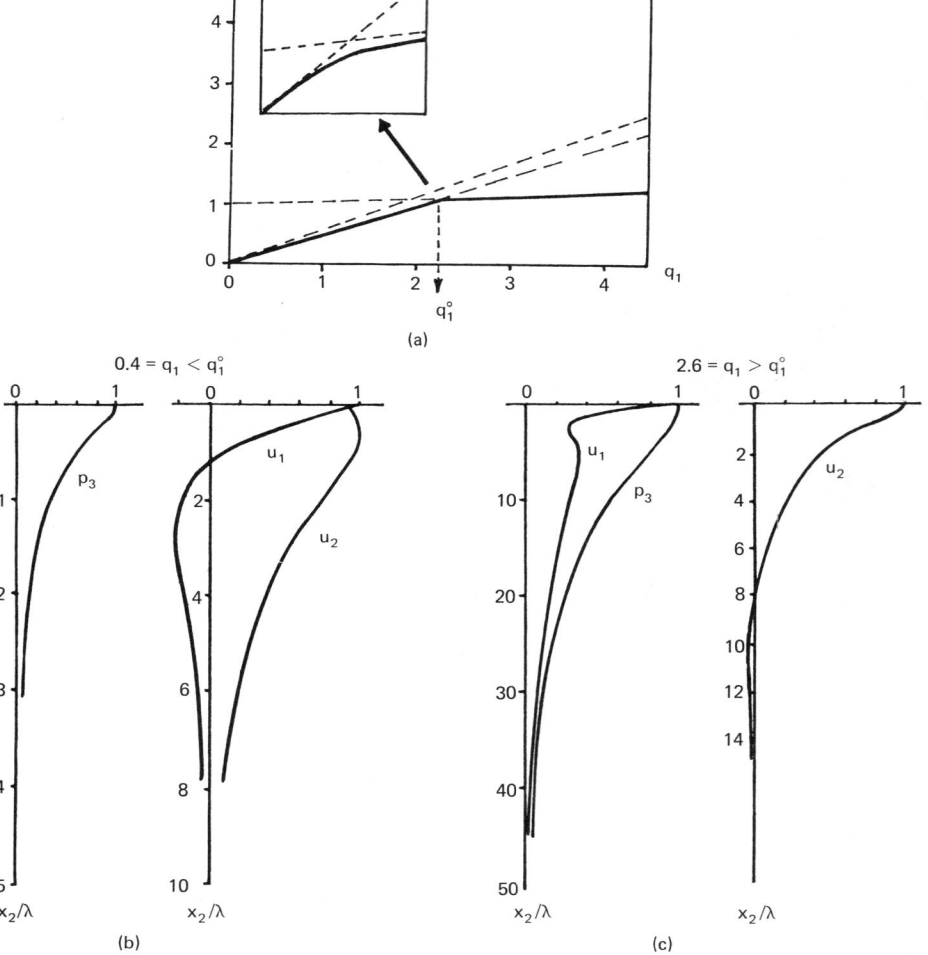

FIG. 22. Piezoelectric Rayleigh waves in $BaTiO_3$ ($\mathbf{P}_0 \perp \mathbf{P}_s$): (a) dispersion relation; (b) behavior with depth of the amplitudes for a wavenumber below electroacoustic resonance; (c) behavior with depth of the amplitudes for a wavenumber above electroacoustic resonance (after Pouget and Maugin, 1981b).

soft-ferroelectric branch (parameter $\varepsilon_2 = \hat{\varepsilon}_1^p$) is obviously small, so that the dispersion branch will be practically flat for $q_1 \gg q_1^0$. We have reproduced in Fig. 22 the numerical dispersion relation relative to piezoelectric Rayleigh waves in $BaTiO_3$, as well as the behavior of the amplitude of these waves in terms of a normalized depth. The amplitude behavior differs, depending

Elastic Surface Waves with Transverse Horizontal Polarization   415

on whether the wavenumber is situated below or above the critical wavenumber q_1^0. In particular, whereas the penetration depth is of the order of a few wavelengths when the surface mode is essentially a transverse elastic mode ($q_1 < q_1^0$), its penetration depth is much larger for the components u_1 and p_3 when the surface wave is essentially a soft ferroelectric mode ($q_1 > q_1^0$): $p \simeq 20\lambda$ for p_3 and $p \simeq 30\lambda$ for u_1.

D. Bleustein–Gulyaev Modes in Elastic Ferroelectrics

After introduction of the Helmholtz decomposition of the electric polarization field,

$$\mathbf{p} = (p_1, p_2, 0) = \nabla \chi + \nabla \times \mathbf{\kappa} = (\chi_{,1} + \kappa_{,2}, \chi_{,2} - \kappa_{,1}, 0), \quad (6.12)$$

system (6.2) for the solution (6.5) takes on the form (see Pouget, 1980, 1981; Pouget and Maugin, 1981a)

$$\begin{aligned}
\rho \ddot{u}_3 &= (C_2 + P_0^2 C_5)\nabla^2 u_3 + 2\rho C_8 P_0 \nabla^2 \chi, \\
d\ddot{\chi} &= -\varphi - 2C_8 P_0 u_3 - 2\rho C_4 \chi + (\bar{C}_{13} + \bar{C}_{14})\nabla^2 \chi, \\
d\ddot{\kappa} &= -2\rho C_4 \kappa + \bar{C}_{14}\nabla^2 \kappa, \\
\nabla^2(\rho \chi - \varphi) &= 0,
\end{aligned} \quad (6.13)$$

for $x_2 > 0$, while the boundary conditions (6.3) at $x_2 = 0$ read

$$\begin{aligned}
\varphi_{,2} - \varphi_{,2}^{\text{out}} - (\chi_{,2} - \kappa_{,1}) &= 0, \\
\bar{C}_2 u_{3,2} + \rho(2C_8 + C_4)P_0(\chi_{,2} - \kappa_{,1}) &= 0, \\
(\bar{C}_{14} + \bar{C}_{13} - \bar{C}_{18})\chi_{,12} + \bar{C}_{14}\kappa_{,22} - (\bar{C}_{13} - \bar{C}_{18})\kappa_{,11} &= 0, \\
(\bar{C}_{13} + \bar{C}_{14})\chi_{,22} + \bar{C}_{18}\chi_{,11} - (\bar{C}_{13} + \bar{C}_{14} - \bar{C}_{18})\kappa_{,12} &= 0,
\end{aligned} \quad (6.14)$$

and

$$\begin{aligned}
\varphi &= \varphi^{\text{out}} \quad \text{(electrically free boundary; } \nabla^2 \varphi^{\text{out}} = 0, x_2 < 0\text{)}, \\
\varphi &= 0 \quad \text{(grounded boundary)}.
\end{aligned} \quad (6.15)$$

The first equation of (6.14) corresponds to the continuity of the normal component of the electric displacement at $x_2 = 0$. In these equations C_2 and C_5 are elastic coefficients, $\bar{C}_2$ is a stiffened elastic coefficient, C_8 is a piezoelectric coefficient, C_4 is related to electric susceptibility, and $\bar{C}_{13}$, $\bar{C}_{14}$, and $\bar{C}_{18}$ refer to polarization gradient effects. From Eq. (6.13) for solutions of the type (2.5), it is found that Eq. (2.9) reads

$$Q(\Gamma_0 + \varepsilon_2' Q - \Gamma)[(1 - \Gamma + \varepsilon_2 Q)(\hat{\varepsilon}_T^e Q - \Gamma) - \varepsilon_3 \mu Q] = 0, \quad (6.16)$$

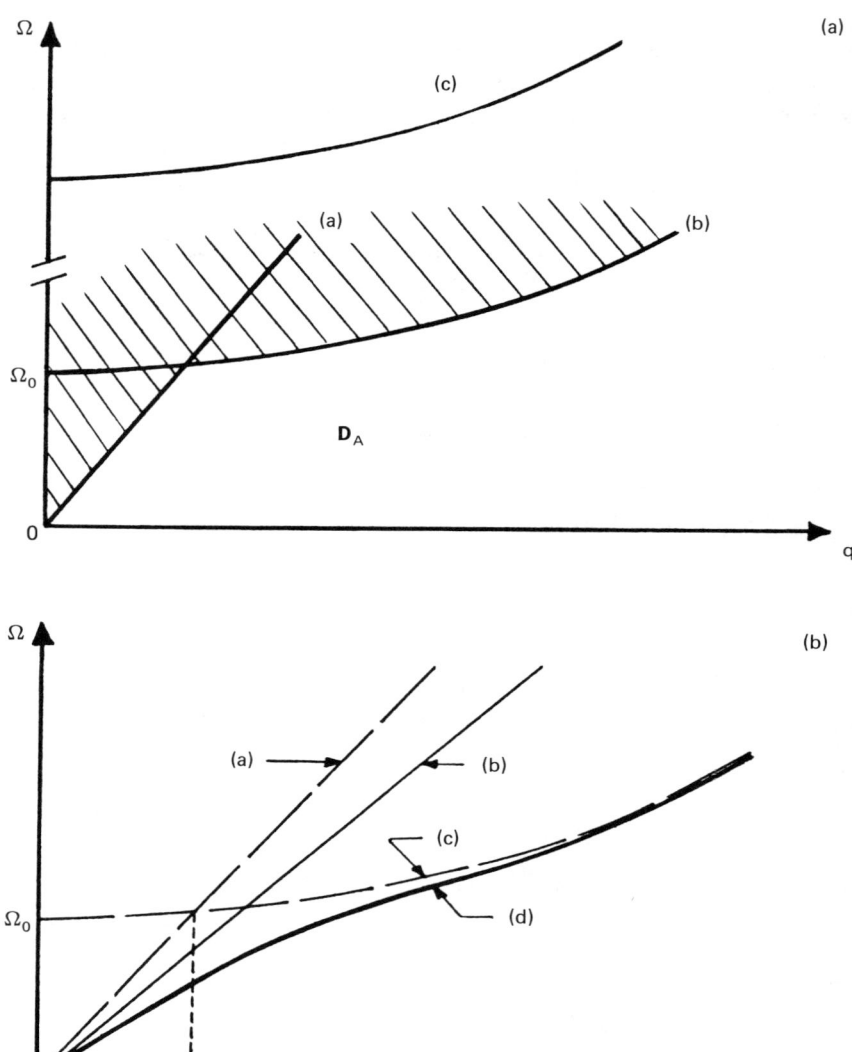

FIG. 23. Bleustein–Gulyaev modes in ferroelectrics ($\mathbf{P}_0 \perp P_s$): (a) the allowed domain of dispersion (the upper branch is a high longitudinal optical mode); (b) qualitative sketch of the dispersion relation (grounded boundary).

wherein

$$Q = q_1^2 + q_2^2, \quad \Gamma = \Omega^2, \quad \Omega = \omega/\omega_1,$$
$$q_1 = ck_1/\omega_1, q_2 = ck_2/\omega_1, \quad (6.17)$$
$$\omega_1^2 = \rho(1 + 2C_4)/d, \quad \Gamma_0 = \Omega_0^2 = 2C_4/(1 + 2C_4),$$

and

$$\varepsilon_2 = (\bar{C}_{13} + \bar{C}_{14})/dc^2, \quad \varepsilon_2' = \bar{C}_{14}/dc^2,$$
$$\hat{\varepsilon}_T^e = (C_2/\rho c^2) + C_5\varepsilon_2, \quad \varepsilon_3 = P_0^2/\rho c^2, \quad (6.18)$$
$$\mu = C_8^2(1 + 2C_4)^{-1}.$$

The admissible domain of dispersion determined in agreement with the reasoning outlined in Eqs. (2.10)–(2.12) is that situated below the hatched region in Fig. 23a, since, on setting $q_2^\alpha = q_{\alpha=J}$, $J = I, \ldots,$ IV, Eqs. (6.16) and (2.11) impose that we consider solutions q_J such that

$$q_I^2 = -q_1^2 \quad (q_I = -iq_1)$$
$$q_{II}^2 = (\Gamma - \Gamma_0 - \varepsilon_2' q_1^2)/\varepsilon_2' \leqq 0,$$
$$q_{III}^2 = (\Gamma - \tilde{\varepsilon}_T^e q_1^2)/\tilde{\varepsilon}_T^e \leqq 0, \quad (6.19)$$
$$q_{IV}^2 = (\Gamma - 1 - \varepsilon_2 q_1^2)/\varepsilon_2 \leqq 0,$$

where

$$\tilde{\varepsilon}_T^e = \hat{\varepsilon}_T^e - \mu\varepsilon_3. \quad (6.20)$$

The real dispersion curve and the amplitude behavior of the surface modes with depth depend critically on the choice of the electric boundary condition (6.15). As was already the case for Bleustein–Gulyaev modes in classic linear piezoelectricity (see Section IV), the case of a grounded or electroded boundary $x_2 = 0$ is easier to handle than the other case. We shall not give the exact dispersion relation for coupled SH acoustic and soft ferroelectric modes here [see Eq. (126) in Pouget and Maugin (1981a)]. A qualitative sketch of the dispersion relation is given in Fig. 23b, where (b) represents the classic (nondispersive) Bleustein–Gulyaev mode [compare Eq. (4.22)]

$$\Omega_{BG}^2 = \tilde{\varepsilon}_T^e(1 - K^4)q_1^2, \quad (6.21)$$

where

$$K^2 = \frac{C_8(2C_8 + C_4)P_0^2}{C_4\bar{C}_2} \quad (6.22)$$

is the electromechanical coupling factor. For small q_1's, the coupled admissible branch, (d) in Fig. 23b, has the following behavior:

$$\Omega = \Omega_{B2}\left[1 - \tilde{\varepsilon}_T^e\left(\frac{K^2}{\Omega_0}\right)^2\left(1 - \frac{2\varepsilon_2}{\tilde{\varepsilon}_T^e(1-K^4)}\right)q_1^2\right.$$
$$\left. - \frac{\tilde{\varepsilon}_T^e\sqrt{\varepsilon_2}}{2\Omega_0}\left(\frac{K^2}{\Omega_0}\right)^2\left(1 - \frac{\varepsilon_2}{\tilde{\varepsilon}_T^e(1-K^4)}\right)q_1^3 + O(q_1^4)\right], \quad (6.23)$$

$$\Omega_{B2} \equiv (c_T/c)(1-K^4)^{1/2}q_1, \qquad c_T^2 = c^2\hat{\varepsilon}_T^e,$$

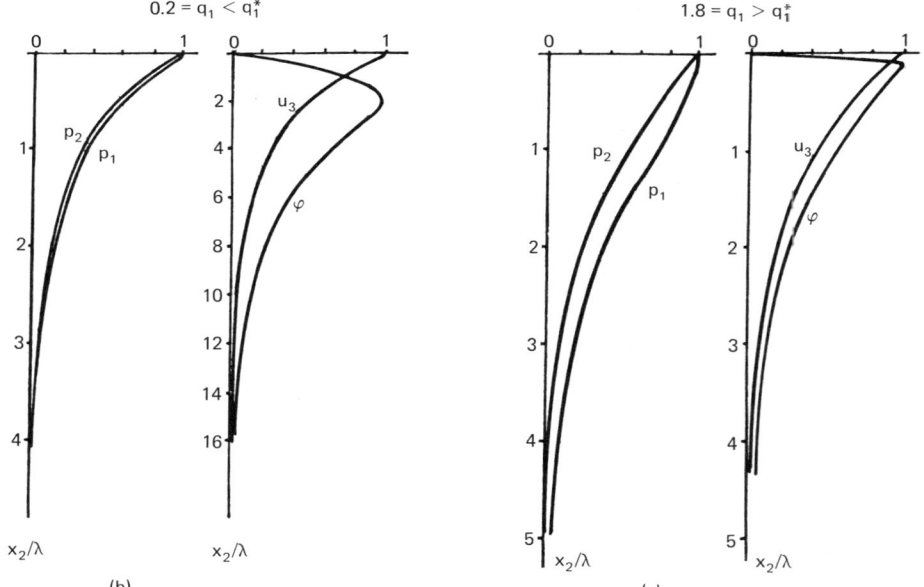

FIG. 24. Bleustein–Gulyaev modes in $BaTiO_3$ ($\mathbf{P}_0 \perp \mathbf{P}_S$; grounded boundary): (a) numerical dispersion relation; (b) behavior with depth of the amplitudes for a wavenumber below electroacoustic resonance; (c) behavior with depth of the amplitudes for a wavenumber above electroacoustic resonance (after Pouget and Maugin, 1981a).

whereas for large q_1's we have

$$\Omega = \Omega_F\left(1 - \frac{1}{2\Omega_F^2}\frac{\varepsilon^2\tilde{\varepsilon}_T^e}{\varepsilon_2(\tilde{\varepsilon}_T^e - \varepsilon_2)q_1^2} + O(q_1^{-4})\right), \qquad (6.24)$$

with

$$\varepsilon = 2C_8(2C_8 + C_4)(1 + 2C_4)P_0^2/\bar{C}_2, \qquad \Omega_F^2 = \Gamma_0 + \varepsilon_2 q_1^2. \qquad (6.25)$$

We have reproduced in Fig. 24 the corresponding numerical dispersion relation for BaTiO$_3$ and the amplitude behavior according to whether the wavenumber q_1 is smaller or greater than the critical wavenumber q_1^* (used in the nondimensionalization). For $q_1 < q_1^*$, both components p_1 and p_2 decrease very quickly with depth (penetration depth of the order of one wavelength), whereas the elastic component penetrates at about 4λ and the electric potential reaches a maximum at about 2λ and thereafter penetrates at about 8λ. The situation is somewhat equivalent for $q_1 > q_1^*$, except for a somewhat greater penetration depth.

The situation is markedly different when the boundary $x_2 = 0$ is not grounded, but it is electrically matched to the exterior electric potential [first of the conditions (6.15)]. What happens is that there is no real resonance coupling at the intersection point of the uncoupled modes (classical Bleustein–Gulyaev and soft ferroelectric modes). The resulting coupled mode presents only a slight deviation (from below) from the uncoupled Bleustein–Gulyaev mode and is accompanied by strong oscillations in the amplitude behavior, as well as rapid damping (i.e., the frequency becomes complex above the "coupling" point). From this there results a dispersion relation of the type reproduced in Fig. 25. Below the critical point q_1^* and for small q_1's, the component p_1 practically behaves like a *face wave* (a degenerate surface wave with practically constant amplitude) for a depth greater than 2λ, whereas u_3 and φ have the same behavior as for the classic Bleustein–Gulyaev mode, with a remarkably important penetration depth (about 2500λ). In addition, the electric potential component exhibits a boundary layer effect with a maximum at about one wavelength as a result of steep polarization gradients near $x_2 = 0$. For $q_1 > q_1^*$, the components p_2 and p_1 present an oscillating behavior, the component p_2 being practically nil in the mean after a few wavelengths, while the component p_1 keeps a constant mean (face wave overall behavior). Here, also, u_3 and φ deeply penetrate the substrate and φ exhibits a boundary layer effect with a maximum at about 0.2λ.

In conclusion, it must be noted that the features of both the Rayleigh and Bleustein–Gulyaev modes placed in evidence in this section are easily *controlled* by the manner (*direction* and *strength*) in which the bias field $\mathbf{E}_0$ is applied. This endows these waves with a particular interest in acoustoelectronics.

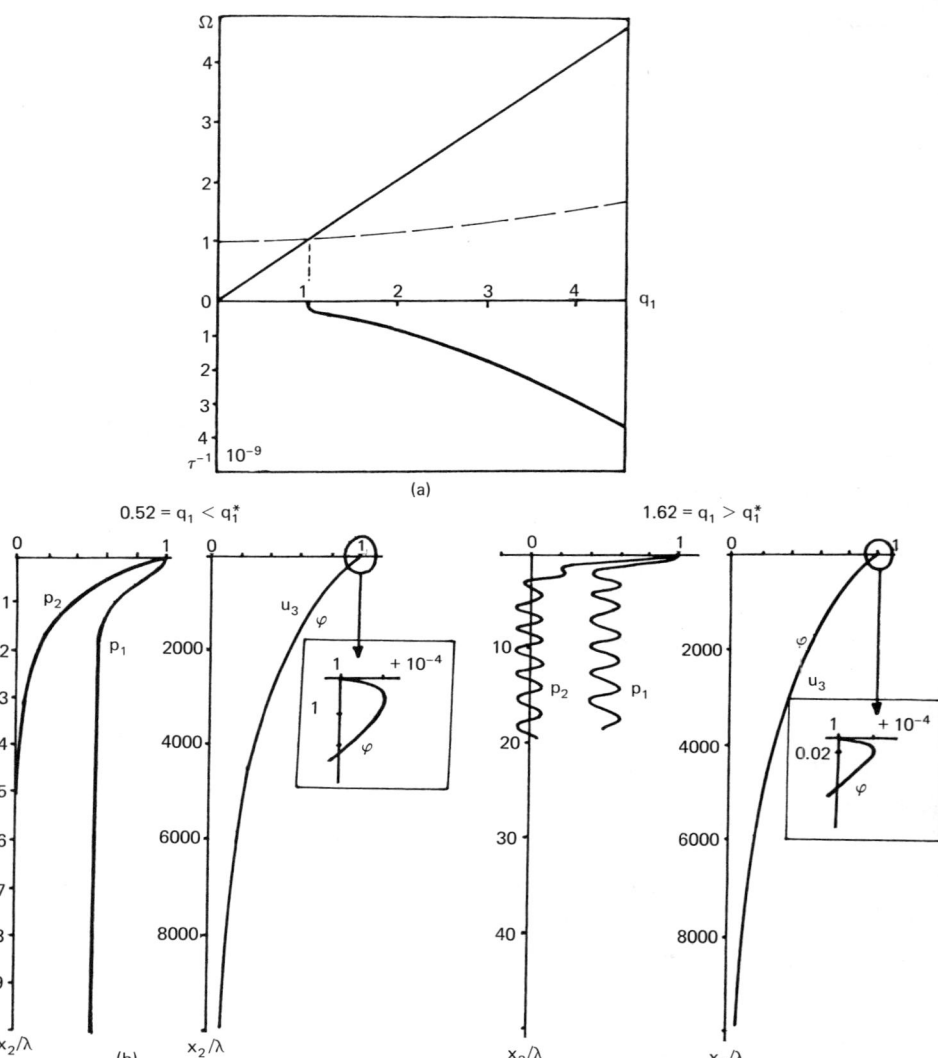

FIG. 25. Bleustein–Gulyaev modes in $BaTiO_3$ ($\mathbf{P}_0 \perp P_S$; free boundary): (a) numerical dispersion relation; (b) behavior with depth of the amplitudes for a wavenumber below electroacoustic resonance; (c) behavior with depth of the amplitudes for a wavenumber above electroacoustic resonance (after Pouget and Maugin, 1981a).

VII. Magnetoacoustic SH Surface Waves in Magnetoelasticity

A. General Features

Most of the preliminary remarks made in Section VI,A hold good in the case of magnetically ordered elastic bodies such as elastic ferromagnetic yttrium–iron garnet (YIG). The methodology for obtaining the linearized bulk field equations and constitutive equations about an initial configuration of intense bias magnetic field follows along the same line (see Maugin, 1979a,b). The same holds true for deduction of the linearized boundary conditions from the exact ones (see Maugin, 1982a). A symmetry breaking is caused by the intense bias magnetic field, a field that also allows one to produce a one-domain, spatially uniformly magnetized configuration from a multidomain ferromagnetic sample. Both acoustic and spin oscillations to be considered are in the long-wave range, and couplings with electromagnetic phenomena are discarded so that a quasi-magnetostatic approximation proves to be sufficient. Two essential features, however, differ from those in the case of ferroelectrics: First, at low temperatures (and hence low levels of energy), temperatures much below the phase transition temperature θ_C that separates the disordered paramagnetic phase of the material and its ordered ferromagnetic one with decreasing temperature, the magnetization is assumed to have reached *saturation*. The applied magnetic field is sufficiently strong to make any perturbation in the magnetic field let the magnetization stay in its saturation regime. From this it first follows that the magnetization density has a constant magnitude at each point in the sample and, next, because of uniformity, the *same* constant magnitude throughout the sample. Therefore the magnetization μ per unit mass has *kinematics* that are necessarily described by an equation of the type

$$\dot{\boldsymbol{\mu}} = \boldsymbol{\omega} \times \boldsymbol{\mu}, \tag{7.1}$$

where $\boldsymbol{\omega}$ is the precessional velocity vector of $\boldsymbol{\mu}$. Obviously, then

$$\boldsymbol{\mu} \cdot \dot{\boldsymbol{\mu}} = 0, \tag{7.2}$$

from which it follows that perturbations $\bar{\boldsymbol{\mu}}$ about an initial value $\boldsymbol{\mu}_0$ satisfy the orthogonality condition

$$\boldsymbol{\mu}_0 \cdot \bar{\boldsymbol{\mu}} = 0. \tag{7.3}$$

Hence *magnetization perturbations in ferromagnets at low temperatures* (as compared to θ_C) *are always orthogonal to the initial magnetization* (this is also true when magnetic spin relaxation is present).

Next, since μ is proportionally related to an angular momentum via the so-called gyromagnetic ratio γ of the material, Eq. (7.1) is also equivalent to a *dynamic* equation, the local balance of angular momentum. On account of this property and the quasi-magnetostatic approximation, in the case of elastic ferromagnets in the absence of damping effects (viscosity and magnetic spin relaxation), the Set of equations (6.1) will be replaced by a set of equations governing perturbations in the bulk of the material in the general form

$$\ddot{\mathbf{u}} = L_u\left[\nabla\nabla\mathbf{u}, \nabla\bar{\mu}, \nabla\nabla\varphi \,|\, \mathbf{M}_0\right],$$
$$\dot{\bar{\mu}} = L_\mu\left[\nabla\mathbf{u}, \nabla\varphi, \bar{\mu}, \nabla\nabla\bar{\mu} \,|\, \mathbf{M}_0\right], \qquad (7.4)$$
$$0 = L_\varphi\left[\nabla^2\varphi, \nabla\nabla\mathbf{u}, \nabla\bar{\mu} \,|\, \mathbf{M}_0\right],$$

where the last equation is none other than the remaining Maxwell static equation after introduction of the magnetostatic potential φ such that $\mathbf{h} = -\nabla\varphi$ if $\mathbf{h}$ is the perturbation magnetic field about $\mathbf{H}_0$. The latter field $\mathbf{H}_0$, and $\mathbf{M}_0 = \rho\mu_0$ are proportionally related by $\mathbf{M}_0 = \chi_0\mathbf{H}_0$. On eliminating φ with the last equation of (7.4), this system yields a *coupled dispersive* system for the perturbations $\mathbf{u}$ and $\bar{\mu}$. The corresponding bulk oscillations are coupled via magnetoelastic couplings and a *magnetoacoustic resonance* coupling of the type illustrated in Fig. 8a results.

The existence of surface waves of the pure spin wave type in ferromagnets was established by Damon and Eshbach (1961) and Filipov (1967). Coupled surface acoustic spin waves were studied on the basis of a discrete model by Kaliski and Kapalewski (1968). Since then many papers have been published by physicists concerning the problem of coupling elastic waves and spin waves by the intermediate of surface effects. Most of these studies are devoted to the case of *generalized Rayleigh waves*, whose main characteristics are modified by the coexistence of spin waves [see, e.g., Parekh and Bertoni (1974a,b), Emtage (1976), Ganguly et al. (1978), Maugin (1983b)]. The method recalled at the end of Section II proves its efficiency in such a complicated scheme. The propagation features obviously depend in a critical manner on the angle made by the direction of propagation and the bias magnetic field in reason of Eq. (7.3), but this remark concerns also bulk waves (see Maugin, 1981). Parekh (1972), however, was the first to recognize that whenever the initial magnetic field is set parallel to the limiting plane $\partial\mathcal{D} = \{x_2 = 0\}$ and, moreover, *orthogonally* to the direction of propagation (Fig. 26a), and hence orthogonally to the sagittal plane, an essentially SH surface wave could be maintained in the same way as the SH surface wave recalled in Sections IV and VI,D. That is, on propagating in a direction orthogonal to the bias magnetic field set parallel to the boundary (so-called *orthogonal* setting), a

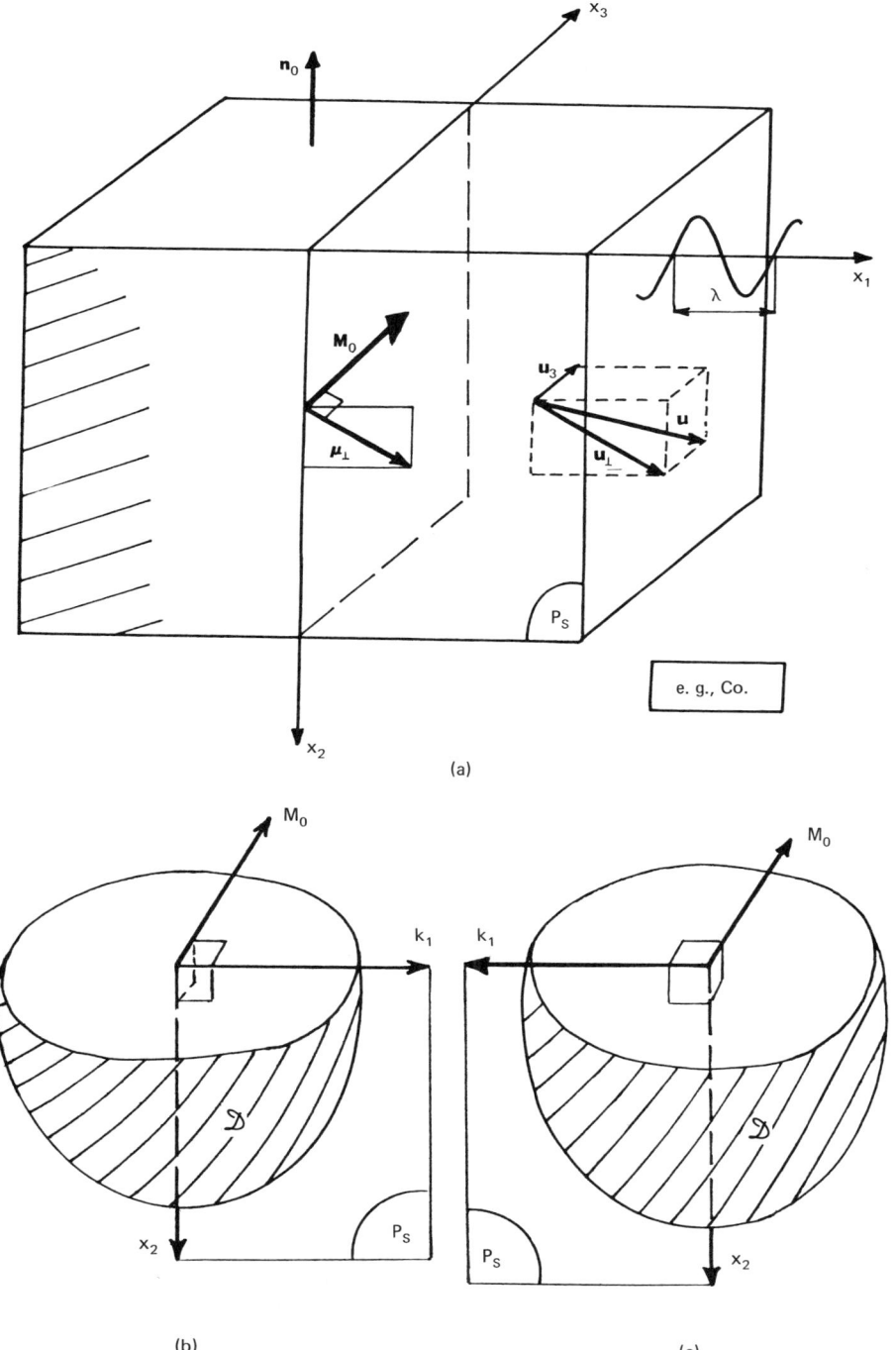

FIG. 26. SH Magnetoacoustic waves ($\mathbf{M}_0 \perp P_S$) in a ferromagnetic crystal with broken symmetry: (a) general setting; (b) ($k_1, \mathbf{M}_0, x_2$) form a nondirect triad; (c) ($k_1, \mathbf{M}_0, x_2$) form a direct triad.

combined magnetoacoustic surface wave of the Bleustein–Gulyaev type can be created. But its properties depend on the direction of propagation relative to the direction of the magnetic field. Such waves are *nonreciprocal*, according to whether they travel from left to right or from right to left. This fact ultimately follows from the *axial* nature of a magnetic field, as well as from the fact that the magnetic solution in the substrate must match a magnetostatic problem of the type

$$\nabla^2 \varphi = 0 \quad \text{for } x_2 < 0, \qquad \varphi \to 0 \quad \text{as } |x_2| \to \infty \tag{7.5}$$

at $x_2 = 0$. This is examined in greater detail in Section VII,B. The boundary conditions [compare Eq. (6.3)] that complement the bulk equations (7.4) valid in the substrate, and Eq. (7.5) valid in the exterior region, have the general form (see Maugin, 1983a)

$$\begin{aligned} B_u \left[\nabla \mathbf{u}, \bar{\boldsymbol{\mu}}, [\![\varphi]\!], [\![\nabla\varphi]\!] \,|\, \mathbf{M}_0 \right] &= 0, \\ B_\mu \left[\nabla \bar{\boldsymbol{\mu}} \,|\, \mathbf{M}_0 \right] &= 0, \\ B_\varphi \left[[\![\varphi]\!], [\![\nabla\varphi]\!], \nabla \mathbf{u}, \bar{\boldsymbol{\mu}} \,|\, \mathbf{M}_0 \right] &= 0 \end{aligned} \tag{7.6}$$

at $x_2 = 0$, where $[\![\Box]\!]$ denotes the jump of the enclosed quantity at $\partial \mathcal{D}$. The second equation of (7.6) is a condition of *pinning* of magnetic spins at the boundary. Whenever the curvature of uncoupled spin waves can be discarded, and hence for long wavelengths, this condition does not play any role in the analysis [compare Scott and Mills (1977)].

B. The Nonreciprocity of Surface Modes

The nonreciprocity of pure surface spin waves and surface magnetoacoustic waves in ferromagnetic bodies can be explained as follows in terms of simple symmetry arguments. Consider a crystal whose rectangular coordinate axes in Fig. 26a are symmetry axes. Let $R_{\alpha\beta}$ denote a reflection operation through a plane spanned by axes x_α and x_β (e.g., R_{12} denotes reflection through the plane $x_3 = 0$). A magnetic field behaves like a *pseudovector* for reflection operations. Thus a magnetic field is left unchanged in a reflection through a plane orthogonal to it, whereas it reverses sign in a reflection through a plane that contains it. For instance, with $\mathbf{M}_0$ or $\mathbf{H}_0$ along the x_3 axis, we have

$$\begin{aligned} R_{12}[\mathbf{H}_0] &= \mathbf{H}_0, & R_{13}[\mathbf{H}_0] &= -\mathbf{H}_0, & R_{23}[\mathbf{H}_0] &= -\mathbf{H}_0, \\ (R_{13} \circ R_{12})[\mathbf{H}_0] &= -\mathbf{H}_0, & (R_{13} \circ R_{23})[\mathbf{H}_0] &= -\mathbf{H}_0. \end{aligned} \tag{7.7}$$

For an *infinite* crystal, we symbolically have

$$R_{12}[\text{crystal}] = \text{crystal}, \quad R_{13}[\text{crystal}] = \text{crystal}, \quad \text{etc.,}$$
$$(R_{13} \circ R_{23})[\text{crystal}] = (R_{13} \circ R_{12})[\text{crystal}] = \text{crystal}, \quad (7.8)$$

whereas for the semi-infinite crystal in Fig. 26a,

$$R_{13}[\mathscr{D} = \{x_2 > 0\}] = \mathscr{D}' = \{x_2 < 0\}. \quad (7.9)$$

For *bulk* waves of the type (2.2) in an *infinite* crystal with dispersion relation $\omega_v(k_1, k_2, k_3, \mathbf{H}_0)$, we have

$$(R_{13} \circ R_{23})[\omega_v(k_1, k_2, k_3; \mathbf{H}_0)] = R_{13}[\omega_v(-k_1, k_2, k_3; -\mathbf{H}_0)]$$
$$= \omega_v(-k_1, -k_2, k_3; \mathbf{H}_0). \quad (7.10)$$

With ω_v even in k_1 and k_2, *the combined reflection operation* $(R_{13} \circ R_{23})$ *is a symmetry operation for the system* (crystal plus bulk modes). However, for *surface modes* of the type (2.5) in the *semi-infinite* crystal $\mathscr{D} = \{x_2 > 0\}$ with dispersion relation $\omega_s(k_1, k_2, k_3; \mathbf{H}_0)$, it cannot be proven by the same argument that ω_s is even in k_1, because R_{13} is no longer a symmetry operation—Eq. (7.9)—and $\omega_s(k_1, k_3; \mathbf{H}_0) \neq \omega_s(-k_1, k_3; \mathbf{H}_0)$ because of the surface. In fact in one case (Fig. 26b,c) (propagation in the direction of positive x_1's), k_1, $\mathbf{M}_0$, and x_2 form a *nondirect* triad, whereas in the other case (propagation in the direction of negative x_1's) these three vectors form a *direct* triad. Therefore in studying the complete surface wave problem for the system (7.4), (7.6), instead of Eq. (2.5) it is convenient to consider solutions of the form

$$g = \tilde{g}\exp[i(\omega t - \sigma k_1 x_1 - k_2 x_2)], \quad (7.11)$$

with $\sigma = +1$ (propagation toward positive x_1's) and $\sigma = -1$ (propagation toward negative x_1's). Whenever damping effects, such as spin relaxation are taken into account, a *nonreciprocal attenuation* will also be observed (see Emtage, 1976). A nonreciprocal attenuation has been reported by Lewis and Patterson (1972) for the case of waves traveling at right angles to the field on a thin film of YIG. We shall leave aside the case of generalized Rayleigh waves [see Parekh and Bertoni (1974a,b), Emtage (1976), Ganguly et al. (1978), Maugin (1983b)]. All we need to note concerning this case is that the boundary conditions prohibit the buildup of any large spin wave, and resonance damping is small. The attenuation of Rayleigh waves is very different from that of bulk waves [see Maugin (1979b) for the latter]. For frequencies above the resonant frequency, spin waves can travel away from the surface and the Rayleigh wave is attenuated through radiation of spin waves into the magnetic substrate. Such waves can be used to conceive a *convolutor of signals* in thin YIG layers (Volluet, 1979).

C. SH Surface Waves of the Bleustein–Gulyaev Type

It is a remarkable feature of the exact nonlinear theory of elastic ferromagnets [see, e.g., Maugin (1979a)] that the surface wave problem associated with Eqs. (7.4) and (7.6) for a body that behaves isotropically with respect to K_R and a setting of $\mathbf{H}_0$ and $\mathbf{M}_0$ orthogonal to the sagittal plane P_S (see Fig. 26a) splits into two surface wave problems, one of the *pure Rayleigh type*, with solution

$$S_{MR} = \{\mathbf{u} = (u_1, u_2, 0) \subset P_S\}, \tag{7.12}$$

and one of the *Bleustein–Gulyaev type*, with solution

$$S_{BG} = \{\mathbf{u} = (0, 0, u_3) \perp P_S, \quad \bar{\boldsymbol{\mu}} = (\bar{\mu}_1, \bar{\mu}_2, 0) \subset P_S, \varphi\}. \tag{7.13}$$

For the first problem, for which we only indicate the dispersion relation, in place of the classic Rayleigh equation (2.24), one has [see Maugin (1983a)]

$$[4(1 + v) + \varepsilon_M \tilde{v}](1 - \xi)^{1/2}\left(1 - \frac{\xi}{\gamma_0}\right)^{1/2}$$
$$= [(2 + v\varepsilon_M) - \xi]\left[\left(2(1 + v) - \frac{\varepsilon_M}{2}\right) + \left(\frac{\gamma_0 + \varepsilon_M \hat{v}}{\gamma_0 + \varepsilon_M v^2}\right)\xi\right], \tag{7.14}$$

with ξ and γ_0 defined as in Eq. (2.25) and

$$\begin{aligned}\varepsilon_M &= M_0^2/\rho c_T^2, & v &= t_1/M_0^2, \\ \hat{v} &= 3v + \tfrac{1}{2}, & \tilde{v} &= 2v(1 + v) - 1,\end{aligned} \tag{7.15}$$

where ε_M is an infinitesimally small parameter and t_1 is an initial (internal) stress due to the presence of the initial magnetic field. Up to terms of the order of ε_M^2, Eq. (7.14) can be shown to reduce to a cubic in ξ, so that the problem of finding the speed c_R is the same as for Eq. (2.24).

For problem (7.13), with $\bar{\boldsymbol{\mu}} \subset P_S$, it is shown that the bulk equations (7.4) take on the form

$$\begin{aligned}\rho \ddot{u}_3 &= \tilde{C}_2 \nabla^2 u_3 + \rho M_0 f(\nabla \cdot \bar{\boldsymbol{\mu}}), \\ \dot{\bar{\boldsymbol{\mu}}} &= \gamma \boldsymbol{\mu}_0 \times (-\nabla \varphi - \rho \bar{b}\bar{\boldsymbol{\mu}} - \bar{b} M_0 \nabla u_3 + \alpha \rho \nabla^2 \bar{\boldsymbol{\mu}}), \\ \nabla^2 \varphi - \rho \nabla \cdot \bar{\boldsymbol{\mu}} &= 0\end{aligned} \tag{7.16}$$

in $\mathscr{D} = \{x_2 > 0\}$, and

$$\nabla^2 \varphi = 0 \tag{7.17}$$

in $\mathscr{D}' = \{x_2 < 0\}$, whereas the boundary conditions (7.6) reduce to

$$\tilde{C}_2 u_{3,2} + \rho M_0 \bar{b} \bar{\mu}_2 = 0,$$
$$\bar{\mu}_{1,2} = \bar{\mu}_{2,2} = 0, \qquad (7.18)$$
$$[\![\varphi_{,2}]\!] + \rho \bar{\mu}_2 = 0, \qquad [\![\varphi]\!] = 0$$

at $x_2 = 0$. In these equations $\tilde{C}_2$ is a stiffened shear elastic modulus, f is a magnetoelastic coupling constant, γ is the gyromagnetic ratio, b is a magnetic anisotropy constant, α is an exchange constant (representation of the interaction between neighboring magnetic spins), and

$$\hat{b} = b + \chi_0^{-1}, \qquad \bar{b} = b + f. \qquad (7.19)$$

For perturbations of the type (7.11) and setting

$$\Omega = \omega/\omega_M, \qquad \Gamma = \Omega^2, \qquad \omega_M = \gamma M_0 = \rho \gamma \mu_0,$$
$$q_{1,2} = k_{1,2}\sqrt{\alpha}, \qquad \bar{q}_2 = \bar{k}_2\sqrt{\alpha}, \qquad Q = \sigma^2 q_1^2 + q_2^2, \qquad Q_S = Q + \hat{b},$$
$$(7.20)$$

and

$$\tilde{\lambda}_T^2 = \tilde{C}_2/\rho\alpha\omega_M^2, \qquad \bar{f} = f/\alpha\gamma^2\rho, \qquad \varphi_0 = \sqrt{\alpha}M_0,$$
$$U_3 = u_3/\sqrt{\alpha}, \qquad \mu = \mu_0 m, \qquad \Phi = \varphi/\varphi_0, \qquad (7.21)$$

system (7.16) yields

$$(\Gamma - \tilde{\lambda}_T^2 Q)\hat{U}_3 - i\bar{f}(\sigma q_1 \hat{m}_1 + q_2 \hat{m}_2) = 0,$$
$$i\Omega \hat{m}_1 - Q_S \hat{m}_2 + i\bar{b}q_2 \hat{U}_3 + iq_2 \hat{\Phi} = 0,$$
$$i\Omega \hat{m}_2 + Q_S \hat{m}_1 - i\bar{b}\sigma q_1 \hat{U}_3 - i\sigma q_1 \hat{\Phi} = 0, \qquad (7.22)$$
$$i(\sigma q_1 \hat{m}_1 + q_2 \hat{m}_2) - Q\hat{\Phi} = 0,$$

system (7.18) yields

$$\tilde{\lambda}_T^2 U_{3,2} + \varepsilon_P(\rho/\bar{f})m_2 = 0, \qquad (\varepsilon_P \equiv \bar{f}\bar{b}),$$
$$m_{1,2} = m_{2,2} = 0, \qquad (7.23)$$
$$\sqrt{\alpha}[\![\Phi_{,2}]\!] + m_2 = 0, \qquad [\![\Phi]\!] = 0,$$

and system (7.17) yields

$$\bar{Q}\tilde{\Phi} = 0, \qquad \bar{Q} = \sigma^2 q_1^2 + \bar{q}_2^2, \qquad (7.24)$$

with

$$\Phi = \tilde{\Phi}\exp[i(\omega t - \sigma k_1 x_1 - \bar{k}_2 x_2)], \qquad x_2 < 0. \qquad (7.25)$$

Eliminating $\hat{\Phi}$ with the help of the last equation of (7.22), we are led to a third-order matrix system that possesses nontrivial solutions for $\hat{U}_3$, $\hat{m}_1$, and $\hat{m}_2$ if and only if the following relation between Γ and Q holds true [this is Eq. (2.9) for the present problem]:

$$\mathscr{H}_{BG}(Q,\Gamma) = Q\{(\Gamma - \tilde{\lambda}_T^2 Q)[\Gamma - Q_S(Q_S + 1)] - \varepsilon_P Q Q_S\} = 0. \quad (7.26)$$

The parameter ε_P is the small parameter ε of Section III and the small parameter ε_3 in the general discussion in Section IV,A. The determination of the admissible domain of dispersion (2.12) can practically be made for $\varepsilon_P = 0$. For the sake of simplicity, we can also discard any curvature in the uncoupled spin wave spectrum. This is equivalent to replacing Q_S by $\hat{b}$ in the second and third equations of (7.22), as well as in Eq. (7.26), and discarding the second and third boundary conditions in Eq. (7.23). The surface wave problem is thus greatly simplified. In the absence of elastic displacement the previously set problem reduces to the surface spin wave problem of Damon and Eshbach (1961), for which it is easily shown that a surface wave solution with penetration depth of the order of one wavelength exists only for $\sigma = -1$, hence for waves traveling toward negative x_1's (configuration of Fig. 26c). The frequency of this *pure surface spin wave* is given by

$$\Omega = \Omega_{DE} \equiv \hat{b} + \tfrac{1}{2}, \quad (7.27)$$

while we recall that for *pure bulk spin waves* under the same conditions we would have (see Maugin, 1979b)

$$\Omega = \Omega_v \equiv [\hat{b}(\hat{b} + 1)]^{1/2}. \quad (7.28)$$

Condition (7.26) is reduced to

$$Q[(\Omega^2 - \tilde{\lambda}_T^2 Q)(\Omega^2 - \Omega_v^2) - \varepsilon_P \hat{b} Q] = 0, \quad (7.29)$$

whose roots are

$$Q_I = 0, \qquad Q_{II} = (\Omega^2/\tilde{\lambda}_T^2)(1 + \varepsilon_P A)^{-1}, \quad (7.30)$$

where

$$A \equiv \hat{b}/(\Omega^2 - \Omega_v^2). \quad (7.31)$$

Equivalently to Eq. (7.30), we have

$$q_{2(I)} = q_I = -iq_1,$$
$$q_{2(II)}^2 = q_{II}^2 = \frac{\Omega^2}{\tilde{\lambda}_T^2}\left(\frac{1}{1 + \varepsilon_P A} - \frac{\tilde{\Omega}_T^2}{\Omega^2}\right), \quad (7.32)$$

Elastic Surface Waves with Transverse Horizontal Polarization 429

where $\tilde{\Omega}_T^2 = \lambda_T^2 q_1^2$. The admissible domain of dispersion (2.12) is determined in the following way: First, Ω must be such that $1 + \varepsilon_p A > 0$ or $\Omega > (\Omega_v^2 - \varepsilon_p \hat{b})^{1/2} \equiv \Omega_v'$, and only those points (Ω, q_1) situated below each curve of the coupled dispersion relation of bulk waves are admissible (discussion of the positiveness of the imaginary part of q_{II}). On account of these conditions, there remains for $\mathscr{D}_A$ the nonhatched domain in Fig. 27. In Fig. 27 all separations are greatly exaggerated for illustrative purposes. The study of the dispersion relation for the coupled magnetoacoustic modes is quite involved, so we refer the reader to original works [Scott and Mills (1977), in the absence of magnetic anisotropy effects and exchange forces; and Maugin (1983a), with these effects taken into account (and therefore the full set (7.23) taken into account)]. For $\sigma = +1$, we already remarked on the absence of pure surface spin waves. The coupled magnetoacoustic branches then obtained consist of a lower branch (SH wave) and an upper branch that go asymptotically to the bulk shear wave (Fig. 28a). In the case where $\sigma = -1$, the lower branch has the same behavior as for $\sigma = +1$, but the upper branch never takes on the character of a shear wave. For large wavenumbers, it goes asymptotically to the Damon–Eshbach pure surface spin wave. The difference in behavior for the upper branches according to whether σ equals plus or minus one illustrates the *nonreciprocity* of surface magnetoacoustic waves in elastic ferromagnets. The case of waves attenuated by magnetic spin damping is also examined in Scott and Mills (1977).

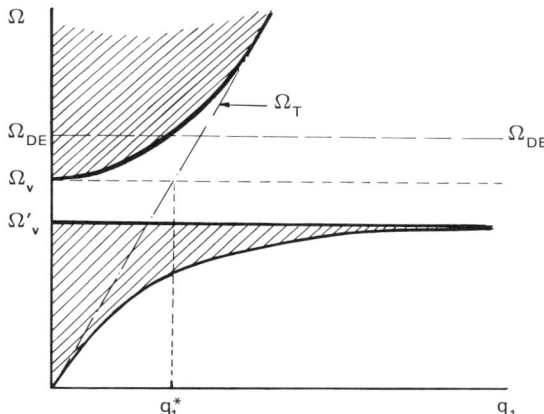

FIG. 27. Admissible domain of dispersion for magnetoacoustic SH surface waves (q_1^*: wavenumber corresponding to bulk magnetoacoustic resonance).

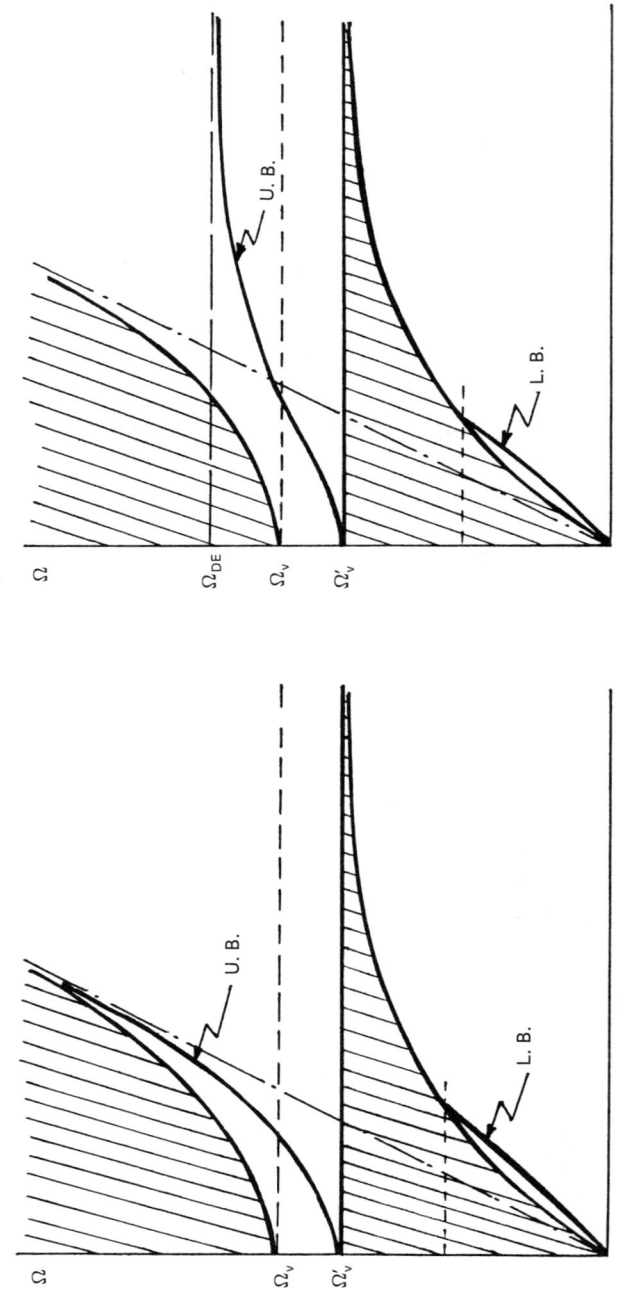

FIG. 28. Dispersion relation for magnetoacoustic SH surface waves of the Bleustein–Gulyaev type: (a) $\sigma = -1$ (backward-traveling waves); (b) $\sigma = +1$ (forward-traveling waves) (U.B. = upper branch, L.B. = lower branch).

ACKNOWLEDGMENTS

The author is indebted to Dr. I. Bakirtaş, Dr. B. Collet, Dr. J. Pouget, and Mr. M. M' Hamdi for their assistance in the preparation of this chapter.

REFERENCES

Abramowitz, M. A., and Stegun, I. A., (1968). "Handbook of Mathematical Functions." Dover, New York.
Achenbach, J. D. (1973), "Wave Propagation in Elastic Solids." North-Holland Publ., Amsterdam.
Balakirev, M. K., and Gorchakov, A. V. (1977). Coupled surface waves in piezoelectrics. *Sov. Phys. Sol. State* (Eng. Tran.), **19**, 355.
Bakirtaş, I., and Maugin G. A. (1982). Ondes de surfaces SH pures en élasticité inhomogène, *J. Mécanique Théor. Appl.* **1**(6), 995–1013.
Bleustein, J. L. (1968), A new surface wave in piezoelectric materials, *Appl. Phys. Lett.* **13**, 412–414.
Buckens, F. (1958). The velocity of Rayleigh waves along a prestressed semi-infinite medium assuming a two-dimensional anisotropy. *Ann. Geofis.* **11**(2), 1–14.
Bullen, K. E. (1965). "Introduction to the Theory of Seismology." Cambridge Univ. Press, London and New York.
Chadwick, P., and Smith, G. D. (1977), Foundations of the theory of surface waves in anisotropic elastic materials. *Adv. Appl. Mech.* **17**, 303–376.
Coddington, E. A., and Lewison, N. (1955). "Theory of Ordinary Differential Equations." McGraw-Hill, New York.
Colloque International Du CNRS No. 111 (1961). "La propagation des ébranlements dans les milieux hétérogènes." Editions du CNRS, Paris.
Coquin, G. A., and Tiersten, H. F. (1967). Rayleigh waves in linear elastic dielectrics. *J. Acoust. Soc. Am.* **41**, 921–939.
Currie, P. K. (1974). Rayleigh waves on elastic crystals. *Q. J. Mech. Appl. Math.* **27**, 489–496.
Damon, R. W., and Eshbach, J. R. (1961). Magnetostatic modes of a ferromagnetic slab. *J. Phys. Chem. Solids* **19**, 308–320.
De Klerk, J. (1972). Elastic surface waves. *Phys. Today* **25**(11), 32–39.
Dieulesaint, E., and Royer, R. (1974). "Ondes élastiques dans les solides: Application au traitement du signal. Masson, Paris.
Emtage, P. R. (1976). Nonreciprocal attenuation of magnetoelastic surface waves. *Phys. Rev. B* **13**, 3063–3070.
Eringen, A. C., and Suhubi, E. S., (1975). "Elastodynamics," Vol. 2, Academic Press, New York.
Ewing, W. M., Jardetzky, W. S., and Press, F. (1957). "Elastic Waves in Layered Media." McGraw-Hill, New York.
Fahmy, A. H., and Adler, E. L. (1973). Propagation of acoustic surface waves in multi-layers: A matrix description. *Appl. Phys. Lett.* **22**, 495–497.
Farnell, G. W. (1970). Properties of elastic surface waves. *In* "Physical Acoustics" W. P. Mason, ed.), Vol. VI, pp. 109–166. Academic Press, New York.
Farnell, G. W. (1978). Types and properties of surface waves. *In* "Acoustic Surface Waves" ("Topics in Applied Physics") (A. A. Oliner ed.), Vol. 24, pp. 13–60. Springer-Verlag, Berlin and New York.

Farnell, G. W., and Adler, E. L. (1972). Elastic wave propagation in thin layers. *In* "Physical Acoustics" (W. P. Mason and R. N. Thurston ed., Vol. IX, pp. 35–127. Academic Press, New York.
Filipov, B. M. (1967). Surface spin waves. *Sov. Phys. Sol. State*, 9, 5.
Fischer, B., and Lagois, J. (1979). Surface excition polaritons. *In* "Topics in Current Physics" (K. Cho, ed.), Vol. 14, pp. 183–210. Springer-Verlag, Berlin and New York.
Fokkema, J. T. (1979). Reflection and transmission of time-harmonic elastic waves by the periodic interface between two elastic media. Ph. D. Thesis, Delft Univ. of Tech., Dept. of Elect. Eng., Report No. 1979-6, Delft, The Netherlands.
Ganguly, A. K., Davis K. L., and Webb, D. C. (1978). Magnetoelastic surface waves on the (110) plane of highly magnetostrictive cubic crystals. *J. Appl. Phys.* **49**, 759–767.
Gérard, A. (1979). Scattering of SH waves by a spherical layer. *Int. J. Eng. Sci.*, **17**, 313–327.
Gérard, A. (1980). Diffraction d'ondes SH par un milieu stratifié sphérique. *Int. J. Eng. Sci.*, **18**, 583–595 [Errata: *Int. J. Eng. Sci.*, **19**, 583–584, 1981].
Graff, K. F. (1975). "Wave Motion in Elastic Solids." Ohio State University Press, Columbus.
Gulyaev, Yu. V. (1969). Electroacoustic surface waves in solids. *Sov. Phys. JETP Lett.* **9**, 37–38.
Ingebrigten, K. A., and Tonning, A. (1969). Elastic surface waves in crystals. *Phys. Rev.* **184**, 942–951.
Kaliski, S., and Kapalewski, J. (1968). Surface waves of the spin-elastic type in a discrete body of cubic structure. *Proc. Vibr. Prob.* **9**, 269–278.
Kiełczynski, P. (1981). Propagation of surface SH waves in nonhomogeneous media. *J. Tech. Phys.* (Warsaw) **22**, 73–78.
Lewis, M. F., and Patterson, E. (1972). Acoustic surface wave isolators. *Appl. Phys. Lett.* **20**, 276–278.
Lifshitz, E. M., and Pitaevskii, L. P. (1979). "Physical Kinetics" (in Russian). Nauka, Moscow.
Maugin, G. A. (1979a). A continuum approach to magnon–phonon couplings—I: General equations, background solution. *Int. J. Eng. Sci.* **17**, 1073–1091.
Maugin, G. A. (1979b). A continuum approach to magnon–phonon couplings—II: Wave propagation for hexagonal symmetry. *Int. J. Eng. Sci.*, **17**, 1093–1108.
Maugin, G. A. (1980). Elastic-electromagnetic resonance couplings in electromagnetically ordered media. *In* "Theoretical and Applied Mechanics" (Sectional Lecture at the 15th ICTAM, Toronto, 1980) (F. P. J. Rimrott and B. Tabarrok, eds.), pp. 345–355. North-Holland Publ., Amsterdam.
Maugin, G. A. (1981). Wave motion in magnetizable deformable solids (review article). *Int. J. Eng. Sci.* **19**, 321–388.
Maugin G. A., (1983a). Surface magnetoelastic waves in elastic ferromagnets—I—Orthogonal setting of the bias magnetic field. *J. Acoust. Soc. Am.* **73** (to be published).
Maugin, G. A., (1983b). Surface magnetoelastic waves in elastic ferromagnets—II—Longitudinal setting of the bias magnetic field. *J. Acoust. Soc. Am.* **73** (to be published).
Maugin, G. A., and Pouget, J., (1980). Electroacoustic equations in one-domain ferroelectric bodies. *J. Acoust. Soc. Am.* **68**, 575–587.
Maugin, G. A., and Pouget, J. (1981). A continuum approach to magnon–phonon couplings—III—Numerical results. *Int. J. Eng. Sci.* **19**, 479–493.
Meunier, J. (1980). Analyse perturbative des caractéristiques de propagation des fibres optiques à gradient d'indice quasi-parabolique. *Optic. Quan. Electron.* **12**, 1.
Musgrave, M. J. P., (1970). "Crystal Acoustics." Holden-Day, San Francisco, California.
Nalamwar, A. L., and Epstein, M. (1976). Surface acoustic waves in strained media. *J. Appl. Phys.* **47**, 43–48.
Pajewski, W. (1977). Transversal Bleustein–Gulyaev (B–G) surface waves on piezoelectric ceramics. *Arch. Acoust.* **2**, 3.

Pajewski, W., (1978). Les propriétés des ondes de surface se propageant sur les cristaux. *In* Proceedings Colloque sur les Ultrasons, pp. 97–100, Editions de Physique, Paris.

Parekh, J. P. (1972). *In* Proceedings of the I.E.E.E. Ultrasonics Symposium (IEEE, Boston, 1972), p. 333. I.E.E.E., New York.

Parekh, J. P., and Bertoni, H. L. (1974a). Magnetoelastic Rayleigh waves propagating along a tangential bias field on a YIG substrate. *J. Appl. Phys.* **45**, 434–445.

Parekh, J. P., and Bertoni, H. L. (1974b). Magnetoelastic Rayleigh waves on a YIG substrate magnetized normal to its surface. *J. Appl. Phys.* **45**, 1860–1868.

Peynircioglu, H. (1980). Geotechnical study of the site of Rasattepe in 1945 and the foundations of Attatürk's mausoleum (in Turkish). *Bull. Comité National Turque Mécanique Sols* **2**(3).

Pouget, J. (1980). Ondes de Bleustein–Gulyaev dans les matériaux ferroélectriques. *C. R. Acad. Sci. Paris* **291 B**, 181–184.

Pouget, J. (1981). Ondes de surface couplées dans les matériaux ferroélectriques. "Symmetries and Broken Symmetries" (Pierre Curie Colloquium, Paris, 1980) (N. Boccara, ed.). IDSET, Paris, pp. 333–341.

Pouget, J., Askar, A., and Maugin, G. A. (1983). Lattice models for elastic ferroelectric bodies: Coupled wave modes and long-wave approximation. *J. Physique* (to be published).

Pouget, J., and Maugin, G. A., (1980). Coupled acoustic-optic modes in deformable ferroelectrics. *J. Acoust. Soc. Am.* **68**, 588–601.

Pouget, J., and Maugin, G. A. (1981a). Bleustein–Gulyaev surface modes in elastic ferroelectrics. *J. Acoust. Soc. Am.* **69**, 1304–1318.

Pouget, J., and Maugin, G. A. (1981b). Piezoelectric Rayleigh waves in elastic ferroelectrics. *J. Acoust. Soc. Am.* **69**, 1319–1325.

Rayleigh, Lord (1885). On waves propagated along the plane surface of an elastic solid. *Proc. London Math. Soc.* **17**, 4–11.

Scott, R. Q., and Mills, D. L., (1977). Propagation of surface magnetoelastic waves on ferromagnetic crystal substrate, *Phys. Rev. B* **15**, 3545–3557.

Sinha, B. R., and Tiersten, H. F. (1973). Elastic and piezoelectric surface waves guided by thin films. *J. Appl. Phys.* **44**, 4831–4854.

Smith, H. I. (1970). The physics and technology of surface elastic waves. *Int. J. Nondestr. Test.* **2**, 31–59.

Stoneley, R. (1955). The propagation of surface elastic waves in a cubic crystal. *Proc. R. Soc. London, Ser. A* **232**, 447–452.

Szabo, T. L. (1974). Nondestructive subsurface gradient determination, *In* "Proceedings of the Ultrasonics Symposium." I.E.E.E., New York.

Synge, J. L. (1956). Flux of energy for elastic waves in anisotropic media. *Proc. R. Irish Acad.* **58A**, 13–21.

Synge, J. L. (1957). Elastic waves in anisotropic media. *J. Math. Phys.* **35**, 323–334.

Taylor, B. D., and Crampin, S. (1978). Surface waves in anisotropic media: Propagation in a homogeneous piezoelectric half space. *Proc. R. Soc. London, Ser. A* **364**, 161–179.

Tiersten, H. F. (1963). Thickness vibrations of piezoelectric plates. *J. Acoust. Soc. Am.* **35**, 53–58.

Tiersten, H. F. (1969). "Linear Piezoelectric Plate Vibrations." Plenum, New York.

Tittman, B. R. (1974). Characterization of subsurface anomalies by elastic surface wave dispersion. In "Proceedings of the Ultrasonics Symposium." I.E.E.E., New York.

Tsutsumi, M., Bhattacharyya, T., and Kumagai, N. (1975). Piezoelectric-magnetoelastic surface wave guided by the interface between semi-infinite piezoelectric and magnetoelastic media. *J. Appl. Phys.* **46**, 5072–5075.

Van den Berg, P. M. (1971). Diffraction theory of a reflection grating. *Appl. Sci. Res.* **24**, 261–293.

Van de Vaart, H. (1971). Magnetoelastic Love-wave propagation in metal-coated layered substrates. *J. Appl. Phys.* **42**, 5305–5312.

Viktorov, I. A. (1967). "Rayleigh and Lamb Waves: Physical Theory and Applications." Plenum, New York.

Viktorov, I. A. (1975). Elastic waves in a solid half-space with a magnetic field. *Sov. Phys. Dokl.*, **20**, 273–274.

Viktorov, I. A. (1978). Surface waves induced by an inhomogeneity in a solid (in Russian). *In* "Proceedings of the 10th All-Union Conference or Quantum Acoustics and Acoustoelectronics," pp. 101–103. Tashkent, USSR.

Viktorov, I. A. (1979). Types of acoustic surface waves in solids. *Sov. Phys. Accoust.* **25**, 1–9.

Volluet, G. (1979). Effets magnétoélastiques nonlinéaires dans une couche mince de grenat d'yttrium-fer, *J. Phys.* **40** (Colloque C8, Suppl. No. 11), C8-256-C8-261.

White, R. M. (1970). Surface elastic waves. *Proc. I.E.E.E.* **58**, 1238–1276.

Wilson, A. J. (1973). Surface and plate waves in biaxially-stressed elastic media. *Pure Appl. Geophys.* **102**, 182–192.

Wilson, A. J. (1974). The anomalous surface waves in uniaxially-stressed elastic media. *Pure Appl. Geophys.* **112**, 665–174.

Author Index

Numbers in italics refer to the pages on which the complete references are listed.

A

Abramowitz, M. A., 399, 403, *431*
Achenbach, J. D., 375, 378, 406, *431*
Achter, M. R., 134, 155, *176*
Adkins, J. E., 325, 355, *366*
Adler, E. L., 384, *431–432*
Aernoudt, E., 99, *114*
Aldag, E., 140, 161–162, *172*
Al-Khozaie, S., 260, *268*
Allen, R. D., 138, 159, *174*
Almroth, B. O., 272, *365*
Alwar, R. S., 237, *262*
Amick, C. J., 255, *262*
Anderson, C. A., 181, *268*
Anderson, R. H., 162, *175*
Antman, S. S., 274, 290, *365*
Argon, A., 123–124, 137, 149, *172*, *175*
Arridge, R. G. C., 236, *264*
Asaro, R. J., 16, 18, 20, 30, 35–36, 40–41, 45–47, 49–53, 64, 66, 69–73, 76–77, 79–87, 90, 93, 97–98, *111–114*
Ashby, M. F., 22–23, *111*, 120–121, 123–125, 127–128, 131, 133, 140–141, 143, 146–147, 149–150, 152–153, 155–157, 159–161, *172–173*, *176–177*
Askar, A., 389, *433*
Atluri, S. N., 274, *365*
Attridge, R. G. C., 237, *262*
Austin, R. J., 134, *173*
Avery, D. H., 99, *111*
Axelrad, E. L., 310, 341, *365*

B

Backofen, W. A., 67, 96, *114*
Baker, G. S., 66, *113*
Bakirtas, I., 402–406, *431*
Balakirev, M. K., 376, *431*
Balluffi, R. W., *172*
Barendreght, J. A., 51, 75, *111*
Barham, P. J., 237, *262*
Barnett, D. M., 16, 18, *111–112*
Barrett, C. R., 134, *172*
Barrett, C. S., 67, *112*
Basinski, Z. S., 28, 30, 32, *113*
Baskes, M. I., 25–26, 106, *115*
Bastow, B. D., 129, *173*
Batra, R. C., 249, *262*
Becher, P. F. 162, *172*
Bechtold, J. H., 159, *172*
Becker, E. B., 279, *365*
Beere, W., 152, *172*
Berdichevskii, V. L., 196, 242, 245–246, 249–250, 261–262, 323, *365*
Berger, N., 319, 323, 325–326, *365*
Berglund, S., 249, *262*
Bergman, S., 208, *262*
Bernstein, B., 242, 245, *262*
Bertoni, H. L., 377, 422, 425, *433*
Berveiller, M., 28, 32, *112*
Betteridge, W., 133, 136, 158, *172*
Bhattacharyya, T., 377, *433*
Bickford, W. B., 239, *268*
Bilde-Sörenson, J. B., 142, *172*
Biollay, Y., 249, *263*
Biot, M., 84, *112*
Bird, J. E., 125, *172*, *176*
Biricikoglu, V., 273, *365*
Bishop, J. F. W., 101, 104, *112*
Bleustein, J. L., 246–247, *268*, 379, 392, *431*
Boas, W., 33, *112*
Bogy, D. B., 231, *263*
Boley, B. A., 257–258, 261, *263*
Bolling, G. F., *172*
Bolton, C. J., 139, *173*
Boone, D. H., 134, *177*
Boussinesq, J., 180, *263*
Boyle, J. T., 341, 344, *365*
Bragg, W. H., 6, *112*
Bragg, W. L., 6, *112*

Brazier, L. G., 341, *365*
Brewer, S., 232, 247, 251, *263*
Brodrick, R. F., 138, 159, *173*
Bromer, D. J., 146, *173*
Brown, A. M., 141, *173*
Brown, K., 67, *112*
Brown, L. M., 149–150, *173–174*
Brown, T. J., 28, *113*
Bruggeman, G., *175*
Brush, D. O., 272, *365*
Buckens, F., 374–375, *431*
Budiansky, B., 106, *112*, 272, 283, 302, 311, 315–316, 318, 324, 350, 354, *365*
Bullen, K. E., 378, *431*
Burgers, J. M., 14, *112*
Burke, M. A., 154, *173*
Burke, P. M., 140–141, *173*
Burton, B., 129, *173*
Bush, A. C., 67, 108–110, *112*
Bushnell, D., 272, 341, *365–366*
Butkovich, T. R., 145–146, *173*
Byrne, J. G., 55, *112*

C

Cahn, R. W., 29–30, 32–33, 67–69, 76–78, 80, *112*, *114*
Campbell, J. D., 62, *112*
Cannaday, J. E., 134, *173*
Cannon, R. M., 144, *173*
Carey, G. F., 279, *365*
Chadwick, P., 375, *431*
Chalmers, B., 30, *114*
Chandra, H., 107, *113*
Chang, K. J., 166, *177*
Chang, Y. W., 30, 64, 66–67, 77, 79, 80–83, 90, 93, 97–98, *112*
Charles, R. J., 164–165, *173*
Chiem, C. Y., 26–27, 55–60, *112*
Childs, S. B., 238, *266*
Choi, I., 237, *263*
Chou, P. C., 189, *263*
Chuang, T., 153, *173*
Clark, A. E., 358–359, *366*
Clark, R. A., 341, *366*
Coble, R. L., 128, *173*
Cocks, A. C. F., 120, 152–153, *173*
Coddington, E. A., 398, *431*
Cohen, M., 148, *176*
Coleman, B. D., 296, *366*
Comer, R. L., 361, *366*
Conrad, H., 123, *173*

Copley, S. M., 142–143, 163–164, *173*
Coquin, G. A., 376, *431*
Corth, R., 159, *176*
Cottrell, A. H., 29–30, 32–33, *114*, 147, *175*
Courant, R., 191, 199, *263*
Cowper, G. R., 278, *366*
Crampin, S., 376, *433*
Crosby, A., 144, 164–165, *173*
Crossman, F. W., 152, *173*
Cullen, G. V., 142–143, *174*
Currie, P. K., 374, *431*

D

Dac Viuana, C. S., 99, 107, *114*
DaDeppo, D., 355, *369*
Dafermos, C. M., 245–246, *263*
Damon, R. W., 380, 422, 428, *431*
Danielson, D. A., 281, 312, 321, 324–327, 341, *366*, *370*
Danneberg, W., *173*
Davidge, R. W., 163–164, *173*
Davies, C. K. L., 164–165, *173*
Davies, G. J., 99, 107, *114*
Davies, P. W., 134, 155, *173*
Davis, K. L., 385, 422, 425, *432*
Davis, L. A., 140, 161–162, *172*
Day, R. B., 142, 163–164, *173*
Day, W. A., 182, *263*
De Klerk, J., 375, *431*
de Meester, B., 123, *173*
Dennison, J. P., 134, 155, *173*
Diaz, J. B., 216, 232, 245, *263*
Dieulesaint, E., 375–376, 379, 394, *431*
Digges, T. G., 134, 155, *175*
Dillamore, I. L., 67, 108–110, *112*
Doner, M., 123, *173*
Dong, S. B., 237, *263*
Donlevy, A., 159, *173*
Donnell, L. H., 348, *366*
Dorn, J. E., 124–125, 129, 139, 141, 161, *172*, *174–176*
Dou, A., 247, *263*
Dougall, J., 246, *263*
Drake, R. L., 189, *263*
Duffy, J. D., 26–27, 55–60, *112*
Duran, S. A., 134, *176*
Dyson, B. F., 152, 158, *173*

E

Edelstein, W. S., 260, *263–264*, *266*
Edward, G., 149–150, *175*

Elam, C. F., 2, 7–11, 20, 28–30, *112*, *115*
Embury, J. D., 24, 26, *112*, 149–150, *173*, *175*
Emtage, P. R., 424–425, *431*
Epstein, M., 375, *432*
Ericksen, J. L., 182, 199, 250, *264*, 296, *366*
Eringen, A. C., 375, 404, *431*
Eshbach, J. R., 380, 422, 428, *431*
Eshelby, J. D., 17, 106, *112*
Eteiche, A. M., 62, *112*
Evans, A. G., 123, 163–164, *173*
Evans, P. E., 144, 164–165, *173*
Everstine, G., 235–236, *264*
Ewing, J. D., 4–5, 7, *112*
Ewing, W. M., 375, *431*
Eyring, H. E., 54–55, *113*

F

Fadle, J., 231, *264*
Fahmy, A. H., 384, *431*
Farnell, G. W., 375, *431–432*
Farrell, C. J., 237, *262*
Farrell, K., 159, *177*
Feng, W. W., 355, *371*
Fichera, G., 242, 248, *264*
Fields, R. J., 140–141, *173*, *177*
Filipov, B. M., 380, 422, *432*
Fine, M. E., 55, *112*
Firestone, R. F., 144, *174–175*
Fischer, B., 380, *432*
Fix, G. J., 279, *370*
Flagella, P. N., 137–139, *174*
Flavin, J. N., 223, 225, 228, 230, 238, 249, *264*
Flinn, J. E., 137–138, *174*
Flynn, P. W., 139, *174*
Fokkema, J. T., 375, *432*
Folkes, M. J., 236–237, *262*, *264*
Folweiler, R. C., 144, *174*
Forrest, P. G., 158, *174*
Franciosi, P., 28, 32, *112*
Frank, F. C., 22, *112*
Frederking, R., 131, 146, *174*
Freiman, S. W., 162, *172*
Frenkel, J., 12, *112*
Friedrichs, K. O., 242, *264*, 322, *366*
Frish-Fay, R., 290, *366*
Fritch, D. J., 138, 159, *173*
Frost, H. J., 121, 123, 125, 127, 129, 131, 133, 141, 143, 146, *172*, *174*
Fung, Y.-C., 181, *264*

G

Galimov, K. Z., 347, *368*
Gandhi, C., 133, 140, 147, 156–157, 159–160, *172*, *174*
Ganuly, A. K., 385, 422, 425, *432*
Gérard, A., 378, *432*
Gibbs, G. B., 139, *174*
Gilbarg, D., 253, *264*
Gilbert, A., 83, *114*
Gilbert, E. R., 137, 138, *174*
Gilling, D., 163, 164, *173*
Gilman, J. J., 140–141, *174*
Glasier, L. F., 138, 159, *174*
Glen, J. W., 146, 166, *175*, *177*
Glenn, J. W., 145, *174*
Glockner, P. G., 273, *368*
Goetschel, D. B., 237, *263*
Gold, L. W., 166, *174*
Goldenveiser, A. L., 273, 322, *366*
Gooch, D. J., 144, *174*
Goodier, J. N., 231, *268*
Goodman, D. J., 146, *174*
Goods, S. H., 149, *174*
Gorchakov, A. V., 376, *431*
Gordon, R. B., 140, 161–162, *172*
Graff, K. F., 375–376, *432*
Grandin, H. T., 261, *264*
Grant, N. J., 134, 151, 155, *176–177*
Graves, N. F., 159, *175*
Green, A. E., 296, 302, 352, 355, *366*
Green, R. E., Jr., 30–31, 66, *113*
Green, W. V., 137–138, 159, *174*
Greenough, A. P., 134, *174*
Greenwood, G. W., 161, *176*
Greenwood, N. N., 141, *174*
Gregory, R. D., 231, *264*
Groves, G. W., 144, *174*
Guard, R. W., 134, *174*
Guitiérrez, A., 247, *264*
Guiu, F., 141, *174*
Gulyaev, Yu. V., 379, *432*
Gurtin, M. E., 181, 213, 223–225, 232, 234, 240–241, 243, 249, 251, *264*, 296, *366*
Guyot, P., 124, *174*

H

Haasen, P., 134, *174*
Hahn, G. T., 83, *114*
Hall, E. O., 24, *113*
Hamel, G., 312, 326, *366*

Hardwick, D., 127, 134, *174*
Hardy, G. H., 227, *264*
Harper, J. G., 129, *174*
Harris, J. E., 139, 153, *175*, *177*
Hart, E. W., 154, *174*
Hasford, W. F., Jr., 67, 96, *114*
Hauser, F. E., 125, 139, 161, *174–175*
Havner, K. S., 36, 41, 66, *113*
Hawkes, I., 166, *174*
Hayward, E. R., 134, *174*
Heard, H. C., 141, *174*
Hedgepeth, J. M., 361–362, *370*
Hensler, J. H., 142–143, *174*
Hermon, E. L., 159, *175*
Hersing, C., 128, *175*
Heslop, J., 133, 136, 158, *172*
Heuer, A. H., 144, *174–175*
Hilbert, D., 191, 199, *263*
Hill, R., 36, 39, 41, 43–44, 70, 84–86, 104, 106–107, *112–113*
Hiram, Y., 355, *369*
Hirschwald, W., *176*
Hirth, J. P., 16, 18, 20, 110, *111*, *113*
Ho, C.-L., 218, 222, *264*
Hobbs, P. V., 145–146, *175*
Hoff, N. J., 255, *264*
Holt, R. T., 158, *175*
Honeycombe, R. W. K., 67, 80, *113*
Horgan, C. O., 210–212, 217–218, 222–223, 234–239, 243, 245–246, 250–255, 260, 263–266, 268–269
Horvay, G., 231, *265*
Huang, N. C., 348, *366*
Hull, A. W., 6, *113*
Hull, D., 153, *175*
Hulse, C. O., 142, 163, *175*
Hum, J. K. Y., 159, *173*
Hutchinson, J. W., 84–86, 102–107, *113*, 272, *366*

I

Im, J., 149, *172*
Ingebrigten, K. A., 374, *432*

J

Jackson, P. J., 28, 30, 32, *113*
Jardetsky, W. S., 375, *431*
Jenkins, W. D., 134–135, *175*
John, F., 191, *265*, 286, 319, 323, 325–326, *366–367*

Johnson, L. R., 134, 155, *175*
Johnson, M. W., 231, 238, *265*
Johnston, T. L., 162, *177*
Johnston, W. G., 141, *175*
Jonas, J. J., 126–128, *175*
Jones, J. J., 127, *177*
Jones, J. P., 278, *367*
Jones, R. B., 139–140, *175*
Jones, S. J., 165–166, *175–176*
Jordan, P. F., 346, *367*
Joshi, N. R., 30–31, 66, *113*

K

Kainins, A., 273, *365*
Kaliski, S., 422, *432*
Kapalewski, J., 422, *432*
Karashima, S., 134, *175*
Kear, B. H., 134, *175–176*
Keeler, J. H., 134, *174*
Keith, H. H., 159, *176*
Keller, A., 237, *262*
Keller, H. B., 181, *265*
Kelly, A., 26, 55, 67, 77, 79–80, *112–114*, 121–122, 147, *175*
Kielczynski, P., 400–402, *432*
King, G. W., 137, *175*
Kingery, W. D., 140–141, 146, *173*, *175*
Klahn, D., 125, 141, *175*
Kleeman, P. W., 337, *368*
Klein, P. H., 162, *172*
Klemm, J. L., 246, *266*
Klepaczko, J., 62, *113*
Klopp, W. D., 137, 159, *175*, *177*
Knowles, J. K., 181, 195, 199, 204, 209–210, 213, 217–218, 222–213, 225–226, 230, 232, 237–238, 245–246, 250–253, 260, *264–266*, 268
Kocks, U. F., 28, 30, 66, 105, 107, *113–114*, 123–124, *175*
Koehler, J. S., 17, 22, *113–114*
Koepke, B. G., 162, *175*
Köster, W., *175*
Koga, T., 361, *367*
Koiter, W. T., 272–273, 285–286, 296, 318–323, 325–327, 329, 332, 338, 340, 345, 347–348, 350–351, *366*, *367*
Koo, R. C., 137, 159, *175–176*
Krausz, A. S., 54–55, *113*
Kreider, K. G., *175*
Kronberg, M. L., 143, *175*

Kroner, E., 105, *113*
Kuhlman-Wilsdorf, D., 26, *113*
Kumagai, N., 377, *433*
Kumar, A., 125, *175*
Kumble, R. G., 125, *175*

L

Ladeveze, P., 323, *367*
Lagois, J., 380, *432*
Landauer, J. K., 145–146, *173*
Landon, P. R., 139, 161, *174*
Langdon, T. G., 141–143, *174*, *175*
Lankford, J., 164–165, *175*
Leknitskii, S. G., 233, *266*
Le Roy, G., 149–150, *175*
Leverant, G R., 134, *175*
Levy, S., 361, *366*
Lewis, M. F., 425, *432*
Lewison, N., 398, *431*
Lezius, D. K., 162, *177*
Libai, A., 286–287, 291, 327, 336–337, 346, 348, 355, 361, *367*, *370*
Librescu, L., 274, *367*
Lifshitz, E. M., 387, *432*
Lifshitz, L. M., 128, *175*
Lindler, R., *175*
Lions, J. L., 243, *266*
Lisiecki, L. L., 81–82, 97, *113*
Little, R. W., 231, 238, 246, 261, *264–266*
Littlewood, J. E., 227, *264*
Llewellyn, R. J., 134, *173*
Loh, B. T. M., 159, *177*
Lomer, W. M., 19, *113*
Lothe, J., 16, 18, 110, *111*, *113*
Love, A. E. H., 14, *114*, 221, *266*
Lundy, T. S., *175*
Luton, M. J., 127, 134, *175*

M

McClintock, F. A., 149, *176*
McConnel, A. J., *368*
McCormick, P. G., 127, *176*
McCoy, H. E., 159, *176–177*
McDonnells, D. L., 159, *176*
McHargue, C. J., *175*
MacKenzie, J. K., 122, *176*
McLean, D., 158, *173*, *176*
McMahon, C. J., 148, *176*
Macmillan, N. H., 147, *176*
Magenes, E., 243, *266*

Maisonneuve, O., 182, 187, 194, 198, *266*
Malcom, D. J., 273, *367*
Maloof, S. R., 137, *172*
Mansfield, E. H., 336–337, 361–362, *368*
Marguerre, K., 347, *368*
Maugin, G. A., 385, 388–389, 402–412, 414–415, 417–418, 420–422, 424–426, 428–429, *431–432*
Mellor, M., 145–146, 166, *174*, *176*
Merrett, G. J., 164–165, *176*
Meunier, J., 397, *432*
Michael, D. J., 144, *177*
Mieth, H. J., 191, 195, 206, 210, 228, 232, *266*
Mikhlin, S. G., 191, 196, 228, *266*
Mills, D. L., 424, 429, *433*
Mitchell, J. B., 163–165, *177*
Mitchell, T. E., 12, 24, 29–32, 66, 76, *114*
Miura, S., 33–35, *114*
Mohammad, F. A., 129, *176*
Monkman, F. C., 151, *176*
Montrone, E. B., 140–141, *175*
Moon, D. M., 137, *176*
Morgan, R. P., 159, *175*
Morii, K., 108–110, *114*
Morris, J. W., Jr., 129, *176*
Morrison, W. B., 25, *114*
Moschetti, A., 165, *176*
Mote, J., 139, *174*
Motomiya, T., 134, *175*
Mukherjee, A. K., 125, 141, *172*, *175–176*
Muller, I., 296, *368*
Muncaster, R. G., 182, 199, 204, 250, *266*
Murty, K. L., 129, *176*
Musgrave, M. J. P., 375, *432*
Mushtari, K. M., 347, *368*

N

Nabarro, F. R. N., 125, 128, *176*
Nachbar, W., 358, *368*
Naghdi, P. M., 272, 296, 302, 347, *366*, *368*
Nakayama, Y., 108–110, *114*
Nalamwar, A. L., 375, *432*
Narayanaswamy, O. S., 358–359, *366*
Neapolitan, R. E., 260, *266*
Needham, N. G., 161, *176*
Needleman, A., 35–36, 41, 46, 64, 66, 84, 86–90, 96, 107, *114*
Nelson, D., 81–82, 97, *113*
Neumann, G. M., *176*
Newman, S. Z., 341, *370*

Nicholls, J. H., 127, *176*
Nicholson, J. W., 286, *368*
Nicholson, R. B., 26, *113*
Niordson, F. I., 284–285, 320, *368*
Nix, W. D., 154, *173*
Noll, W., 296, *366*
Norman, E. C., 134, *176*
Norton, F. H., 164, 165, *177*
Novozhilov, V. V., 273, 320–321, *368*
Nunziato, J. W., 260, *266*

O

Oden, T. J., 279, *365*
Ogilvie, G. J., 33, *112*
Oikawa, H., 134, *175*
Oleinik, O. A., 191, 195, 223, 228, 230, 249, 255, 260, *266–267*
Olmstead, W. E., 250, *265*
Orowan, E., 12–13, 67, *114*

P

Pajewski, W., 375, 379, *432*, *433*
Palamà, A., 249, *267*
Pan, J., 61, *114*
Papkovich, P. F., 231, *267*
Parameswaran, V. R., 165–166, *176*
Parekh, J. P., 377, 422, 425, *433*
Paren, J. G., 146, *177*
Parfitt, G. D., *175*
Parker, E. R., 142, *176*
Pask, J. A., 142–143, 163–164, *173*, *175*, *177*
Passmore, E., 165, *176*
Paterson, M. S., 142, *176*
Patterson, E., 425, *432*
Pawel, R. E., *175*
Payne, L. E., 216, 232, 242, 245–246, *263*, *267*
Peach, M., 17, *114*
Peirce, D., 35–36, 41, 46, 64, 66, 84, 86–90, 96, 107, *114*
Petch, N. J., 24, *114*
Petrovsky, I. G., 355, *368*
Peynircioğlu, H., 397, *433*
Pflüger, A., 290, *368*
Piearcey, B. J., 134, *175–177*
Piercy, G. R., 29–30, 32–33, *114*
Pietraszkiewicz, W., 274, 281, 310, 312, 321, 325, 327, *368*
Pipkin, A. C., 235–236, *264*, *267*

Pipkin, A. D., 357, *368*
Pitaevskii, L. P., 387, *432*
Poirier, J. P., 140–141, *176*
Polan, N. W., 99, *111*
Polanyi, Von, M., 8, 13–14, *114*
Pólya, G., 227, *264*
Pouget, J., 388–389, 407, 409–412, 414–415, 417–418, 420, *432*, *433*
Prager, W., 321, *368*
Press, F., 375, *431*
Price, R. J., 67, 77, 79–80, *114*
Protter, M. H., 211–212, *267*
Pugh, J. W., 137, 159, *176*

Q

Quarrell, A. G., 134, *176*

R

Rabenstein, A. S., 159, *176*
Raffo, P. L., 137, 159, *175–176*
Raj, R., 128–129, 153, *176*
Ramaswami, B., 30, *114*
Rao, N. R., 237, *267*
Rawlings, R. D., 123, *173*
Rayleigh, Lord, 374, *433*
Read, W. T., Jr., 22, *112*
Reid, C. N., 83, *114*
Reinhardt, G., 159, *175*
Reissner, E., 238, 245, *265*, *267*, 280–281, 284, 310–312, 319, 321, 338, 341, 344, 346, 347, 358–359, 361–362, *366*, *368–369*
Reiter, S. F., 134, *174*
Rice, J. R., 18, 36, 39–41, 43–44, 47, 49–53, 61, 69–73, 76, 79, *111*, *113–114*, 149, 153, *173*, *176*
Rice, R. W., 162, *172*
Richardson, G. J., 134, *176*
Richman, R. H., *172*
Richmond, O., 52, 76, *114*
Rigbi, Z., 355, *369*
Rimmer, D. E., 153, *175*
Ripling, E. J., 139, 161, *177*
Roark, A. L., 239, *268*
Roberts, C. S., 139, *176*
Roberts, J. G., 67, 108–110, *112*
Robinson, A., 181, *267*
Robinson, S. L., 137, *176*
Rodgers, M. J., 158, *173*
Roseman, J. J., 232, 247, 249–251, *263*, *267*

Author Index 441

Rosenhain, W., 4–5, 7, *112*
Rossettos, J. N., 346, *369*
Royer, R., 375–376, 379, 394, *431*
Rozenberg, V. M., 134, *176*

S

Saeki, Y., 33–35, *114*
Safoglu, R., 149, *172*
Saint-Venant, A.-J.-C. B de, 180, 204, *267*
Saiomoto, S., 67, 96, *114*
Saldinger, I. L., 138–139, *174*
Sanders, J. L., Jr., 311, 318, 321, 346–348, 365, *369*
Saxer, R. K., 134, *173*
Schiffer, M., 208, *262*
Schmid, E., 7, 8, *114*
Schmidt, R., 355, *369*
Schnitzel, R. H., 159, *176*
Schwarz, R., 166, *177*
Schwerin, E., 358, *369*
Scott, R. Q., 424, 429, *433*
Sell, H., 137, 159, *175–176*
Sellars, C. M., 126–128, 134, *174–176*
Sevillano, J. G., 99, *114*
Shaffer, P. T. B., 163, 165, *176*
Shahinian, P., 134, 155, *176–177*
Shalaby, A. H., 66, *113*
Shamina, V. A., 321, *368*
Sharpe, W. N., Jr., 51, 75, *111*
Shelley, P. E., 346, *367*
Shepard, L. A., 129, *174*
Sherby, O. D., 134, 137, *172, 176*
Shewmon, P. G., 159, *172*
Shield, R. T., 181, 204, 209, *267—268*
Sigillito, V. G., 260, *268*
Signorelli, R. A., 159, *176*
Simmonds, J. G., 181, 218, 256–257, *266, 268*, 273, 281, 285–287, 291, 293, 295, 312, 318–327, 332, –334, 336–337, 340–341, 359, *367—370*
Simpson, L. A., 164–165, *176*
Sinha, B. R., 378, *433*
Smashey, R. W., *176*
Smith, E., 34, *115*
Smith, G. D., 375, *431*
Smith, H. I., 375, *433*
Smith, P. A., 158, *174*
Snow, J. D., 144, *175*
Snowden, W. E., 142, *177*
Sobel, L. H., 341, *370*

Sober, R. J., 52, 76, *114*
Sokolnikoff, I. S., 213, 215, 244, *268*
Sowerby, R., 99, 107, *114*
Speight, M. V., 152–153, *172, 177*
Spencer, A. J. M., 235, *268*
Spitzig, W. A., 52, 76, *114*
Spriggs, R. L., 163, *177*
Spriggs, R. M., 164–165, *177*
Stanley, R. M., 246–247, *268*
Stavrolakis, J. A., 164–165, *177*
Stegun, I. A., 399, 403, *431*
Stein, M., 272, 361–362, *370*
Sternberg, E., 180–181, 199, 213, 217, 260, 266, *268*
Stickler, R., 137, *176*
Stiegler, J. O., 159, *177*
Stobbs, W. M., 149, *173*
Stocker, R. L., 143, *177*
Stokes, R. J., 140–142, 161–164, *173, 175, 177*
Stoloff, N. S., 162, *177*
Stoneley, R., 374, *433*
Strang, G., 279, *370*
Struik, D. J., 336, *370*
Stüwe, H. P., 127, 134, *177*
Suhubi, E. S., 375, 404, *431*
Suiter, J. W., 139, 161, *177*
Sutherland, E. C., 159, *177*
Swearengen, J. C., 34, *115*
Sweeting, T. B., 163, *177*
Synge, J. L., 246, *268*, 321, *368*, 374, *433*
Szabo, T. L., 397, *433*
Szwabowicz, M. L., 325, *368*

T

Taggart, R., 34, *115*
Taplin, D. M. R., 133, 147, 156, 157, *172*
Tavernelli, J. F., 159, *177*
Taylor, B. D., 376, *433*
Taylor, G. I., 2, 7–11, 13–16, 19–20, 28, 30, 36, 99–102, 105, *115*
Taylor, J. L., 134, *177*
Tegart, W. J. McG., 126–128, 134, 139, *174–177*
Teh, S. K., 164–165, *177*
Testa, R., 145–146, *176*
Thomason, P. F., 150, *177*
Thompson, A. W., 25–26, 106, *115*
Thornton, P. R., 29–32, 66, *114*
Thurston, G. A., 341, *370*

Tiersten, H. F., 376, 378, *431, 433*
Timoshenko, S. P., 231, *268*
Timpe, A., 14, *115*
Titchmarsh, E. C., 186–187, 208, *268*
Tittman, B. R., 397, *433*
Toaz, M. W., 139, 161, *177*
Tonning, A., 374, *432*
Toupin, R. A., 181, 192, 194, 205, 239, 241, 245, 248, *262, 268,* 299, 363, *370*
Tracey, D. M., 149, *176*
Tradinger, N. S., 253, *264*
Tressler, R. E., 144, *177*
Truesdell, C. A., *268,* 296, 299, 311, 313, 363, *370*
Tsao, M. C. C., 62, *112*
Tsutsumi, M., 377, *433*
Tullis, J. D., 144, *175*
Turteltaub, M. J., 210–212, *269*
Tyson, W. R., 121–122, 147, *175, 177*

V

Valsarajan, K. V., 237, *267*
Van de Berg, P. M., 406, *433*
Van der Heijden, A. M. A., 273, 322, *370*
Van de Vaart, H., 378, *434*
van Houtle, P., 99, *114*
Vasilos, T., 163–165, *176–177*
Vause, R. F., 66, *113*
Vekua, I. N., 320, *370*
Verrall, R. A., 140–141, *177*
Ver Synder, F. L., 134, *177*
Viktorov, I. A., 375, 377, 399, *434*
Vlasov, V. Z., 347, *370*
Volluet, G., 425, *434*
Volterra, V., 14, 16, *115*
Von Karman, T., 341, *370*
von Mises, R., 180, *266*

W

Wagner, H., 361, *370*
Wallace, W., 158, *175*
Walles, K. F. A., *177*
Wan, F. Y. M., 256–257, 260, *268,* 321, 359, *369–370*

Warren, W. E., 239, *268*
Weaver, C. W., 142, 158, *176–177*
Webb, D. C., 385, 422, 425, *432*
Webster, G. A., 134, *177*
Weck, N., 191, 249, *268*
Weerasootiya, T., *173*
Weertman, J., 125, 134, 146, 155, *177*
Weinberger, H. F., 211–212, 242, 245–246, *267*
Weinitschke, H. J., 341, 358, *359, 371*
Wempner, G., *371*
Westbrook, J. H., 142, 144, *177*
Wheatley, J. E., 161, *176*
Wheeler, L. T., 210–212, 218, 222–223, 250, 254, 260, 265, *268–269*
White, R. M., 375, *434*
Whitworth, R. W., 146, *177*
Wilshire, B., 134, 155, *173*
Wilson, A. J., 375, *434*
Winslow, F. R., *175*
Wittrick, W. H., 346, *371*
Witzke, W. R., 137, 159, *175*
Wong, W. A., 127, *177*
Wood, J. D., 341, *371*
Wood, W. A., 139, 161, *177*
Woodford, D. A., 155, *177*
Worthington, P. J., 34, *115*
Wu, C.-H., 195, 210, *269,* 352, 355–357, 361–362, *371*
Wu, H. C., 166, *177*
Wu, T. Y., 106, *112*

Y

Yaggee, F. L., 137–138, *174*
Yang, W. H., 355, *371*
Yim, W. M., 134, 155, *177*
Yin, C., 123, *173*
Yosifian, G. A., 195, 223, 228, 230, 249, 255, 260, *267, 269*

Z

Zanaboni, O., 181, *269*
Zaoui, A., 28, 32, *112*

Subject Index

A

Acceleration gradient tensor, 329
Activation volume, 55
  versus sheer strain, 58–59
Adiabatic heating, 130
Alkali halides
  deformation mechanism map, 140–141
  fracture mechanism map, 161–162
Al_2O_3
  deformation mechanism map, 143–144
  fracture mechanism map, 164–165
α-Alumina, deformation mechanism map, 143–144
Aluminum
  bicrystal, slip bands, 33, 35
  dislocation cell structure, 26–27
  flow strength and cell diameter, 26–27
  polycrystal, slip-line pattern, 33
  single crystal
    coarse slip lines, 77–79
    dislocation density versus shear strain, 55, 57
    flow strength and dislocation cell diameter, 55, 57
    kink bands, 67–68
    resolved shear stress versus shear strain curves, 55–56
    strain rate jump tests, 55–56
    strain rate sensitivity parameter m, 60
    thermal rate analysis, 55, 58
  slip planes, 7
  strain rate and overshoot, 30
Aluminum–copper alloy, see also Copper–aluminum alloy
  age hardened, macroscopic shear bands, 81–82
  orientation contrast, 83
  polycrystal, shear band, 108–109
  precipitation hardened
    activation volume, 55
    coarse slip bands, 77–78
    macroscopic shear bands, 77–78, 81–82
Anisotropy inducement. 375
Attenuation, nonreciprocal, 425

B

$BaTiO_3$
  Bleustein–Gulyaev modes in, 418–420
  piezoelectric Rayleigh waves in, 414
Beamshell, cylindrical motion, 274–302
  alternate stresses and strains, 282–283
  basic assumption, 283
  constitutive laws, 284–287
  defined, 283
  differential equations, 278–279
  elastodynamics, 292–296
  elastostatics, 287–292
  external force, 277
  geometry, 274–276
  integral equations, 276–278
  mechanical theory, 273, 283–284
  mechanical work identity, 280
  planar motion, 274–276
  strain energy density, 283–287
  strain measures, 280–282
  thermodynamics, see Thermodynamics, beamshell
  virtual work identity, 279
Bishop–Hill maximum work principle, 104
Bleustein–Gulyaev waves, 379, 391–395
  amplitude variation with depth, 394
  Cauchy stress tensor, 391
  constitutive equations, 391–392
  electroacoustic, 415–420
    amplitude behavior, 418–419
    in $BaTiO_3$, 418–420

dispersion relation, 416–419
Helmholtz decomposition, 415
electromechanical coupling factor, 392
magnetoacoustic, 426–430
  dispersion relation, 429–430
  surface wave problem, 426–427
piezoelectrically stiffened, 392–393
speed, 393, 395
surface wave problem, 411
Boundary value problem, 182–212
  axisymmetric theory, 220
  characteristic decay length, 187
  cross-sectional estimates, 201–204
  eigenvalues, 185
  energy decay inequality, see Energy, decay inequality
  exact solution representation, 185–189
  flow in a cylinder, 182–185
  formal solution, 185–186
  integral inequality, 190–191, 193
  Neumann problem, 187–188, 203
  pointwise decay, see Pointwise decay
  two-dimensional problem, 187–188
  uniform heat flow, 184, 187
α-Brass
  compression texture, 101
  deformed in tension, 29, 32
  overshoot, 29–30
Burgers vector, 14, 17

C

Cataclastic flow, 155
Cauchy stress tensor, 391
Clausius–Duhem inequality, 297–298, 364
Cleavage, 147–149
Coble creep, 129
Composite material, fiber-reinforced, energy decay inequality, 235–236
Copper, single crystal
  macroscopic shear bands, 81–82
  shear band, 108–109
Copper–aluminum alloy
  overshoot, 29–30
  slip bands, 33–34
Crack
  advance
    by ductile tearing, 169–171
    by rupture, 171
  extension mechanisms, 167–168
  wedge, 160

Creep
  Coble, 129
  diffusional flow, 128–129
  Harper–Dorn, 129
  high-temperature, tungsten, 137–138
  Nabarro–Herring, 129
  power-law, 125–128
    breakdown, 126–127
    constitutive law, 125
    dynamic recrystallization, 127–128
    rate equation, 127
  steady-state rate, 151
  transgranular, 149–150
Cross-slip, 18–19
  and coarse slip band formation, 79
  double, 22–23
Crystal, micromechanics, 1–115
  elastic–plastic, 2–3
    constitutive laws, 39–41, 61
    deformation applications, see Kink band; Shear band; Slip band
    deformation gradient, 37
    deformation kinematics, 36–39
    dislocation, see Dislocation
    elastic law, 39
    hardening modulus, 63–66
    Kirchhoff stress, 40
    plastic deformation, 38–39
    rate-insensitive, strain-hardening laws, see Strain hardening laws, rate-insensitive
    residual deformation, 37
    Schmid stress, 40–41
    slip, see Slip
    velocity gradient, 38
  notation, 3–4
  plastic flow, 2
  rate-insensitive, strain-hardening laws, see Strain-hardening laws, rate- insensitive
  strengthening mechanisms, 24–27
    flow stress, and grain diameter, 24–25
    Hall–Petch relation, 24–25
    unyielded grain, 26
    yield strength and grain diameter, 24–25
  structure, 6–7
  and slip plane, 17

D

Deformation, mechanisms, 117–177
  constitutive laws, 118–120, 146–147

Subject Index

mechanism maps, 121, 130–147, 171–172
  Al_2O_3, 143–144
  ceramics, 140–147
  construction, 130–132
  ice, 144–146
  magnesium, 138–140
  metals, 132–140
  MgO, 141–143
  NaCl, 140–141
  nickel, 130–131, 133–136
  tungsten, 136–138
plasticity, see Plasticity
power density, 281
Diffusional flow, 128–129
Diffusional fracture, 153
Dislocation, 13–24
  activation volume, 55
  average velocity, 54
  basic properties, 16–19
  Burgers vector, 14, 17
  cell diameter and flow strength, 55, 57
  cell structure, 26–27
  conjugate, 19–20
  coordination frame, 18
  cross-slip, 18–19
  density, 22–24, 35
    versus shear strain, 55, 57
  double cross-slip, 22–23
  drift velocity, ice, 146
  edge, 13–14, 20
    cross-slip, idealized model, 47–48
  elastic theory, 16
  flow stress versus average density, 24
  forest, 1
  glide, 14, 20
  glide force, 17
  interactions, 19–21
  Koehler's model, 23
  latent hardening, 21
  Lomer, 19–20
  merging reaction, 19
  multiplication, 19
  Orowan's model, 13
  partial, dissociation, 18
  Polanyi's model, 14
  resistance to, 53
  screw, 14, 17, 19
    cross-slip, idealized model, 47–48
    segment movement, idealized model, 22
  strain hardening, 14–16, 19, 21
  Taylor's flow stress equation, 15–16, 24

Taylor's idealized arrangement, 15
Taylor's model, 13–15
Dispersion relation, 382, 385
  Bleustein–Gulyaev modes, 416–420
  Love wave problem, 404
  magnetoacoustic Bleustein–Gulyaev waves, 429–430
  piezoelectric Rayleigh waves in ferroelectrics, 412–414
  resonance coupling, 385, 388–390
  uncoupled modes, 386
Donnell–Mushtari–Vlasov theory, 348–349
Ductile tearing, crack advance, 169–171

E

Elastic collapse, 121–122
Elastic energy, 16
Elastic energy function, mixed, 284
Elastic law, 39
Elastic moduli, 240
Elastic shell theory, nonlinear, 271–371
  approximate, 344–362
    comparable bending and extensional strains, 345
    Connell–Mushtari–Vlasov theory, 348–349
    introduction, 344–347
    membrane theory, see Membrane theory, nonlinear
    nonlinear membrane-inextensional theory, 346–347
    order-of-magnitude estimates, 346–347
    quasi-shallow, 347–349
  beamshell, cylindrical motion, see Beamshell, cylindrical motion
  constitutive relations and boundary conditions, 318–323
    Kirchoff boundary conditions, 319–321
    membrane-inextensional bending theory, 320
    modified Kirchoff, 321–323
    stress–strain relations, 320
  direct approach, 272–273, 284–285, 319
  equations, 302–332
    alternate strains, rational in the displacements, see Strain, alternate, rational in the displacements
    assumptions, 317
    compatibility conditions, 309
    covariant base vectors, 303–304

differential equations of motion, 307
field, see Field equations, formulation
finite rotation vector, 311–312
Gauss–Weingarten, 303, 309
integral equations of motion, 305–306
internal power density, 311, 315–318
load-deflection curves, 304
mechanical work identity, 308
strain measures, 308–311
undeformed, geometry, 303–305
virtual work identity, 307–308
mixed approach, 273
reduction approach, 273, 285–286, 319–320
static one-dimensional strain fields, 337–340
curved tubes, pure bending, 340–344
torsionless axisymmetric deformation of shells of revolution, 338–340
Elastic surface waves, 373–434
Bleustein–Gulyaev, see Bleustein–Gulyaev waves
electroacoustics, 407–421
coupled bulk modes, 409–411
dynamical fields, superimposition on bias fields, 408
general features, 407–409
piezoelectric Rayleigh modes, 411–415
quasi-electrostatic version, 410
splitting of surface wave problem, 409–411
general features, 380–385
dispersion relation, 382, 385
matrix method, 384–385
Love waves, 378–379
magnetically stiffened waves, 377
magnetoacoustic, 421–430
combined reflection operation, 425
coupled spin waves, 422
ferromagnetic crystal with broken symmetry, 422–423
general features, 421–424
nonreciprocity, 424–425
perturbation equations, 422
temperature effects, 421–422
nonhomogeneous elastic media, 395–406
compliance, dependence on depth, 395–396
free inhomogeneous layer problem, 404–406
Love wave problem, 402–406
other problems, 402–406
Rasattepe's site, variation with depth, 396–397

Sturm–Liouville problem, see Sturm–Liouville problem
piezoelectrically stiffened waves, 376–377
Rayleigh waves, 374–375
resonance coupling, between modes, 385–391
crossover regions, 386–387
dispersion relation, 385, 388–390
electroacoustic, 388–390
magetoacoustic, 388–390
repulsion point, 388
Stoneley waves, 376
Elastodynamics, 260–261
beamshell
classical flexural motion, 295–296
displacement shear strain form, 293–294
stress resultant rotation form, 294–295
Elastostatic problems, linear, 213–250
axisymmetric theory, 213–223
boundary conditions, 215, 219
decay estimates, 213–217
mean twist, 217
meridional cross section, 213–214, 218
shear stress, estimates, 216–217, 221
stored energy, 215
elastic cylinder, see Elastostatic problems, linear, three-dimensional
plane strain, 223–237
energy decay inequality, see Energy decay inequality
isotropic materials, 223–233
analog, energy decay inequality, 241–242
Berdichevskii's result, 243–245
boundary value problem, 240
cross-sectional properties, decay estimate, 242–247
three-dimensional problems
elasticity operator, 249
elastic moduli, 240
energy decay, slice argument, 239–242
isotropic cylinder, 249
Korn's inequality, 242
strain energy density, 239
stresses, pointwise estimates, 247–248
total energy, pointwise estimates, 247–249
warping function, theory of torsion, 244
torsionless axisymmetric problems, 237–239
hollow circular cylinder, 238
isotropic solid circular cylinder, 237–238
transversely isotropic circular cylinder, 238–239

Elastostatics
  nonlinear effects, 250–253
    average shear traction, 252
    finite antiplane shear, 251
    neo-Hookean material, 253
    plane strain problem, 250–251
  pressure-loaded beamshells, 287–292
    qualitative discussion, 288–290
    quantitative discussion, 290–292
Electroacoustic elastic surface waves, *see* Elastic surface waves, electroacoustic
Electromechanical coupling factor, 293
Energy
  balance, reduced differential equation, 298
  conservation, 297
  decay rate, 253, 255
  stacking fault, 110
  strain inequality, 225
  total
    pointwise estimates, 247–249
    upper bound, 196–198
Energy decay inequality, 189–196
  acoustic eigenvalue problem, 193–194
  alternate procedure, 192–195
  Dirichlet problem, 191
  energy distribution, 190
  estimate, using total energy upper bound, 198–199
  exponential, 194
  higher-order energies, 199–201, 203–204
  hollow circular cylinder, 238
  in terms of cross-sectional properties, 242–247
    circular cross section, 246
    general cross section, 246–247
    hollow cross section, 246
  isotropic materials, 223–228
  isotropic solid circular cylinder, 237–238
  linear elastostatic problems, 216
  other elliptic second-order problems, 195–196
  plane strain, 223–233
    alternative argument, 228–231
    anisotropic case, 233–235
    application to composite materials, 235–237
    biharmonic problem, 232
    constitutive law, 233
    discussion, 231–233
    eigenvalue problems, 231
    lemma, 226–228
    results, 231–233

  semi-infinite-strip, 231
  slow stress decay, 236–237
  strain energy density, 234
  second-order energy, 200–201
  slice argument, 239–242
  strain energy density, 234, 239
  third-order energy, 201
  total energy, upper bound, 196–198
  transversely isotropic circular cylinder, 238–239
  two-dimensional problem, 191, 194–195, 203
Equipresence, principle of, 299
Euler equations, 323–324

F

Ferroelectrics, elastic, *see also* Elastic surface waves, electroacoustic
  crystal, broken symmetry, 410
  defined, 407
  soft modes, 379–380
Field equations, formulation, 323–332
  curved tube bending, 343
  displacement form, 323–324
  Euler equations, 323–324
  intrinsic form, 326–331
    dynamic equations, 327–331
    equation of motion, 329–330
    initial and boundary conditions, 330–331
    preliminary kinematics, 328–329
    longitudinal rod dynamics, 328
    static equations, 326–327
  rotation form, 324–326
    dynamic equations, 325–326
    static equations, 324–325
  velocity form, 331–332
Flexural motion, classical, 295–296
Flow law, 64
  rate-dependent, 107
  strain rate-dependent, 53–63
    activation volume, 55, 58–59
    average dislocation velocity, 54
    dislocation density versus shear strain, 55, 57
    flow strength and dislocation cell diameter, 55, 57
    isothermal relation, 61
    Orowan relation, 54
    power-law hardening, 61
    rate sensitivity, 54–55

resolved shear stress versus shear strain
  curves, 55–56
 sensitivity parameter β, 59–60
 sensitivity parameter m, 54–55, 60–62
 strain rate jump tests, 55–56, 61–62
 thermal fluctuation, 54
Flow stress
 and grain diameter, 24–25
 latent and active primary systems, 30
 versus average dislocation density, 24
Föppl–Hencky theory, 358
Forest dislocations, 21
Fracture, mechanism
 constitutive laws, 120
 dynamics, 156
 by general damage, 147–155
  cleavage, 147–149
  ductile, low temperature, 149–151
  ideal fracture strength, 147
  intergranular creep-controlled, 151–152
  other mechanisms, 154–155
  pure diffusional, 153
  rupture, 154
  transgranular creep, 149–151
 mechanism maps, 121, 155–172
  ceramics, 161–166
  construction, 155–157
  failure mechanism, 156
  metals, 155–161
  and order of magnitude of K_c and G_c, 166–171
  cleavage without plasticity, 168
  crack advance by ductile tearing, 169–171
  crack advance by rupture, 171
  crack extension by cleavage, 168
  G_c and relationship to crack advance, 170
  toughness versus mechanism, 167
  void sheet linkage, 171
 plasticity and damage accumulation, 120
 strength, ideal, 147
 wedge cracks, 160
Free inhomogeneous layer problem, 404–406
 phase velocity, 406
Frene–Serret equation, 336

G

Gauss–Codazzi equations, 316, 351
Gauss–Weingarten equations, 303, 309
Glide force, 17

H

Hall–Petch relation, 24–25
Hardening
 latent
  dislocation, 21
  effect on elastic–plastic response, 63
  estimates, 65–66
  measurements, 26, 28–32
  nonuniform slip mode, 32–33
  patchy slip, 35
  and polycrystalline slip, 8–10
 power-law, 61
 strain
  and macroscopic shear band formation, 83
  rate, and onset of coarse slip band formation, 79–80
  role of kink bands, 69
  slip, 8–10
  slip mode uniqueness, 46
  symmetry and hardening law, 66
Hardening law, 66
Hardening modulus, 63–66
 imposing symmetry, 66
 lattice rotation, 64–66
 plane double-slip model, 42–46
 resolved shear stress, 65
Hardness, parameters, 63
Heat, three-dimensional equations, 363
Helmholtz decomposition
 displacement field, 411
 electric polarization field, 415
Hill model, 106–107
Hill's minimum principle, 104
H_2O
 deformation mechanism map, 144–146
 fracture mechanism map, 165–166
Hooke's law, 285–286
Hutchinson's model, 102

I

Ice
 deformation mechanism map, 144–146
 fracture mechanism map, 165–166
Inextensional beamshell model, 288, 290
Integral inequality, 190–191, 193
Iron–silicon alloy, slip bands, 33–34

K

Kink band, formation, 67–77
 bifurcation modes, 69

constitutive laws, 69–70
examples, 67–68
localization conditions, 70–73
localized deformation, 70–71
non-Schmid effects, 73–74
orientation, 67–69
particular cases, 75–77
pressure sensitivity, yielding, 76
role in strain hardening, 69
specific results, 73–75
velocity gradient field, 70
yielding, stress dependent, 75–76
Kirchhoff boundary conditions, 319–321
  modified, 321–323
Kirchhoff hypothesis, 273, 285
Koehler's model, double cross-slip, 23
Korn's inequality, 242

L

Lagrange strain tensor, 313
Lagrangian strain, 282
Lattice, resistance-limited plasticity, 124
Lattice rotation, 64
  and dislocation tilt boundaries, 98
  jumps
    macroscopic shear bands, 88
  and macroscopic shear bands, 96–97
  Taylor's model, 100–101
  versus shear strain, 65–66
  versus tensile strain, 31–32
Lead, deformed polycrystalline, 5
Linearly extensional model, 291
Lomer dislocation, 19–20
Love wave problem, 402–405
  dispersion relation, 404
  fundamental mode, dispersion, 404–405
Love waves, 378–379, 400–401

M

Magnesium
  deformation mechanism map, 138–140
  fracture mechanism map, 160–161
Magnons, see Spin waves
Mar-M200, 134–135
Mechanical theory, shells, 273, 283–284
Mechanical work identity, 280, 308
Membrane beamshell model, 288, 290–291
Membrane theory, nonlinear, 349–362
  annular membranes, 357–359
  axisymmetric membranes, 355–357

boundary conditions, 354–355
compatibility, 351–352, 360
constitutive laws, 352–353
equilibrium, 350–351, 359
general cylindrical membranes, 359–361
membrane stress function, 360
nonlinear membrane problem, 354–357
small-strain approximation, 359–360
wrinkled membrane, 361–362
Metals, see also specific metals
  bcc
    deformation mechanism map, 136–138
    fracture mechanism map, 159–160
  fcc
    deformation mechanism map, 130–131, 133–136
    fracture mechanism map, 155, 157–158
  hexagonal
    deformation mechanism map, 138–140
    fracture mechanism map, 160–161
  polycrystalline, slip steps, 4–5
MgO
  deformation mechanism map, 141–143
  fracture mechanism map, 163–164
Momentless beamshell model, 288
Monge–Ampere equation, 335–336
Motion, equations of
  classical flexural motion, 295–296
  differential, 306
  displacement shear strain form, 293–294
  integral, 305–306, 362–363
  intrinsic form, 329–330
  stress resultant rotation form, 294–295
  weak, 279, 307–308

N

Nabarro–Herring creep, 129
NaCl
  deformation mechanism map, 140–141
  fracture mechanism map, 161–162
  impurities, defect structure effects, 141
Neo-Hookean material, elastostatics, 253
Neumann problem
  kernel, 206, 208–209
  three-dimensional, 208–210
  two-dimensional, 187–188, 203, 206–208
Nickel
  deformation mechanism map, 130–131, 133–136
  fracture mechanism map, 155, 157–158
  polycrystals, stress–strain curves, 105–106

Nickel-based superalloys
 deformation mechanism map, 133–135
 fracture mechanism map, 158
Nimonic 80A, fracture mechanism map, 158
Niobium, single crystal, macroscopic shear bands, 81–83
Nonlinear membrane-inextensional theory, 346–347

O

Orowan relation, 54
Orowan's model, 13
Overshoot
 aluminum, 30
 α-brass, 29–30
 latent hardening measurement, 29–30
 and strain rate, 30, 32
Oxide
 refractory
  deformation mechanism map, 143–144
  fracture mechanism map, 164–165
 simple
  deformation mechanism map, 141–143
  fracture mechanism map, 163–164

P

Patchy slip
 and latent hardening, 96
 polycryalline models, 107
Peach–Koehler relation, 17, 48
Phase velocity, free inhomogeneous layer, 406
Pipe
 elbow routines, 341
 pure bending, see Strain field, static one-dimensional, curved tubes, pure bending
Plastic flow, 2, 122
Plasticity, 121–130
 adiabatic heating, 130
 diffusional flow, 128–129
 elastic collapse, 121–122
 Harper–Dorn creep, 129
 ideal shear strength, 121–122
 low-temperature, dislocation glide, 122–125
  athermal flow strength, 123
  explosive loading, 125
  lattice-resistance-limited, 124
  obstacle characteristics, 123
  obstacle-limited, 122–123

Taylor factor, 124
 other mechanisms, 129–130
 power-law creep, 125–128
 twinning, 129
Plate theory, nonlinear, 332–337
 extended von Karman equations, 332–335
  Cartesian coordinates, 334–335
  polar coordinates, 335
  simplifications, 333–334
 inextensional deformation, 335–337
  Frene–Serret equations, 336
Pointwise decay, 204–210
 estimates valid up to boundary, 206–210
  Neumann kernel, 208–209
  three-dimensional Neumann problem, 208–210
  two-dimensional Neumann problem, 206–208
 interior estimates, 204–206
 maximum principles, 210–212
Polanyi's model, 14
Polar decomposition theorem, 313
Polycrystal, micromechanics, see also Crystal
 dislocation, see Dislocation
 grain size effects, 26
 models, problems, 97–108
  Bishop–Hill maximum work principle, 104
  Chang and Asaro's model, 98
  dislocation tilt boundaries, 98
  equal strain restriction, 105
  five independent slip systems, 99–100
  grain size effects, 105–106
  Hill model, 106–107
  Hill's minimum principle, 104
  Hutchinson's models, 102–105
  lattice rotation, 98, 100–101
  patchy slip, 107
  rigid lattice spin, 100
  self-consistent model, 106
  small-strain rate-dependent, 102–105
  stress-strain curves, 105–106
  Taylor's model, 99–100
  tensor of creep compliance, 102–103
  texture development, 99–101, 107
  x-ray beam trajectories and lattice orientation, 107–108
 slip, see Slip
 Taylor's model, 2
Power-law creep, see Creep, power-law
Power-law materials, decay rate, 253

Q

Quasi-shallow shell theory, 347–349

R

Rayleigh waves, 374–375
  magnetically stiffened, 377
  piezoelectric in elastic ferroelectrics, 411–415
    dispersion relation, 412–414
  piezoelectrically stiffened, 376–377
  surface wave problem, 411, 426
Recrystallization, dynamic, 127–128
Resolved shear stress, 7–8, 10, 65
  and onset of coarse slip band formation, 79–80
  basal slip system, 51
  critical, 12, see also Schmid stress
    dislocation segment movement, 22
    temperature effects, 12
  Hall–Petch form, 58
  non-Schmid effects, 49–50
  versus shear strain, 55–56
Resonance coupling, see Elastic surface waves, resonance coupling
Rotational momentum equation, 293–294
Rotation vector, finite, 311–312
Rupture, 154, 171

S

Saint-Venant's principle, 179–269
  boundary value problem, see Boundary value problem
  history, 180–181
  linear elastostatic problems, see Elastostatic problems, linear
  nonlinear effects, 250–255
    difference velocity fields, 254
    elastostatics, 250–253
    viscous flows, pipes and channels, 254–255
  slow stress diffusion, shells, 255–257
  stress field comparison, 181
  time-dependent problems, 257–261
    boundary-initial value problem, 259–260
    elastodynamics, 260–261
    parabolic equations, 257–260
    viscoelasticity, 260
  von Mises–Sternberg formulation, 257–260
Schmid factor, 10, 97

Schmid law
  and plastic normality, 42–46
  critical system, 43
  stress changes, 43
  stress state, 45–46
  uniqueness, 44–46
  yield locus, 43
  and slip, 7–8
Schmid rule, 2, 14
Schmid stress, 40–41, see also Resolved shear stress, critical
  rate, 51–52
Shear band
  deformation, 107–111
    examples, 109
    geometrically soft, 110
    rolling, 110–111
    stacking fault energy, 110
    Taylor factor, 111
    twinning, 110
  macroscopic
    aluminum–copper alloys, and strain-hardening law, 90–93
    bifurcation and uniqueness, 84–89
    bifurcation stress, 86
    constitutive assumptions, 88–89
    constitutive laws, 84
    differentiation from coarse slip bands, 83
    dislocation substructure, 97
    examples, 81–82
    geometrically softened, 83
    inclined to tensile axis, 87–88
    lattice kinematics, 96
    lattice rotation contours, 95–96
    lattice rotation jumps, 88
    load-engineering strain curve, 90, 92
    numerical results, 89–97
    orientation contrast, 83
    patchy slip, 96
    plane strain tension regimes, 86–87
    prebifurcation stress state, 84–85
    Schmid factors, 97
    shearing rate ratio, 88
    slip plane hardening matrix, 89
    strain-hardening law, 90–93
      softer crystal, 92–97
    vertex softening, 88
Shear
  finite antiplane, 251
  mode, 8
  simple, 252

strength, ideal, 121–122
traction, 252
Shear stress, estimates, 216–217, 221
Shell
    defined, 272
    elastic, theory, see Elastic shell theory, nonlinear
    infinite cylindrical, see Beamshell
    in torsion, axisymmetric theory, error estimates, 217–223
    mechanical theory, 273, 283–284
    slow stress diffusion, 255–257
    thermodynamic theory, 273
SH surface waves, see Elastic surface waves
Slip
    at modest strains, 32–35
    direction, 31, 37
    double primary-conjugate, 36, 45
    flow law, 48–49
    geometrical softening, 11–12
    grain reorientation, 33
    -line pattern, aluminum, 33
    multiple mode, 35
    nonuniform mode, 32–35
    patchy, 33, 35
    planes, 7, 17
    plastic, 37
    polycrystals, 4-35
        conjugate, 10
        early observations, 4–12
        latent hardening, 8–10
        Schmid factor, 10
        Schmid law, 7–8
        steps, 4–5
        strain-hardening, 10–11
    rate of rotation, 10
    resolved shear stress, 7–8, 10, 12
    rigid lattice spin rate, 11
    single plane model, 36
Slip band, 33–35
    coarse, formation, 77–81
    and cross-slip, 79
    and macroscopic shear bands, 77–78
    non-Schmid effects, 79
    onset and strain-hardening rate, 79–80
    shearing mode, 79
Spin wave, 379–380
    nonreciprocity, 424
    pure, frequency, 428
Steel, strain jump tests, 62
Stoneley waves, 376

Strain
    alternate, rational in the displacements, 312–318
    bending strain, 318
    Gauss–Codazzi equations, 316
    internal power density, 315–318
    Lagrange strain tensor, 313
    polar decomposition theorem, 313
    stress resultant, 318
    beamshell measures, 280–282, 308–311
    bending, 318
    comparable bending and extensional, 345
    compatibility equation, 342
    energy density, 234, 239, 283–284
        differentiation, 287
        direct method, 284–285, 319
        incompressible Mooney material, 286
        reduction method, 285–286, 319–320
    Lagrangian, 282
    tensile, versus lattice rotation, 31–32
Strain field, static one-dimensional, 337–344
    curved tubes, pure bending, 340–344
        boundary conditions, 343–344
        constitutive relations, 343
        equilibrium, 342–343
        field equations, 343
        kinematics, 342
    torsionless axisymmetric deformation of shells of revolution, 338–340
Strain hardening
    dislocation, 14–16, 21
    slip, 10–11
Strain-hardening laws, rate-insensitive, 52–53
    cross-slip process, 47–48
    hypothetical yield surfaces, 48
    non-Schmid effects, 47–53
    non-Schmid terms, estimates, 51–53
    plastic normality, deviations from, 47–53
    resolved shear stress, 49–50
    Schmid law, see Schmid law
    strain rate-dependent flow laws, see Flow laws, strain rate-dependent
    strength differential, 52
    yield strength, dependence on biaxial tension, 51
Strain rate, 119
    jump tests, 55–56
Strain tensor, Lagrange, 313
Stress
    rate, 118
    slow decay, 236–237

slow diffusion, 255–257
Stress resultant, beamshell, 277, 306, 318
Stress–strain law, 16
Stress tensor
  Cauchy, 391
  nominal, components, 364–365
Sturm–Liouville problem
  generalized, 398–399
  simple, 400–402
    amplitude dependence on depth, 400–401
    dispersion curves, 400–402
    finite element method, 400
Surface wave problem
  magnetoacoustic, 426
  splitting, 409–411

T

Taylor factor, 111, 124
Taylor's flow stress equation, 15–16, 24
Taylor's model, 13, 99–101
Temperature, effects
  flow strength, 123
  ideal shear strength, 122
  magnetoacoustic waves, 421–422
  resolved shear stress, 12
Temperature field, heat equation, 257–260
Thermal fluctuations, probability of experiencing, 54
Thermal rate analysis, 55, 58
Thermodynamics, beamshell, 296–302
  Clausius–Duhem inequality, 297–298
  conservation of energy, 297
  constitutive variables, changes, 300–301
  reduced differential equation of energy balance, 298
Thermodynamic theory, shells, 273
Thermoelasticity, beamshell, 297–298
Tube, pure bending, *see* Strain field, static one-dimensional, curved tubes, pure bending

Tungsten
  deformation mechanism map, 136–138
  fracture mechanism map, 159–160
Twinning, 110, 129

V

Velocity gradient tensor, 328
Virtual work identity, 279, 307–308
Viscoelasticity, 260
Viscous flow, pipes and channels, 254–255
Void sheeting, 154
von Karman equations, extended, 332–335
  Cartesian coordinates, 334–335
  polar coordinates, 335
  simplifications, 333–334
von Mises–Sternberg formulation, Saint-Venant's principle, 257–260

W

Warping function, theory of torsion, 244
Waves
  Bleustein–Gulyaev, 379
  elastic surface, *see* Elastic surface waves
  Love, 378–379
  magnetically stiffened, 377
  piezoelectrically stiffened, 376–377
  Rayleigh, 374–375
  Stoneley, 376

Y

Yield strength, 24–25, 51

Z

Zinc
  slip, 8
  yield strength, 51